Distributed Strategic
Learning
for Wireless Engineers

Distributed Strategic Learning
for Wireless Engineers

Hamidou Tembine

CRC Press
Taylor & Francis Group
Boca Raton London New York

CRC Press is an imprint of the
Taylor & Francis Group, an **informa** business

CRC Press
Taylor & Francis Group
6000 Broken Sound Parkway NW, Suite 300
Boca Raton, FL 33487-2742

First issued in paperback 2017

© 2012 by Taylor & Francis Group, LLC
CRC Press is an imprint of Taylor & Francis Group, an Informa business

No claim to original U.S. Government works

ISBN-13: 978-1-4398-7637-4 (hbk)
ISBN-13: 978-1-138-07702-7 (pbk)

Visit the Taylor & Francis Web site at
http://www.taylorandfrancis.com

and the CRC Press Web site at
http://www.crcpress.com

Dedicated to Bourere Siguipily

Contents

9 Learning in Risk-Sensitive Games

List of Figures

List of Tables

Foreword

We live today in a truly interconnected world. Viewed as a network of decision making agents, decisions taken and information generated in one part, or one node, rapidly propagate to other nodes, and have impact on the well being (as captured by utilities) of agents at those other nodes. Hence, it is not only the information flow that connects different agents (or players, in the parlance of game theory), but also the cross-impact of individual actions. Individual players therefore know that their performance will be affected by decisions taken by at least a subset of other players, just as their decisions will affect others. To expect a collaborative effort toward picking the "best" decisions is generally unreasonable, and for various reasons, among which are nonalignment of individual objectives, limits on communication, incompatibility of beliefs, and lack of a mechanism to enforce a stable cooperative solution. Sometimes a player will not even know the objective or utility functions of other players, their motivations, and the possible cross-impacts of decisions.

How can one define an equilibrium solution concept that will accommodate different elements of such an uncertain decision making environment? How can such an equilibrium be reached when players operate under incomplete information? Can players learn through an iterative process and with strategic plays the equilibrium-relevant part of the game? Would such an iterative process converge, and to the desired equilibrium, when players learn at different rates, employ heterogeneous learning schemes, receive information at different rates, and adopt different attitudes toward risk (some being risk-neutral, other being risk-sensitive)?

The questions listed above all relate to issues that sit right in the heart of multi-agent networked systems research. And this comprehensive book meets the challenge of addressing them all, in the nine chapters to follow.

Professor Tamer Başar,
Urbana-Champaign,
Illinois, 11-11-11.

Preface

Preface to the book Distributed Strategic Learning for Wireless Engineers

Much of Game Theory has developed within the community of Economists, starting from the book "Theory of Games and Economic behavior" by Morgenstern and Von Neumann (1944). To a lesser extent, it has had an impact on biology (with the development of evolutionary games) and on road traffic Engineering (triggered by the concept of Wardrop equilibrium introduced already in 1952 along with the Beckmann potential approach introduced in 1956). Since 1999 game theory has had a remarkable penetration into computer science with the formation of the community of Algorithmic game theory.

I am convinced that game theory will play a much more central role in many fields in the future including telecommunication network engineering. I use the term Network Engineering Games (NEGs) to call games that arise within the latter context. NEG is the young brother of the Algorithmic game theory. NEG is concerned with competition that arises at all levels of a network. This includes aspects related to information theory, to power control and energy management, to routing, to the transport and application layers of communication networks. It also includes competition arising in spread of information over a network as well as issues related to the economy of networks. Finally, it includes security issues, service denial attacks, spread of virus in computers and measures to fight it.

This book is the first to consider a systematic analysis of games arising in all network layers and is thus an important contribution to NEGs.

The word "game" may have connotations to "toys" or of "playing" (as opposed to decision making). But in fact it stands for decision making by several decision makers, each having her (or his) own individual objectives. Is game theory a relevant tool for research in communication networks? On 20/12/2011 I searched on Scholar Google the documents containing "wireless networks" together with "power control". 20500 documents were found. Of these, 3380 appeared in 2011, and 1680 dated from 2000 or earlier. I then repeated the experience restricting further to documents containing "game theory". 2600 documents were found. Of these, 20 dated from prior to 2001 and 580 dated from the single year 2011. The share of documents containing "game theory" thus increased from 1.2% to 17% within 10 years.

Is game theory relevant in wireless Engineering?

A user that changes some protocols in his cellular telephone may find out that a more aggressive behavior is quite beneficial and allows to obtain better performances. Yet if all the population tried to act selfishly and use more aggressive protocols, then everyone may loose in performance. But in practice we do not bother to change the protocols in our cellular phones. Making such changes would require access to the hardware, skills and training which is too much to invest. This may suggest that game theory should be used for other networking issues, perhaps in other scales (such as auctions over bandwidth, competition between service providers etc). So is there a need in NEG? Here are two different angles that one can use to look at this problem. First, we made here the assumption that decisions are taken concerning how to use equipment. But we can instead consider the decisions as being which equipment to buy. The user's decisions concerning which protocol to use are taken when one purchases a telephone terminal. One prefers a telephone that is known to perform better. The game is then between equipment constructors. Secondly, not all decisions require special skills and knowhow. The service providers and/or the equipment constructors can often gain considerably by delegating to the users to take decisions. For example, when you wish to connect to the Internet from your laptop, you often go to a menu that provides you with a list of available connections along with some of their properties. The equipment provider has decided to leave us, the users, this choice. It also decides what information to let us have when we take the decision.

Leaving the decisions to the users is beneficial for service providers because of scalability issues: decentralizing a network may reduce signaling, computations and costs. When designing a network that relies on decisions taken by users, one needs to predict the users behavior. Learning is part of their behavior. Much of the theory of learning in games has been developed by biologists that used mathematical tools to model learning and adaptation within competing species. In NEG one need not restrict to describing existing learning approaches, one can propose and design learning procedures.

Why learn to play an equilibrium?

Sometimes it's better not to learn. For example, assume that there are two players, one chooses x and the other chooses y, where both x and y lye in the half closed unit interval $[0, 1[$ Assume that both have the same utility to maximize, which is given by $r(x, y) = xy$. It is easily seen that this game has an equilibrium which is unique: $(0, 0)$. This is the worst possible choice for both players. Any values of x and y that are strictly different from the equilibrium value give both a strictly better utility!

When a service provider delegates to users some decisions, it can control what parameter to let them control and what information to let them have so as to avoid such situations. Learning to play equilibrium may then be in the interest of the players, and exploring learning algorithms enrich the tools available in designing networks.

This book is unique among the books on learning in game theory in focusing on problems relevant to games in wireless engineering. It is a masterpiece bringing the state-of-the art foundations of learning in games to wireless.

Professor Eitan Altman
INRIA Sophia Antipolis
February 3rd, 2012

Strategic learning has made substantial progress since the early 1950s and has become a central element in economics, engineering, and computer science. One of the most significant accomplishments in strategic decision making during the last decades have been the development of game dynamics. Learning and dynamics are necessary when the problem to be solved is under uncertainty, time-variant and depends on the structure of the dynamic environment. This book develops distributed strategic learning schemes in games [15, 16, 17]. It offers several examples in networking, communications, economics and evolutionary biology in which learning and dynamics play an important role in understanding the behavior of the system.

As a first example, consider a spectrum access problem where the secondary users can sense a subset of channels. If there are unused channels by primary users at a given time slot, then the secondary users which sensed can access to the free channels. The problem is that even under slotted time and frames, several secondary users can simultaneously sense the same channels at the same time. We can explicitly describe this problem depending the channel conditions, the throughput, the set of primary users, the set of malicious users, the set of altruistic users (relays), the set of secondary users, their arrivals, departure rates, their past activities, but we are unable to explain how the secondary users do it if they sensed the same channel at the same time. Thus, it is useful to find a learning mechanism that allows an access allocation in the long-run.

As a second example, consider a routing packet over a wireless ad hoc network. The wireless path maximizing the quality of service with minimal end-to-end delay from a source to a destination changes continuously as the network traffic and the topology change. A learning-based routing protocol is therefore needed to estimate the network traffic and to predict the best stochastic path.

Already there are many successful applications of learning in networked games but also in many other domains: robotics, machine learning, bioinformatics, economics, finance, cloud computing, network security and reliability, social networks, etc. A great many textbooks have been written about learning in dynamic game theory. Most of them adopt either an economic perspective or a mathematical perspective. In the past several years, though, the application of game theory to problems in networking and communication systems has become more important. Specifically, game-theoretic models have been developed to better understand flow control, congestion control, power control, admission control, access control, network security, quality of service, quality of experience management and other issues in wireline and wireless systems. By modeling interdependent decision makers such as users, transmitters, radio devices, nodes, designer, operators, etc, game theory allows us to model scenarios in which there is no centralized entity with a full picture of the system conditions. It allows also teams, collaborations, and coalitional behaviors among the participants. The challenges in applying game theory to networking systems has attracted a lot of attention in the last decade. Most of the game-theoretic models can abstract away important assumptions and mask critical unanswered questions. In absence of observation of the actions of the other participants and under unknown dynamic environment, the prediction of the outcome are less clear. It is our hope that this book will illuminate both the promise of learning in dynamic games as a tool for analyzing network evolution and the potential pitfalls and difficulties likely to be encountered when game theory is applied by practicing engineers, undergraduate, graduate students, and researchers. We have not attempted to cover either learning in games or its applications to networking and communications. We have severely restricted our exposition to those topics that we feel are necessary to give the reader a grounding in the fundamentals of learning in games under uncertainty or robust games and their applications to networking and communications.

As most of wireless networks are dynamic and evolve in time, we are seeing a tendency toward decentralized networks, in which each node may play multiple roles at different times without relying on an access point or a base station (small base station, femto-cell BS or macro-cell BS) to make decisions such as in what frequency band to operate, how much power to use during transmission frame, when to transmit, when to go in sleep mode, when to upgrade, etc. Examples include cognitive radio networks, opportunistic mobile ad hoc networks, and sensor networks that are autonomous and self-organizing and support multihop communications. These characteristics lead to the need for distributed decision making that potentially takes into account network conditions as well as channel conditions. In such distributed systems, an individual terminal may not have access to control information regarding other terminal's actions and network congestion may occur. We address the following questions:

- Question One: How much information is enough for effective distributed decision making?

- Question Two: Is having more information always useful in terms of system performance (value/price of information)?

- Question Three: What are the individual learning performance bounds under outdated and imperfect measurement?

- Question Four: What are the possible dynamics and outcomes if the players adopt different learning patterns?

- Question Five: If convergence occurs, what is the convergence time of heterogeneous learning (at least two of the players use different learning patterns)?

- Question Six: What are the issues (solution concepts, non-convergence, convergence rate, convergence time etc) of hybrid learning (at least one player changes its learning pattern during the interaction)?

- Question Seven: How to develop very fast and efficient learning schemes in scenarios where some players have more information than the others?

- Question Eight: What is the impact of risk-sensitivity in strategic learning systems?

- Question Nine: How do we construct learning schemes in a dynamic environment in which one of the players does not observe a numerical value of its own-payoffs but only a signal of it?

- Question Ten: How to learn "unstable" equilibria and global optima in a fully distributed manner?

These questions are discussed through this book. There is an explicit description of how players attempt to learn over time about the game and about the behavior of others (e.g. through reinforcement, adaptation, imitation, belief updating, estimations or combination of these etc.). The focus is both on finite and infinite systems, where the interplay among the individual adjustments undertaken by the different players generate different learning dynamics, heterogeneous learning, risk-sensitive learning, and hybrid dynamics.

How to use this book?

This Guide is designed to assist instructors in helping students grasp the main ideas and concepts of *distributed strategic learning*. It can serve as the text for learning algorithm courses with a variety of different goals, and for courses that are organized in a variety of different manners. The Instructor's note and supporting materials is developed for use in a course using *distributed strategic learning* with the following goals for students:

Students will be better able to think about iterative process for engineering problems;

Students will be better able to make use of their algorithmic, graphing, and computational skills in real wireless networks based on data;

Students will be better able to independently read, study and understand the topics that are new to the students such as solution concepts in robust games;

Students will be better able to explain and describe the learning outcomes and notions orally and to discuss both qualitative and quantitative topics with others;

We would like to make the following remarks. The investigations of various solutions are almost independent of each other. For example, you may study the strategy dynamics by reading Chapter 2 and payoff dynamics by reading Chapter 3. If you are interested only in the risk-sensitive learning, you should read Chapter 8. Similar possibilities exist for the random updates, heterogeneous learning, and hybrid learning (see the Table of Contents).

If you plan an introductory course on robust game theory, then you may use Chapter 1 for introducing robust games in strategic-form. Remark. Chapters 2 - 8 may be used for a one-semester course on distributed strategic learning.

Each chapter contains some exercises. The reader is advised to solve at least those exercises that are used in the text to complete the proofs of various results.

This book can be used for a one semester course by sampling from the chapters and possibly by discussing extra research papers; in that case, I hope that the references at the end of the book are useful. I welcome your feedback via email to *tembineh(at)gmail.com*. I very much enjoyed writing this course, I hope you will enjoy reading it.

Notation and Terminology

The book is comprised of nine chapters and one appendix. Each chapter is divided into sections, and sections occasionally into subsections. Section 2.3, for example, refers to the third section of Chapter 2, while Subsection 2.3.1 is the first section of Subsection 2.3.

Items like theorems, propositions, lemmas, etc, are identified within each chapter according to the standard numbering; Equation (7.1) would be the first equation of Chapter 7.

Organization of the book

The manuscript comprises nine chapters.

- Chapter one introduces basic strategic decision-making and robust games. State-dependent games with different level of information are formulated and the associated solution concepts are discussed. Later, distributed strategic learning approaches in different layers of the open systems interconnection model (OSI) including physical layer (PHY), medium access control (MAC) layer, network layer, transport layer, and application layer are presented.

- In Chapter two, we overview classical distributed learning schemes. We start with partially distributed strategy-learning algorithms and their possible implementation in wireless networks. Generically, partially distributed learning schemes, sometimes called semi-distributed schemes, assume that all players knew their own-payoff functions and, observe others' actions in previous stages. This is clearly not the case in many networking and communication problems of interest. Under this strong assumption, several game-theoretic formulations are possible for uncertain situations. Then, the question of how to learn the system characteristics in presence of incomplete information and imperfect measurements is addressed. Both convergence and nonconvergence results are provided. In the other chapters of this book, we develop strategic learning framework by assuming that each player is able to learn progressively its own-action space, knows his or her current action, and observes a numerical (possibly noisy) value of her (delayed) payoff (the mathematical structure of the payoff functions are unknown as well as the actions of the other players). This class of learning procedures is called fully distributed strategy-learning algorithm or model-free strategy-learning and is presented in section 2.3.

- Chapter 3 focuses on *payoff learning and dynamics*. The goal of Payoff Learning is to learn the payoff functions, the expected payoffs and the risk-sensitive payoffs. In many cases, the exact payoff functions may not be known by the players. The players try to learn the unknown data through the long-run interactions. This chapter complements the Chapter two.

- Chapter 4 studies *combined fully distributed payoff and strategy learning* (CODIPAS). The core chapter examines how can evolutionary game theory be used as a framework to analyze multi-player reinforcement learning algorithms in an heterogeneous setting. In addition, equilibrium seeking algorithms, learning in multi-armed bandit problems and algorithms for solving variational inequality are presented. CODIPAS combines both strategy-learning and payoff-learning.

- Chapter 5 examines *combined learning under delayed and unknown payoffs*. Based on outdated and noisy measurements, combining learning schemes that incorporates the delays, as well as schemes that avoid the delays, are investigated. Relevant applications to wireless networks are presented.

- Chapter 6 analyzes *combined learning in constrained-like games*. The core of the chapter comprises two parts. The first part introduces constrained games and the associated solution concepts. Then, we address the challenging question of how such a game can be played? How player can choose their actions in constrained games? The second part of the chapter focuses on satisfactory solutions. Instead of robust optimization framework, we propose a robust satisfaction theory which is relevant quality-of-experience (QoE, application layer) and quality-of-service (QoS, network layer) problems. The feasibility conditions as well as satisfactory solutions are investigated. The last part of the chapter is concerned about *random matrix games* (RMGs) with variance criterion.

- Chapter 7 extends the heterogeneous learning to *hybrid learning*. Uncertainty, random updates and switching between different learning procedures are presented.

- Chapter 8 develops *learning schemes for global optima*. The chapter provides specific class of games in which global optimum can be found in a fully distributed manner. Selection of larger sets, Pareto optimal solutions, are discussed. A detailed MATLAB code associated to the example of resource selection games is provided.

- Chapter 9 presents *risk-sensitivity aspects in learning*. The classical game-theoretic approach to modeling multi-player interaction assumes that players in a game want to maximize their expected payoff. But in many settings, players instead often want to maximize some more complicated function of their payoff. The expected payoff framework for games is obviously very general, but it does exclude the possibility that players in the game have preferences that depend on the entire distribution of payoff, and not just on its expectation. For example, if a player is sensitive to risk, her objective might be to balance the variance to be closer to the expectation. Indeed, this is the recommendation of modern portfolio theory, and a version of the mean-variance objective is widely used by investors in financial markets as well as in network economics. The chapter also addresses the generalization of familiar notions of Nash and correlated equilibria to settings where players are sensitive to the risk. We especially examine the impact of risk-sensitivity in the outcome.

- Background materials on dynamical systems and stochastic approximations are provided in appendices.

Acknowledgments

I would like to thank everyone who made this book possible. I owe a special debt of gratitude to those colleagues who gave up their time to referee the Chapters. I would like to thank Professors Eitan Altman, Anatoli Iouditski, and Sylvain Sorin who initiated my interest in learning under uncertainty. The development of this book has spanned many years. The material as well as its presentation has benefited greatly from the inputs of many bright undergraduate and graduate students who worked on this topic. I would like to thank my colleagues from Ecole Polytechnique for their comments. It is a pleasure to thank my collaborators and coauthors of articles and papers on which part of the chapters of this book is based. They have played an important role in shaping my thinking for so many years. Their direct and indirect contributions to this work are significant. They are, of course, not responsible for the way I have assembled material, especially the parts I have added to and subtracted from our joint works to try to make the manuscript more coherent.

Special thank to Professor Tamer Başar who kindly accepted my invitation to write a foreword to the book, to Professor Eitan Altman who kindly accepted to write a preface. My thanks go to Professors Vivek Borkar, Mérouane Debbah, Samson Lasaulce, David Leslie, Galina Schwartz, Mihaela van der Schaar, Thanos Vasilakos, and Peyton H. Young for fruitful interactions or collaborations. I am grateful to the anonymous reviewers for assistance in proofreading the manuscript.

I thank seminar participants at the University of California at Los Angeles (UCLA), Ecole Polytechnique, University of California at Berkeley, University of Avignon, the National Institute for Research in Computer Science and Control (INRIA), University of Illinois at Urbana Champaign (UIUC), McGill University, Ecole Polytechnique Fédérale de Lausanne (EPFL), Ecole Supérieure d'Electricité (Supelec), University of California at Santa Cruz (UCSC), etc.

Artwork

The scientific graphs in the book are generated using the MATLAB software by Mathworks Inc. and the mean field package for simulation of large population games. The two figures in the cover of the book are examples of cycling learning processes.

The Author Bio

Hamidou Tembine has two master degrees, one in applied mathematics and one in pure mathematics from respectively Ecole Polytechnique and University Joseph Fourier, France. He received a PhD degree from Avignon University, France. He is an assistant professor at Ecole Supérieure d'Electricité, Supelec, France. His current research interests include evolutionary games, mean field stochastic games and applications. He was the recipient of many student travel grant awards, and best paper awards (ACM Valuetools 2007, IFIP Networking 2008, IEEE/ACM WiOpt 2009, IEEE Infocom Workshop 2011).

Contributors

Below is the list of contributors of the preface of the book and the foreword.

Eitan Altman, Research Director, Institut de Recherche en Informatique et en Automatique, INRIA Sophia Antipolis, France

Tamer Başar, Professor, Coordinated Science Laboratory, University of Illinois at Urbana-Champaign, Illinois, US.

Symbol Description

$\mathbb{R}^k$ $k-$dimensional Euclidean space, $k \geq 2$.

$\mathcal{N}$ Set of players (finite or infinite).

$\mathcal{B}(t)$ Random set of active players at time t.

$\mathcal{A}_j$ Set of actions of player j.

$s_j \in \mathcal{A}_j$ An element of $\mathcal{A}_j$.

$\Delta(\mathcal{A}_j)$ Set of probability distributions over $\mathcal{A}_j$.

$\mathcal{X}_j$ Mixed actions $\Delta(\mathcal{A}_j)$.

$a_{j,t}$ Action of the player j at time t. Element of $\mathcal{A}_j$.

$\mathbf{x}_{j,t}$ Randomized action of the player j at t. Element of $\mathcal{X}_j$.

$r_{j,t}$ Perceived payoff by player j at t.

$\hat{\mathbf{r}}_{j,t}$ Estimated payoff vector of player j at t. Element of $\mathbb{R}^{|\mathcal{A}_j|}$.

$\tilde{\beta}_{j,\epsilon}(\hat{r}_{j,t})$ Boltzmann-Gibbs strategy of player j. Element of $\mathcal{X}_j$.

$\tilde{\sigma}_{j,\epsilon}(\hat{r}_{j,t})$ Imitative Boltzmann-Gibbs strategy of player j. Element of $\mathcal{X}_j$.

$\mathbb{1}_{\{.\}}$ Indicator function.

l^2 Space of sequences $\{\lambda_t\}_{t \geq 0}$ such that $\sum_t |\lambda_t|^2 < +\infty$.

l^1 Space of sequences $\{\lambda_t\}_{t \geq 0}$ such that $\sum_t |\lambda_t| < +\infty$

$\lambda_{j,t}$ Learning rate of player j at t.

$\mathbf{e}_{s_j} \in \mathcal{X}_j$ Unit vector with 1 at the position of s_j, and zero otherwise.

$\| \cdot \|_2$ $\| x \|_2 = (\sum_k |x_k|^2)^{\frac{1}{2}}$.

$\langle ., . \rangle$ Inner product.

$\mathcal{W}$ State space, environment state.

$\mathbf{w} \in \mathcal{W}$ A scalar, a vector or a matrix (finite dimension).

$2^{\mathcal{D}}$ The set of the all the subsets of $\mathcal{D}$.

$C^0(A, B)$ Space of continuous functions from A to B.

$\mathbb{N}$ Set of natural numbers (non-negative integers).

$\mathbb{Z}$ Set of integers.

M_t Martingale.

1

Introduction to Learning in Games

One of the objectives in distributed interacting multi-player systems is to enable a collection of selfish players to achieve a desirable objective. There are two overriding challenges to achieving this objective. The first one is related to the complexity of finding the optimal solution. A centralized algorithm may be prohibitively complex when there are a large number of interacting players. This motivates the use of adaptive methods that enable players to self-organize into suitable, if not optimal, alternative solutions. The second challenge is limited information. Players may have limited knowledge about the status of other players, except perhaps for a small subset of neighboring players. The limitations in terms of information induce robust stochastic optimization, bounded rationality and inconsistent beliefs. As a consequence, there are many simple games in which the beliefs may converge but not the strategies. The outcome is sensitive to

- how much signalling is available to the players,

- the nature of the learning scheme,

- the way the information is exploited.

Common Knowledge

An event is part of the common knowledge [13, 14] if all the players know it; and all the players know that all other players know it; and all other players know that all other players know that all other players know it, and so on ad infinitum. This is much more than simply saying that something is known by all.

Rationality

If every player always maximizes his payoff, thus being able to perfectly calculate the possible probabilistic result of every action, we say that the player is *rational*.

Rationality under uncertainty

We use the notation $\mathbf{x}_1 \succeq \mathbf{x}_2$ to say that $\mathbf{x}_1$ is preferred to $\mathbf{x}_2$. A single decision maker (player) is called rational under uncertainty if its preferences satisfy the following conditions:

- (*Completeness*) Either $\mathbf{x}_1 \succeq \mathbf{x}_2$ or $\mathbf{x}_2 \succeq \mathbf{x}_1$ for any pair $(\mathbf{x}_1, \mathbf{x}_2)$.

- (*Transitivity*) If $\mathbf{x}_1 \succeq \mathbf{x}_2$ and $\mathbf{x}_2 \succeq \mathbf{x}_3$, then $\mathbf{x}_1 \succeq \mathbf{x}_3$.

- (*Monotonicity*) If $\mathbf{x}_1 \succeq \mathbf{x}_2$ and $\alpha_1 > \alpha_2, (\alpha_1, \alpha_2) \in [0,1]^2$, then $\alpha_1 \mathbf{x}_1 + (1-\alpha_1)\mathbf{x}_2 \succeq \alpha_2 \mathbf{x}_1 + (1-\alpha_2)\mathbf{x}_2$.

- (*Continuity*) If $\mathbf{x}_1 \succeq \mathbf{x}_2$ and $\mathbf{x}_2 \succeq \mathbf{x}_3$, then there exists a probability α such that $\mathbf{x}_2 = \alpha \mathbf{x}_1 + (1-\alpha)\mathbf{x}_3$.

- (*Independence*) If $\mathbf{x}_1 \succeq \mathbf{x}_2$, then $\alpha \mathbf{x}_1 + (1-\alpha)\mathbf{x}_3 \succeq \alpha \mathbf{x}_2 + (1-\alpha)\mathbf{x}_3$ for any $\alpha \in [0,1]$ and $\mathbf{x}_3$.

As above, the completeness condition ensures that all distributions can be compared with each other, and the transitivity condition implies that the distributions can be listed in order of preference. The monotonicity and continuity conditions assert that a lottery gets better smoothly as the probability of a preferred outcome increases. The independence condition implies that preferences only depend on the differences between lotteries; components that are the same can be ignored.

We will focus on payoffs representing the above properties. Expected payoffs and other type of payoffs such as risk-sensitive payoffs will be considered.

In the context of strategic decision-making, a strong property such as *Common Knowledge of Rationality* has been widely used.

Common Knowledge of Rationality (CKR)

CKR is informally defined as follows:

- The players are rational.

- The players all know that the other players are rational.

- The players all know that the other players know that they are rational.

- ... (in principle) ad infinitum.

However, such maximization is often quite difficult, and even if they wanted to, most players would be unable to carry them out in practice due to computational limitations, communication complexity, and anticipation errors [14].

Why should we be interested in distributed strategic learning?

Rationality does not imply that the outcome of a game must be an equilibrium, and neither does the common knowledge that players are rational, as equilibrium requires all players to coordinate on the same equilibrium. However, game theory results show that the outcome after multiple stages of moves is often much closer to equilibrium predictions than initial moves, which supports the idea that equilibrium arises as a result of players learning from experiences. The learning approach in games formalizes this idea, and examines how, which and what kind of equilibrium might arise as a consequence of a long-run non-equilibrium process of learning, adaptation, and/or imitation. Learning approaches are not necessarily procedures that are *"trying to reach Nash equilibrium"* but ask also the following questions:

- How can players optimize their own payoff while simultaneously learning about the strategies of other players?

- How can players optimize their own unknown payoff functions while simultaneously learning about their payoff functions and the behavior of the other players?

If players' strategies are completely observed at the end of each time slot, and players interact in the long-term, fairly simple rules perform well in terms of the players worst-case payoffs, and also guarantee that steady state of the process must be closer to an equilibrium point (saddle point, Cournot/Nash equilibrium, Hannan set, correlated equilibrium, etc.) than at the starting point. Now, if players cannot observe the strategies chosen by the other players, then steady states of learning process are not necessarily Nash equilibria because players can maintain incorrect beliefs about off-path play. Beliefs can also be incorrect due to cognitive limitations, errors, and anticipation mistakes.

The classical learning procedures in games (see [59, 60, 194, 193] and the references therein) assume that:

C1 The number of players is small and fixed,

C2 The rules of the game are known by every player, At least the distribution of possible payoffs, is known by all the players, and the action spaces are fully known.

C3 Every player is rational.

Even under these "strong" classical assumptions (C1-C3) and hyperrationality of players, learning in games is difficult because it is interactive:

> Each player's learning process correlates and affects what has to be learned by every other player over time.

Note that the correlations of the learning processes in games make a clear difference with what is usually done in classical single player decision-making.

In many practical scenarios, the above assumptions on information about the other players may not hold or may need a central authority or a coordinator to feedback some signals/messages. We will discuss how these assumptions can be relaxed. We specially examine the case where players can adjust their behavior only in response to their own realized payoffs, they have no knowledge of the overall structure of the game, the number of players that interact can be random, they cannot observe the actions or payoffs of the other players, they occasionally mistakes on their choices and readapt themselves based on past experiences, regret, reinforcement, and imitation.

In many situations in engineering, economics, and finance, one would like to have a learning procedure that does not require any information about the other players actions or payoffs and less memory (small number of parameters in terms of past own-actions and past perceived own-payoffs) as possible. Such a rule is said to be *uncoupled* or *fully distributed*. However, for a large class of games, no such general algorithm causes the players' period-by-period behavior to converge to Nash equilibrium. Hence, there is no guarantee for uncoupled dynamics that behaviors come close to Nash equilibrium most of the time.

A common correlation scheme in a one-shot game is the following: each player chooses his/her action according to his/her observation of the value of the same public signal. A strategy assigns an action to every possible observation a player can make. If no player would want to deviate from the recommended strategy (assuming the others don't deviate), the distribution is called a *correlated* Cournot/Nash equilibrium. A set is said to be *convex* if it is empty or it contains all the convex combinations of its elements. Convex sets and functions are very useful in robust optimization models, and have a structure that is convenient for learning algorithms. The set of correlated equilibria is convex and includes the Nash equilibria but is frequently much larger.

The works in [58, 70, 46] show that *regret minimizing procedures* can cause the empirical frequency distribution of play to converge to the set of correlated Nash equilibria. But there is no arbitrator, and no random device in these models. So,

What is the origin of this correlation?

It is clear that the correlation comes from the interactions via the perceived payoff which is incorporated in the regret term. Thus, the perceived own-payoffs over several stages have played the role of private signals of the process.

> Under these correlations during the interactions, it is important to know *when (rational or not) players can learn to play equilibrium starting from out of an equilibrium point.*

To answer to this question, we will distinguish two main classes of learning algorithms.

> A learning algorithm is *fully distributed or uncoupled if it does not need explicit information about the other players.*

Each player constructs his/her strategies and updates them by using the knowledge of his set of own-actions and measurement of own-payoffs.

> A learning scheme is *partially distributed* if each player implements its updating rules after explicitly receiving an explicit signal (e.g., action profile) or an information exchange about the other players.

This last class of learning procedures is more sophisticated than the class of fully distributed learning algorithms. These algorithms are still decentralized but they use explicit information from the others. In some scenarios, it may be possible to deduce useful information in terms of payoffs and/or strategies by combining the observed signal and the measured payoff if external knowledge is available to the player. These situations are referred as implicit information. Each of these learning classes is important in games and will be discussed in detail in this book.

1.1 Basic Elements of Games

Our approach in this book is based on some elements of dynamic robust game theory. It will be useful to present at the outset some of the basic notions of robust game theory and some general results on the existence and characterization of robust equilibria and maximin points. This section is dedicated to readers who are not familiar with strategic form games.

1.1.1 Basic Components of One-Shot Game

The four basic components of a one-shot game are

- The set of players, $\mathcal{N} = \{1, ..., n\}$, where n is the total number of players,

- The action spaces $\mathcal{A}_1, ..., \mathcal{A}_n$. These sets can be discrete or continuous, finite or infinite and assumed to be non-empty. Most of the time we will work with compact[1] action spaces (in Euclidean space).

- The preference structure of the players. In this book we focus on the case where the preference relation can be represented by a certain function: utility, payoff, cost, loss, reward, benefit, etc. The functions $\tilde{R}_1, ..., \tilde{R}_n$ are real-valued functions defined over the product space $\prod_{j=1}^{n} \mathcal{A}_j$; the outcome.

- The information structure and order of play of the game.

If the game is played once, the game is called a one-shot game, and its *strategic form* (also called normal-form) is represented by a collection[2]:

$$\tilde{\mathcal{G}} = \left(\mathcal{N}, \{\mathcal{A}_j\}_{j \in \mathcal{N}}, \{\tilde{R}_j\}_{j \in \mathcal{N}} \right).$$

A generic element of $\mathcal{A}_j$ is denoted by s_j. If player $j \in \mathcal{N}$ chooses an action s_j in $\mathcal{A}_j$ according to a probability distribution $\mathbf{x}_j = (x_j(s_j))_{s_j \in \mathcal{A}_j}$ over $\mathcal{A}_j$, the choice of $\mathbf{x}_j$ is called a randomized action or mixed strategy of the one-shot game. We denote by $\mathcal{X}_j = \Delta(\mathcal{A}_j)$ the set of probability distributions over $\mathcal{A}_j$. We will work with the canonical probability space. When the vector $\mathbf{x}_j \in \mathcal{X}_j$ is on a vertex of the simplex $\mathcal{X}_j$, the mixed strategy boils down to a pure strategy i.e., the deterministic choice of an action. We use the canonical inclusion that maps an element of $\mathcal{A}_j$ to the simplex by considering the mapping $s_j \longmapsto e_{s_j}$ where e_{s_j} is the unit vector with same size as the cardinality of $\mathcal{A}_j$, with 1 at the position of s_j and 0 otherwise.

⚠ Note that the notation e_{s_j} is to represent the vector $e_{j,s_j} \in \mathbb{R}^{|\mathcal{A}_j|}$. We do not use a specific notation for scalars, vectors, matrices or matrix of vectors because the action can be of any of these types or even a generic object.

The mixed extension payoff is $R_j : \prod_j \mathcal{X}_j \longrightarrow \mathbb{R}$ defined by

$$R_j(\mathbf{x}_1, \ldots, \mathbf{x}_n) \quad := \quad \mathbb{E}_{\mathbf{x}_1, \ldots, \mathbf{x}_n} \tilde{R}_j \tag{1.1}$$

$$= \quad \sum_{\mathbf{a} \in \prod_j \mathcal{A}_j} \tilde{R}_j(\mathbf{a}) \left(\prod_{j' \in \mathcal{N}} x_{j'}(a_{j'}) \right), \tag{1.2}$$

where $\mathbb{E}$ denotes the mathematical expectation operator. This is well defined by continuity and compactness assumption.

---

[1] A topological space $\mathcal{X}$ is called compact if every collection of open subsets of $\mathcal{X}$ that covers $\mathcal{X}$ contains a finite subcollection covering $\mathcal{X}$. It is well-known that $\mathcal{X} \subseteq \mathbb{R}^k$ is compact if and only if it is bounded and closed.

[2] Note that one-shot games in extensive form can be transformed in normal form but one may lose some information if the information sets are not added.

What is the difference between action and pure strategy?

For a static game in normal form as above, there is no real distinction between pure strategies and actions. However, the distinction will become important when we consider state-dependent one-shot games, robust games with partial information and dynamic games. This distinction will be clarified in Chapters 2 and 4. The next example explains the difference between actions and pure strategies in the context of access control protocols. The key idea is that

Information is crucial in the process of decision making.

Example 1.1.2. *Consider a finite number of users over a common shared medium [98, 112, 165]. Each user has its own queue with arrival process described over a probability space. Since the medium is common, there is interference if several users transmit simultaneously. This means the probability of success of each user depends on the other users; thus there is an interaction between them.*

An action in this game is a transmit power to be chosen among a list. There is also a "null" action that corresponds to the case where the user does not transmit, i.e, power zero. However, a user will not transmit if there is no packet to transmit. This implies that the knowledge of the size of own-queue or the knowledge of the state of the queue (empty or not) is necessary in order to construct a realistic model. Suppose that each user is able to observe the size of its queue or backoff. Then, a pure strategy is a mapping from the queue sizes to a set of transmit powers, which is completely different from just selecting a power from the list independently of everything else. Thus,

- *An action is an element of the list of transmit powers.*

- *A pure strategy is a mapping from the set of queue sizes to the list of transmit powers.*

Here an output of a pure strategy (function of queues) corresponds to a transmit power.

In the case of nonrandom payoffs, we will denote the instantaneous payoff function by $r_j(\mathbf{x})$ where $\mathbf{x} = (\mathbf{x}_j)_{j \in \mathcal{N}} \in \prod_{j \in \mathcal{N}} \mathcal{X}_j$.

Next, we define a Cournot/Nash equilibrium.

Definition 1.1.2.1 (Cournot/Nash equilibrium). *A strategy profile* $\mathbf{x}^*$ *is a Cournot/Nash equilibrium of the one-shot game*

$$\mathcal{G} = (\mathcal{N}, \{\mathcal{X}_j\}_{j \in \mathcal{N}}, \{r_j\}_{j \in \mathcal{N}}),$$

if it satisfies, for all $j \in \mathcal{N}$ and for all $\mathbf{x}_j \in \mathcal{X}_j$,

$$r_j(\mathbf{x}_j^*, \mathbf{x}_{-j}^*) \geq r_j(\mathbf{x}_j, \mathbf{x}_{-j}^*). \tag{1.3}$$

where $\mathbf{x}_{-j}^* = (\mathbf{x}_k^*)_{k \neq j}$.

We will often wish to focus on the value of $\mathbf{x}$ at which the maximum is achieved rather than the maximum value of the function itself, so we introduce a symbol argmax.

> **About the notation arg max.**

Suppose $\mathbf{x}$ is an arbitrary member of the set $\mathcal{X}$. Let $r(\mathbf{x})$ be some function that is defined for all $\mathbf{x} \in \mathcal{X}$. The symbol arg max is defined by the following equivalence:

$$\mathbf{x}^* \in \arg\max_{\mathbf{x} \in \mathcal{X}} \ r(\mathbf{x}) \iff r(\mathbf{x}^*) = \max_{\mathbf{x} \in \mathcal{X}} r(\mathbf{x}).$$

This suppose that the function has a maximum (the sup is attained).

⚠ Note that arg max returns a set. For this reason we do not write $\mathbf{x}^* = \arg\max r(\mathbf{x})$ because a function may take its maximum value for more than one element in the set $\mathcal{X}$. Because the symbol arg max may return a set of values rather than a unique value, it is called a correspondence rather than a function. Even in the case of unique maximizer, one has to write $\{\mathbf{x}^*\} = \arg\max r(\mathbf{x})$.

Example 1.1.3. *Consider the function defined on $x \in [-1, 1]$, by $r(x) = x^2$. This function achieves its maximum at $x^* \in \{-1, +1\}$. So,*

$$\arg\max_{x \in [-1, +1]} x^2 = \{-1, 1\}.$$

The following subproblem is a part of Definition 1.1.2.1.

> For each $j \in \mathcal{N}$, and $\mathbf{x}_{-j}$, one would like to solve the optimization problem $OP_j(\mathbf{x}_{-j})$:
>
> $$OP_j(\mathbf{x}_{-j}): \quad \sup_{\mathbf{x}_j \in \mathcal{X}_j} \quad r_j(\mathbf{x}_j, \mathbf{x}_{-j}) \tag{1.4}$$
>
> An interesting question that we will investigate is the following: Is it possible to solve the optimization problem $OP_j(\mathbf{x}_{-j})$ without knowledge of the mathematical structure function r_j?

The solutions to the problem $OP_j(\mathbf{x}_{-j})$ are sometimes referred as **best-responses** of player j to the strategies $\mathbf{x}_{-j}$ of the others. A best response $BR_j(.)$ is a correspondence (multi-valued mapping) from $\prod_{j' \in \mathcal{N}, \ j' \neq j} \mathcal{X}_{j'} \longrightarrow 2^{\mathcal{X}_j}$, defined by the set

$$BR_j(\mathbf{x}_{-j}) = \arg\max_{\mathcal{X}_j} r_j(., \mathbf{x}_{-j}).$$

In the next section, we address the question of non-emptiness of this set for a given strategy profile of the other players.

Exercise 1.1. *Show that the Definition (1.1.2.1) is equivalent to the following characterization: For any player j, any action $s_j \in \mathcal{A}_j$,*

$$\mathbf{x}_j^*(s_j) > 0 \Longrightarrow r_j(\mathbf{e}_{s_j}, \mathbf{x}_{-j}^*) = \max_{s_j' \in \mathcal{A}_j} r_j(\mathbf{e}_{s_j'}, \mathbf{x}_{-j}^*) = \max_{\mathbf{x}_j' \in \mathcal{X}_j} r_j(\mathbf{x}_j', \mathbf{x}_{-j}^*) \quad (1.5)$$

which says exactly that

$$\forall\, j \in \mathcal{N}, \ support(\mathbf{x}_j^*) := \{s_j \mid x_j^*(s_j) > 0\} \subseteq \arg\max_{\mathcal{X}_j} r_j(., \mathbf{x}_{-j}^*). \quad (1.6)$$

Hint: To prove the last equality in (1.5), use the concavity of the mixed extension with the respect to x_j, continuity and compactness [37].

These relations have been established in Nash (1951, [119]) in his paper entitled Non-Cooperative Games. Note that this property is related to the Wardrop first principle (Wardrop 1952, [187], Beckmann et al. 1956, [21]). Typically, if $\mathcal{P} = \{1, 2, \dots, K\}$, $K \geq 1$ denotes a set of subpopulations and $\mathcal{A}^p$ a finite set of actions in subpopulation, the population profile $\mathbf{z} = (\mathbf{z}^p)_{p \in \mathcal{P}}$, $\mathbf{z}^p \in \Delta(\mathcal{A}^p)$, is an equilibrium state if for any subpopulation p, and any action $a^p \in \mathcal{A}^p$ such that

$$z^p(a^p) > 0 \Longrightarrow r_{a^p}^p(\mathbf{z}) = \max_{b^p \in \mathcal{A}^p} r_{b^p}^p(\mathbf{z}) = \max_{\mathbf{y}^p \in \Delta(\mathcal{A}^p)} \int_{b^p} \mathbf{y}^p(db^p) r_{b^p}^p(\mathbf{z}) \quad (1.7)$$

where $r_{b^p}^p(z)$ if the payoff of a generic player in subpopulation p with the action b^p. As we can see this definition of *equilibrium state* in the context of multiple subpopulations is more general than the definition in (1.5). Note that, in the multiple subpopulation game, for a fixed action a^p, the corresponding payoff $r_{a^p}^p(.)$ may depend on $\mathbf{z}^{-p} = (\mathbf{z}^{p'})_{p' \in \mathcal{P},\ p' \neq p}$ but also on z^p. Also, considering macro-players, one gets the first definition (1.1.2.1). However, the function $r_{a^p}^p(\mathbf{z})$ do not need to be concave/convex in z^p.

1.1.4 State-Dependent One-Shot Game

If there are independent random variables that determine the one-shot game, we add a state space $\mathcal{W}$ and the payoff function will be defined on product space: $\mathcal{W} \times \prod_{j \in \mathcal{N}} \mathcal{X}_j$. A strategy in this game depends on the information of the realized state.

> A pure strategy of player j in the one-shot game is a mapping from the set of information available to player j to its action space.

1.1.4.1 Perfectly-Known State One-Shot Games

Given a state $\mathbf{w} \in \mathcal{W}$ that is known by all the players, we define a normal-form game

$$\tilde{\mathcal{G}}^1(\mathbf{w}) = (\{\mathbf{w}\}, \mathcal{N}, (\mathcal{A}_j^{\mathbf{w}}, \tilde{R}_j(\mathbf{w}, \mathbf{a}))_{j \in \mathcal{N}}),$$

where $\mathcal{A}_j^{\mathbf{w}}$ is the set of mapping from $\{\mathbf{w}\}$ to the set $\mathcal{A}_j$ and the payoff takes an element of $\prod_{j \in \mathcal{N}} \mathcal{A}_j^{\mathbf{w}}$ to a real value which we simply denote by $\tilde{R}_j(\mathbf{w}, \mathbf{a})$. This is a parameterized game with a given parameter $\mathbf{w}$ that is common knowledge and which influences both strategies and payoffs.

1.1.4.2   One-Shot Games with Partially-Known State

Given a state $\mathbf{w} \in \mathcal{W}$, each player knows only some components of the state. For example, player j observes/measures w_j but does not know the current value of w_{-j}. We assume that the distribution μ_{-j} over state w_{-j} is known by player j. This defines a one-shot game with incomplete information. Each player j is able to compute the payoff function given by

$$\tilde{R}_j^2(w_j, \mu_{-j}; a) = \mathbb{E}_{w_{-j} \sim \mu_{-j}} \left[\tilde{R}_j(w, a) \mid w_j, \mu_{-j} \right].$$

Let $\Delta(W_{-j})$ be the set of probability measures over W_{-j}, equipped with the canonical sigma-algebra. Then, $\mu_{-j} \in \Delta(W_{-j})$. Note that $\tilde{R}_j^2$ is defined over $W_j \times \left(\prod_j \mathcal{A}_j \right) \times \Delta(W_{-j})$, where W_j denotes the j-th component state space observed by player j. A pure strategy of player j is this game with partial state observation is a mapping from the given own-state w_j, and the distribution μ_{-j} to $\mathcal{A}_j$. We denote the game by

$$G^2 = (\mathcal{N}, \mathcal{A}, (w_j, \mu_{-j}), (\tilde{R}_j^2)_{j \in \mathcal{N}}).$$

A pure equilibrium for the game G^2 can be seen as a Bayesian (Nash) equilibrium, and it is characterized by

$$\sigma_j : \; W_j \times \Delta(W_{-j}) \longrightarrow \mathcal{A}_j,$$

$$\sigma_j(w_j, \mu_{-j}) \in \arg \max_{\tilde{a}_j \in \mathcal{A}_j} \tilde{R}_j^2(w_j, \mu_{-j}; \tilde{a}_j, \sigma_{-j}(w_{-j})),$$

$\forall w_j$ and $\forall j \in \mathcal{N}$.

> Note that, a pure strategy for player j in this one-shot game is a mapping from the set $\{(w_j, \mu_{-j})\}$ to the set $\mathcal{A}_j$

1.1.4.3   State Component is Unknown

The players do not know any component of the current state. The distributions over the full states are known in addition to the mathematical structure of the payoffs. Now each player j is able to compute the following payoff function

$$\tilde{R}_j^3(\mu; a) = \mathbb{E}_{w \sim \mu} \tilde{R}_j(w, a),$$

where $\mu \in \Delta(\mathcal{W})$. This leads to an expected robust game that we denote by

$$G^3 = (\mathcal{N}, \mathcal{A}, \mathcal{W}, \mu, (\tilde{R}_j^3)_{j \in \mathcal{N}}).$$

A pure strategy for player j in the game G^3 corresponds to a function of μ that gives an element of $\mathcal{A}_j$.

1.1.4.4 Only the State Space Is Known

Now, we assume that the distribution over the state is also unknown but the state space $\mathcal{W}$ is known. Then, each player can adopt different behaviors depending on how he/she sees the state space. The well-known approaches in that case are the maximin robust and the maximax approaches (pessimistic or optimistic approaches) and their variants. The payoff of player j in the maximin robust game is given by

$$\tilde{R}_j^4(a) = \inf_{w \in \mathcal{W}} \tilde{R}_j(w, a).$$

We denote the game by

$$G^4 = (\mathcal{N}, (\mathcal{A}_j)_{j \in \mathcal{N}}, \mathcal{W}, (\tilde{R}_j^4)_{j \in \mathcal{N}}).$$

The equilibria of G^4 are called *maximin robust equilibria*.

Definition 1.1.4.5. *Worse-case Cournot/Nash equilibrium if (i) there exists* $w^*(x)$ *such that*

$$r_j(w^*(x), x) = \inf_{\mathbf{w}} r_j(w, x),$$

and (ii):

$$r_j(w^*(x), x) = \max_{x_j'} r_j(w^*(x_j', x_{-j}), x_j', x_{-j})$$

⚠ The choice of the set $\mathcal{W}$ plays an important role in the Worse-case Cournot/Nash equilibrium. For example, if $\tilde{R}_j(H, A) = \log \det \left(I + \frac{1}{N_0} \sum_{j' \in \mathcal{N}} H_{j'}^* P_{j'} H_{j'} \right)$, and $H_j \in \mathbb{C}^{n \times n}$, then

$$\inf_H \tilde{R}_j(H, A) = 0.$$

Therefore everything is optimal, which does not reflect the solutions to the mutual information optimization.

Similarly, the payoff of player j in the maximax robust game is given by

$$\bar{R}_j^5(a) = \sup_{w \in \mathcal{W}} \tilde{R}_j(w, a).$$

1.1.5 Perfectly Known State Dynamic Game

At each time-step t, a state is randomly picked according to some distribution $\nu(.)$ defined in a probability space. The selected state $\mathbf{w}_t$ is perfectly and publicly known. Then, the perfectly-known state simultaneous-act game $\tilde{\mathcal{G}}(\mathbf{w}_t)$ is played. Note that, in this context, the actions at time t are dependent on the state $\mathbf{w}_t$.

1.1.6 Unknown State Dynamic Games

We consider dynamic games where the players have an opportunity to act several times. Additionally, the dynamic game can be under uncertainty; that is, the game can have a state which changes randomly from time slot to another.

> In network scenarios, the state will be typically *a backoff state, a battery state, a vector of channel states, the availability of the resources etc.*

The games under state uncertainty are sometimes called *robust games*, that is in the class of games with incomplete information and imperfect payoffs. If $R_j(\mathbf{w}, \mathbf{x})$ denotes the non-random payoff functions of player j, one way to analyze the equilibrium issue is to consider the expected game defined by

$$\left(\mathcal{N}, (\mathcal{X}_j)_{j \in \mathcal{N}}, \mathbb{E}_{\mathbf{w}} R_j(\mathbf{w}, .)\right).$$

We denote the set of states by $\mathcal{W}$. The set $\mathcal{W}$ is equipped with a probability measure ν.

> **Problem formulation:** For each $j \in \mathcal{N}$, and $\mathbf{x}_{-j}$, one would like to solve the problem $ROP_j(\mathbf{x}_{-j})$:
>
> $$ROP_j(\mathbf{x}_{-j}) : \sup_{\mathbf{x}_j \in \mathcal{X}_j} \quad \mathbb{E}_{\mathbf{w}} R_j(\mathbf{w}, \mathbf{x}_j, \mathbf{x}_{-j}) \qquad (1.8)$$
>
> The key question that we will address in this manuscript is: how to solve the robust optimization problem $ROP_j(\mathbf{x}_{-j})$ without knowledge of the function $R_j(.)$?

The Nash equilibrium of the expected robust game is therefore defined as follows:

Definition 1.1.6.1 (Equilibrium of the expected robust game). *A strategy profile* $(\mathbf{x}_j)_{j \in \mathcal{N}} \in \prod_{j=1}^{n} \mathcal{X}_j$ *is a (mixed) state-independent equilibrium for the expected robust game if and only if* $\forall j \in \mathcal{N}, \ \forall \mathbf{y}_j \in \mathcal{X}_j$,

$$\int_{\mathbf{w} \in \mathcal{W}} R_j(\mathbf{w}, \mathbf{y}_j, \mathbf{x}_{-j}) \ \nu(d\mathbf{w}) \le \int_{\mathbf{w} \in \mathcal{W}} R_j(\mathbf{w}, \mathbf{x}_j, \mathbf{x}_{-j}) \ \nu(d\mathbf{w}). \qquad (1.9)$$

which is exactly the definition 1.1.2.1 with the payoff function $r_j : \prod_{j \in \mathcal{N}} \mathcal{X}_j \longrightarrow \mathbb{R}$, defined by

$$r_j(\mathbf{x}_j, \mathbf{x}_{-j}) = \int_{\mathbf{w} \in \mathcal{W}} R_j(\mathbf{w}, \mathbf{x}_j, \mathbf{x}_{-j}) \ \nu(d\mathbf{w}) = \mathbb{E}_{\mathbf{w}} R_j(\mathbf{w}, \mathbf{x}_j, \mathbf{x}_{-j}).$$

Note that the strategy profile is state-free, i.e., independent of the environment state $\mathbf{w}$.

Two-by-two expected robust games

Consider two players. The action spaces of each player are independent on the state. Player I has two actions $\{Up, Down\}$ in any state, and Player II's action set is $\{Left, Right\}$ in any state. We transform the robust game into expected game by taking the expectation of the payoff. Table 6.1 represents the payoff entries. where

Decisions of player I \ Decisions of player II	*Left*	*Right*
Up	$\bar{a}_1, \bar{a}_2$	$\bar{b}_1, b_2$
Down	$\bar{c}_1, \bar{c}_2$	d_1, d_2

TABLE 1.1

2×2 expected robust game.

- $\bar{a}_1 = \mathbb{E}_{\mathbf{w}}\tilde{R}_1(\mathbf{w}, Up, Left),\ \bar{a}_2 = \mathbb{E}_{\mathbf{w}}\tilde{R}_2(\mathbf{w}, Up, Left)$

- $\bar{b}_1 = \mathbb{E}_{\mathbf{w}}\tilde{R}_1(\mathbf{w}, Up, Right),\ \bar{b}_2 = \mathbb{E}_{\mathbf{w}}\tilde{R}_2(\mathbf{w}, Up, Right)$

- $\bar{c}_1 = \mathbb{E}_{\mathbf{w}}\tilde{R}_1(\mathbf{w}, Down, Left),\ \bar{c}_2 = \mathbb{E}_{\mathbf{w}}\tilde{R}_2(\mathbf{w}, Down, Left)$

- $\bar{d}_1 = \mathbb{E}_{\mathbf{w}}\tilde{R}_1(\mathbf{w}, Down, Right),\ \bar{d}_2 = \mathbb{E}_{\mathbf{w}}\tilde{R}_2(\mathbf{w}, Down, Right)$.

It is easy to show that the optimal completely mixed state-independent strategy (if it exists) is given by

$$x^* = \frac{\bar{d}_1 - \bar{b}_1}{\bar{d}_1 - \bar{b}_1 + \bar{a}_1 - \bar{c}_1},$$

$$y^* = \frac{\bar{d}_2 - \bar{b}_2}{\bar{d}_2 - \bar{b}_2 + \bar{a}_2 - \bar{c}_2}.$$

Convexity and Quasi-Convexity

We briefly introduce the notion of convexity. Let $\mathcal{E}$ be a linear space. A subset $\mathcal{C}$ of $\mathcal{E}$ is *convex* if for all

$$\alpha c_1 + (1 - \alpha)c_2 \in \mathcal{C},\ \forall (c_1, c_2) \in \mathcal{C}^2, \alpha \in [0, 1].$$

The following properties hold:

- The closure and the interior of a convex set are convex.

- The intersection of two convex sets is a convex set.

A function $f : \mathcal{C} \longrightarrow \mathbb{R}$ is convex if

$$f(\alpha c_1 + (1 - \alpha)c_2) \leq \alpha f(c_1) + (1 - \alpha)f(c_2), \ \forall (c_1, c_2) \in \mathcal{C}^2, \alpha \in [0, 1]$$

Note that the convexity of the domain $\mathcal{C}$ is a prerequisite for saying that a function is convex.

A function $f : \mathcal{C} \longrightarrow \mathbb{R}$ where the set $\mathcal{C}$ is a convex set is called *concave* if the function $(-f)$ is convex.

A function $f : \mathcal{C} \longrightarrow \mathbb{R}$ where the set $\mathcal{C}$ is a convex set is called *strictly convex* if

$$f(\alpha c_1 + (1 - \alpha)c_2) < \alpha f(c_1) + (1 - \alpha)f(c_2),$$

$$\forall (c_1, c_2) \in \mathcal{C}^2, c_1 \neq c_2, \ \alpha \in (0, 1) = [0, 1] \backslash \{0, 1\}.$$

If $f : \mathcal{C} \longrightarrow \mathbb{R}$ is a function and η is a scalar, the set

$$\{c \in \mathcal{C} \mid f(c) \leq \eta\}$$

is called level set of f. One nice property is that if f is a convex function, then all its level sets are convex.

⚠ Note, however, that convexity of the level sets does not imply convexity of the function. A counterexample is obtained with the function $f : c \longmapsto \sqrt{|c|}$ is not convex. The level sets are the empty set or the set $\{c \mid c^2 \leq \eta^4\}$.

What is a topological space?

A topological space is a set $\mathcal{E}$ together with $\mathcal{T}$ (a collection of subsets of $\mathcal{E}$) satisfying the following axioms:

- The empty set $\emptyset$ and the whole space $\mathcal{E}$ are in $\mathcal{T}$.

- $\mathcal{T}$ is closed under arbitrary union.

- $\mathcal{T}$ is closed under finite intersection.

The collection $\mathcal{T}$ is called a topology on $\mathcal{E}$. The sets in $\mathcal{T}$ are called the open sets, and their complements in $\mathcal{E}$ are called closed sets. A subset of $\mathcal{E}$ may be neither closed nor open, either closed or open, or both.

⚠ A topological set is not like a DOOR!

Compactness

A topological space $\mathcal{E}$ is compact if, whenever a collection of open sets covers the space, some subcollection consisting only of finitely many open sets also covers the space. For Euclidean space, a subset is compact if it is closed and bounded subset.

A function between topological spaces is called *continuous* if the inverse image of every open set is open. A *Hausdorff space* or separated space is a topological space in which distinct points can have disjoint neighborhoods.

Semicontinuity

A real-valued function f is lower semicontinuous if $\{e \in \mathcal{E} \mid f(e) > \alpha\}$ is an open set for every $\alpha \in \mathbb{R}$. For real-valued functions, this is equivalent to say that a mapping $f : A \longrightarrow \mathbb{R} \cup \{+\infty\}$ is lower semicontinuous at a_0 if $\liminf_{a \longrightarrow a_0} f(a) \geq f(a_0)$. f is lower-semicontinuous if it is lower semicontinuous at every point of its domain.

Upper semicontinuity is defined in a similar way by considering $\limsup$.

An important property that we will use in robust games for the emptiness of the best-response set is the following:

> If $\mathcal{C}$ is a compact space and $f : \mathcal{C} \longrightarrow \mathbb{R}$ is upper semicontinuous, then f has a maximum on $\mathcal{C}$.

A function $f : \mathcal{C} \longrightarrow \mathbb{R}$ is quasiconvex if

$$f(\alpha c_1 + (1 - \alpha)c_2) \leq \max(f(c_1), f(c_2)), \ \forall (c_1, c_2) \in \mathcal{C}^2, \ \alpha \in [0, 1]$$

⚠ The floor function $x \longmapsto \lfloor x \rfloor$ is a quasiconvex function that is neither convex nor continuous.

A quasiconcave function is a function whose negative is quasiconvex. The probability density function of the normal distribution is quasiconcave but not concave.

Next, we present some well-known results in fixed point theory.

Fixed-point theorem

Theorem 1.1.6.2 (Kakutani, 1941). *Let $\mathcal{D} \subseteq \mathbb{R}^d$, $d \geq 1$ be a nonempty, compact and convex set. Let the correspondence $\phi : \mathcal{D} \longrightarrow 2^{\mathcal{D}}$ be upper hemicontinuous (its graph is closed) with nonempty convex values. Then ϕ has a fixed point, i.e,*

$$\exists \, x^* \in \mathcal{D}, \ x^* \in \phi(x^*) \in 2^{\mathcal{D}}.$$

Theorem 1.1.6.2 provides sufficiency conditions for existence of equilibria. It is a generalization of Brouwer's fixed-point theorem, Schauder's fixed point theorem, etc.

Theorem 1.1.6.3 (Existence of equilibrium, Kakutani-Glicksberg-Debreu–Fan). *Suppose that*

- *for each $j \in \mathcal{N}$, $\mathcal{X}_j$ is a nonempty, convex, and compact subset of a Haussdorf topological vector space,*

- *for each j, r_j is upper semicontinuous on $\prod_{j \in \mathcal{N}} \mathcal{X}_j$, for each j and*

- *each $x'_j \in \mathcal{X}_j$ the functions $x_{-j} \longmapsto r_j(x'_j, x_{-j})$ is lower semicontinuous and,*

- *for each j and $x_{-j} \in \prod_{i \neq j} \mathcal{X}_i$, $x'_j \in \mathcal{X}_j \longmapsto r_j(x'_j, x_{-j})$ is quasi-concave.*

Then, the one-shot game

$$(\mathcal{N}, (\mathcal{X}_j, r_j(.))_{j \in \mathcal{N}}),$$

has at least one (Cournot/Nash) equilibrium.

⚠️ The above result provides a sufficiency condition for existence of an equilibrium. It does not address the question of how an equilibrium can be reached or can be played, which will be formulated in terms of partial information and bounded reasoning. Remember that you don't have to know you are playing a game to be in a game!

Theorem 1.1.6.4. *The following results hold.*

- *If the action spaces are nonempty and finite and $\mathcal{W}$ is compact and nonempty, then the expected robust game has at least one equilibrium in mixed (state-independent) strategies.*

- *If the action spaces and $\mathcal{W}$ are compact and non-empty, and the functions $\tilde{R}_j(\mathbf{w}, a_j, \mathbf{a}_{-j})$ are continuous then the expected robust game has at least one equilibrium in mixed strategies (measures over a compact set).*

- *If the action spaces are convex, compact and non-empty, $\mathcal{W}$ compact and non-empty and $\tilde{R}_j(\mathbf{w}, a_j, \mathbf{a}_{-j})$ continuous in $\mathbf{a}$, quasi-concave in a_j, then the existence of pure equilibria holds in the expected robust game.*

Proof. • To prove the first statement, we show that for every player j, the payoff function

$$\mathbf{x} \longmapsto r_j(\mathbf{x}) = \mathbb{E}_{\mathbf{w}}[R_j(\mathbf{w}, \mathbf{x})] = \int_{\mathbf{w}} R_j(\mathbf{w}, \mathbf{x}) \ \nu(d\mathbf{w})$$

is continuous. Moreover, $\mathbf{x}_j \longmapsto r_j(\mathbf{x})$ is quasi-concave for any vector

$\mathbf{x}_{-j}$. The strategy spaces $\mathcal{X}_j$ are convex, compact, and non-empty. We can therefore use one of the fixed point theorems (Brouwer, Schauder, Kakutani, etc.) which guarantee the existence of least one equilibrium in mixed strategies. To use the Brouwer theorem, we may need to construct Brown-von Neumann excess payoff ratio or Nash map.

- The proofs of the second statement and the last statement follow the same line using the Ky Fan fixed point theorem (which is a generalization of the Kakutani fixed point theorem).

- The proof of the last statement follows from Kakutani-Glicksberg-Debreu-Fan's theorem applied to the payoff function $\mathbb{E}\tilde{R}_j(\mathbf{w}, a_j, \mathbf{a}_{-j})$.

$\square$

As we can observe, fixed-point theorems play important roles in establishing existence of equilibria. But there are many classes of games (namely, weakly acyclic game, potential games, supermodular games, robust potential games, population games with monotone payoffs, etc, see [99] and the references therein) where the existence of Nash equilibrium can be established without recourse to fixed-point theorems.

Another way to analyze the equilibrium of robust games is to consider the worst-state case game or Maximin robust game defined by

$$\left(\mathcal{N}, (\mathcal{X}_j)_{j\in\mathcal{N}}, \inf_{\mathbf{w}} R_j(\mathbf{w}, .)\right).$$

The Nash equilibrium of the maximin robust game is therefore defined as follows:

Definition 1.1.6.5 (Maximin Robust Equilibrium). *A strategy profile* $(\mathbf{x}_j)_{j\in\mathcal{N}} \in \prod_{j\in\mathcal{N}} \mathcal{X}_j$ *is a (mixed) state-independent maximin robust equilibrium if and only if*

$$\forall\, j \in \mathcal{N}, \quad \inf_{\mathbf{w}\in\mathcal{W}} R_j(\mathbf{w}, \mathbf{y}_j, \mathbf{x}_{-j}) \leq \inf_{\mathbf{w}\in\mathcal{W}} R_j(\mathbf{w}, \mathbf{x}_j, \mathbf{x}_{-j}) \,, \; \forall\mathbf{y}_j \in \mathcal{X}_j. \quad (1.10)$$

which is exactly the Definition 1.1.2.1 with the payoff function $r_j : \prod_{j\in\mathcal{N}} \mathcal{X}_j \longrightarrow \mathbb{R}$, defined by

$$r_j(\mathbf{x}_j, \mathbf{x}_{-j}) = \inf_{\mathbf{w}\in\mathcal{W}} R_j(\mathbf{w}, \mathbf{x}_j, \mathbf{x}_{-j}).$$

Again, the strategy profile is distribution-free. The players do not need to know the state $\mathbf{w}$ since they focus on the worst-state case. Note that Theorem 1.1.6.4 extends to this maximin robust game under similar sufficient conditions.

Other Notions of Robust Equilibria

We describe the robust game $\mathcal{G}$ the collection $\mathcal{G} = (\mathcal{G}(\mathbf{w}))_{\mathbf{w}\in\mathcal{W}}$ where $\mathcal{G}(\mathbf{w})$ is a normal form game. All the games "$\mathcal{G}(\mathbf{w})$" have the same set of action spaces (independent of the state "$\mathbf{w}$"). Denote by $E(\mathcal{G}(\mathbf{w}))$ the set of Nash equilibria of the normal-form game $\mathcal{G}(\mathbf{w})$ for a fixed $\mathbf{w}$.

Definition 1.1.6.6. *An action profile a^* is state-robust equilibrium of $\mathcal{G}$ if*

$$\forall \mathbf{w} \in \mathcal{W}, \; a^* \in E(\mathcal{G}(\mathbf{w})).$$

Denote by $E(\mathcal{G})$ the set of equilibria of $\mathcal{G}$. The interpretation of this state-robust equilibrium is the following: a state-robust equilibrium is independent of $\mathbf{w}$ and it is an equilibrium for "all" the games $\mathcal{G}(\mathbf{w})$, $\mathbf{w}$ describing $\mathcal{W}$. This is a distribution-free equilibrium. This notion of equilibrium may be too demanding since one has an existence problem even for simple cases. Below, we provide examples of state-robust equilibrium as well as examples where there is no such equilibrium.

An alternative to this definition of quantifying the set of states for which a given profile a is equilibrium. It consists of finding the probability for a given action profile $\mathbf{a} \in \prod_{j \in \mathcal{N}} \mathcal{A}_j$ to be an equilibrium of $\mathcal{G}$.

Similarly, the analogue of the uniqueness problem becomes: What is the probability that $(a, a') \in E(\mathcal{G})^2$, $a \neq a'$?

This last problem consists to quantify the measure of the set: $\mathbb{P}\left(\{\mathbf{w} \mid (a, a') \in E(\mathcal{G}(\mathbf{w}))^2\}\right)$ for a given pair $(a, a'), a \neq a'$. Here, we have a random variable over $\mathcal{W}$, associated with the collection $(\mathcal{W}, \mathcal{P}(\mathcal{W}), \mathbb{P})$ where $\mathcal{P}(\mathcal{W})$ is a $\sigma-$algebra.

We say that the probability of having more than one pure equilibrium in $\mathcal{G}$ is zero if $\forall a \neq a'$,

$$\mathbb{P}\left(\{\mathbf{w} \mid (a, a') \in E(\mathcal{G}(\mathbf{w}))^2\}\right) = 0.$$

Example 1.1.6.7 (Many equilibria versus domination). *Consider the following two-player, two-state, two-action per state game with symmetric payoff.*

P1\P2	s_1' of P2	s_2' of P2
s_1 of P1	1	1
s_2 of P1	1	1
state "w = 0"		

P1\P2	s_1' of P2	s_2' of P2
s_1 of P1	1	1/4
s_2 of P1	1/4	0
state "w = 1"		

There are two states represented by $\{0, 1\}$ indexed by w. The set of actions of player 1 $\mathcal{A}_1(w = 0) = \{s_1, s_2\} = \mathcal{A}_1(w = 1)$, and the actions space of player 2 is $\mathcal{A}_2(w = 1) = \{s_1', s_2'\} = \mathcal{A}_2(w = 0)$. Player 1 chooses a row and player 2 chooses a column. Lets examine the equilibria in each state. In state $w = 0$, any action is an equilibrium, i.e, the set of pure equilibria is $\{1, 2\}^2$ and the set of mixed equilibria is

$$\{((x, 1 - x), (y, 1 - y)) \mid x \in [0, 1], y \in [0, 1]\}.$$

In state $w = 1$, the first action is strictly dominant in the sense that the payoff of the action s_1 is strictly greater than the payoff of the action s_2, independently of the choice of the other player. Thus, the equilibrium is $(1, 1)$. We deduce that the action profile which consists of playing the first action for both players leads to a state-robust equilibrium.

Example 1.1.6.8 (No pure distribution-independent equilibrium). *Consider the following state-dependent one-shot game. There are two states. In state 0, the first action dominates the second one, and in state 1 the second action is dominant. This means that (s_1, s_1) is a pure equilibrium of the matrix game $\mathcal{G}(0)$, and (s_2, s_2) is an equilibrium of the game $\mathcal{G}(1)$. We verify that there is no pure distribution-independent equilibrium in $\mathcal{G}$.*

P1\P2	s_1' of P2	s_2' of P2
s_1 of P1	1	1/4
s_2 of P1	1/4	0

state"$w = 0$"

P1\P2	s_1' of P2	s_2' of P2
s_1 of P1	0	1/4
s_2 of P1	1/4	1

state"$w = 1$"

Example 1.1.6.9 (Three states). *We consider a three-player game. There are three states. In each state, each player has two actions. In the table below, Player 1 chooses a row, player 2 chooses a column and player 3 chooses one of the two blocks.*

1	1
1	1

$w = 0$

continuum of MEs

1	1/4
1/4	0

$w = 1$

$1 - E$

0	1/4
1/4	1

$w = 2$

$1 - E$

Assume that the states are drawn according the probability distribution $\tilde{w} \hookrightarrow (\mu_1, \mu_2, 1 - \mu_1 - \mu_2)$. We compute $\mathbb{P}(\{\mathbf{w} \mid (1,1) \in E(\mathcal{G}(\mathbf{w}))\})$ in order to quantify the probability for $(1,1)$ to be an equilibrium. It is easy to see that this probability is at least $\mu_1 + \mu_2$, which can be arbitrary close to 1. Furthermore, there is no pure (distribution-independent) equilibrium in $\mathcal{G}$ but each of the games $\mathcal{G}(\mathbf{w})$ has at least one pure equilibrium.

This leads naturally to the question of existence in distribution-dependent pure equilibrium.

Example 1.1.6.10. *To illustrate an example of distribution-dependent equilibrium, we consider the above state-dependent scenario where the states are drawn according to a distribution $(\mu, 1 - \mu)$ for some $1 > \mu > 0$. In addition to the state-dependent game, we represent also the expected robust game:*

1	1/4
1/4	0

$w = 0$

0	1/4
1/4	1

$w = 1$

μ	1/4
1/4	$1 - \mu$

expected

We can take a strategy in the form:

$$x_1^*(\mu) = \begin{cases} 1 & if \ \mu > \mu^* \\ 0 & otherwise \end{cases}$$

independent of w but depends on the law μ.

Note that $\mathcal{G}$ has a distribution-dependent pure equilibrium.

Example 1.1.6.11 (Continuum of mixed equilibria in each game). *Let* $\Delta(\{1,2,3,4\}) = \{x = (x_1, x_2, x_3, x_4) \in \mathbb{R}_+^4 \mid x_1 + x_2 + x_3 + x_4 = 1\}$ *be the three-dimensional simplex of* $\mathbb{R}^4$ *and consider the two family of games*

0	0	−1	−1		−1	−1	−1	−1
0	0	−1	−1		−1	−1	−1	−1
−1	−1	−1	−1		−1	−1	0	0
−1	−1	−1	−1		−1	−1	0	0

Each game has a continuum of mixed equilibria: $\Delta(\{1,2\})^2$ *for the first game and* $\Delta(\{3,4\})^2$ *for the second game. It is not difficult to see that the robust game* $\mathcal{G}$ *has no pure equilibrium in distribution-independent strategies. This means that the existence of pure equilibria in each state-dependent game does not necessarily implies the existence of pure equilibria in the expected robust game.*

1.1.7   State-Dependent Equilibrium

Let $\tilde{S}_j$ be the set of maps from the state space $\mathcal{W}$ to the action set $\mathcal{A}_j$ of player j. A state-dependent pure strategy of player j is an element of $\tilde{S}_j$. We define a state-dependent Cournot/Nash pure equilibrium as follows.

Definition 1.1.7.1 (Pure state-dependent equilibrium). *A strategy profile* σ^* *is a pure state-dependent equilibrium of the robust game*

$$\mathcal{G} = \left(\mathcal{W}, \mathcal{N}, \{\mathcal{A}_j\}_{j\in\mathcal{N}}, \{\tilde{R}_j\}_{j\in\mathcal{N}} \right),$$

if it satisfies, for all $j \in \mathcal{N}$ *and for all* $\sigma_j \in \tilde{S}_j$ *such that* $\sigma_j(\mathbf{w}) \in \mathcal{A}_j$,

$$\int_{\mathbf{w}\in\mathcal{W}} \tilde{R}_j(\mathbf{w}, \tilde{\sigma}_j^*(\mathbf{w}), \tilde{\sigma}_{-j}^*(\mathbf{w}))\ \nu(d\mathbf{w}) \geq \int_{\mathbf{w}\in\mathcal{W}} \tilde{R}_j(\mathbf{w}, \tilde{\sigma}_j(\mathbf{w}), \tilde{\sigma}_{-j}^*(\mathbf{w}))\nu(d\mathbf{w})$$

(1.11)

where $\tilde{\sigma}_{-j}^* = (\tilde{\sigma}_k^*)_{k\neq j} \in \prod_{j'\neq j} \tilde{S}_{j'}$.

Let S_j be the set of maps from the state space $\mathcal{W}$ to the randomized action set $\mathcal{X}_j$.

We define a mixed state-dependent Cournot/Nash equilibrium as follows.

Definition 1.1.7.2 (Mixed state-dependent equilibrium). *A strategy profile* σ^* *is a state-dependent equilibrium of the robust game*

$$\mathcal{G} = \left(\mathcal{W}, \mathcal{N}, \{\mathcal{X}_j\}_{j\in\mathcal{N}}, \{\tilde{R}_j\}_{j\in\mathcal{N}} \right),$$

if it satisfies, for all $j \in \mathcal{N}$ *and for all* $\sigma_j \in S_j$ *such that* $\sigma_j(\mathbf{w}) \in \mathcal{X}_j$,

$$\int_{\mathbf{w}\in\mathcal{W}} R_j(\mathbf{w}, \sigma_j^*(\mathbf{w}), \sigma_{-j}^*(\mathbf{w}))\nu(d\mathbf{w}) \geq \int_{\mathbf{w}\in\mathcal{W}} R_j(\mathbf{w}, \sigma_j^*(\mathbf{w}), \sigma_{-j}^*(\mathbf{w}))\nu(d\mathbf{w})$$

(1.12)

where $\sigma_{-j}^* = (\sigma_k^*)_{k\neq j} \in \prod_{j'\neq j} S_{j'}$.

Note that we cannot talk about uniqueness in robust games without specification. The reason is that the uniqueness is in the probabilistic sense, but it is also sensitive to the class of strategies that is under consideration. For example, a game that has a unique state-independent strategy equilibrium, it does not necessarily have a unique *state-dependent strategy* equilibrium. To see this statement, consider a modified strategy

$$\sigma_j(\mathbf{w}) = \begin{cases} \sigma_j^* & \text{if } \mathbf{w} \in \tilde{\mathcal{W}} \\ \sigma_j \in S_j \text{ arbitrary} & \text{otherwise} \end{cases}$$

where $\mathcal{W} \backslash \tilde{\mathcal{W}}$ is a $\nu-$null set.

1.1.8 Random Matrix Games

A random matrix game is a particular case of robust game where the interaction payoffs are given by matrices with random entries. Below we describe the case of two players. Player 1 has $l_1 \geq 2$ choices and Player 2 has $l_2 \geq 2$ choices. The rows are labeled with the pure strategies for Player 1, the columns with the pure strategies of Player 2. The payoff matrix of player 1 is a random matrix $M_1 \in \mathbb{R}^{l_1 \times l_2}$ and the payoff of player 2 is a random matrix $M_2 \in \mathbb{R}^{l_1 \times l_2}$. The matrix entries $M_{1,ij}, M_{2,ij}$ are the pairs of payoffs for the pure strategy pair (s_i, s_j'). Note that the labels of the actions or pure strategies (rows or columns) do not matter. We write these matrices as follows

$$M_1 = \bar{M}_1 + (M_1 - \bar{M}_1) = \bar{M}_1 + N_1 \qquad (1.13)$$
$$M_2 = \bar{M}_2 + (M_2 - \bar{M}_2) = \bar{M}_2 + N_2 \qquad (1.14)$$

where $\bar{M}_j = \mathbb{E}M_j$ is the expected matrix for $j \in \{1, 2\}$ and $N_j = M_j - \bar{M}_j$, is a zero-mean random variable (a noise). More details on random matrix games (RMGs) can be found in Chapter 6.

1.1.9 Dynamic Robust Game

We describe a generic model of robust difference games, i.e, games under uncertainty and in discrete time. The dynamic game is described by

$$\Gamma = (\mathcal{W}, \mathcal{N}, \mathcal{T}, (\mathcal{A}_j, \Theta_{j,t}, h_{j,t}, N_{j,t}, \Sigma_{j,t})_{j \in \mathcal{N}, t \in \mathcal{T}}, (F_j)_{j \in \mathcal{N}})$$

where

- $\mathcal{T}$ is the time space. This set is assumed to be nonempty. For difference game, the set $\mathcal{T} = \{1, 2, 3, 4, \ldots\}$ represents the game stages.

- A finite or infinite set $\mathcal{W}$, w_t is the state of the environment at stage t. A state may represent a joint channel state, collection of individual locations, collection of queue sizes, weather conditions or activity/inactivity of the players, etc.

- A set of players $\mathcal{N} = \{1, 2, \ldots, n\}$. There is at least one potential player in the game, $1 \leq n < +\infty$. Sometimes "Nature" can be considered as a player.

- $\mathcal{A}_j$ is an action set of player $j \in \mathcal{N}$ at time $t \in \mathcal{T}$. Its elements are allowed action $a_{j,t}$ of player j at stage t.

- A set $\Theta_{j,t}$ is the observation set of the player j at stage t, $\theta_{j,t} \in \Theta_{j,t}$.

- The function $h_{j,t} : \mathcal{W} \times \prod_{j' \in \mathcal{N}} \mathcal{A}_{j'} \longrightarrow \Theta_{j,t}$ defined by

$$\theta_{j,t} = h_{j,t}(w_t, a_t),$$

  is the observation of player j concerning the current state of the game.

- A set $\eta_{j,t}$ defined as a subset of

$$\left(\prod_{1 \leq t' \leq t-1} \prod_{j \in \mathcal{N}} \theta_{j,t'} \times \mathcal{A}_{j,t'} \right) \times \prod_{j \in \mathcal{N}} \theta_{j,t},$$

  which determines the information set of player j at stage t. The specification of $\eta_{j,t}$ for the stage of the game characterizes the information structure of player k and the set $\{\eta_{j,t}\}_{t \in \mathcal{T}, j \in \mathcal{N}}$ is the information structure of the game.

- $N_{j,t}$ is the information space at stage t induced by $\eta_{j,t}$. This set contains the action observation structure of the player j.

- A class $\Sigma_{j,t}$ of mappings $\sigma_{j,t} : N_{j,t} \longrightarrow \mathcal{A}_j$, are the admissible decisions at stage t. The collection $\sigma_j = (\sigma_{j,t})_{t \in \mathcal{T}}$ is the strategy of player j and Σ_j is the strategy set of player j.

- A function

$$F_j : \prod_{t \in \mathcal{T}} \left(\mathcal{W} \times \prod_{j' \in \mathcal{N}} \mathcal{A}_{j'} \right) \longrightarrow \mathbb{R},$$

  called payoff function of player j in the dynamic game. An intermediary payoff is not specified. In the additive payoff function the long-term payoff is the cumulative stage payoffs. An example of nonadditive payoff F_j is the norm-sup of all the stage payoffs obtained along the trajectory.

1.2 Robust Games in Networks

Emerging wireless networks such as cognitive radio, ad hoc, sensor networks, disruption/delay tolerant networks, and distributed opportunistic networks have been found to be more and more important in the network community. One common point between the corresponding scenarios is that nodes/mobile devices/base stations/users have to take some decisions in a quasi-autonomous manner and interact with each other under the form of interdependency such as multiuser interference, congestion cost, or capacity region in wireless networks.

> Game theory, which is a branch of mathematics analyzing interactions between (inter-dependent) decision makers, appears as a natural paradigm to understand and design these types of wireless networks.

Below we mention some pioneer works on networking games. Starting from

(i) The paper entitled Datagram Network as a Game by Nagle (RFC 970, 1985),

(ii) *Multi-Objective Routing In Integrated Services Networks: A Game Theory Approach* by Economides, A.A. and Silvester J.A. in 1991,

(iii) A Game Theoretic Perspective to Flow Control in Telecommunication Networks by Christos Douligeris & Ravi Mazumdar (1992),

(iv) Pricing in Computer Networks: Motivation, Formulation, and Example by Ron Cocchi, Scott Shenker, Deborah Estrin, Lixia Zhang, and

(v) Competitive Routing Game by Orda & Shimkin in 1993 in the first volume of IEEE Transaction in Networking,

the networking community is much interested in game theoretic modeling to understand strategic behaviors in wired and wireless networks (stochastic or deterministic).

In the last two decades, game theory has been applied more and more intensively to wireless networking and communications [5, 147, 78]. For instance, many distributed congestion control, routing algorithms, power control, and resource allocation schemes have been proposed by exploiting concepts of game theory. There is a quite general consensus to say that the corresponding analysis are useful but rely on strong information and behavior assumptions on the part of users (e.g., games with complete information and rational players, where rationality is common knowledge and predetermined beliefs). One of the consequences of this is that the predicted game outcomes (most often Cournot/Nash equilibria, Hannan set, correlated equilibria, communication equilibria, bargaining solutions, hierarchical solutions, etc.) will rarely be observed. One of the many purposes of learning procedures is to design fast and efficient algorithms exploiting partial information on the game. This chapter presents a unified view of learning algorithms with different levels of information. The goal is also to contribute to the understanding of learning in strategic decision making that will become more acute with the increased deployment of wireless devices and limited resources.

The uncertainty/variability of the environment and the restrictive natures of wireless communications, for example, limited energy, battery, limited spectrum, scarce and shared communication medium, together with the resulting economical issues, make the distributed strategic learning approach a reasonable fit of approach to the wireless problems.

The framework covers scenarios where players are users/mobile devices/transmitters/base stations/network designers who are able to choose their actions by themselves. The actions can be typically

- to upgrade or not a technology, turn ON/OFF at device,

- to sense or not a channel,

- vector of powers in the case of a distributed power allocation problem or just a power level in the case of distributed power control in one dimension.

- to choose a subset of channels in the case of spectrum sharing and resource selection problems

- rate vectors in the case of rate allocation,

- choice of routes (concatenation of different nodes) from sources to destinations in the case routing and flow control, etc.

The payoff function can be, for example,

- the transmission rate or energy-efficiency, the cost of energy consumption per capacity, for power control/allocation problems [6].

- the individual/total throughput or mean delays, for the spectrum sharing problem,

- the cost, the queueing length, or the end-to-end delay for the routing problem.

Example 1.2.1. *As an example of interaction between nodes we consider multiple access problem in wireless networks [169]. Suppose that there are two mobile nodes tx1 and tx2 who want to access to a shared wireless channel to send information to the corresponding receivers rx1 and rx2. Both receivers are within the transmission range of both the transmitters. Each transmitter has one packet to transmit in each time step, and it can either choose to transmit during a time step or wait. If tx1 transmits, the packet is successfully* transmitted if *tx2* chooses *not* to transmit during that time step (and hence there is no collision). For successful packet transmission, *tx1* obtains a benefit at the cost of transmit power. It is of interest to analyze the interactions between the transmitters under different network settings and performance objectives.

Example 1.2.2 (Energy management). *In wireless networks, most devices work on their own batteries. The limitation of battery power gets worse when devices communicate because much more power is consumed to transmit than to compute. As a result, optimal power consuming networks have been studied for various scenarios such as broadcasting, multicasting, routing, and target coverage (for wireless sensor networks). The power control problem for wireless data is to assign each device's transmission power to promote the quality and efficiency of the wireless networking system. In this manuscript, one of the main goals is the development of combined learning schemes such that the iterative power control algorithm converges both synchronously and asynchronously when outdated or incorrect interferences are used.*

Example 1.2.3. *Wireless sensor network is a network consisting of a large number of sensor devices with processing and wireless transmission capability. Wireless Sensor Networks (WSN) is considered as a class of ad hoc networks with the salient feature that it could have a centralized entity for the entire network called base station or sink. The sensors in the network sense different types of occurrences (e.g., temperature, magnetism, sound, light) from the surrounding environments and send the data to the base station. However, acquiring data from the deployment area of sensors is not always easy, and multiple issues are to be considered in dealing with a wireless sensor network. The major issue is the limited resources of the tiny sensor devices, which are run with one-time batteries. Even if rechargeable batteries are used, it is not always easy to replace them once the sensors are deployed over a geographical area. Most of the time, the application requires the network to be unattended, which means it is free from human intervention. Other problems might come from the external environment; heat, rain, wind condition, etc. Routing, data aggregation, ensuring Quality of Experience (QoE), Quality of Service (QoS), security, all these are various facets of wireless sensor networks. The interaction between the sensors can be modeled as an evolving with variable and nonreciprocal set of players. A typical objective function could be an end-to-end delay under long-term energy constraint (with renewable energy).*

The above payoff functions may depend on

- queue lengths,

- channel gains,

- location of the players,

- environment state, etc.

which are uncertain quantities. Thus, the resulting games are games under uncertainty, also called *robust games*. This class of games is frequently met in networks.

Below we provide examples where the class of robust games in wireless networking and communications is relevant.

- Outage probability

$$\mathbb{P}(\tilde{R}_j(\mathbf{w}, \mathbf{a}) < \beta_j) = \mathbb{E}_w \left(\mathbb{1}_{\{\tilde{R}_j(\mathbf{w}, \mathbf{a}) < \beta_j\}} \right)$$

- Success probability (Psucc) $\mathbb{P}(\tilde{R}_j(\mathbf{w}, \mathbf{a}) \geq \beta_j) = \mathbb{E}_w \left(\mathbb{1}_{\{\tilde{R}_j(\mathbf{w}, \mathbf{a}) \geq \beta_j\}} \right) = \mathbb{E}_w \left(\mathbb{1}_{\{ACK\}} \right)$

  Energy-consumption based variants: $\frac{Psucc}{c_j(a_j)}$, $Psucc - c_j(a_j)$,

- Coverage probability

$$\mathbb{P}(\bigcup_{SBS} \tilde{R}_{j,SBS}(\mathbf{w}, \mathbf{a}) \geq \beta_j) = \mathbb{E}_w \left(\mathbb{1}_{\{\bigcup_{SBS} \tilde{R}_{j,SBS}(\mathbf{w}, \mathbf{a}) \geq \beta_j\}} \right)$$

- Ergodic Shannon-like payoff:

  - $\mathbb{E}_w \left(\log(1 + SINR(\mathbf{w}, \mathbf{a})) \right)$
  - $\mathbb{E}_w \left(\log(1 + SINR(\mathbf{w}, \mathbf{a})) \mathbb{1}_{SINR_j \geq \beta_j} \right)$
  - Energy-based payoff:

$$\mathbb{E}_w \left(\sum_{j \in \mathcal{N}} \log(1 + SINR_j(\mathbf{w}, \mathbf{a})) - c_j(a_j) \right)$$

- Cognitive radio networks

  - Primary users: $\mathbb{P}(SINR_{j'}(\mathbf{w}, \mathbf{a}) \geq \beta_{j'})$
  - Secondary users: $\mathbb{P} \left(\frac{a_j |w_{j,rxPU}|^2}{(\epsilon^2 + d_{j,RxPU}^2)^{\alpha/2}} \leq \eta, \ SINR_j(\mathbf{w}, \mathbf{a}) \geq \beta_j \right)$ where $rxPU$ denotes the receiver of primary user and $d_{j,RxPU}$ is the current distance between user j and the receiver of the primary user $RxPU$.

- Energy market as a multi-level robust game

  Consider a generic expected robust games in the energy market with three classes of players: producers of energy, distributors and users/customers with the following specifications

  - Producers/generators (of different types of renewable energy)
    * Each producer knows only her own production cost
    * does not know the costs of the other producers

> ∗ does not know the (exact) weather condition in advance (before making the decision on how much to produce)

- Distributors
  - ∗ Each distributor has a contract with a subset of producers
  - ∗ Contract with the users
- Users/Consumers
  - ∗ Each user sees its realized prices per period/slot
  - ∗ Users interact also through a network, application in the Cloud

We analyze the system by considering the expectation over type distribution, over environment distribution (weather), leading to expected robust games.

⚠ In the above example, there are *robust games* within the robust game.

1.3 Basic Robust Games

> What is the difference between the classical one-shot games and robust games?

We have defined in Subsection 1.1.1 the basic components of a one-shot game. If the description involves in addition a state parameter that can be deterministic or stochastic (uncertain), and which can be correlated with the action profile or independent, then one gets a robust game. In particular, if the uncertainty set is reduced to a singleton or an empty set, then we get a standard one-shot game.

Below we provide basic examples of robust games.

Example 1.3.1 (Robust games with dominated strategies). *An action is dominated if, regardless of what any other players do, the action earns a player a smaller payoff than some other action. Hence, an action is dominated if it is always better to play some other strategy, regardless of what opponents may do.*

> *In the context of robust games, a strategy is dominated if it is always better to play some other strategy, regardless of the states and what opponents may do.*

Consider a two-player robust game. Let $n^i_{aa'}$ be an independent and identically distributed (i.i.d) random variable with zero mean and finite variance. Table 1.2, the action profile (s_1, s'_1) is dominant.

1\ 2	s'_1	s'_2			
s_1	$(100, 100)$	$(3, 10)$	$+$	$n^1_{1,1}, n^2_{1,1}$	$n^1_{1,2}, n^2_{1,2}$
s_2	$(3, 10)$	$(0, 0)$		$n^1_{2,1}, n^2_{2,1}$	$n^1_{2,2}, n^2_{2,2}$

TABLE 1.2
Robust game with dominant strategy.

> *Robust games with dominant strategies can be solved using the so-called Iterated elimination of dominated strategies.*

An example of such a game is a power control game in a multiple-input-single-output (MISO) system, where the maximum power is a dominant action.

Note that for some states of the robust game, the action may not be dominant but it can be a dominant of the expected payoff. This allow ones to extend the definition of robust game to expected robust game with dominant strategies.

Example 1.3.2 (Robust team game). *The payoff function has the form*

$$R_j(w, a) = \bar{r}(w, a).$$

Theorem 3.1.0.3 of Chapter 3 characterizes the equilibrium payoffs of such team problems. This class of games can be generalized to potential robust games. See Chapter 4.

Example 1.3.3 (Anti-coordination robust games). *Consider the following two-player game. Each player can choose one of the two options. When they use the same action at the same time and same area, there is congestion or a collision (e.g, in same frequency band or same medium) and the data is lost. Table 1.3 represents the two-by-two game. The actions are denoted by s_1 and s_2. Player 1 chooses a row, and player 2 chooses a column. A successful configuration gives a payoff 1 to the corresponding player. In addition, we look at it in an independent and identically distributed (i.i.d) noisy environment. Let $0 < \alpha \leq 1$. In table 1.3, n^i_{ab} : is a real-valued i.i.d noise. This class of game can be extended to robust crowd game and to robust congestion game, etc.*

1\ 2	s_1	s_2			
s_1	$(0,0)$	(α,α)	+	$n_{1,1}^1, n_{1,1}^2$	$n_{1,2}^1, n_{1,2}^2$
s_2	(α,α)	$(0,0)$		$n_{2,1}^1, n_{2,1}^2$	$n_{2,2}^1, n_{2,2}^2$

TABLE 1.3
Anti-coordination robust game.

1.4 Basics of Robust Cooperative Games

Cooperative game theory is concerned primarily with coalitions, that is, groups of players, who coordinate their actions and pool their payoffs. Consequently, one of the problems here is how to divide the payoff among the members of the formed coalition. The basis of this theory was laid by John von Neumann & Oskar Morgenstern [120] with coalitional games in characteristic function form, known also as transferable utility games (TU-games). The theory has been extended to non-transferable utility games (NTU-games). Robust cooperative games are cooperative games under uncertainties.

1.4.0.1 Preliminaries

Let $\mathcal{N}$ be a non-empty finite set of players who consider different cooperation possibilities. Each subset $\mathcal{N}' \subseteq \mathcal{N}$ is referred to as a crisp coalition. The set $\mathcal{N}$ is called the grand coalition and $\emptyset$ is called the empty coalition.

Definition 1.4.0.2. *A cooperative game in characteristic function form is an ordered pair $(\mathcal{N}, v)$ consisting of the player set $\mathcal{N}$ and the characteristic function $v : 2^{\mathcal{N}} \longrightarrow \mathbb{R}$, with $v(\emptyset) = 0$ and $2^{\mathcal{N}}$ is the set of subsets of $\mathcal{N}$.*

A cooperative game in characteristic function form is usually referred to as a transferable utility game (TU-game). The real number $v(\mathcal{N}')$ can be interpreted as the maximal payoff that the members of $\mathcal{N}'$ can obtain when they cooperate.

We now address one of the basic questions in the theory of cooperative TU-games:

If the grand coalition forms, how to divide the utility $v(\mathcal{N})$?

This question is approached using different solution concepts in cooperative game theory like cores, stable sets, bargaining sets, the Shapley value, the nucleolus. A solution concept gives an answer to the question of how the payoff obtained when all players in $\mathcal{N}$ cooperate should be distributed among the individual players while taking account of the potential utility of all different coalitions of players. Hence, a solution concept assigns to a coalitional game at least one payoff vector $r = (r_j)_{j \in \mathcal{N}}$, where r_j is the payoff allocated to player $j \in \mathcal{N}$.

Denote by $\mathcal{G}_{\mathcal{N}}$ the set of characteristic functions of coalitional games with player set $\mathcal{N}$ forms with the usual operations of addition and scalar multiplication of functions a $(2^{|\mathcal{N}|} - 1)$-dimensional linear space. A basis of this linear space is obtained by value function $v^*_{\mathcal{N}'}(\mathcal{N}'') = \mathbf{1}_{\{\mathcal{N}'' \subseteq \mathcal{N}'\}}$ for $\mathcal{N}' \in 2^{\mathcal{N}} \backslash \{\emptyset\}$. This class cooperative games with value v^* are called *unanimity games*. An interpretation is unanimity game is the following: In the game $v^*_{\mathcal{K}'}$, the payoff one can be obtained if and only if all players in coalition $\mathcal{N}'$ are involved in cooperation.

Definition 1.4.0.3. *Let* $r : \mathcal{G}_{\mathcal{N}} \longrightarrow \mathbb{R}^n$. *Then,* r *satisfies*

- *Individual rationality if* $r_j(v) \geq v(\{j\})$, $\forall (v, j) \in \mathcal{G}_{\mathcal{N}} \times \mathcal{N}$

- *Efficiency if* $\sum_{j=1}^{n} r_j(v) = v(\mathcal{N})$, $\forall v \in \mathcal{G}_{\mathcal{N}}$.

- *Additivity if* $r(v + v') = r(v) + r(v')$, $\forall (v, v')$

- *the dummy player property if* $r_j(v) = v(\{j\})$ *for all* v *and for all dummy players* k *such that* $v(\mathcal{N}' \cup \{j\}) = v(\mathcal{N}') + v(\{j\})$ *for all* $\mathcal{N}' \in 2^{\mathcal{N} \backslash \{j\}}$.

- *the anonymity property if* $r(v_\pi) = \pi^*(r(v))$ *for all* $\pi \in S_n$. *Here* v_π *is the game with* $v_\pi(\pi(\mathcal{N}')) := v(\mathcal{N}')$ *for all* $\mathcal{N}' \in 2^{\mathcal{N}}$ *and* $\pi^* : \mathbb{R}^n \longrightarrow \mathbb{R}^n$ *is defined by* $(\pi^*(r))_{\pi(j)} := r_j$ *for all* $r \in \mathbb{R}^n$ *and* $j \in \mathcal{N}$.

1.4.0.4   Cooperative Solution Concepts

Definition 1.4.0.5. *A payoff vector* $r \in \mathbb{R}^n$ *is an imputation for the game* $v \in \mathcal{G}_{\mathcal{N}}$ *if it is efficient and individually rational, i.e.*

- $\sum_{j \in \mathcal{N}} r_j = v(\mathcal{N})$,

- $r_j \geq v(\{j\})$, $\forall\, j \in \mathcal{N}$

We denote by $I(v)$ the set of imputations of $v \in \mathcal{G}_{\mathcal{N}}$. Clearly, $I(v)$ is empty if and only if $v(\mathcal{N}) < \sum_{j \in \mathcal{N}} v(\{j\})$.

We now define set-valued solution concepts called **core** [64].

Definition 1.4.0.6 (Core). *The core* $C(v)$ *of a TU-game* $v \in \mathcal{G}_{\mathcal{N}}$ *is the set*

$$C(v) = \left\{ r \in I(v), \ \sum_{j' \in \mathcal{N}'} r_{j'} \geq v(\mathcal{N}'), \ \forall\, \mathcal{N}' \in 2^{\mathcal{N}} \backslash \{\emptyset\} \right\}$$

If $r \in C(v)$, then no coalition $\mathcal{N}'$ has an incentive to split off if r is the proposed payoff allocation in $\mathcal{N}$, because the total amount $\sum_{j' \in \mathcal{N}'} r_{j'}$ allocated to $\mathcal{N}'$ is not smaller than the amount $v(\mathcal{N}')$ which the players can obtain by forming the sub-coalition. If the core is non-empty, then the core is a polytope because bounded and it is a finite system of linear inequalities.

Definition 1.4.0.7 (Balanced game). *Denote by $e_{\mathcal{N}'}$ the characteristic vector of $\mathcal{N}'$ i.e $e_{\mathcal{N}',i} = \mathbf{1}_{\{i \in \mathcal{N}'\}}$.*

- *Balanced map: $\lambda : 2^{\mathcal{N}} \backslash \{\emptyset\} \longrightarrow \mathbb{R}_+$ is a balanced map if*

$$\sum_{\mathcal{N}' \in 2^{\mathcal{N}} \backslash \{\emptyset\}} \lambda_{\mathcal{N}'} e_{\mathcal{N}'} = e_{\mathcal{N}}$$

- *A collection $\mathcal{N}'$ of coalitions is called balanced if there is a balanced map λ such that*

$$\mathcal{N}' = \{\mathcal{N}'' \in 2^{\mathcal{N}} \mid \lambda_{\mathcal{N}''} > 0\}$$

(the support of λ).

- *A game v is balanced if for each balanced map $\lambda : 2^{\mathcal{N}} \backslash \{\emptyset\} \longrightarrow \mathbb{R}_+$ one gets*

$$\sum_{\mathcal{N}' \in 2^{\mathcal{N}} \backslash \{\emptyset\}} \lambda_{\mathcal{N}'} v(\mathcal{N}') = v(\mathcal{N}).$$

The next theorem gives a characterization of games with a nonempty core.

Theorem 1.4.0.8. *Let $v \in \mathcal{G}_{\mathcal{N}}$. Then the following two assertions are equivalent:*

- *$C(v) \neq \emptyset$.*

- *The game v is balanced.*

We now focus on the notion of stable set[120].

We say that an imputation r dominates another imputation r' via coalition $\mathcal{N}'$, if

- $r_j > r'_j$ for all $j \in \mathcal{N}'$, this says that the payoff r is better than the payoff r' for all members of $\mathcal{N}'$.

- $\sum_{j \in \mathcal{N}'} r_j \leq v(\mathcal{N}')$. This guarantees that the payoff r is reachable for the coalition $\mathcal{N}'$.

For $v \in \mathcal{G}_{\mathcal{N}}$, we denote by $dom(V)$ the set consisting of all imputations that are dominated by some element in V. The dominance core $DC(v)$ is defined by $DC(v) = I(v) \backslash dom(I(v))$.

Definition 1.4.0.9. *For $v \in \mathcal{G}_{\mathcal{N}}$ a subset $\mathcal{N}'$ of $I(v)$ is called a stable set if the following conditions hold:*

- *(Internal stability) $\mathcal{N}' \cap dom(\mathcal{N}') = \emptyset$.*

- *(External stability) $I(v) \backslash \mathcal{N}' \subset dom(\mathcal{N}')$.*

Note that for a game v the set $\mathcal{N}'$ is a stable set if and only if $\mathcal{N}'$ and $dom(\mathcal{N}')$ form a partition of $I(v)$.

Definition 1.4.0.10. • *A game v is superadditive if $v(\mathcal{N}' \cap \mathcal{N}'') \geq v(\mathcal{N}') + v(\mathcal{N}'')$ for all $(\mathcal{N}', \mathcal{N}'') \in 2^{\mathcal{N}}$ with $\mathcal{N}' \cap \mathcal{N}'' = \emptyset$*
  • *A game v is subadditive if $-v$ is superadditive.*

Theorem 1.4.0.11. *Let $v \in \mathcal{G}_{\mathcal{N}}$ and $\mathcal{N}'$ be a stable set of v. Then*
  • $C(v) \subseteq DC(v) \subseteq \mathcal{N}'$;
  • *If v is superadditive, then $DC(v) = C(v)$;*
  • *If $DC(v)$ is a stable set, then there is no other stable set.*

We now focus on the Shapley value. The Shapley value uses the marginal vectors of a cooperative TU-game. First we define a marginal vector.

Definition 1.4.0.12 (Shapley value). • *Let S_n is the set of permutation $\mathcal{N}$. Let $\pi \in S_n$. The set $p_{\pi,j} = \{j' \in \mathcal{N}, \mid \pi^{-1}(j') < \pi^{-1}(j)\}$ consists of all predecessors of j with respect to the permutation. The marginal contribution vector $m_{\pi,j} \in \mathbb{R}^n$ with respect to π and v has the j-th coordinate*

$$m_{\pi,j}(v) = v(p_{\pi,j} \cup \{j\}) - v(p_{\pi,j}), \ j \in \mathcal{N}$$

  • *The Shapley value Φ of a game v is the average of the marginal vectors of the game: $\Phi(v) = \frac{1}{n!} \sum_{\pi \in S_n} m_{\pi}(v)$*

Theorem 1.4.0.13. *A solution $r : \mathcal{G}_{\mathcal{N}} \longrightarrow \mathbb{R}^n$ satisfies additivity, anonymity, the dummy player property, and efficiency if and only if it is the Shapley value.*

For cooperative robust games, we define the value as $v : \mathcal{W} \times 2^{\mathcal{N}}$, defined by $v(w, J)$ where w is the uncertainty parameter and J is a coalition. Then, the robust cooperative game is given by $(\mathcal{N}, v, \mathcal{W})$, where $\mathcal{W}$ is the uncertainty set. For a cooperative expected robust game, the allocated expected payoff will be defined as $\bar{r}_j = \mathbb{E}_w r_j(w)$. The notions of expected robust core and expected robust Shapley value are defined in a similarly way.

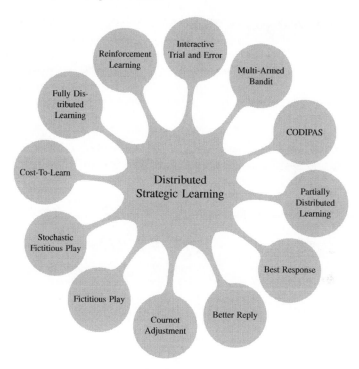

FIGURE 1.1
Strategic Learning.

1.5 Distributed Strategic Learning

What is distributed strategic learning?

Distributed strategic learning is learning in the presence of other players (i.e., from, with, or about others) or (from, with or about) Nature/environment. The term *strategic learning* stands for how the players can learn from/with/about the others (players, environment) under their complex strategies/beliefs. A strategic learning consists of acting (eventually optimally or for satisfactory criterion) when outcomes are interdependent and interactive (depend on others).

In this book, distributed strategic learning is a learning-based process for creating and executing breakthrough payoff functions, strategies, and distribution laws over time. The strategic learning aspect includes learning how to communicate better but also communicating to learn better. When the players can act locally with joint decisions during the interactions, we refer to it as *coalitional strategic learning*. In figure 1.1, we represent some of the learning topics related to distributed strategic learning.

Strategy learning and dynamics

We use the term *strategy learning* or *strategy dynamics* to emphasize to learning process for strategies, optimal strategies, better response strategies, satisfactory strategies, equilibrium strategies, etc.

Belief learning and dynamics

We use the term *belief learning* to refer to a process of learning the other players beliefs or one's own beliefs.

Payoff learning and dynamics

We use the term payoff learning or payoff dynamics to refer to a process of learning the payoff functions, the optimal payoffs, the equilibrium payoffs, etc. See Chapter 3.

Combined learning

The term *combined learning* is used to emphasis a joint learning process. For example, learning simultaneously the payoff functions and the optimal (or satisfactory) strategies is a combined learning.

Figure 1.2 represents a generic combined learning scheme. There are n entities (players) in a dynamic interactive system. Each player can make a decision and is able to measure a reaction from the system. Each player updates her or his decision based on their own-measurements without information exchange among the players. The environment parameters can be stochastic or deterministic and outcomes depend on the players decisions.

Heterogeneous learning

The term *heterogeneous learning* is used to emphasize that it refers to differences in the ways players update and learn. It does not necessarily refer to the structural heterogeneity of the model. For example, heterogeneous learning can be considered in symmetric robust games with simultaneous moves.

We will examine five types of heterogeneity: players that

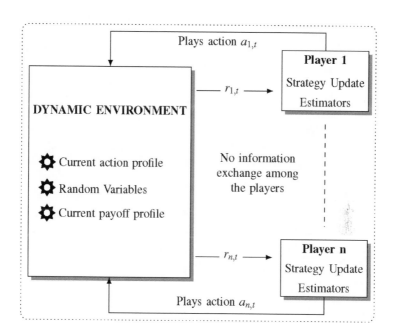

FIGURE 1.2
A generic combined learning scheme.

- have different perceptions
- have different temperature-like parameters and degrees of inertia in updating and
- use different learning patterns
- use different learning rates (slow learners, medium learners, fast learners, etc.)
- different updating times, synchronicity, and activities.

In the next chapters, if at least two of the players have distinct learning patterns, we will use the term heterogeneous learning.

The asymptotic results of convergence of heterogeneous learning are examined in Chapter 4. Moreover, we do not neglect the issue of how heterogeneity affects the speed of convergence of the interacting system towards equilibrium.

Hybrid learning

Different from heterogeneous learning, the term *hybrid learning* will be used in the case where at least one of the players changes the learning pattern during the interactions. At each time, each active player will be able to select from a list of learning patterns. The resulting dynamics are related to *hybrid game dynamics* in which each player can switch between different rules called *revision protocols*.

CODIPAS: COmbined fully DIstributed PAyoff and Strategy

CODIPAS is a combined learning. It consists of learning jointly and in an interdependent manner payoffs and strategies in long-run games.

Players	Information	CODIPAS: Summary
Actions	Own-action	Make a decision based on own-measurement and strategies.
Observation	Numerical measurement	New numerical payoff values are perceived.
Response	Learning pattern	Update the strategies and the estimates based on the recent reaction.

CODIPAS-RL : COmbined fully DIstributed PAyoff and Strategy Reinforcement Learning

CODIAPS-RL is a reinforcement-based CODIPAS. This class of learning will be developed in Chapter 4. The basic idea of CODIPAS can be summarized as follows:

- The consequences influence behavior.

- The behaviors influence the outcomes.

The idea of CODIPAS-RL goes back at least to Bellman (1957, [23]) for joint computation of optimal value and optimal strategies. The idea has been followed in [19] with less information requirement. Since then, different versions of CODIPAS-RLs has been studied. These are:

- Bush-Mosteller-based CODIPAS-RLs: combine the scheme of Bush-Mosteller with the perceived payoff averaging (per action).

- Boltzman-Gibbs-based CODIPAS-RLs: construct Gibbs distribution scheme based on estimated payoffs (per action). The main difference between Boltzman-Gibbs CODIPAS-RLs and the classical Boltzman-Gibbs dynamics or logit dynamics is in terms of information. In logit dynamics the vector payoff corresponding to all the actions of the player are required in order to update. In the Boltzman-Gibbs-based CODIPAS-RLs, the payoff corresponding to the others actions than the current action are not observed. Each player needs to estimate these alternatives.

- Imitative Boltzman-Gibbs-based CODIPAS-RLs: Combine imitative rules that are proportional to the Gibbs distribution.

- Multiplicative weighted Imitative CODIPAS-RLs: Instead of the exponential function, an imitation rule based on multiplicative weights is considered.

- Weakened fictitious play based CODIPAS-RLs: Construct a stochastic fictitious play scheme based only on the payoff estimations. The actions of the other players are not observed. The frequency of others play is also unknown.

- Logit-based payoff learning: The payoff dynamics is based on logit map.

- Imitative Boltzmann-Gibbs-based payoff learning.

- Multiplicative weighted payoff learning.

- No-regret based CODIPAS-RLs.

- Pairwise comparison based CODIPAS-RLs: Construct a strategy learning based on a rate that is proportional to the gap between two estimates.

- Projection-based CODIPAS-RLs.

- Excess-payoff-based CODIPAS-RL.

- Risk-sensitive CODIPAS.

- Cost-to-Learn CODIPAS, etc

Mean field learning

The term *mean field learning* refers to a learning framework for large populations of players. It is different from *learning the mean field* which is related to transport equations and Fokker-Planck-Kolmogorov (FPK) forward equations.

Risk-sensitive learning

The term *risk-sensitive learning* refers to learning in the presence of risk-sensitive players. It consists of learning the risk-sensitive payoffs, risk-sensitive equilibrium, and or other risk-sensitive solutions. See Chapter 9.

Risk-sensitive payoff dynamics

We use the term *risk-sensitive payoff dynamics* to refer to a process of learning the risk-sensitive payoff functions, the optimal risk-sensitive payoffs, the risk-sensitive equilibrium payoffs, etc. The risk-neutral payoff dynamics is simply called payoff dynamics. See Chapter 3.

Risk-sensitive CODIPAS

Joint payoff and strategy learning in the risk-sensitive context.

Slow, Medium, or Fast Learners

We refer to players with negligible learning rates compared to the others as *slow learners*. The fast learners have high learning rates. The medium learners are the players with learning rates between the slow learners' rates and the fast learners' rates. The other configurations are considered as *proportional learning rates*.

Evolutionary game dynamics

Evolutionary games in a large population provide a simple framework for describing strategic interactions among large numbers of players. Traditionally, predictions of behavior in game theory are based on some notion of equilibrium, typically Cournot equilibrium, Bertrand equilibrium, Nash equilibrium, Stackelberg solution, Wardrop equilibrium, mean field equilibrium, resilient by invasion, or some refinement/extension thereof. These notions require the assumption of equilibrium knowledge, which posits that each user correctly anticipates how the other players will act or react. The equilibrium knowledge assumption is too strong and is difficult to justify, in particular, in contexts with large numbers of players.

As an alternative to the equilibrium approach, an explicitly dynamic updating choice is proposed, a model in which players myopically update their behavior in response to their current strategic environment. This dynamic procedure does not assume the automatic coordination of players actions and beliefs, and it can derive many specifications of players' choice procedures. These procedures are specified formally by defining a revision of actions called *revision protocol*. A revision protocol takes current payoffs (also called *fitness* in behavioral ecology) and aggregate behavior as inputs; its outputs are conditional switch rates, which describe how frequently players in some class playing action a who are considering switching action switch to another action b, given the current payoff vector and subpopulation state. Revision protocols are flexible enough to incorporate a wide variety of paradigms, including ones based on imitation, adaptation, learning, optimization, etc. The revision protocols describe the procedures players follow in adapting their behavior to the dynamic evolving environment such as evolving networks. The methodology is extended to delayed evolutionary game dynamics as well as stochastic evolutionary game dynamics. For recent developments of evolutionary dynamics (deterministic or stochastic), we refer the reader to [140, 158, 160] and the references therein.

1.5.1 Convergence Issue

Evolutionary game dynamics provides a powerful tool for prediction of the outcome of a game. Convergence to equilibria has been established in many classes of games. These include generic two-players-two-action games, common interest games, potential games, S-modular games, many classes of aggregative games, games with unique evolutionarily stable strategy, stable games, games with monotone payoffs, etc. As a consequence, convergence issue of networking problems including parallel routing, routing with M/M/1 cost (Poisson arrival process, exponentially distributed service time and single server queue), network congestion games, network selection games, power allocation games, rate control, spectrum access games, resource sharing games, and many others, can be investigated through evolutionary game dynamics for both linear and non-linear payoff functions.

1.5.2 Selection Issue

A fundamental question we address is the selection problem (selection of equilibrium, Pareto optimal solutions, global optimum, etc.) in a *fully distributed way* (minimal signaling to the users, no message exchange, no recommendation, etc.).

1.5.2.1   How to Select an Efficient Outcome?

The problem of selection of a global optimum in a fully decentralized way is a very challenging problem. Very little is known about this. In Chapter 8 we provide examples of games where fully distributed reinforcement learning algorithms lead to evolutionary game dynamics that converge to global optima. However, in general, evolutionary game dynamics may not converge.

1.5.2.2   How to Select a Stable Outcome ?

Another important question is the stability/instability of the system. How to design fast algorithms such that a network will behave well after some iterations? This question will be translated by the fact that when time horizon is large, the behavior of the system looks like at a stationary point or set, and any small perturbation around this point or set will quickly come back to that point or set due to losses. It should be noticed that the two properties, *Stability* and *Efficiency*, may not be compatible in some situations.

Simulated Annealing

There are certain optimization problems that become unmanageable using combinatorial methods as the number of objects (players, resources, etc.) becomes large. This is the case for NP-complete problem, a problem which is both NP (verifiable in nondeterministic polynomial time) and NP-hard. The optimization community has been using *simulated annealing* for many years. Annealing is a metallurgical process in which a material is heated and then slowly brought to a lower temperature. The crystalline structure of the material can reach a global minimum in this way. The high temperature excites all atoms or molecules, but later, during the cooling phase, they have enough time to assume their optimal position in the crystalline lattice and the result is fewer fractures and fewer irregularities in the crystal.

Example 1.5.3. *A typical example is the traveling salesman problem, which belongs to the NP-complete class of problems. For these problems, there is a very effective practical algorithm called simulated annealing. While this technique is unlikely to find the optimum solution, it can often find a very good solution, even in the presence of noisy data.*

The traveling salesman problem can be used as an example application of simulated annealing. In this problem, a salesman must visit some large number of cities while minimizing the total mileage traveled. If the salesman starts with a random itinerary, he can then pairwise trade the order of visits to cities, hoping to reduce the mileage with each exchange. The difficulty with this approach is that while it rapidly finds a local minimum, it cannot get from there to the global minimum.

1.6 Distributed Strategic Learning in Wireless Networks

"It can be very costly to learn quickly, and learning can take some time".

1.6.1 Physical Layer

Distributed strategic learning has recently been applied to determine the transmission power of mobile nodes. In [84], the authors model the distributed power control problem in an infrastructure wireless network as a nonzero-sum game between the mobile nodes. Distributed learning algorithms for distributed solution of the power control game are proposed. Each node operates a learning algorithm, which determines the probability of choosing a certain transmission power based on the feedback received from the Base Station (BS). This feedback is essentially the environmental response for choosing a certain power level and expresses the amount of information that a mobile node can transfer during the lifetime of its battery. However, the proposed algorithms can also operate for other definitions of the environmental response as well.

In [191], the authors considered Bush-Mosteller-based reinforcement learning to analyze power control in wireless network. They conducted a detailed stability analysis for the two-player-two-action case. However, the stability analysis for arbitrary number of players and or the associated performance metrics remains a challenging open problem.

Rate control

Rate control is one of the most important problems in networks. If input rates to a network are left uncontrolled, then the offered load may exceed the network capacity causing congestion, i.e., buffers fill up and packets have to be dropped. It is well known that network congestion has a negative impact on performance, potentially leading to a throughput collapse. Rate control schemes should therefore

- produce a predictable and stable throughput,

- allow a distributed implementation, and

- allocate network bandwidth among applications with a certain fairness criterion (proportional fairness, maxmin fairness, $\alpha-$fairness, Bargaining solutions, etc.).

In [84, 68] the authors proposed a Stochastic Automata Rate Adaptation

Algorithm in order to achieve rate adaptation. According to this scheme, each transmitter has a list of possible modulation schemes to use, each one yielding a different transmission rate.

1.6.2   MAC Layer

Distributed strategic learning has found use in both infrastructure and ad-hoc wireless local area networks (WLANs). In [122] the authors proposed distributed algorithms and learning the parameters of the operating environment is exploited in order to provide efficient MAC protocols for bursty-traffic wireless networks. More information can be found in the survey [121].

In wireless networks, it is widely observed that the bandwidth utilization of a single channel-based wireless networks decreases due to congestion and interference from other sources. Therefore, the transmission on multiple channels are needed. The authors of [55], proposed a distributed dynamic channel allocation scheme for wireless networks using adaptive learning whose nodes are equipped with single radio interfaces so that a more suitable channel can be selected.

The authors in [87] studied a dynamic channel selection in multichannel wireless systems. Dynamic channel selection allows a transmitter to identify the channel offering the best radio conditions and to avoid interference created by other transmitters. In the absence of interference, remarkable regret bounds can be obtained in the channel selection problem. With interference, the problem is complicated by the fact that transmitters interact, and that a given transmitter experiencing a transmission failure cannot identify whether the failure is due to a channel error or to interference. The authors examined the convergence to pure Nash equilibria.

1.6.3   Network Layer

Routing in wireless networks

Besides its hardware, the energy consumption of a wireless node is also dependent on its position in the network. Nodes that are closer to the sink have to forward more messages and therefore need to listen more, while those far away from the sink could spend more time sleeping. Thus, the behavior of users cannot be the same for all users, and each node needs to learn what behavior is energy efficient in the network. To achieve that, we make nodes adopt an algorithm for optimization in order to improve the performance of the whole system. Each node in the wireless network uses a strategic learning algorithm to learn an optimal schedule that will maximize the energy efficiency and minimize the latency of the system in a distributed manner.

The authors in [114] use the reinforcement learning algorithm with the aim of increasing the autonomous lifetime of a Wireless Sensor Network (WSN) and decreasing latency in a decentralized manner.

In [181], distributed learning has been used in the context of multicast routing in wireless adhoc networks. Based on predictions of node mobility, the authors proposed to find the routes with the higher lifetimes, based on which a stochastic graph representing the virtual multicast backbone of the ad-hoc network is built. Then a distributed algorithm is applied to this graph to solve the multicast routing problem.

The work in [18] considered a routing problem over multiple parallel links. The cost function is assumed to be increasing with the congestion. Considering noisy measurement from each user side, it was shown that an equilibrium payoff can obtained using Boltzmann-Gibbs-like strategies. The details of the algorithm will be given in Chapter 4.

Resource allocation in wireless networks

The wireless channel is a shared medium over which many users compete for resources. Since there are many users, it is important to allocate this shared resource in a fair manner among the users. Further, since the available spectrum is limited, it is important to efficiently use the channel. However, the stochastic and time-varying nature of the wireless environment, coupled with different channel conditions for different users, poses significant challenges to accomplishing these goals. Moreover, the lack of availability of channel, queue sizes, and arrival statistics further complicates the solution. The authors in [183] apply strategic learning to spectrum access games. Therein, each wireless user will try to maximize its payoff by acquiring as much spectrum resources as possible unless a preemptive mechanism exists in the network. Using a stochastic game framework, where secondary wireless users can compete over time for the dynamically available transmission opportunities (spectrum holes), the authors explain how selfish, autonomous agents that strategically interact can adapt and learn their optimal strategies in order to acquire the necessary spectrum opportunities. The authors in [48] consider allocation games and show that by modifying the payoff functions, if the players select their strategy according to a stochastic approximation of the replicator dynamics, the strategy learning procedure converges to a Nash equilibrium of the game that is a locally optimal for the initial game. The authors in [113] consider the behavior of a large number of selfish users that are able to switch dynamically between multiple wireless access points. The authors show that users start out completely naive but, by using a fixed set of strategies to process a broadcasted training signal, they quickly evolve and converge to an evolutionarily stable state.

Evolutionary coalitional learning

Coalitional learning is a strategy/payoff learning in which different teams and coalitions, each with different heterogeneous/hybrid schemes and different learning ability, use a variety of learning activities to improve their understanding of the dynamic environment over long-run interaction. Each member of a coalition is responsible not only for learning jointly the optimal payoff/strategy for the coalition but also for helping teammates learn, thus creating an environment with a common goal. The group members work through the internal solution concepts for value allocation until all coalition members successfully learn and achieve a target.

Cooperative efforts result in mutual benefit so that:

- Each member of the coalition may gain from each other's efforts. Your success benefits me, and my success benefits you.

- All members of the same coalition may share a common payoff and know that one's performance is mutually caused by oneself and one's team members. This means that a subgroup cannot implement a specific scheme without the other members of the coalition.

How can we design networks to automatically learn and to improve with experience in a dynamic and interactive environment when nodes may form local coalitions? Below we provide a partial answer to this question with an explicit coalitional scheme. The widespread use of heterogeneous wireless technologies, their integration, advent of multimode terminals, and the envisioned user (service) centricity enable users to get associated with the best available networks according to user preferences and applications specific requirements. When it comes to user-centric network selection, operators with a view to increasing their user pool may offer different incentives to users to motivate them to form coalitions. The mechanism of evolutionary coalition learning proposed in [90] is described as follows:

- The procedure starts with an initial coalitional structure.

- At each frame unit, each user can decide to stay in a coalition or try to join another coalition (including singleton). The user migrates depending on its payoff which is based on the cooperative solution concept used inside its current coalition (it can be the Shapley value allocated to that user in the specific coalition).

- A migration may take a certain time unit and a learning cost.

- The process is iterated many times until the emergence of a stable coalitional or fuzzy coalitional structure.

The authors in [91, 157, 90] analyzed this learning algorithm to form new coalitions. The preliminary results are promising in the sense of speed of convergence as well as in terms of average performance. A negative part of the algorithm is that it may be very costly to make a coalition or to re-form a coalition due delay, negotiation issue, co-existence, reputation, patience, etc. Example of evolutionary coalition without considering cost of coalition formation is illustrated below.

Example of evolutionary coalition game without cost of coalition formation

Most issues of cooperation among the cognitive radio (CR) nodes are based on partially competitive behavior of the nodes for available resources in the decentralized network. In practice, one needs to introduce quantitative metrics of the cost of selfishness (the Price of Anarchy - PoA), of having limited knowledge of the competitors (the Price of Ignorance - PoI), and of limited time for learning the network environment (the Price of Impatience - PoIm) in dynamically changing radio channels.

Consider a random access network composed by power-limited users communicating to a common access point in a coordinated way. Each user can be in one of the two states: active or dormant. In an active state, a user attempts a communication. In a dormant state, a user shuts off in order to maximize the battery lifetime. Transmitted packets are successful only if there is a single active user in the range, otherwise collision occurs. In order to reduce collisions, users are allowed to form coalitions and the users that are in the same coalition can share their activity plans. Define the value of coalition J without cost of moves as

$$v(J) = \left(1 - \prod_{j \in J}(1 - x_j)\right) \prod_{j' \in \mathcal{N} \setminus J}(1 - x_{j'}),$$

where $\mathcal{N} = \{1, 2, \ldots, n\}$ is the grand coalition, $J \subseteq \mathcal{N}$, and x_j is the probability for user j to be active. One can show that the pair $(\mathcal{N}, v)$ defines a cooperative game in the sense of von Neumann-Morgenstern (1944). A payoff vector $r \in \mathbb{R}^n$ is an imputation for the game if it is efficient and individually rational. One of the set-valued solution concepts in that context is the core. Define the *core* and specify the difference (if any) between the core and the capacity region.

However, in practice, it is not trivial to form a coalition.

- How to form a coalition? Is it possible to form a coalition using a distributed strategic learning?

- What are the costs, benefits, and delays associated to the coalition formation?

- Propose an evolutionary coalitional learning framework for the above game.

- Extend the methodology to cognitive radio network with finitely many primary users and a coalitions of secondary users.

Learning and dynamics in network economics

In a wide range of network economics situations, users, designers, and operators make decisions without being fully informed about the rewards from different options. In many of these instances, the decision problems are of a recurring nature, and it is natural that players use their past experience and the experience of others in making current decisions. The experience of others is important for two reasons: one, it may yield information on different actions (as in the case of choice of new technology or quality of experience of user per subscription), and two, in many settings the rewards from an action depend on the choices made by others, and so there is a direct value to knowing about other's actions (as in the case of which operator to choose, or which technology to use, or whether to buy a novel battery or not). This suggests that the precise way in which individuals interact can influence the generation and dissemination of useful information and that this could shape players choices and network outcomes. In recent years, these considerations have motivated a substantial body of work on learning in economics, which takes explicit account of the structure of interaction among entities. In [66] the authors consider a set of individuals who are located on the nodes of a network; the arcs of the network reflect relations between these individuals. At regular intervals, individuals choose an action from a set of alternatives. They are uncertain about the rewards from different actions. They use their own past experience as well as gather information from their neighbors (individuals who are linked to them) and then choose an action that maximizes own-payoffs. The model is shown to be well-adapted to network formation, emergence of collaborative behavior and cooperation [38, 94, 82].

Learning in wireless network security

Wireless network security is a challenging problem due to the complexity of underlying network entities, their interdependencies as well as designers and exogenous factors. It involves decision making at multiple levels and multiple time scales, given the limited resources available to altruistic users, jammers and malicious attackers, regular users, designers, and administrators defending networked systems. The book in [3] develops a decision and game theoretic approach to network security. Various aspects of learning and controls in antagonistic problems are examined. In particular the Chapter 9 is on machine learning for intrusion and anomaly detection.

1.6.4 Transport Layer

The transport layer provides end-to-end communication services for applications within a layered architecture of network components and protocols. Improving congestion control and queue management in the Internet is a very active research area. The most well-known transport protocol is the Transmission Control Protocol (TCP). When transferring data between nodes, flow control protocols are needed to regulate the transmission rates so as to adapt to the available resources. A connection that loses data units has to retransmit them later. In the absence of adaptation to the congestion, the ongoing transmissions along with the retransmissions can cause increased congestion in the network resulting in losses and further retransmissions by this and/or by other connections. This type of phenomenon, which had lead to several congestion collapses, motivated the evolution of the Internet transport protocol, TCP, to a protocol that reduces dramatically the connection's throughput upon congestion detection. The possibilities of deploying freely new versions of congestion protocols (TCP, additive-increase multiplicative-decrease -AIMD-, multiplicative-increase multiplicative-decrease -MIMD-) on terminals connected to the Internet creates a competitive environment between protocols. Much work has been devoted analyzing such competition and predicting its consequences.

Learning which transport protocol will emerge.

The works in [7, 160, 8] address the question of protocol dominance with the evolutionary game paradigm and provide an alternative answer along with a more detailed analysis of this competition scenario.

Evolutionary games have been developed in biological sciences with the aim of studying the evolution and equilibrium behavior (called evolutionarily stable state, ESS, see also Section 8.3.2) of large populations. While the rich theoretical foundations of evolutionary games allow biologists to explain past and present evolution and predict future evolution, it can be further used in engineering to architect evolution. One of the components of evolutionary games is evolutionary game dynamics. As we will see the next chapters, there are close connections between learning and game dynamics. The evolutionary game approach helps to better understand the evolution and the design of congestion control protocols in wireless networks containing large number of individual non-cooperative terminals or sensors. The authors studied the evolution of congestion control protocols, and show how the evolution and the evolutionarily state states are influenced by the characteristics of the wireless channel.

> **Strategic learning in cross-layer PHY-MAC problem**

One of the fundamental aspects of wireless networking is the problem of access control, where it must be decided which wireless nodes should access the wireless channel at any given time and at what transmission powers and rates. The simultaneous consideration of access, which is a medium access control (MAC) layer issue, and of power and rate control, which are physical layer (PHY) issues, make this a cross-layer problem. In a game-theoretic setting, this problem can be seen as a multilevel interacting system under uncertainty. The two layers are interdependent and it is not necessarily a hierarchical game with fixed ordering. Due to the uncertainty of channel states, one gets a multilevel robust game. One of the challenging questions is the design of very fast fully distributed learning algorithms for such cross-layer problems. A particular case of such a problem will be examined in Section 4.8.

1.6.5   Application Layer

Below, we consider a network and application layer example. The Transport layer handles a lot of interfacing between the network and applications. The application layer focuses more on network services, utilities, and operating system environments. It supports network access, as well as providing services for user applications. Here we use the concept of user quality-of-experience (QoE), which formalizes the users' experiences with a service (phone call, video downloading files). Quality of experience systems will try to measure metrics that the user will directly perceive as a quality parameter: for example,

- time for new switching codecs of videos

- time for a new channel to be played when changing channel on TV

> **IP TV service user-centric network selection**

It is envisioned that in Internet Protocol television (IPTV) of services will be run on the top, a heterogeneous mix of network infrastructure. The existence of multimode terminals enables users get associated to the best available networks according to user preferences and application specific requirements. The work in [92] proposed a user-centric service selection for IPTV services. User quality of experience and user satisfaction function for video services are studied under measurements. The users select the service based on their own-satisfaction function. The authors introduced the notion of *cost of learning* which incorporates the cost to switch to an alternate IPTV service provider (the idea cost of learning received the best paper award at INFOCOM Workshop 2011 on Future Media Networks and IP-based TV) and have shown that the resulting iterative process with switching cost converges to a global optimum in a basic user-centric selection problem.

In order to demonstrate the user-centric IPTV service selection, and the effect of learning in such a telecommunication landscape, extensive rounds of simulation runs are required. The simulation scenario dictates that IPTV consumers are under the coverage of heterogeneous technologies owned by different infrastructure providers. Considering Long Term Evolution (LTE) and WLAN access network technologies, the authors implemented the integration of these two technologies following 3GPP standards for intra-operator heterogeneous technologies integration. Intra-operator mobility management is carried out using Mobile IPv6. Furthermore in total there are four IPTV service providers, who are considered as potential candidate service providers (competitors) to extend IPTV services to the consumers. Service requests of different quality classes, and content types are generated by consumers. The arrival of requests is modeled by random process, and the service quality class is chosen randomly. In order to capture the different consumer preferences we assume that the sizes of different quality class requests are assumed to be static and classified as low, medium, and high video quality respectively. Under this setting, it is shown that the stochastic learning algorithm with switching cost converges and stability properties are studied in the user-centric IPTV service selection problem using *evolutionary game dynamics*. See Chapter 2 for more details.

1.6.6 Compressed Sensing

In this subsection we present a robust game theoretic and learning formulation in *compressed sensing*.

What is compressed sensing?

Given matrices $A \in \mathbb{R}^{l_1 \times l_2}$, and matrices B, C, D. Compressed Sensing focuses on recovery of a sparse signal $Bx \in \mathbb{R}^m$ from its noisy observations $y = Ax + Cu + D\xi$ where ξ is an observation noise such that $\|\xi\| \leq \epsilon$ for a certain norm. u is an unknown parameter known to belong to a given set.

Compressive Sensing is based on the empirical observation that many types of real-world signals and images have a sparse expansion in terms of a suitable basis or frame, for instance a wavelet expansion. This means that the expansion has only a small number of significant terms, or in other words, that the coefficient vector can be well-approximated with one having only a small number of nonvanishing entries.

The support of a vector (signal) $x \in \mathbb{C}^N$ is denoted $supp(x) = \{s, \ x_s \neq 0\}$. Note that $x \longmapsto |supp(x)| = \sum_{s=1}^{N} \mathbb{1}_{\{x_s \neq 0\}}$ is not a quasi-norm. A vector is k-sparse if $|supp(x)| \leq k$. We denote by S_k the set of k-sparse vectors. The best k-term approximation error of a vector Bx is defined in $l^p, p > 0$ as

$$\epsilon_{k,p}(x) := \inf_{z \in S_k} \|Bx - z\|_p$$

where $\|x\|_p = \left(\sum_s |x_s|^p\right)^{1/p}$ denotes the l^p−norm.

If $\epsilon_{k,p}(x)$ decays quickly in k then Bx is called compressible. Indeed, in order to compress Bx one may simply store only the k largest entries. When reconstructing Bx from its compressed version the nonstored entries are simply set to zero, and the reconstruction error is $\epsilon_{k,p}(x)$. It is emphasized at this point that the procedure of obtaining the compressed version of Bx is adaptive and nonlinear since it requires the search of the largest entries of Bx in absolute value. In particular, the location of the non-zeros is a nonlinear type of information. The best k−term approximation of Bx can be obtained using the nonincreasing rearrangement of the component. However this approach of compressing a signal Bx by only keeping its largest coefficients is certainly valid when full information on Bx is available. If, however, the signal first has to be acquired or measured by a somewhat costly or lengthy procedure then this seems to be a waste of resources because large efforts are made to acquire the full signal and then most of the information is thrown away when compressing it.

Compressive sensing nevertheless predicts that reconstruction from vastly undersampled nonadaptive measurements is possible under suitable conditions. The approach for a recovery procedure that probably comes first to mind is to search for the sparsest vector Bx which is consistent with the measurement of y.

This leads in solving one of the following formulations:

- (P1) $p > 0$

$$\inf_{z \in S_k} \|B(x - z)\|_p \text{ such that } y = Az + Cu + D\xi$$

- (P2)

$$\inf_{z \in S_k} \|Az - y\|_p \text{ such that } y = Az + Cu + D\xi$$

Due to the combinatorial nature of S_k, an algorithm that the above problem for any matrix A, B, C, D and any right hand side y is computationally intractable and many authors has suggested a relaxed problem with the l^1−norm instead of the support. However the solutions of these relaxations are different than the original problem. Now, we formulate this problem as a zero-sum robust game.

$$\inf_{z \in S_k} \sup_{\|\theta\|_{p^*} \leq 1} \langle \theta, B(x - z) \rangle \text{ such that } y = Az + Cu + D\xi$$

where $\frac{1}{p^*} + \frac{1}{p} = 1$. Here we have used the fact that the norm itself is a sup of a linear form:

$$\sup_{\|\theta\|_{p^*} \leq 1} \langle \theta, z \rangle = \|z\|_p$$

In this robust game, player 1 chooses z and player 2 chooses θ which are continuous vector variables. In Chapter 4 we explain how very fast algorithms can be constructed for this type robust game problem using CODIPAS.

In the next chapter we present strategy-learning and dynamics.

2

Strategy Learning

2.1 Introduction

This chapter presents basic learning schemes in games. There are two main parts. The first part focuses on strategy learning in games under perfect monitoring of past actions. This class of schemes is referred to partially distributed learning algorithms or semi-distributed schemes, which basically introduces learning mechanisms that use information about the other players (information about their past actions or their past payoffs via some signals or messages, public or private), and in this context, learning different solution concepts such as Nash equilibrium, and bargaining solution in both pure and mixed strategies are introduced. Convergence, nonconvergence properties and features of learning algorithms are thoroughly investigated.

The framework is used also for bargaining problems in long-run interactions. Then, we focus on learning schemes that does not require the monitoring assumption. The learning schemes that are between the two are not specially discussed.

2.2 Strategy Learning under Perfect Action Monitoring

In this section, we present strategy learning algorithms for engineering games with action monitoring and one-step recall. Partially distributed learning algorithms are characterized by the following property:

> *In their learning patterns, these learning schemes require some explicit information about the choices of the other players*, i.e, each player observes the previous actions picked by the other players (perfect action-monitoring at the last round).

These distributed learning algorithms are illustrated in wireless contexts including power allocation, spectrum access in small cell networks [100], cognitive radio networks, access control, resource selection in wireless networks, rate bargaining in additive white Gaussian noise channel, etc. We discuss also the limitation of these learning algorithms and their possible implementations in wireless communications.

Noncooperative solution concepts (Cournot/Nash equilibrium, correlated equilibrium, subgame perfect equilibrium) as well as cooperative solution concepts (Bargaining solution, Pareto optimality) are addressed.

We start with sophisticated learning algorithms in which each player adapts its strategy based on observations of the other players' actions. These algorithms are called *partially distributed learning algorithms* and include classical fictitious play (FP), best response, classical logit learning, adaptive play, joint strategy FP, stochastic FP, regret matching, sequential asynchronous FP learning, cost-to-learn, learning bargaining solution, etc. The term *partially distributed* is used because these procedures require information about the other players. We provide detailed classification of the learning schemes in terms of information needed to update the learning patterns as well as their efficiency and stability. In order to improve the efficiency of the outcomes, new learning algorithms are proposed such as *learning bargaining solutions*. The convergence time of some of the learning algorithms is presented. This allows us to address a comparative view between some of the learning algorithms in the same category [99]. The limitation of each learning scheme in terms of implementation is addressed.

2.2.1 Fictitious Play-Based Algorithms

Fictitious play and its variants (stochastic fictitious play, smooth fictitious play, etc) are simple models of learning extensively used in the literature of game theory. This subsection complements the description we provided in [99].

Brown's Fictitious Play Model

The standard *fictitious play* algorithm introduced by Brown (1949-1951,[40, 41, 137]) assume that each player observes his or her actions, its payoffs but also the last actions played by the other players. In it, each player presumes that the other players are playing stationary (possibly mixed) strategies. At each time slot, each player thus best responds to the empirical frequency of play of his or her opponent (this presumes that he or she observes the last actions chosen by the other players).

The next Lemma 2.2.1.1 gives a recursive equation satisfied by the frequencies of strategies and payoffs. This equation is fundamental in the construction of learning algorithms.

Lemma 2.2.1.1. *The empirical process*

$$\bar{f}_{j,t}(s_j) = \frac{1}{t}\sum_{t'=1}^{t}\mathbb{1}_{\{a_{j,t'}=s_j\}}$$

satisfies the recursive equation

$$\bar{f}_{j,t+1}(s_j) = \bar{f}_{j,t}(s_j) + \lambda_{t+1}\left(\mathbb{1}_{\{a_{j,t+1}=s_j\}} - \bar{f}_{j,t}(s_j)\right) \tag{2.1}$$

where $\lambda_{t+1} = \frac{1}{t+1}$, *the term* $\mathbb{1}_{\{a_{j,t'}=s_j\}}$ *denotes the indicator of* $\{a_{j,t'} = s_j\}$; *i.e,* $\mathbb{1}_{\{a_{j,t'}=s_j\}} = 1$ *if* $a_{j,t'} = s_j$, *and 0 otherwise.*

Proof.

$$
\begin{aligned}
(t+1)\bar{f}_{j,t+1}(s_j) &= \mathbb{1}_{\{a_{j,t+1}=s_j\}} + \sum_{t'=1}^{t}\mathbb{1}_{\{a_{j,t'}=s_j\}} \\
&= \mathbb{1}_{\{a_{j,t+1}=s_j\}} + t\bar{f}_{j,t}(s_j)
\end{aligned}
$$

By dividing both sides by $(t+1)$ one gets,

$$\bar{f}_{j,t+1}(s_j) = \frac{1}{t+1}\mathbb{1}_{\{a_{j,t+1}=s_j\}} + \frac{t}{t+1}\bar{f}_{j,t}(s_j)$$

$$
\begin{aligned}
\bar{f}_{j,t+1}(s_j) &= \frac{1}{t+1}\mathbb{1}_{\{a_{j,t+1}=s_j\}} + \left(1 - \frac{1}{t+1}\right)\bar{f}_{j,t}(s_j) \\
&= \bar{f}_{j,t}(s_j) + \frac{1}{t+1}\left(\mathbb{1}_{\{a_{j,t+1}=s_j\}} - \bar{f}_{j,t}(s_j)\right)
\end{aligned}
$$

which is the announced recursive equation:

$$\bar{f}_{j,t+1}(s_j) = \bar{f}_{j,t}(s_j) + \lambda_{t+1}\left(\mathbb{1}_{\{a_{j,t+1}=s_j\}} - \bar{f}_{j,t}(s_j)\right).$$

$\square$

Fundamental equation for learning processes

The recursive equation (2.1) will be fundamental of the learning algorithms studied in book.

Exercise 2.1. *Verify that the sequence* λ_t *defined by* $\lambda_t = \frac{1}{t+1}$, $t \geq 0$, *is in* $l^2\backslash l^1$, *i.e., it satisfies*

$$\lambda_t \geq 0, \quad \sum_{t\in\mathbb{Z}_+}\lambda_t = +\infty, \quad \sum_{t\in\mathbb{Z}_+}\lambda_t^2 < +\infty.$$

From its definition, the vector $\bar{f}_{j,t+1} = (\bar{f}_{j,t+1}(s_j))_{s_j \in \mathcal{A}_j}$ gives a probability measure over $\mathcal{A}_j$, equipped with the canonical sigma-algebra. Each player j chooses an action at time $t + 1$ to maximize his/her payoff in response to the frequencies $\bar{f}_{-j,t}$.

Brown's fictitious play algorithm is determined by

$$a_{j,t+1} \in \arg \max_{s_j \in \mathcal{A}_j} \ r_j(\mathbf{e}_{s_j}, \bar{f}_{-j,t}). \qquad (2.2)$$

This choice is myopic, i.e., players are trying to maximize current payoff without considering their future payoffs. Players only need to know their own payoff or utility function and observe the actions of the others at the end of each play. Thus, it is in the class of a partially distributed learning scheme.

Exercise 2.2. *Suppose that each player follows the above fictitious play procedure. Verify that the frequencies generated by the plays satisfy: for any $t \geq 2$,*

$$\bar{f}_{j,t}(s_j) \geq 0, \ \forall s_j \in \mathcal{A}_j,$$

$$\sum_{s'_j \in \mathcal{A}_j} \bar{f}_{j,t}(s'_j) = 1.$$

Definition 2.2.1.2. *Let $\{a_{j,t}\}_{t \in \mathbb{Z}_+}$ be a sequence of action profiles generated by the fictitious play learning algorithm. We say that the fictitious play algorithm converges in frequencies if the sequence of empirical frequencies of play $\{\bar{f}_{j,t}\}_{j \in \mathcal{N}, \ t \in \mathbb{Z}_+}$ converges to some point $\{\bar{f}_j^*\}_{j \in \mathcal{N}} \in \prod_j \mathcal{X}_j$ (componentwise).*

Next, we give a simple class of games that has the convergence in frequency property.

Definition 2.2.1.3 (Exact finite potential game). *A finite game in normal form $\mathcal{G}$ is an exact potential game if it admits a potential function $\tilde{V}^n : \prod_{j \in \mathcal{N}} \mathcal{A}_j \longrightarrow \mathbb{R}$, which is a function satisfying $\forall j, \ s_j \in \mathcal{A}_j, \ a \in \prod_j \mathcal{A}_j,$*

$$\tilde{R}_j(s_j, a_{-j}) - \tilde{R}_j(a) = \tilde{V}^n(s_j, a_{-j}) - \tilde{V}^n(a).$$

In terms of frequencies, the convergence of fictitious play has been proved in the following configurations:

- Both players have only a finite number of strategies and the game is two-player zero sum (Robinson 1951, [137])

- The game is solvable by iterated elimination of strictly dominated strategies (Nachbar 1990, [118]). A simple example of such a game is studied in [100] in the context of coverage of small-cell networks.

- The game is a potential game (Monderer and Shapley 1996). This includes two-player finite games with common payoffs, common interest games, etc.

- The game has generic payoffs and is $2 \times m$ (Berger 2005, [30]).

> ⚠ However, fictitious play does not always converge. Shapley (1964 , [143]) proved that the play can cycle indefinitely in Rock-Paper-Scissor games. Note that, in the above formulation, the players update their beliefs simultaneously, but they can do it sequentially (alternating update of choices).

Proposition 2.2.1.4. *Suppose a (Brown) fictitious play sequence $\{a_{j,t}\}_{j,t}$ generated a sequence with empirical frequencies $\{\bar{f}_{j,t}\}_{j \in \mathcal{N},\ t \in \mathbb{Z}_+}$ converging to $\{\bar{f}_j^*\}_{j \in \mathcal{N}}$. Then $\{\bar{f}_j^*\}_{j \in \mathcal{N}}$ is an (Nash) equilibrium point of the one-shot game.*

The result follows immediately from Equation (2.2) and the continuity of the mappings $(r_j)_{j \in \mathcal{N}}$.

Remark 2.2.1.5. *Note that for continuous action space $\mathcal{A}_j$, the average :* $\bar{f}_{j,t} = \frac{1}{t} \sum_{t'=1}^{t} a_{j,t'}$ *satisfies also the recursive equation*

$$\bar{f}_{j,t+1} = \bar{f}_{j,t} + \lambda_{t+1} \left(a_{j,t+1} - \bar{f}_{j,t} \right).$$

Similarly, the average payoff $\bar{R}_{j,t} = \frac{1}{t} \sum_{t'=1}^{t} r_{j,t'}$ satisfies

$$\bar{R}_{j,t+1} = \bar{R}_{j,t} + \lambda_{t+1} \left(r_{j,t+1} - \bar{R}_{j,t} \right).$$

Arthur (1993, [11]), Borger and Sarin (1993), [179] studied variants of these iterative equations (in terms of strategy reinforcement and average payoff learning etc) by changing the step-size λ_{t+1}.

For a constant step size, i.e., $\lambda_{t+1} = b$, $b \in \mathbb{R}_+$, one gets

$$x_{j,t+1}(s_j) = \mathbf{x}_{j,t}(s_j) + b \left(\mathbb{1}_{\{a_{j,t+1}=s_j\}} - x_{j,t}(s_j) \right) \tag{2.3}$$

If the step-size is proportional to the current payoff $\lambda_{t+1} = b r_{j,t+1}$, one gets

$$x_{j,t+1}(s_j) = x_{j,t}(s_j) + b r_{j,t+1} \left(\mathbb{1}_{a_{j,t+1}=s_j} - x_{j,t}(s_j) \right). \tag{2.4}$$

Other Versions of Fictitious Play

We now present one of the most used version of fictitious play in game theory literature (see [60], [194]). Adaptive learning means that players consider past play as the best predictor of their opponents' future play and they best-respond to their forecast. Fictitious play and its variants are *adaptive learning algorithms* used in standard repeated games (the same game is played many times). Assume that players play a standard repeated game with a fixed strategic form game $\mathcal{G}$ many times. Each player is assumed to know the joint strategy space, his or her own-payoff function and observe the actions of the other players at the end of each time slot (perfect monitoring of past actions of the others). Players play as Bayesian players:

> The players believe that the strategies of the other players correspond to some unknown mixed strategy.

The beliefs are updated as follows:

A Generic Fictitious Play Procedure:
Start with some initial references: For each player j, $\alpha_j(0)$: $\prod_{i \neq j} \mathcal{A}_i \longrightarrow \mathbb{R}_+$, which is updated by adding 1 if an action (profile) is played.

$$\alpha_{j,t+1}(s_{-j}) = \alpha_{j,t}(s_{-j}) + \mathbb{1}_{\{a_{-j,t} = s_{-j}\}}, \ s_{-j} \in \prod_{j' \neq j} \mathcal{A}_{j'}.$$

The probability that the player j assigns to the action profile s_{-j} of the other players at time slot $t + 1$ is

$$\tilde{x}_{j,t+1}(s_{-j}) = \frac{\alpha_{j,t+1}(s_{-j})^{\gamma_j}}{\sum_{s'_{-j}} \alpha_{j,t+1}(s'_{-j})^{\gamma_j}},$$

where $\gamma_j \geq 1$.
Fictitious play assigns actions to histories by
Computing the probabilities $\{\tilde{x}_{j,t}\}_t$ after computing the parameters $\{\alpha_{j,t}\}_t$.
Choosing an action

$$a_{j,t+1} \in \mathrm{BR}_j\left(\tilde{x}_{j,t+1}\right),$$

at time slot t.

The above inclusion defines a Hannan set (Hannan'1957), also known as coarse correlated equilibrium or weak correlated equilibrium.

Below we give an illustrative example that shows the limitation of this learning scheme [115].

Example 2.2.1.6 (Channel selection [129, 125, 146]). *Consider two receivers $Rx1, Rx2$ and two mobile terminals $MT1, MT2$. Each mobile terminal can transmit to one of the two receivers. If both MTs transmit at the same receiver at the time slot, there is collision and the packets are lost and both MTs get zero as payoff. If they choose different receivers, the probability $\mu > 0$ of having a successful transmission is very high. One can represent this scenario by the following table: Mobile terminal 1 (MT1) chooses a row, and MT2 chooses a column. The first component of the payoff vector is for MT1 and the second component is for MT2.*

$MT1 \backslash MT2$	$Rx1$	$Rx2$
$Rx1$	$(0,0)$	(μ, μ)
$Rx2$	(μ, μ)	$(0,0)$

Suppose now that the players have implemented the fictitious play algorithm in their mobile terminal, and the algorithm starts with the initial references $\alpha_j(0) = (1, \bar{\epsilon})$, $j \in \{1, 2\}$ with $0 < \bar{\epsilon} < 1$. At the first time slot, the MTs compute

$$x_{1,1}(Rx1) = x_{2,1}(Rx1) = \frac{1}{1 + \bar{\epsilon}},$$

$$x_{1,1}(Rx2) = x_{2,1}(Rx2) = \frac{\bar{\epsilon}}{1 + \bar{\epsilon}}.$$

The best response to $(\frac{1}{1+\bar{\epsilon}}, \frac{\bar{\epsilon}}{1+\bar{\epsilon}})$ is reduced to the choice of receiver $Rx2$. Both MTs transmit at receiver $Rx2$ at time slot 1. This leads to a collision, and the payoffs are $(0, 0)$. At time slot 2, the updated parameters become $\alpha_2^1(Rx1) = 1 = \alpha_2^2(Rx1)$, and $\alpha_2^1(Rx2) = 1 + \bar{\epsilon} = \alpha_2^2(Rx2)$. The new probabilities are

$$x_{j,2}(Rx1) = \frac{1}{2 + \bar{\epsilon}}, \quad x_{j,2}(Rx2) = \frac{1 + \bar{\epsilon}}{2 + \bar{\epsilon}}.$$

So both mobile terminals transmit at receiver $Rx1$ at time slot 2 and the outcome is again a collision. We then have an alternating sequence $(Rx2, Rx2), (Rx1, Rx1), (Rx2, Rx2), (Rx1, Rx1), (Rx2, Rx2), (Rx1, Rx1), \ldots$ and at each time slot there is a collision. This outcome cannot be a (Nash) equilibrium. But if we take the empirical frequencies of use of each receiver

$$\bar{f}_{j,t}(Rx1) = \frac{1}{t} \sum_{t'=1}^{t} \mathbb{1}_{\{a_{j,t'}=Rx1\}},$$

converges to $\frac{1}{2}$, i.e, a mixed (Nash) equilibrium of the one-shot game. As you can see, the realized payoff of each MT via the fictitious play algorithm is zero (because there is a collision at each time slot). Note that under the fictitious play algorithm

$$\frac{1}{t} \sum_{t'=1}^{t} \mathbb{1}_{\{a_{t'}=(Rx1,Rx1)\}} \neq \bar{f}_{1,t}(Rx1)\bar{f}_{2,t}(Rx1).$$

   This property is sometimes referred to as violation of propagation of chaos [160, 163]. The reason is that, even if the players start playing independently, they will be correlated during the interactions and the frequencies are not decoupled. Thus, it does not propagate the initial independence.

   In this channel selection example, even if the mobile terminals act on their best choices at each time slot based on their best estimations and beliefs, it does not necessarily lead to a better outcome. The cumulated realized payoff under the algorithm is worse than the worst (Nash) equilibrium payoff of the one-shot game. We conclude that the equilibrium payoff is not obtained by the players. This says that the equilibrium is not played. We will see in Section 4 how to learn the equilibrium payoffs simultaneously with optimal strategies using CODIPAS.

   The next example studies a two-player game with fixed (and nonzero) channel state parameters.

Example 2.2.1.7. *Consider two mobile stations (MSs) and two base stations (BSs). Each mobile station can transmit to one of the base stations. If mobile station $MS1$ chooses a different base station than mobile station $MS2$*

then, mobile station $MS1$ gets the payoff $\log_2\left(1 + \frac{p_1|h_1|^2}{N_0}\right)$, and mobile station 2 gets $\log_2\left(1 + \frac{p_2|h_2|^2}{N_0}\right)$ where p_1 and p_2 are positive powers, h_1, h_2 are channel states with nonzero modulus, and N_0 is a background noise parameter. If both MSs transmit at the same base station, there is an interference; mobile station $MS1$ gets $\log_2\left(1 + \frac{p_1|h_1|^2}{N_0 + p_2|h_2|^2}\right)$, and mobile station $MS2$ gets $\log_2\left(1 + \frac{p_2|h_2|^2}{N_0 + p_1|h_1|^2}\right)$. The following table summarizes the different configurations. Mobile station $MS1$ chooses a row and $MS2$ chooses a column. The first component of the payoff is for $MS1$ and the second component is for $MS2$.

$MS1\backslash MS2$	$BS1$	$BS2$
$BS1$	$\log_2\left(1 + \frac{p_1|h_1|^2}{N_0 + p_2|h_2|^2}\right),$ $\log_2\left(1 + \frac{p_2|h_2|^2}{N_0 + p_1|h_1|^2}\right)$	$\log_2\left(1 + \frac{p_1|h_1|^2}{N_0}\right),$ $\log_2\left(1 + \frac{p_2|h_2|^2}{N_0}\right)$
$BS2$	$\log_2\left(1 + \frac{p_1|h_1|^2}{N_0}\right),$ $\log_2\left(1 + \frac{p_2|h_2|^2}{N_0}\right)$	$\log_2\left(1 + \frac{p_1|h_1|^2}{N_0 + p_2|h_2|^2}\right)$ $\log_2\left(1 + \frac{p_2|h_2|^2}{N_0 + p_1|h_1|^2}\right)$

- *Under the assumptions $|h_j|^2 > 0, p_j > 0, N_0 > 0$, it is easy to see that this game has the same equilibrium structure as the game described in Example 2.2.1.6.*

- *We apply the fictitious play algorithm with the initial point $(\frac{\bar{\epsilon}}{1+\bar{\epsilon}}, \frac{1}{1+\bar{\epsilon}})$, $\bar{\epsilon} \in (0, 1)$.*

- *Since the two alternating cycling configurations lead to worse payoffs, the Cesaro-limit payoff obtained under the fictitious play algorithm is again worse than the worst Nash equilibrium payoff.*

- *We now change the initial point to $(\frac{1}{2}, \frac{1}{2})$, then the $\arg\max$ is a set. The pure best response set is equal to the set of actions. The outcome of the fictitious play algorithm is the completely mixed Nash equilibrium.*

- *If $|h_1|^2 = 0 = |h_2|^2$ or $p_1 = p_2 = 0$, then every strategy profile is an equilibrium and the set of equilibria coincides with the set of global optima.*

For non-zero modulus in the above example, we distinguish four configurations:

- *$(h_1, h_2) \neq (0, 0)$*

- *$(0, h_2), \ h_2 \neq 0.$*

- *$(h_1, 0), \ h_1 \neq 0.$*

- *$(0, 0)$*

The tables below represent the four configurations.

$MS1\backslash MS2$	$BS1$	$BS2$
$BS1$	$\log_2\left(1+\frac{p_1\|h_1\|^2}{N_0+p_2\|h_2\|^2}\right),$ $\log_2\left(1+\frac{p_2\|h_2\|^2}{N_0+p_1\|h_1\|^2}\right)$	$\log_2\left(1+\frac{p_1\|h_1\|^2}{N_0}\right),$ $\log_2\left(1+\frac{p_2\|h_2\|^2}{N_0}\right)$
$BS2$	$\log_2\left(1+\frac{p_1\|h_1\|^2}{N_0}\right),$ $\log_2\left(1+\frac{p_2\|h_2\|^2}{N_0}\right)$	$\log_2\left(1+\frac{p_1\|h_1\|^2}{N_0+p_2\|h_2\|^2}\right)$ $\log_2\left(1+\frac{p_2\|h_2\|^2}{N_0+p_1\|h_1\|^2}\right)$

$MS1\backslash MS2$	$BS1$	$BS2$
$BS1$	$0,$ $\log_2\left(1+\frac{p_2\|h_2\|^2}{N_0}\right)$	$0,$ $\log_2\left(1+\frac{p_2\|h_2\|^2}{N_0}\right)$
$BS2$	$0,$ $\log_2\left(1+\frac{p_2\|h_2\|^2}{N_0}\right)$	0 $\log_2\left(1+\frac{p_2\|h_2\|^2}{N_0}\right)$

$MS1\backslash MS2$	$BS1$	$BS2$
$BS1$	$\log_2\left(1+\frac{p_1\|h_1\|^2}{N_0}\right),$ 0	$\log_2\left(1+\frac{p_1\|h_1\|^2}{N_0}\right),$ 0
$BS2$	$\log_2\left(1+\frac{p_1\|h_1\|^2}{N_0}\right),$ 0	$\log_2\left(1+\frac{p_1\|h_1\|^2}{N_0}\right)$ 0

$MS1\backslash MS2$	$BS1$	$BS2$
$BS1$	$(0,0)$	$(0,0)$
$BS2$	$(0,0)$	$(0,0)$

On the Notion of Convergence in Frequencies

As illustrated in Example 2.2.1.6, even when the frequencies converge, the average payoff under fictitious play algorithms can be worse than the worst Nash equilibrium payoff. This leads naturally to the following question:

> *Is convergence of the frequencies/beliefs a natural notion of convergence in adaptive learning algorithms?*

From the game theory point of view, the game given in Example 2.2.1.6 is in the class of anticoordination games. The game has a unique completely (fully) mixed (Nash) equilibrium and two pure equilibria. As we remarked earlier, if the standard fictitious play algorithm is used in this game, the players may never take opposite actions at the same time slot (they may never anticoordinate) but there is a convergence in frequencies. This example proves that the "convergence to an equilibrium" does not mean that this equilibrium will be played.

The next result characterizes strict Nash equilibria and absorbing states of the stochastic process generated by fictitious play learning.

Proposition 2.2.1.8 ([59]). *Under fictitious play learning, the following results hold:*

- *Strict Nash equilibria are absorbing for the fictitious play learning process.*

- *Any pure strategy steady state of fictitious play (not joint) learning must be a Nash equilibrium.*

Asynchronous Clocks in Stochastic Fictitious Play

The model that we previously presented supposes that all the players play at every time slot (all the players are always active) so that every player knows that the actions played by the opponents. This is typically the case in some *atomic games*, but in many settings; players play at varying frequencies, and do not know how many times their current opponents have played the game. To take into account such situations, [61] proposed an asynchronous stochastic fictitious play model that we describe below.

Consider a system with n players. Time is discrete. At each time slot, two players are drawn randomly from the n players; we denote the set of selected players by $\mathcal{B}^n(t)$. When a player j is in $\mathcal{B}^n(t)$, i.e., he or she is selected at time t, he chooses a smoothed best response in the stage-game given his assessment about his current opponents' actions and beliefs. After the meeting, he updates his assessment based on the realized strategy of the other players. If the player j is not in $\mathcal{B}^n(t)$, player j keeps his old assessment.

As in the stochastic fictitious play with synchronous clocks described above, at each time slot, each player j has a parameter $\alpha_{j,t}$ and forms a probability $\mathbf{x}_{j,t}$ that he assigns to the other player. If player j is drawn from the system, one of the two active players at time slot t (i.e. $j \in \mathcal{B}^n(t)$.), then he chooses an action $a_{j,t}$ that is a pure best response to $\mathbf{x}_t$.

The learning algorithm can be described as follows:

Start with some initial references. At time 0, each player j has a mapping $\alpha_{j,0} : \prod_{i \neq j} \mathcal{A}_i \longrightarrow \mathbb{R}_+$, which is updated by adding $+1$ if an action is played $$\alpha_{j,t+1}(s_{-j}) = \alpha_{j,t}(s_{-j}) + \mathbb{1}_{\{j \in \mathcal{B}^n(t), a_{-j,t} = s_{-j}\}}, \quad (2.5)$$ $$s_{-j} \in \prod_{j' \neq j} \mathcal{A}_{j'}. \quad (2.6)$$

The main difference between this model and the stochastic fictitious play is that in this model, each player only updates his assessment when he or she has opportunities.

The term $\alpha_{j,t+1}$ is equal to the initial weight on s_{-j}, $\alpha_{j,0}$ plus the number of time slots in which player j was drawn and his or her opponents have chosen s_{-j} up to the time slot t. The number $\frac{1}{\sum_{s_{-j}} \alpha_{j,t}(s_{-j})}$ is player j's step size at time t which governs the size of the influence of new observations.

Let $y_{j,t}$ be the ratio between step sizes of synchronous and asynchronous clocks, $y_{j,t} = \frac{1+\bar{\alpha}_{j,t}}{1+t}$ where $\bar{\alpha}_{j,t} = \sum_{s_{-j}} \alpha_{j,t}(s_{-j})$. The updating rule becomes

$$\mathbf{x}_{j,t+1} - \mathbf{x}_{j,t} = \mathbb{1}_{\{j \in \mathcal{B}^n(t)\}} \frac{1}{t+1} \frac{1}{y_{j,t}} \left(\mathbb{1}_{\{a_{j,t+1}=s_j\}} - \mathbf{x}_{j,t} \right)$$

$$= \frac{1}{t+1} \left[\frac{2}{n} \frac{1}{y_{j,t}} \left(\frac{1}{n-1} \sum_{i \neq j} \beta_i(\mathbf{x}_{-i,t}) - \mathbf{x}_{j,t} \right) + M_{j,t+1} \right],$$

where

$$M_{j,t+1} = \frac{1}{y_{j,t}} \left[\mathbb{1}_{\{j \in \mathcal{B}^n(t)\}} \left(\mathbb{1}_{\{a_{j,t}=s_j\}} - \mathbf{x}_{j,t} \right) \right.$$

$$\left. - \frac{2}{n} \left(\frac{1}{n-1} \sum_{i \neq j} \beta_i(\mathbf{x}_{-i,t}) - \mathbf{x}_{j,t} \right) \right]$$

The evolution of y_j is given by

$$y_{j,t+1} - y_{j,t} = \frac{1}{t+1} \left(\frac{2}{n} - y_{j,t} + \epsilon^1_{j,t+1} + \epsilon^2_{j,t+1} \right),$$

where $\epsilon^1_{j,t+1} = \mathbb{1}_{\{j \in \mathcal{B}^n(t)\}} - \frac{2}{n}$, $\epsilon^2_{j,t+1} = -\frac{1}{t+2} \left[\mathbb{1}_{\{j \in \mathcal{B}^n(t)\}} - y_{j,t} \right]$.

The stochastic process $\{(\mathbf{x}_{j,t}, y_{j,t})\}_t$ satisfies the conditions in Benaïm (1999). We then use it and approximate the stochastic process by the following system of ordinary differential equations:

$$\begin{cases} \dot{\mathbf{x}}_j(t) = \frac{2}{n} \frac{1}{y_j(t)} \left(\frac{1}{n-1} \sum_{i \neq j} \beta_i(\mathbf{x}_i(t)) - \mathbf{x}_j(t) \right) \\ \dot{y}_j(t) = \frac{2}{n} - y_j(t) \end{cases}$$

A rest point $\mathbf{x}^*$ is *linearly unstable* with respect to a dynamic process $\dot{\mathbf{x}} = K(\mathbf{x})$, if its linearization $DK(\mathbf{x})$ evaluated at $\mathbf{x}^*$ has at least one eigenvalue with positive real part.

> The asymptotic pseudo-trajectory of the learning dynamics does not converge to unstable points. The probability of going to such a point is zero.

Theorem 2.2.1.9 ([61, 124]). *If $\mathbf{x}^*$ is a linearly unstable equilibrium for the system $\dot{\mathbf{x}} = \beta(\mathbf{x}) - \mathbf{x}$, then*

$$\lim_{t \to \infty} \mathbf{x}_t = \mathbf{x}^*,$$

with probability zero.

Lemma 2.2.1.10. *Let*

$$\bar{f}_{j,t}(s_j) = \frac{\sum_{t'=1}^{t} w_{j,t'} \, 1\!\!1_{\{a_{j,t'}=s_j\}}}{\sum_{t'=1}^{t} w_{j,t'}}$$

the weighted frequency of the player j up to t. Then, the vector $\bar{f}_{j,t}$ satisfies the recursive equation:

$$\bar{f}_{j,t+1}(s_j) = \Big(1 - \frac{w_{j,t+1}}{\sum_{t'=1}^{t+1} w_{j,t'}}\Big)\bar{f}_{j,t}(s_j) + \frac{w_{j,t+1}}{\sum_{t'=1}^{t+1} w_{j,t'}} \, 1\!\!1_{\{a_{j,t+1}=s_j\}}$$

Assume that the coefficients $\lambda_{j,t+1} = \frac{w_{j,t+1}}{\sum_{t'=1}^{t+1} w_{j,t'}}$ satisfies

$$\lambda_{j,t} \geq 0, \; \sum_t \lambda_{j,t+1} = \infty, \; \sum_t \lambda_{j,t+1}^2 < \infty.$$

The asymptotic pseudo-trajectory is given by

$$\frac{d}{dt}\bar{f}_{j,t}(s_j) \in \mathbb{E}\left(1\!\!1_{\{a_{j,t+1}=s_j\}} \mid \mathcal{F}_t\right) - \bar{f}_{j,t}(s_j),$$

where $\mathcal{F}_t$ is the filtration generated by the past play of actions and the frequencies.

The proof of this lemma follows the line of the Robbins-Monro's algorithm given in the appendix. Note that for w constant, i.e., λ uniform, one gets the standard recursive equations satisfied by the frequencies.

Nonconvergence (to a point) of fictitious play

The above negative results (Examples 2.2.1.6 and 2.2.1.7) do not imply that fictitious play is a poor learning pattern for all robust games. Fictitious play does converge to equilibrium in a variety of interesting cases if the full and perfect observation assumptions hold. The examples of coordinations and anti-coordination games show, however, that fictitious play has its limitations as a procedure for learning in such games. This leads to the question of whether a simple variant of fictitious play might get around the difficulty. The variants considered in the literature generally have the feature that best replies to the historical distribution of play are perturbed in some way, and/or the historical distribution is truncated, exponentially smoothed, imitative best-reply, etc. A combination of these features can indeed guarantee convergence in coordination and anticoordination games.

Once convergence is ensured, the second common question is the convergence time. At this point we should mention

> Why should we care about the convergence time of learning patterns?

This has been and is still a challenging open question. The answer to this question is not clear at all, and it is motivated by the fact that there are many scenarios where the average performance along the nonconvergent trajectory (limit cycle for example) can be greater than the expected performance/payoff at the stationary point which the algorithm will never reach except if it started from there.

This opens the question of alternative solution concepts such as *nonequilibrium solution, dynamic equilibrium concepts*, not a point but a whole *set as solution concept*, etc.

> Is fictitious play applicable in wireless networked games?

In the standard fictitious play algorithm and many of its variants, each player needs to know its own-actions, own-payoff functions, and the actions of other players at the end of each time slot. There are some wireless scenarios in which these assumptions hold. However, there are many others in which these assumptions may not hold. The observation of actions of the other players is one of the critical points. In many wireless scenarios, the players may not be able to observe the actions outside some range (neighborhood), and hidden nodes may be present as in opportunistic adhoc networks. Hence, observations about the actions of the others may not hold.

The second critical point is *cost of learning* and the time-consuming aspect. Even when the fictitious play algorithm converges in specific games, there may be some cost attached to having the information about the actions of the other players or need help from a local arbitrator (such as base station broadcasting, central coordinator, etc). Thus, the applicability of this procedure is limited; it needs too much information for decentralized implementations.

The next section describes how to relax the observation assumptions.

> Rate of convergence of fictitious play: a negative result

Consider the following basic two-player game. Player 1 chooses rows, and player 2 chooses columns. Outcomes are denoted by the payoff of player 1 and player 2, respectively. Let $L \geq 2$ be a natural number, $0 < \epsilon = \frac{1}{2^L} < 1$, and

$c > 0$.

	s_1	s_2	s_3
s_1	$(0, c)$	$(-1, 1+c)$	$(-\epsilon, \epsilon+c)$
s_2	$(1, -1+c)$	$(0, c)$	$(-\epsilon, \epsilon+c)$
s_3	$(\epsilon, -\epsilon+c)$	$(\epsilon, -\epsilon+c)$	$(0, c)$

Theorem 2.2.1.11. *In the above two-player game, Fictitious Play may require exponentially many rounds before an equilibrium action is eventually played.*

Proof. Let start the algorithm with (s_1, s_1). In the second round, the players will choose (s_2, s_2) and the frequencies are $(1/2, 1/2, 0)$. In the third stage, each player chooses a best reply to $(1/2, 1/2, 0)$, i.e., the action s_2 will be played again. At iteration $1 < l < L$, the players are still playing (s_2, s_2) because the frequencies are $(\frac{1}{l}, 1 - \frac{1}{l})$. It is easy to see (s_2, s_2) is not an equilibrium of the game. This gives rise to a learning sequence that is exponentially long in L and in which no equilibrium action is played. We have the sequence

$$(s_1, s_1), \underbrace{(s_2, s_2), \ldots, (s_2, s_2)}_{(2^L - 1) - times}.$$

Since the game can clearly be encoded using $O(L)$ bits in this case, the result follows.

$\square$

2.2.2 Best Response-Based Learning Algorithms

Asynchronous update or sequential update: The iterative best reply dynamics is often termed the Cournot adjustment model or Cournot tâtonnement after Augustin Cournot (1838, [49]) who was one of the first to propose it in the context of a duopoly model.

The Cournot adjustment process is as follows:

> Each of the players select actions sequentially. In each time slot, a player selects the action that is best response to the action chosen by the other players in the previous time slot.

In the original formulation, the best response is assumed to be a singleton. However, when there are multiple best-response actions in the iterative best reply dynamics, one player moves at a time and chooses at random a best reply (if any) to the current strategies of the others. Cournot noted that this process converges to the equilibrium (so-called Nash equilibrium today) in some duopoly games in which players act sequentially.

Let us illustrate the Cournot tâtonnement in the two player setting.

Consider two players engaged in the game described by $\mathcal{G} = (\mathcal{N} = \{1,2\}, \mathcal{A}_1, \mathcal{A}_2, \tilde{r}_1, \tilde{r}_2)$ where the action spaces $\mathcal{A}_j$ are compact and convex sets, and the functions $\tilde{r}_1, \tilde{r}_2$ are continuous, $\tilde{r}_j(., a_{-j})$ strict concave for each a_{-j}

As a consequence, the Cournot-Nash reaction sets are reduced to singleton, unique, and continuous. $b_j(a_{-j}) \in \arg\max_{y_j} \tilde{r}_j(y_j, a_{-j})$ is uniquely determined.

Time is discrete and the horizon is infinite. Let $\Phi : \mathcal{A}_1 \times \mathcal{A}_2 \longrightarrow \mathcal{A}_1 \times \mathcal{A}_2$, defined by $\Phi(a_1, a_2) = (b_1(a_2), b_2(a_1))$. The best-response algorithm is given by the dynamical adjustment $a_{t+1} = \Phi(a_t)$, i.e

$$a_t = \Phi^{\circ t}(a_0), \ a_0 \in \mathcal{A}_1 \times \mathcal{A}_2.$$

The main difficulty in the Cournot adjustment is the composition of mapping $b_1 \circ b_2$,

$$\Phi^{\circ 2t}(a_0) = ((\tilde{r}_1 \circ \tilde{r}_2)^{\circ t}(a_1), (\tilde{r}_2 \circ \tilde{r}_1)^{\circ t}(a_2)).$$

Depending on the scenario, one can observe nonconvergent or convergent behavior.

- nonconvergence: cycling phenomena (even if the equilibrium is unique!)

- sufficiency condition of convergence: monotonicity, strict contraction, etc.

Nonconvergence of Cournot adjustment

Example 2.2.3 (Convergent trajectory). *Consider a simultaneous moves Cournot tâtonnement with two players. Let $\mathcal{A}_1 = [0,1]$,*

$$b_1(a_2) = a_2, b_2(a_1) = sa_1(1 - a_1),$$

$$\tilde{r}_2(a) = -sa_2a_1^2 + sa_1a_2 - a_2^2/2.$$

Let $s = 2$. Using the fact that the fixed-point iteration (Picard iteration) of the logistic function $2x(1 - x)$ converges, we deduce a convergent Cournot adjustment for $s = 2$. Figure 2.1 illustrates a trajectory of the dynamics.

Example 2.2.4 (Non-convergence). *Consider a simultaneous moves Cournot tâtonnement with two players. Let $\mathcal{A}_1 = [0,1]$,*

$$b_1(a_2) = a_2, b_2(a_1) = sa_1(1 - a_1),$$

$$\tilde{r}_2(a) = -sa_2a_1^2 + sa_1a_2 - a_2^2/2.$$

Let $s = 4$. Using the fact that the fixed-point iteration of the logistic function $4x(1 - x)$ does not converge, we deduce a non-convergent Cournot adjustment for $s = 4$. Figure 2.2 illustrates the trajectory starting from $a_0 = x_0 = 0.2$.

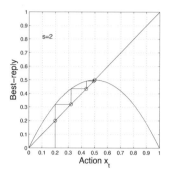

FIGURE 2.1

Convergence of best-reply.

Since $s = 2$ one gets a strict contraction, the best-response procedure
converges globally to the unique fixed point of the logistic map.

The MATLAB code is provided below:

Initialization

```
%%% MAKES A BEST-REPLY PLOT

% Compute trajectory
s=4.0;           % parameter
x0=0.2;           % Initial condition
N=40;            % Number of iterations
x(1) = x0;
```

Trajectory of best reply

```
for ic=1:N
 x(ic+1) = s*x(ic)*(1-x(ic));
end
```

Plot the best reply the segment $y = x$

```
% Plot the best-response correspondence and the line y=x
```

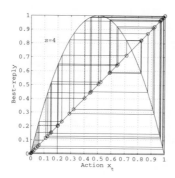

FIGURE 2.2

Nonconvergence of best-reply

    The best-response procedure does not converge to the equilibrium.

```
clf;
t = 0:0.01:1;
plot(t,s*(t.*(1-t)),'-k'); hold on;
 axis('square'); axis([0 1 0 1]);
 set(gca,'XTick',(0:0.1:1),'YTick',(0:0.1:1))
 grid on;

 fplot('1*y',[0 1],'-k');
```

Plot the iterates of the best-response dynamics

```
%%%%% STEP 3: PLOT THE ITERATES
 line([x(1) x(1)],[0 x(2)],'Color','k')
 plot(x(1), x(1),'ko');
 for ic=1:N-1
  line([x(ic) x(ic+1)],[x(ic+1) x(ic+1)],'Color','k')
  line([x(ic+1) x(ic+1)],[x(ic+1) x(ic+2)],'Color','k')
  plot(x(ic+1), x(ic+1),'ko');
 end
 line([x(N) x(N+1)],[x(N+1) x(N+1)],'Color','k')
```

Plot the parameter

```
%%%%% STEP 4: SIGN THE PLOT
at=text(0.1,0.82,['s=',num2str(s)]); set(at,'FontSize',12);
xlabel('Action x_t','FontSize',16), ylabel('Best-reply','FontSize',16
```

The figure 2.2 plots a nonconvergent trajectory as in example 2.2.4.

Sequential Cournot tâtonnement

Consider Player I and Player II engaged in the following sequential game:

- Time is discrete. The players moves sequentially.

- In odd time slots $t \in \{1, 3, 5, 7, 9, \ldots\}$ Player I chooses an action that is committed for two time slots, $a_{1,2t+2} = a_{1,2t+1}$, $t \geq 0$

- In even time slots $t \in 2\mathbb{Z}_+$ Player II chooses an action that is committed for two time slots $a_{2,2t+1} = a_{2,2t}$, $t \geq 0$

- Dynamical reaction functions:

$$a_{1,2t+1} = b_1(a_{2,2t}), \; a_{2,2t+2} = b_2(a_{1,2t+1})$$

The above process can be easily extended to n players. Consider n players engaged in the Cournot tâtonnement. Denote the one-shot game by

- $\mathcal{G} = (\mathcal{N}, \mathcal{A}_j, \tilde{r}_j)$,

- $\mathcal{A}_j$ compact convex set, $\tilde{r}_j$ are continuous, $\tilde{r}_j(., a_{-j})$ strict concave for each $a_{-j} = (a_i)_{i \neq j}$

As a consequence the Cournot-Nash reaction sets are reduced to singleton, unique, and continuous. $b_j(a_{-j}) \in \arg\max_{y_j} \tilde{r}_j(y_j, a_{-j})$.

Define the mapping $\Phi : a \longmapsto (b_j(a_{-j}))_{j \in \mathcal{N}}$ Then, the Cournot dynamical adjustment is given by $a_{t+1} = \Phi(a_t)$

Specific class of games and convergence results

Below we define a specific class of games.

A game is *dominance solvable* if the set that remains after iterated elimination of dominated strategies is a singleton. Moulin (1984) showed that if a game is dominance solvable, its (unique) pure Nash equilibrium is Cournot stable.

A sequential best-reply path (Cournot adjustment) is *admissible* if whenever n successive periods have passed, all n players have moved.

The game $\mathcal{G}$ is a best-reply potential game if there exists a function V such that $\arg\max_{a_j} \tilde{r}_j(a) = \arg\max_{a_j} V(a)$. A game is a best-reply potential game if each player instead of maximizing his payoff function can maximize the *common objective*.

Theorem 2.2.4.1 (Monderer-Shapley'96). *Let $\mathcal{G}$ a best-reply potential game with V a best-response potential. If the action a maximizes V, then a is a (pure) Nash equilibrium.*

The function $f(x,y)$ has increasing differences in (x,y) if for all

$$x' \succ x, y' \succ y$$

$$f(x',y') - f(x,y') \geq f(x',y) - f(x,y).$$

$\mathcal{G}$ is a supermodular game if all j, $\tilde{r}_j$ has increasing differences in (a_j, a_{-j}).

Theorem 2.2.4.2. *Consider a game with finite number of players.*
* *Let V be a continuous function over a compact and nonempty domain. Then, a pseudo-potential game with potential function V has a at least one pure strategy Nash equilibrium.*
* *Let $\mathcal{G}$ be a best-reply potential game with single-valued, continuous best-reply functions and compact strategy sets. Then any admissible sequential best reply path converges to the set of pure strategy Nash equilibria.*
* *Moreover, assume that $\mathcal{G}$ has a unique pure Nash equilibrium a^*. Then $\mathcal{G}$ is a best-reply potential game if and only if every admissible sequential best-reply path converges to a^*.*

Sequential best reply

Consider a finite game pseudo-potential game. Assume that all best reply correspondences are single-valued (functions):

* Start with an arbitrary strategy profile,

* Let each player, one at a time, unilaterally deviate from his or her unique best reply (if he does not happen to be there already). $(a_{1,t}, \ldots, a_{n,t})_{t \geq 1}$ the sequence generated.

* At every time slot t, if $a_t^J \neq b_J(a_{-J,t-1})$ for some nonempty set of players in J Then, one of them (say $j \in J$) is required to move and choose its best-reply $a_{j,t} = b_j(a_{-j,t-1})$. All the other players stay put at their previous strategies, i.e., $a_{-j,t} = a_{-j,t-1}$. Since the game is pseudo-potential, and since the monotone sequence $V(a_t)$ cannot have infinitely many strict increases on its finite domain $\prod_{j \in \mathcal{N}} \mathcal{A}_j$, the algorithm is convergent.

From supermodular to best-reply potential

- Any *dominance solvable* game with continuous best-reply functions and compact strategy sets, is a best-reply potential game.

- Any *supermodular game* with continuous best-reply functions, compact strategy sets, and a unique pure Nash equilibrium is a best-reply potential game.

2.2.5 Better Reply-Based Learning Algorithms

Better Reply versus Best Reply

We mention briefly the difference between Best-Response-Dynamics and the Best-Reply-Dynamics. The Best-Reply-Process is defined as follows.

Each of the players select actions sequentially. In each time slot, a player selects the action that is its better-reply to the action chosen by the other players in the previous time slot. When there are multiple best-reply actions in the iterative better-reply dynamics, one player moves at a time and chooses at random a better reply (if any) to the current strategies of the others.

The stochastic process generated by the better-reply is called *better-reply dynamics*, because players move from their current actions to a better reply, not necessarily a *best reply*; even though players move in the direction of their best replies, they can overshoot or undershoot them.

The better-reply dynamics is consistent with a player not having precise knowledge, or memory, of his or her own and his or her opponents' payoff functions and past actions.

Acyclicity and Existence of Pure Equilibria

Let $\mathcal{G}^*$ be the graph $(\mathcal{A}, E)$ where $\mathcal{A} = \prod_{j \in \mathcal{N}} \mathcal{A}_j$ is the set of action profiles and E is the set of edges defined by

$$E = \{(\mathbf{a}, \mathbf{b}) \mid \exists j, \ \mathbf{b} = (\mathbf{b}_j, \mathbf{a}_{-j}), \tilde{r}_j(\mathbf{a}) < \tilde{r}_j(\mathbf{b})\}.$$

Lemma 2.2.5.1. *If the graph $\mathcal{G}^*$ is acyclic, then the corresponding game $\mathcal{G}$ possesses at least one pure equilibrium.*

Proof. Every acyclic graph has a sink (vertex without outgoing edges): since there are a finite number of vertices if there is no sink there will be a cycle. Every sink on the graph $\mathcal{G}^*$ is a pure equilibrium (follows immediately from the definition of the graph $\mathcal{G}^*$). $\square$

Consider any finite game with a set $\mathcal{A}$ of action profiles. A better reply path is a sequence of action profiles $\mathbf{a}_1, \mathbf{a}_2, \ldots, \mathbf{a}_T$ such that for each successive pair $(\mathbf{a}_t, \mathbf{a}_{t+1})$ there is exactly one player such that $a_{j,t} \neq a_{j,t+1}$ and for that player the payoff $\tilde{r}_j(\mathbf{a}_t) < \tilde{r}_j(\mathbf{a}_{t+1})$. This says that only one player moves at a time, and each time a player moves he or she increases his own payoff.

Definition 2.2.5.2. *The finite game $\mathcal{G}$ is weakly acyclic if for any action profile $\mathbf{a} \in \mathcal{A}$, there exists a better reply path starting at a and ending at some pure Nash equilibrium of $\mathcal{G}$.*

One of the nice properties of weakly acyclic games is that they have at least one pure Nash equilibrium. The following theorem gives a convergence to pure Nash equilibria under better reply dynamics.

Theorem 2.2.5.3 ([193]). *In finite weakly acyclic games, the iterative better reply dynamics converges with probability one to a pure Nash equilibrium.*

Since weakly acyclic games include exact potential games, the following result holds.

Corollary 2.2.5.4. *In potential games, the iterative better reply dynamics converges with probability one to a pure Nash equilibrium.*

Best-Reply in Aggregative Games

We focus on a particular class of aggregative, n-player strategic games with an additive aggregation term.

Definition 2.2.5.5. *In an aggregative game with an additive aggregation term, the payoff of each player is a function of the players own action and of the sum of the actions (or, equivalently, the mean action) of all players.*

Let $\mathcal{G}$ be a game with nonempty, convex, and compact action spaces on the real line. There exists a function $\tilde{r}_j : \prod_{i \in \mathcal{N}} \mathcal{A}_i \longrightarrow \mathbb{R}$ such that, for all action profile, the payoff function of player j can be written as

$$\tilde{r}_j(a) = \phi_j(a_j, \Sigma),$$

where $\Sigma = \sum_{i=1}^{n} a_i \in \mathbb{R}$.
- HA0: We assume that $\tilde{r}_j$ is twice continuously differentiable in all its arguments and assume that a player payoff is a strictly quasi-concave function of her own action.
- HA1: We restrict attention to games in which, for each player, the slope of the best-reply function is bounded below by -1. That is,

$$\frac{\partial^2}{\partial a_j^2} \phi_j(a_j, \Sigma) + \frac{\partial^2}{\partial \Sigma \partial a_j} \phi_j(a_j, \Sigma) < 0$$

This condition does not imply, and is not implied by, the concavity of $\tilde{r}_j$ with respect to a_j. For example, if $\tilde{r}_j(a) = \alpha a_j^2 + \beta \Sigma^2$.

If equation

$$\frac{\partial}{\partial a_j} \tilde{r}_j(a) = 0 = \frac{\partial}{\partial a_j} \phi_j(a_j, \Sigma) + \frac{\partial}{\partial \Sigma} \phi_j(a_j, \Sigma)$$

holds, then for any given

$$\Sigma \in \{\sum_{i=1}^{n} a_i \mid a_j \in \mathcal{A}_j, \ j \in \mathcal{N}\}$$

there is (at most) a unique solution $b_j(\Sigma)$ to the implicit equation:

$$\frac{\partial}{\partial a_j} \tilde{r}_j(b_j(\Sigma), \Sigma) = 0,$$

up to the projection at the ending points.

We assume that

- HA2: Denote by $BR_j(\Sigma^*_{-j})$ be the best reply to a^*_{-j}. We assume that

$$BR'_j(\Sigma^*_{-j}) BR'_i(\Sigma^*_{-i}) \neq 1, \ j \neq i.$$

Consider the following system of differential equations as best-response dynamics in continuous time:

$$\dot{a}_j = b_j(\Sigma) - a_j, \ j \in \mathcal{N}. \tag{2.7}$$

For all $j \in \mathcal{N}$, $(b_j(\Sigma) - a_j)(BR_j(\Sigma_{-j}) - a_j) \geq 0$ and $BR_j(\Sigma_{-j}) = a_j$ if and only if $b_j(\Sigma) = a_j$.

Theorem 2.2.5.6 ([53]). *Under the hypothesis HA0-HA2, the best-response dynamics defined in (2.7) globally converges to some Nash equilibrium of the aggregative game. Moreover, for almost all aggregative games satisfying $HA0 - HA2$ there is a finite sequence of single-player improvements that ends arbitrarily close to a Nash equilibrium.*

Non-convergence in aggregative games

Next, we consider the continuous kernel game with n players and the payoff function of each player j is

$$\tilde{r}_j(x) = \frac{x_j}{\sum_{k=1}^{n} x_k} - x_j.$$

We provide the code of the best reply dynamics.

```
%%% MAKES A BEST-REPLY PLOT

% Compute trajectory
n=4.0;            % parameter
x0=0.3;           % Initial condition
N=50;             % Number of iterations
x(1) = x0;
for ic=1:N
 x(ic+1) = sqrt((n-1)*x(ic))-(n-1)*x(ic);
end


% Plot the best-response correspondence and the line y=x
clf;
t = 0:0.01:0.5;
plot(t,sqrt((n-1)*t)-(n-1)*t,'-k'); hold on;
 axis('square'); axis([0 0.5 0 0.5]);
 set(gca,'XTick',(0:0.1:0.5),'YTick',(0:0.1:0.5))
 grid on;

 fplot('1*y',[0 0.5],'-k');




%%%%% STEP 3: PLOT THE ITERATES
 line([x(1) x(1)],[0 x(2)],'Color','k')
  plot(x(1), x(1),'ko');
 for ic=1:N-1
  line([x(ic) x(ic+1)],[x(ic+1) x(ic+1)],'Color','k')
  line([x(ic+1) x(ic+1)],[x(ic+1) x(ic+2)],'Color','k')
  plot(x(ic+1), x(ic+1),'ko');
 end
line([x(N) x(N+1)],[x(N+1) x(N+1)],'Color','k')

   %%%%% STEP 4: SIGN THE PLOT
at=text(0.1,0.42,['n=',num2str(n)]); set(at,'FontSize',12);
xlabel('Action x_t','FontSize',16), ylabel('Best-reply','FontSize',16)
```

The code is illustrated in Figure 2.3. Based on Figure 2.3, we propose to prove a clear statement of cycling behavior in Exercise 2.3.

Exercise 2.3. *Consider a continuous-kernel game with n players with $n \geq 4$. Each player has a compact action space given by $[\epsilon, 1]$ with $\epsilon > 0$. The payoff*

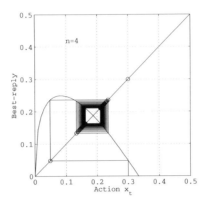

FIGURE 2.3
Nonconvergent aggregative game.
The best response procedure goes to limit cycle. The equilibrium cannot be
reached if not started there.

function of each player j is

$$\tilde{r}_j(x) = \frac{x_j}{\sum_{k=1}^{n} x_k} - x_j.$$

- *Justify why this is an aggregative game.*

- *Verify that the payoff function is welldefined.*

- *Compute the best-response correspondence.*

- *Compute the set of equilibria.*

- *Show that the following iterative scheme can be obtained as best-reply path:*

$$x_{t+1} = \sqrt{(n-1)x_t} - (n-1)x_t$$

- *Is the best-response dynamics convergent? Divergent?*

- *Conclude.*

> Limitation of the best-reply dynamics

As we can see, the iterative better reply dynamics gives nice convergence
properties in weakly acyclic games and other particular class of games such
as S-modular games (supermodular games and submodular games in small
dimensions) but this assumes that each player knows what actions the other

players are taking. In general, the player may not observe the actions taken by the other players. This occurs in many wireless scenarios (sensor networks with visibility constraints, multiple random access channels, resource allocation problems, etc.). As a consequence, a learning mechanism that relaxes this observation assumption is needed. If players observe only their *own payoffs*, one can consider a simple learning scheme based experimentation which assumes only that players react to their own recent payoffs and own-actions. When there exists a fully distributed learning algorithm that has nice properties of convergence, then this learning algorithm is well-adapted for completely decentralized implementations because it does not requires any information from the other players. It is important to mention that in many engineering games, the payoff functions have specific structure such as *aggregative games* which include many power allocation games, random access games, Cournot-like competition between network operators, etc. In these cases, a specific signal that allows one to compute the aggregative term will be sufficient as information (the actions of the others are not needed for the best reply).

Best Response-like Iterative Methods

2.2.6 Fixed Point Iterations

Let $\mathcal{X}$ be a convex, compact and non-empty subset of an Hilbert space.

Best response dynamics and Picard iteration

In the case where the best response is uniquely defined, the simultaneous best-response dynamics can be seen as a the Picard iteration and it is given by

$$x_{t+1} = \mathrm{br}(x_t), \ t \geq 1 \tag{2.8}$$
$$x_0 \in \mathcal{X}. \tag{2.9}$$

where $\{\mathrm{br}\} = BR$ is the best-response set and $\mathrm{br}: \mathcal{X} \longrightarrow \mathcal{X}$. We study the behavior of (2.8), we look at the properties of the function br.

Definition 2.2.6.1. *We say that* $f: \mathcal{X} \longrightarrow \mathcal{X}$ *is (strictly) contractive if there exists* $k < 1$

$$\| f(x) - f(y) \| \leq k \| x - y \|, \ \forall (x,y) \in \mathcal{X}^2. \tag{2.10}$$

For contractive map, the Picard iterations are known to be convergent (Banach-Picard algorithm) with a geometric decay with rate $k < 1$.

Definition 2.2.6.2. *We say that* $f: \mathcal{X} \longrightarrow \mathcal{X}$ *is non-expansive if*

$$\| f(x) - f(y) \| \leq \| x - y \|, \ \forall (x,y) \in \mathcal{X}^2.$$

For nonexpansive map, the Picard iteration may not converge, generally, or even if it converges, its limit is not a fixed point of f.

Definition 2.2.6.3. *We say that* $f : \mathcal{X} \longrightarrow \mathcal{X}$ *is pseudo-contractive if*

$$\| f(x) - f(y) \| \leq \| x - y \| + \| x - y + f(y) - f(x) \|, \; \forall (x, y) \in \mathcal{X}^2.$$

Best response and Krasnoselskij iteration

$$x_{t+1} \in (1 - \lambda)x_t + \lambda BR(x_t), \; t \geq 1 \qquad (2.11)$$

$$x_0 \in \mathcal{X}. \qquad (2.12)$$

$$\lambda \in (0, 1) \qquad (2.13)$$

We have seen above two examples where the best reply dynamics has a cycling behavior. We explain how the algorithm can be designed such that the size of the limit cycle diminishes.

First consider the logistic map example. We have seen that for $s = 4$ there a cycling behavior of the best reply dynamics. Is there a way to reduce the size of the limit cycle?

To answer to this question, we start by modifying the best reply with a convex combination. Figure 2.4 illustrates a Mann/Krasnoselskij iteration based best reply algorithm for step size equal to 0.3 and 0.9 respectively. We observe that the convergence or non-convergence depends mainly on the learning rate λ.

In Figure 2.5, we plot the trajectory of the Mann-based best replay dynamics in the allocation game for $\lambda_t \in \{0.3, 0.9\}$.

Best response and Mann iteration

$$x_{t+1} \in (1 - \lambda_t)x_t + \lambda_t BR(x_t), \; \forall t \geq 1 \qquad (2.14)$$

$$x_0 \in \mathcal{X}. \qquad (2.15)$$

$$\lambda_t \in (0, 1), \; \forall t \geq 1 \qquad (2.16)$$

Best response and Ishikawa iteration

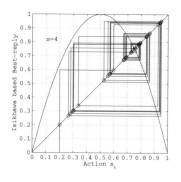

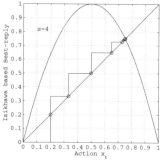

FIGURE 2.4

Design of step size.

Mann-based best reply iteration for $\lambda_t \in \{0.9, 0.3\}$

$$x_{t+1} \in x_t + \lambda_t[BR(y_t) - x_t] \tag{2.17}$$

$$y_t \in x_t + \mu_t[BR(x_t) - x_t], \tag{2.18}$$

$$x_0 \in \mathcal{X}, \tag{2.19}$$

$$0 \le \lambda_t \le \mu_t < 1 \tag{2.20}$$

$$\lim_{t \longrightarrow +\infty} \mu_t = 0 \tag{2.21}$$

$$\sum_{t \ge 1} \lambda_t \mu_t = +\infty \tag{2.22}$$

In Section 2.3, we present some fully distributed learning algorithms that relax the knowledge of the mathematical expression of the payoff functions or specific signals/measurements are observed by the players.

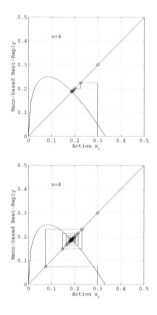

FIGURE 2.5
Mann iteration. Design of step size.
    Allocation game: Mann-based best reply iteration for $\lambda_t \in \{0.3, 0.9\}$

2.2.7 Cost-To-Learn

In this subsection we present an algorithm called "Cost-to-Learn" in simple games.

> The basic idea is that *it can be very costly to learn quickly and learning can take some time.* When a player changes his or her action, there is a cost termed as cost of moving.

We can think about the cost of adopting a new technology or the cost of producing a specific product. We illustrate this in a two-player case.
    Consider the two-player game with cost functions for player one

$$\tilde{R}_1(a_1, a_2) = \underline{f}_1(a_1) + \underline{\beta}_1 \underline{c}(a_1, a_2),$$

$$\tilde{R}_2(a_1, a_2) = \underline{f}_2(a_2) + \underline{\beta}_2 \underline{c}(a_1, a_2).$$

The coupling term defines the more or less competitive part of the game. The coefficients $\{\underline{\beta}_j\}_j$ represent how each player gets from the joint interaction. In addition, we assume that there is a cost of learning and a cost to move from some action to another. Player 1 has a cost of learning of $\underline{\alpha}_1 \underline{h}_1(a_1, a_1')$. Similarly, Player 2 has to pay $\underline{\alpha}_2 \underline{h}_2(a_2, a_2')$ to change its action from a_2 to a_2'. The motivations behind this *"cost of learning"* approach is that, in many situations, changing, improving the performance, guaranteeing a certain quality of service has cost. If only player 1 moves, its objective function becomes $\tilde{R}_1(a_1', a_2) + \underline{\alpha}_1 \underline{h}_1(a_1, a_1')$. Similarly if player 2 is the only mover, he or she gets $\tilde{R}_2(a_1, a_2') + \underline{\alpha}_2 \underline{h}_2(a_2, a_2')$. Assume that $\underline{h}_j(x, x) = 0$, $j \in \{1, 2\}$ (there is no cost to learn if the action remains the same).

We now construct the sequence of actions of the cost of moves.

- Denote by $(a_{1,t}, a_{2,t})$ the actions profile at time slot t. Start with $(a_{1,0}, a_{2,0})$. At each time slot t, there are two stages of moves (sequential updates): Player 1 moves from $a_{1,t}$ to $a_{1,t+1}$ by solving the following problem:

$$a_{1,t+1} \in \arg \min_{a_1' \in \mathcal{A}_1} \left\{ \underline{f}_1(a_1') + \underline{\beta}_{1,t} \underline{c}(a_1', a_{2,t}) + \underline{\alpha}_1(t) \underline{h}_1(a_{1,t}, a_1') \right\},$$

and then given $a_{1,t+1}$, Player 2 moves from $a_{2,t}$ to $a_{2,t+1}$ by solving the following problem:

$$a_{2,t+1} \in \arg \min_{a_2' \in \mathcal{A}_2} \left\{ \underline{f}_2(a_2') + \underline{\beta}_{2,t} \underline{c}(a_{1,t+1}, a_2') + \underline{\alpha}_{2,t} \underline{h}_2(a_{2,t}, a_2') \right\}$$

The learning algorithm *Cost-To-Learn* is then given by:

$$\begin{cases} a_{1,t+1} \in \arg \min_{a_1' \in \mathcal{A}_1} \left\{ \underline{f}_1(a_1') + \underline{\beta}_{1,t} \underline{c}(a_1', a_{2,t}) + \underline{\alpha}_{1,t} \underline{h}_1(a_{1,t}, a_1') \right\} \\ a_{2,t+1} \in \arg \min_{a_2' \in \mathcal{A}_2} \left\{ \underline{f}_2(a_2') + \underline{\beta}_{2,t} \underline{c}(a_{1,t+1}, a_2') + \underline{\alpha}_{2,t} \underline{h}_2(a_{2,t}, a_2') \right\} \end{cases}$$

Definition 2.2.7.1. *A mapping* $f : A \longrightarrow \mathbb{R} \cup \{+\infty\}$ *is lower semicontinuous at* a_0 *if* $\liminf_{a \longrightarrow a_0} f(a) \geq f(a_0)$. *f is lower semicontinuous if it is lower semicontinuous at every point of its domain.*

This is equivalent to saying that $\{a \in A \mid f(a) > \alpha\}$ is an open set for every $\alpha \in \mathbb{R}$. Assume that the mapping $(f_i)_i$ is proper, bounded below, and lower semicontinuous.

Lemma 2.2.7.2. *If the algorithm Cost-to-Learn converges and if*

$$(\underline{\alpha}_{1,t}, \underline{\alpha}_{2,t}) \longrightarrow (\underline{\alpha}_1, \underline{\alpha}_2),$$

$$(\underline{\beta}_{1,t}, \underline{\beta}_{2,t}) \longrightarrow (\underline{\beta}_1, \underline{\beta}_2),$$

when t *goes to infinity then, the limit* (a_1^*, a_2^*) *of any subsequence* $(a_{1,\phi(t)}, a_{2,\phi(t)})$ *generated by the algorithm leads to an equilibrium of the game with cost of learning.*

Proof. By definition of $a_{1,t+1}$, one has,

$$\underline{f}_1(a_1') + \underline{\beta}_{1,t} \underline{c}(a_1', a_{2,t}) + \underline{\alpha}_{1,t} \underline{h}_1(a_{1,t}, a_1')$$

$$\geq \underline{f}_1(a_{1,t+1}) + \underline{\beta}_{1,t}\underline{c}(a_{1,t+1}, a_{2,t}) + \underline{\alpha}_{1,t}\underline{h}_1(a_{1,t}, a_{1,t+1}), \ \forall a_1'.$$

The first term of the inequality goes to $\underline{f}_1(a_1') + \underline{\beta}_1\underline{c}(a_1', a_2^*) + \underline{\alpha}_1\underline{h}_1(a_1^*, a_1')$ when t goes to infinity. By taking limit inferior, the second term gives

$$
\begin{aligned}
\geq \ & \liminf_t \left[\underline{f}_1(a_{1,t+1}) + \underline{\beta}_{1,t}\underline{c}(a_{1,t+1}, a_{2,t}) + \underline{\alpha}_{1,t}\underline{h}_1(a_{1,t}, a_{1,t+1}) \right] \\
\geq \ & \liminf_t \underline{f}_1(a_{1,t+1}) + \liminf_t \left[\underline{\beta}_{1,t}\underline{c}(a_{1,t+1}, a_{2,t}) \right. \\
& \left. + [\underline{\alpha}_{1,t}\underline{h}_1(a_{1,t}, a_{1,t+1})] \right. \\
\geq \ & \underline{f}_1(a_1^*) + \left[\underline{\beta}_1\underline{c}(a_1^*, a_2^*) \right] + [\underline{\alpha}_1\underline{h}_1(a_1^*, a_1^*)] \\
\geq \ & \underline{f}_1(a_1^*) + \underline{\beta}_1\underline{c}(a_1^*, a_2^*)
\end{aligned}
$$

This means that $\forall a_1'$, $\underline{f}_1(a_1') + \underline{\beta}_1\underline{c}(a_1', a_2^*) + \underline{\alpha}_1\underline{h}_1(a_1^*, a_1') \geq \underline{f}_1(a_1^*) + \underline{\beta}_1\underline{c}(a_1^*, a_2^*)$. Similarly, $\forall a_2'$, $\underline{f}_2(a_2') + \underline{\beta}_2\underline{c}(a_1^*, a_2') + \underline{\alpha}_2\underline{h}_2(a_2^*, a_2') \geq \underline{f}_2(a_2^*) + \underline{\beta}_2\underline{c}(a_1^*, a_2^*)$. This implies that (a_1^*, a_2^*) is an equilibrium point. $\square$

Theorem 2.2.7.3 (Convergence). *Assume that $\underline{h}_j$, $j \in \{1, 2\}$ are positive coefficients and all the coefficients $\underline{\alpha}_{j,t}$ and $\underline{\beta}_j(t)$ are positive, and $\underline{\beta}_{1,t} = \underline{\beta}_{2,t} = \underline{\beta}(t)$ is nonincreasing, and $\sum_j \inf_{A_j} \underline{f}_j > -\infty$ then, for any sequence of $(a_{1,t}, a_{2,t})_t$ generated the algorithm of Cost-to-Learn, one has*

- $\left\{ \xi_t := \underline{f}_1(a_{1,t}) + \underline{f}_2(a_{2,t}) + \underline{\beta}(t)\underline{c}(a_{1,t}, a_{2,t}) \right\}_t$ *is nonincreasing and has a limit when t goes to infinity.*

- *The partial sum of the series defined by the sequence of cost of learning, i.e.,*

$$\sum_{t=0}^{T} \left[\underline{\alpha}_{1,t}\underline{h}_1(a_{1,t}, a_{1,t+1}) + \underline{\alpha}_{2,t}\underline{h}_2(a_{2,t}, a_{2,t+1}) \right]$$

is convergent.

Proof. We write the two inequalities defined by the algorithm:

$$\underline{f}_1(a_{1,t+1}) + \underline{\beta}(t)\underline{c}(a_{1,t+1}, a_{2,t}) + \underline{\alpha}_{1,t}\underline{h}_1(a_{1,t}, a_{1,t+1}) \leq \underline{f}_1(a_{1,t}) + \underline{\beta}(t)\underline{c}(a_{1,t}, a_{2,t}),$$

$$\underline{f}_2(a_{2,t+1}) + \underline{\beta}(t)\underline{c}(a_{1,t+1}, a_{2,t+1}) + \underline{\alpha}_{2,t}\underline{h}_2(a_{2,t}, a_{2,t+1}) \leq \underline{f}_2(a_{2,t}) + \underline{\beta}(t)\underline{c}(a_{1,t+1}, a_{2,t}).$$

By adding the two inequalities, one has,

$$
\begin{aligned}
\xi_{t+1} \ := \ & \underline{f}_1(a_{1,t+1}) + \underline{f}_2(a_{2,t+1}) + \underline{\beta}(t+1)\underline{c}(a_{1,t+1}, a_{2,t+1}) \\
& + \underline{\alpha}_{1,t}\underline{h}_1(a_{1,t}, a_{1,t+1}) + \underline{\alpha}_{2,t}\underline{h}_2(a_{2,t}, a_{2,t+1}) \\
\leq \ & \underline{f}_1(a_{1,t+1}) + \underline{f}_2(a_{2,t+1}) + \underline{\beta}(t)\underline{c}(a_{1,t+1}, a_{2,t+1}) \\
& + \underline{\alpha}_{1,t}\underline{h}_1(a_{1,t}, a_{1,t+1}) + \underline{\alpha}_{2,t}\underline{h}_2(a_{2,t}, a_{2,t+1}) \\
\leq \ & \underline{f}_1(a_{1,t}) + \underline{f}_2(a_{2,t}) + \underline{\beta}(t)\underline{c}(a_{1,t}, a_{2,t}) \\
=: \ & \xi_t
\end{aligned}
$$

$$(2.23)$$

This implies that the sequence $\left\{\xi_t := \underline{f}_1(a_{1,t}) + \underline{f}_2(a_{2,t}) + \underline{\beta}(t)\underline{c}(a_{1,t}, a_{2,t})\right\}_t$ is non-increasing and has a limit. We now prove that the partial sum of the series defined by the sequence of cost of learning is convergent.

$$
\begin{aligned}
\xi_{T+1} - \xi_0 &= \sum_{t=0}^{T} \left[\underline{\alpha}_{1,t}\underline{h}_1(a_{1,t}, a_{1,t+1}) + \underline{\alpha}_{2,t}\underline{h}_2(a_{2,t}, a_{2,t+1})\right] \\
&= \underline{f}_1(a_1(0)) + \underline{f}_2(a_2(0)) + \underline{\beta}(0)c(a_1(0), a_2(0)) \\
&\quad -\underline{f}_1(a_{1,t+1}) - \underline{f}_2(a_{2,t+1}) - \underline{\beta}(T+1)\underline{c}(a_{1,t+1}, a_{2,t+1}) \\
&\leq \underline{f}_1(a_1(0)) + \underline{f}_2(a_2(0)) + \underline{\beta}(0)c(a_1(0), a_2(0)) - \left[\sum_{j=1}^{2} \inf_{\mathcal{A}_j} \underline{f}_j\right]
\end{aligned}
$$

This completes the proof.

$\square$

Note that the quantity $\underline{f}_1 + \underline{f}_2 + \beta\underline{c}$ defines a potential function of the game without cost of learning.

Exercise 2.4. *Generalize the convergence proof of the cost-to-learn algorithm in generic robust potential games.*

Next, we present the concepts of cost of learning and cost to move in heterogeneous wireless networks. The work received the best paper award at INFOCOM Workshop 2011 on Future Media Networks and IP-based TV.

Handover and Codec Switching Cost in 4G Networks

This example illustrates the importance of switching costs (codec, handover) in heterogeneous wireless networks.

The business models of telecommunication operators have traditionally been based on the concept of the so-called closed garden: they operate strictly in closed infrastructures and base their revenue-generating models on their capacity to retain a set of customers and effectively establish technological and economical barriers to prevent or discourage users from being able to utilize services and resources offered by other operators. After the initial monopoly-like era, an increasing number of (real and virtual) network operators have been observed on the market in most countries. Users benefit from the resulting competition by having much wider spectrum choices for more competitive prices.

In its most generic sense, the user-centric view in telecommunications considers that the users are free from subscription to any one network operator and can instead dynamically choose the most suitable transport infrastructure from the available network providers for their terminal and application requirements. While bandwidth offered by these technologies is increasing, the link layer properties are also exhibiting time-varying characteristics. These factors may force users, who have different preferences for quality, service cost, and service security, to react according to the network conditions. For instance, due to adverse nature of the wireless medium it is imperative to change speech bandwidth in order to maintain a certain quality of service in VoIP like services. This action sometimes results in changing of speech codec during the ongoing call, i.e., *codec changeover* and/or change of network point of attachment, i.e., *network handover*.

In this approach, the decision of interface selection is delegated to the mobile terminal, enabling end users to exploit the best available characteristics of different network technologies and network providers, with the objective of increased satisfaction. In order to more accurately express the user satisfaction in telecommunications, the term QoS has been extended to include more subjective and also application specific measures beyond traditional technical parameters, giving rise to the quality of experience (QoE) concept. Such a telecommunication landscape provisions an efficient resource usage on operator part that does not only concentrate on operator profit, but also provides high quality to service users (increase the user QoE).

Thus there is a need to model an analytical function that captures user satisfaction for both *real-time* and *non-real-time* applications, taking into account the new conditions of handover and codec switchovers in addition to other QoE affecting parameters (delay, packet loss, service pricing, etc.). Modeling such an analytical function is the key focus of this work.

The usage of *real-time* applications such as voice services over IP and video streaming has grown significantly. *Non-real-time* applications have also become an integrated important part of the future telecommunication market. On the one hand, the availability of high-speed access technologies such as High-Speed Downlink Packet Access (HSDPA), Worldwide Interoperability for Microwave Access (WiMAX), and 3GPP Long Term Evolution (LTE) have encouraged users to employ the diverse applications while they are on move, and on the other hand the physical layer implementation of these wireless technologies have different impact these applications. For instance, most of the cellular access networks like code division multiple access (CDMA), HSDPAm and LTE link exhibit more fluctuations than WiFi due to shared downlink. The primary reason for this is base station scheduling, which allocate resources only to users having good wireless link conditions. WiFi is an interference limited technology and link transmission is greatly dependent on the simultaneous demand of the wireless channel resources. However, WiFi as compare to HSDPA is better with the respect to throughput and latency. These difference in access technologies are the key factors of different quality of experience of users for different services.

Thus call quality monitoring is really crucial for future mobile service providers. The call quality monitoring may be based on network transmission metrics for different types of real and non-real time traffic, which can further be mapped into the quality experienced by the end user. Real user perception measurement requires time-consuming and expensive subjective tests but also provide the ground truth and the most accurate Mean Opinion Score (MOS) in case of real-time applications, which expresses the mean quality score of a group of users according to International Telecommunication Union (ITU), ITU-T Recommendation P.800. The test complexity limits the applicability of such a quality measurement. Due to these limitations, an alternate approach of instrumental quality prediction is used that can be employed for online optimization.

The functional principle of quality prediction models can rely on evaluation of transmission parameters measured in the networking layer e.g., for Voice over Internet Protocol (VoIP) applications, the E-model, or on the analysis of the signal waveform measured at the output of the transmission system e.g. perceptual evaluation of speech quality (PESQ). This approach allows to derive the quality index without performing subjective tests and is based on either a parametric approach or signal-based approach. However these models have been tested for certain conditions and their applicability for future mobile typical networking conditions is still limited. This underscore is the strong need for a another modeling approach. In addition the user satisfaction function will lead to a false network selection decision, if the satisfaction function does not capture user reaction to the nontechnical parameters e.g., the service pricing, service security, and operator reputation, etc. For instance the users willing to pay lesser will prefer comparatively degraded services rather than higher quality of service and paying more. These facts justify the need for an analytical user satisfaction model that can be used on the one hand by users for the network selection and on the other hand by operators for online optimization of their resources and increasing their satisfied user pool, which in turn increases operator profit.

In a single criterion utility function, the utility function captures user satisfaction for the possible degradation effect introduced by the criterion. Consider an ideal scenario, where a static user experiences *zero* delay, no vertical handover, and no codec switchovers. In such a case, the user satisfaction is fully dependent on the packet loss rate.

$$\tilde{r}_j(pl) := \mu_w^{co} e^{-\sigma \cdot pl} \; \forall \; pl \in [pl_{ac}, pl_{max}] \tag{2.24}$$

Let us consider that the user is associated with a specific codec say $co = G711$, and with network technology $w = HSDPA$, where μ_w^{co} represents the maximum achievable MOS value in this scenario, which is achievable for packet loss less than pl_{ac}. In addition we also investigate single attribute utility function for a different network technology $w = WLAN$, and broadband codec $G722$. One has that $\mu_{WLAN}^{G722} > \mu_{HSDPA}^{G711}$. This strengthens the importance of the codec and the technology with the respect to user payoff. One more point to be noticed here is that $\sigma-$value in a wireless local area network (WLAN) is greater than in HSDPA, which results in greater decay of MOS value with

No.	Network	Codec	Ppl	Subj.	Model	Diff.
1	W→H	722.2	0	4.27	4.27	0.00
2	W	722.2	0	4.49	4.49	0.00
3	W→H(b.)	722.2→711	0	2.34	2.35	-0.01
4	H→W(a.)	711→722.2	0	2.55	3.98	-1.43
5	H→W	711	10	2.42	2.158	0.26
6	W	722.2	10	2.02	2.23	-0.21
7	H→W(b.)	711→722.2	20	1.81	1.339	0.47
8	W→H(a.)	722.2→711	0	2.32	2.33	-0.01
9	W→H	722.2	10	2.34	2.12	0.22
10	H	711	0	2.95	2.95	0.00
11	H→W	711	0	3.28	3.10	0.18
12	H	711	10	1.96	1.95	0.01
13	W→H(b.)	722.2→711	20	1.38	1.28	0.1
14	W	722.2	20	1.27	1.11	0.16
15	H	711	20	1.45	1.3	0.15

TABLE 2.1
Comparison of analytical model estimates and auditory judgments (MOS).
Ppl: Packet loss in %; Diff.: Difference of subjective MOS and Utility-based
Model MOS; W: WLAN; H: HSDPA; →: Handover/Changeover, a./b. :
Changeover before/after handover.

increasing packet loss. The experiments are summarized in Table 2.1 where
different values of packet loss $(0\%, 10\%, 20\%)$ are examined

2.2.8 Learning Bargaining Solutions

The bargaining solution is a natural framework that allows us to define and
design fair assignment between players which will play the role of bargainers.
It is characterized by a set of axioms that are appealing in defining fairness or
by a maximization of log-concave function on the set of feasible utilities. We
are interested in Nash bargaining since it can be seen as a natural extension
of the proportional fairness criterion, which is probably the most popular fairness notion in networks. Users are faced with the problem to negotiate for a
fair point in the set of feasible utilities. If no agreement can be achieved by the
users, the disagreement utilities occur. In contrast to most of the noncooperative solutions (Nash equilibrium, Stackelberg solution, etc.), the bargaining
solutions are known to be Pareto optimal (defined below). We present below
a fictitious play-based learning algorithm for dynamic Nash bargaining.

Nash Bargaining

Various solutions have been proposed based on slightly different assumptions about what properties are desired for the final agreement point between the players. John Nash proposed that a solution should satisfy certain axioms:

- Invariance to affine transformations or Invariance to equivalent utility representations

- Pareto optimality

- Independence of irrelevant alternatives

- Symmetry

There are many other type of Bargaining solutions: Egalitarian bargaining solution, Kalai-Smorodinsky bargaining solution, Rubinstein bargaining model, etc. See [72] for more details on these notions and their applications to wireless engineering.

Pareto optimal solution

Pareto efficiency, or Pareto optimality, is a concept in decision-making problems with applications in economics, engineering, computer science, etc. The term is named after Vilfredo Pareto, an Italian economist who used the concept in his studies of economic efficiency and income distribution. Given an initial allocation of payoffs among a set of players, a change to a different allocation that makes at least one player better off without making any other player worse off is called a Pareto improvement. An allocation is defined as *Pareto efficient* or *Pareto optimal* when no further Pareto improvements can be made. Pareto efficiency is a minimal notion of efficiency and does not necessarily result in a global optimum or socially desirable distribution of payoffs: it makes no statement about equality, or the overall well-being of an environment.

The Pareto optimality concept is used in noncooperative as well as in cooperative games.

Example of Pareto optimality

An example of typical situations in which Pareto optimality is involved is the service selection or network selection problem. Consider a population of users in a given city. Suppose that the city suggests to us three operators O_1,

O_2, and O_3 of the same price that fits our budget. We set four main criteria to evaluate the offers: quality of experience (QoE), energy consumption, throughput and delay and range the score from -10 to 10. Here is the table of the evaluation:

	O_1	O_2	O_3
QoE	4	3	3
Energy	-2	-3	-4
Throughput	40	50	45
Delay	-10	-3	-5

By looking at this table, we can eliminate the offer O_3 from our choice because it is worse than O_2 from all points of view. As to the remaining offers O_1 and O_2, we observe that O_1 is better than O_2 with regard to the quality of experience of the users but worse than O_2 with regard to the throughput criterion. At this stage, it is impossible to say which one is the best with regard to the four criteria. Actually they are both optimal according to the Pareto criterion. The main idea of this concept is that the population is enjoying maximum ophelimity when no one can be made better off without making someone else worse off.

FP Learning for Bargaining Solutions

Consider a dynamic bargaining game with n players interacting infinitely many times. Time is discrete, and time space is $\mathbb{Z}_+$. At each time slot t,

- Each player j selects an action $a_j \in \mathcal{A}_j$.

- If the action profile $(a_j, \mathbf{a}_{-j})$ meets the constraints set $\mathcal{C}$, then player j receives $\tilde{R}_j(a_j)$, where $\tilde{R}_j$ is strictly increasing function, $\tilde{R}_j(0) = 0$.

- If the constraints are not satisfied, player j receives $\tilde{R}_j(0)$. There is a discount factor δ between periods.

The system goes to the next slot $t + 1$. We define discounted payoff as

$$F_{j,\delta} = (1 - \delta) \sum_{t=1}^{\infty} \delta^{t-1} \tilde{R}_j(a_{j,t}) \mathbb{1}_{\{(a_{j,t}, \mathbf{a}_{-j,t}) \in \mathcal{C}\}}.$$

Learning in Two-Player Dynamic Bargaining

We introduce a learning rule to play this dynamic bargaining game using Nash bargaining formulation. We consider a strategic Nash bargaining described as follows:

- Two players $1, 2$. At $t = 0$, players bargain for the expected discounted payoff.

- At time $t > 0$, each player j reacts to the frequency of use of action of the other player. From the time $t = 2$, both players use a simple learning rule and make a decision according to the fictitious play.

- The discounted payoff are

$$F_{\delta,1} = (1-\delta) \sum_{t=1}^{\infty} \delta^{t-1} \tilde{R}_1(a_{1,t}) \mathbb{1}_{(a_{1,t}, a_{2,t}) \in \mathcal{C}},$$

$$F_{\delta,2} = (1-\delta) \sum_{t=1}^{\infty} \delta^{t-1} \tilde{R}_2(a_{2,t}) \mathbb{1}_{(a_{1,t}, a_{2,t}) \in \mathcal{C}}$$

- At time $t > 0$, each player j reacts to the frequency use of action of the other player.

- Denote $\bar{f}_{j,t}$ be the relative mean use of actions by player j up to $t-1$.

- According to the fictitious play, players choose, $a_{1,t}$ (resp. $a_{2,t}$)

$$a_{1,t} \in \arg\max_{a_1} \sum_{a_2, \bar{f}_{2,t}(a_2) > 0} \bar{f}_{2,t}(a_2) \tilde{R}_1(a_1) \mathbb{1}_{\mathcal{C}}(a_1, a_2)$$

- That is, in each period each player chooses his best response to the observed historical frequency of his opponent's choices.

We analyze the fictitious play outcome for the constraint set

$$\mathcal{C} = \{a = (a_1, a_2) \mid 0 \le a_1 \le \bar{a}_1, \ 0 \le a_2 \le \bar{a}_2, \ 0 \le a_1 + a_2 \le v(\{1,2\})\},$$

with

$$\bar{a}_1 + \bar{a}_2 > v(\{1,2\}), \bar{a}_j < v(\{1,2\}).$$

Theorem 2.2.8.1. • *For any $(a_{1,1}, a_{2,1}) \in \mathcal{C}$? the following hold:*

(i)

$$a_{1,2} = v(\{1,2\}) - a_{2,1},$$

(ii)

$$a_{2,2} = v(\{1,2\}) - a_{1,1}.$$

- *For any $t \ge 3$, $a_{1,t}$ must be either $a_{1,1}$ or $(v(\{1,2\}) - a_{2,1})$, and*

$$a_{2,t} \in \{a_{2,1}, v(\{1,2\}) - a_{1,1}\}.$$

-

$$\bar{f}_{1,t}(a_{1,1}) + \bar{f}_{1,t}(v(\{1,2\}) - a_{2,1}) = 1,$$
$$\bar{f}_{2,t}(a_{2,1}) + \bar{f}_{2,t}(v(\{1,2\}) - a_{1,1}) = 1.$$

$$(a_1, a_2) \in \arg\max_{a_1'} \left[\tilde{R}_1(a_1')\right]^{w_1} \left(\tilde{R}_2(v(\{1,2\}) - a_1')\right)^{w_2},$$

$a_2 = \phi(a_1)$, ϕ defines the Pareto frontier and satisfies the implicit equation

$$\left[\tilde{R}_1(a_1)\right]^{w_1} \left[\tilde{R}_2(v(\{1,2\}) - a_1)\right]^{w_2} = \left[\tilde{R}_1(v(\{1,2\}) - \phi(a_1))\right]^{w_1} \left[\tilde{R}_2(\phi(a_1))\right]^{w_2}.$$

Example 2.2.8.2 (Rate control in AWGN channel). *We consider a system consisting of one receiver and its uplink additive white Gaussian noise (AWGN) multiple access channel with two senders. The signal at the receiver is given by*

$$Y = \xi + \sum_{j=1}^{m} h_j \tilde{s}_j,$$

where X_j is a transmitted signal of user j and ξ is zero mean Gaussian noise with variance N_0. Each user has an individual power constraint $\mathbb{E}(\tilde{s}_j^2) \leq p_j$. The optimal power allocation scheme for Shannon rate (see [50] and the references therein) is to transmit at the maximum power available, i.e., p_j, for each user. Hence, we consider the case in which maximum power is used. The decisions of the users then consist of choosing their communication rates, and the receiver's role is to decode, if possible. The capacity region is the set of all vectors $\mathbf{a} \in \mathbb{R}_+^2$ such that senders $j \in \{1,2\}$ can reliably communicate at rate a_j, $j \in \{1,2\}$. The capacity region $\mathcal{C}$ for this channel is the set

$$\mathcal{C} = \left\{ \mathbf{a} \in \mathbb{R}_+^2 \;\middle|\; \sum_{j \in J} a_j \leq \log\left(1 + \frac{\sum_{j \in J} p_j w_j}{N_0}\right), \; \forall \, \emptyset \subsetneq J \subseteq \{1,2\} \right\}$$

where $w_j = |h_j|^2$. Let

$$\bar{a}_1 = \log\left(1 + \frac{p_1 w_1}{N_0}\right),$$

$$\bar{a}_2 = \log\left(1 + \frac{p_2 w_2}{N_0}\right),$$

$$v(\{1,2\}) = \log\left(1 + \sum_{j \in \{1,2\}} \frac{p_j w_j}{N_0}\right).$$

Under the constraint $\mathcal{C}$ if a sender j wants to communicate at a higher rate, one of the other senders has to lower his rate; otherwise, the capacity constraint is violated. We let

$$r_j := \log\left(1 + \frac{p_j w_j}{N_0 + \sum_{j' \in \{1,2\}, j' \neq j} p_{j'} w_{j'}}\right),$$

be the rate of a sender when the signals of the other sender are treated as noise.

The strategic bargaining game is given by

$$\left(\mathcal{N} = \{1,2\}, \mathcal{C}, (\mathcal{A}_j)_{j \in \mathcal{N}}, (\tilde{R}_j)_{j \in \mathcal{N}}\right),$$

where the set of senders $\mathcal{N}$ is also the set of players $\mathcal{A}_j$, $j \in \mathcal{N}$, is the set *of actions; $\mathcal{C}$ is the bargaining constraint set and $\tilde{R}_j$, $j \in \mathcal{N}$, are the payoff functions. We define $\tilde{R}_j : \prod_{j \in \mathcal{N}} \mathcal{A}_j \to \mathbb{R}_+$ as follows.*

$$\tilde{R}_j(a_j, a_{-j}) = \mathbb{1}_{\mathcal{C}}(a)\tilde{g}_j(a_j) \tag{2.25}$$

$$= \begin{cases} \tilde{g}_j(a_j) & \text{if } (a_j, a_{-j}) \in \mathcal{C} \\ 0 & \text{otherwise} \end{cases}, \tag{2.26}$$

where $\tilde{g}_j : \mathbb{R}_+ \to \mathbb{R}_+$ is a positive and strictly increasing function.

Using the fictitious play learning algorithm and Theorem 2.2.8.1, the following results are obtained:

- *For any $(a_{1,1}, a_{2,1}) \in \mathcal{C}$, the following hold: The "optimal" rates at the first iteration are in the maximal face of rate region $\mathcal{C}$, i.e*

  (i)
  $$a_{1,2} = v(\{1,2\}) - a_{2,1} \geq r_1,$$

  (ii)
  $$a_{2,2} = v(\{1,2\}) - a_{1,1} \geq r_2.$$

- *When the third step of the fictitious play algorithm, the rates chosen by sender one alternate between two values of $a_{1,t}$. The action $a_{1,t}$ must be either $a_{1,1}$ or $(v(\{1,2\}) - a_{2,1})$, and similarly for sender 2:*

  $$a_{2,t} \in \{a_{2,1}, v(\{1,2\}) - a_{1,1}\}.$$

- *The frequencies of use of the rate satisfy*

  $$\bar{f}_{1,t}(a_{1,1}) + \bar{f}_{1,t}(v(\{1,2\}) - a_{2,1}) = 1,$$

  $$\bar{f}_{2,t}(a_{2,1}) + \bar{f}_{2,t}(v(\{1,2\}) - a_{1,1}) = 1.$$

This is equivalent to the time sharing solution between the two boundary points $(a_{1,1}, v(\{1,2\}) - a_{1,1})$ and $v(\{1,2\}) - a_{2,1}, a_{2,1})$. Note that this solution is known to be a maxmin fair solution and is Pareto optimal. The corner points $(r_1, \bar{a}_2)$ and $(\bar{a}_1, r_2)$ are particular Pareto optimal solutions.

2.2.9 Learning and Conjectural Variations

In this section, we assume that the players do not know the payoff functions of the other players. They do, however, observe the outcomes of past actions.

The players form conjectures about variations of the opponent choice, and they have the ability to revise their beliefs as a function of the discrepancy between the actual conjectural variation deduced from the observed actions of the opponents and their current conjectural variations.

At time t, player i forms a conjecture about player j, captured by the value $\hat{c}_{ij,t}$. Based on the observed behavior of the opponent, player i will revise this conjecture over time.

Let $c_{ij,t}$ the conjecture that player i thinks he or she should have used, based on his/her observations up to time t. He or She then updates his or her conjecture based on the discrepancy between the current value $\hat{c}_{ij,t}$ and the ideal value of i's conjecture $c_{ij,t}$.

$$\hat{c}_{ij,t+1} = \hat{c}_{ij,t} + \rho_i(c_{ij,t} - \hat{c}_{ij,t}),$$

where ρ_i is a speed adjustment parameter.

In an n-player game, player i seeks to maximize her payoff $r_i(.)$, not knowing what the strategies of the other players are, nor even what their payoff functions are. For the purpose of his/her maximization, he/she makes the conjecture about the behavior of player j that his conjectured strategy $\hat{x}_j$ is

$$\hat{x}_j = x_j + \hat{c}_{ij}(-x_i + \hat{x}_i),$$

$$\hat{c}_{ij} \in \mathbb{R}, \ (i,j) \in \mathcal{N}^2.$$

The value x_i represents a *reference point*. In other words, player i assumes that other players will observe her deviation from her reference point x_i and deviate from their own reference points x_j by a quantity proportional to this deviation.

The reference point is considered as exogenously given and of common knowledge. The only adjustable quantities will be the coefficients $\hat{c}_{ij}$ of the affine conjecture.

Each player maximizes her payoff function taking into account the conjectures that she has made about the other players and solves the following problem:

$$\max_{\hat{x}_i} r_i(\hat{x}_i, \{x_j + \hat{c}_{ij}(-x_i + \hat{x}_i)\}_{j \neq i}),$$

this results in a conjectured reaction function or set $\hat{x}_i^* = cr_i(x, \hat{c}_i)$ where $\hat{c}_i = (\hat{c}_{ij})_{j \neq i}$ and x is the profile of reference strategies.

For an interior solution (whenever it exists), cr_i solves the first-order conditions:

$$\frac{\partial}{\partial \hat{x}_i} r_i(\hat{x}_i, \{x_j + \hat{c}_{ij}(-x_i + \hat{x}_i)\}_{j \neq i}) + \sum_{j \neq i} \frac{\partial}{\partial \hat{x}_j} \hat{c}_{ij} r_i(\hat{x}_i, \{x_j + \hat{c}_{ij}(-x_i + \hat{x}_i)\}_{j \neq i})$$

At the end of each time slot, the moves of the players are observed by everyone. In particular, player i observes that player j has played x_j. Player i concludes that if his conjectural scheme is realistic, her conjecture should have been c_{ij} such that

$$\hat{x}_j = x_j + c_{ij}(-x_i + \hat{x}_i),$$

i.e $c_{ij} = \frac{\hat{x}_j - x_j}{-x_i + \hat{x}_i}, \ x_i \neq \hat{x}_i.$

Since player i revises his conjectures at each step, one gets

$$\hat{c}_{ij,t+1} = \hat{c}_{ij,t} + \rho_i \left(\frac{\hat{x}_{j,t} - x_j}{\hat{x}_{i,t} - x_i} - \hat{c}_{ij,t} \right),$$

where $\hat{x}_{i,t} = cr_i(x, c_{i,t})$

The set of all conjectures will evolve according to the recurrence:

$$\hat{c}_{ij,t+1} = \hat{c}_{ij,t} + \rho_i \left(\frac{cr_j(x, c_{j,t}) - x_j}{cr_i(x, c_{i,t}) - x_i} - \hat{c}_{ij,t} \right),$$

where $cr_i(x, c_{i,t})$ and $cr_j(x, c_{j,t})$ are nonempty and $cr_i(x, c_{i,t}) \neq x_i$,

$$cr_i(x, \hat{c}_{i,t}) = \arg\max_{\hat{x}_i} r_i(\hat{x}_i, \{x_j + \hat{c}_{ij,t}(-x_i + \hat{x}_i)\}_{j\neq i}),$$

$\rho_i \in (0, 1)$.

Now, we assume that the reference strategy also changes in time denoted by x_t defined by a recurrence relation $x_{i,t+1} = g_i(x_{i,t}, \hat{c}_{i,t})$ for some continuous function g_i.

Here, we have implicitly assumed that the argmax is reduced to a singleton and we have identified the singleton to its element. Further assume that cr_i is continuous in both variables.

Next, we show that the stationary point of the above scheme is a candidate for Pareto optimality.

Theorem 2.2.9.1. *Suppose that* $(x_{i,t}, \hat{c}_{ij,t}) \longrightarrow (x_i, \hat{c}_{ij})$ *when* $t \longrightarrow +\infty$, *for all* i, j. *Assume that* $cr_i(x, \hat{c}_i) = x_i + \lambda_i$ *for some* $\lambda_i \neq 0$. *Then, for all* i, *the vector* $\hat{c}_i$ *is proportional to* λ.

Proof. We provide a short proof. By continuity of cr and convergence assumptions, we have $cr_i(x, \hat{c}_i) - x_i = \lambda_i \neq 0$. This means that the fixed point equation becomes

$$\hat{c}_{ij} = \hat{c}_{ij} + \rho_i \left(\frac{cr_j(x, c_j) - x_j}{cr_i(x, c_i) - x_i} - \hat{c}_{ij} \right) = \hat{c}_{ij} + \rho_i \left(\frac{\lambda_j}{\lambda_i} - \hat{c}_{ij} \right).$$

Using the fact that $\rho_i \neq 0$, we have

$$\hat{c}_{ij} = \frac{\lambda_j}{\lambda_i}.$$

Thus,

$$\hat{c}_i = \frac{1}{\lambda_i}(\lambda_1, \ldots, \lambda_n).$$

$\square$

This result allows one to address the link with Pareto optimal solutions. As we know, the maximizers of the weighted sum of payoffs are Pareto optimal. The first order conditions for interior Pareto solutions are given by

$$\sum_{j \in \mathcal{N}} \bar{\lambda}_j \frac{\partial}{\partial x_i} r_j(x) = 0, \ \bar{\lambda}_j \geq 0,$$

i.e., $\bar{\lambda}.Dr(x) = 0$ where $(Dr(x))_{ij} = [\frac{\partial}{\partial x_i} r_j(x)]$.

We conclude that both the limits of the learning scheme and the Pareto optima are such that gradient of $r(x)$ is singular. From the equation,

$$\frac{\partial}{\partial \hat{x}_i} r_i(\hat{x}_i, \{x_{j,t} + \hat{c}_{ij,t}(-x_{i,t} + \hat{x}_i)\}_{j \neq i})$$

$$+ \sum_{j \neq i} \frac{\partial}{\partial \hat{x}_j} \hat{c}_{ij,t} r_i(\hat{x}_i, \{x_{j,t} + \hat{c}_{ij,t}(-x_{i,t} + \hat{x}_i)\}_{j \neq i}),$$

we let $t \longrightarrow \infty$. Since $\hat{c}_{ij,t} \longrightarrow \frac{\lambda_j}{\lambda_i}$, By multiplying by $\lambda_i \neq 0$, the limiting becomes

$$\lambda_i \frac{\partial}{\partial \hat{x}_i} r_i(\hat{x}_i, \{x_j + \hat{c}_{ij}(-x_i + \hat{x}_i)\}_{j \neq i})$$

$$+ \sum_{j \neq i} \lambda_j \frac{\partial}{\partial \hat{x}_j} r_i(\hat{x}_i, \{x_j + \hat{c}_{ij}(-x_i + \hat{x}_i)\}_{j \neq i}).$$

This leads to $Dr(x).\lambda' = 0$, i.e., λ' is the (column) vector of λ. In many cases the kernel of the matrix Dr and the kernel of its transposition may be related. If the vector λ happens to be positive, then x is a candidate solution for the maximization problem by analogy with λ and $\bar{\lambda}_j$. In that case, the conjectural variation outcome is a candidate for Pareto optimal. To complete the analysis, it should be checked that x is a maximizer (the first order condition may not be sufficient).

2.2.10 Bayesian Learning in Games

A central result in learning theory is that Bayesian forecasts eventually become accurate under suitable conditions, such as absolute continuity of the data generating process with respect to the player's beliefs. Hence, multiple repetitions of Bayes'rule may transform the historical data into a near perfect guide for the future. Here "Bayesian" refers to the 18th century mathematician and theologian Thomas Bayes (1702-1761).

In the previous sections, we provided the properties and benefits of Bayesian techniques for reinforcement learning and illustrated them with case studies. Below we provide the basic ideas based on priors and/or observations.

For each player j, let $\sigma_j(s_j|\, h_t)$ be the probability that $a_{j,t+1} = s_j$ conditionally on $h_t \in H_t$, i.e., the Bayes rule is adopted:

$$\sigma_j(s_j \mid h_t) = \frac{\sigma_j(s_j, h_t)}{\sigma_j(h_t)}.$$

where $\sigma_j(h_t) > 0$. The marginal distribution should be compatible with the belief and s_j maximizes

$$\sum_{s_{-j}} r_j(., s_{-j}) \tilde{\sigma}_{-j}(s_{-j} \mid h_t),$$

on $\mathcal{A}_j$, where

$$\tilde{\sigma}_{-j}(s_{-j} \mid h_t) = \frac{\sigma_j(s_{-j}, h_t)}{\sigma_j(h_t)}.$$

Generically, the Bayesian-based iterative procedure has the following form:

$$\mu_{j,t+1} = BU_j(\mu_{j,t}, r_{j,t}) + \bar{\lambda}_{j,t}(\mu_{j,t} - BU_j(\mu_{j,t}, r_{j,t+1})),$$

where BU_j is the Bayesian rule for player j and $\bar{\lambda}_{j,t}$ is a parameter. If $\bar{\lambda}_{j,t+1} = 0$ then the model reduces to standard Bayesian model. If $\bar{\lambda}_{j,t+1} > 0$ then the updating rule can be interpreted as attaching too much weight to the prior $\mu_{j,t}$ and hence underreacting to measurement/observations. Conversely, if $\bar{\lambda}_{j,t+1} < 0$, then the updating rule can be interpreted as overreacting to measurement/observations.

> ⚠ The beliefs of the players may converge or not after multiple overreactions to observations, but possibly to incorrect forecasts.

Infinite hierarchy of beliefs

In the presence of the state space $\mathcal{W}$ naturally led to handle infinite hierarchies of beliefs. Adopting a Bayesian approach, each player will make her decision on some subjective beliefs (i.e., probability measure) on $\mathcal{W}$. Since the outcome is determined not only by the player's own actions but the other players' actions as well, and those are influenced by their beliefs about the set $\mathcal{W}$, each player must also have beliefs on other players' beliefs about $\mathcal{W}$. By the same argument she must have

- beliefs about other players' beliefs about her own beliefs about $\mathcal{W}$,

- beliefs about other players' beliefs about her beliefs on their beliefs about the set $\mathcal{W}$, etc.

Thus, it seems that one needs to have an infinite hierarchy of beliefs for each player. These hierarchies are linked together by the fact that each belief of a player is also the subject of beliefs of the other players. Such an approach leads to the so-called *Universal Belief Space*.

2.2.11   Non-Bayesian Learning

A series of experiments suggest that players may repeatedly process [131] information using non-Bayesian heuristics (as in experimental games). These experiments contributed to a growing interest in the properties of non-Bayesian learning.

Departures from Bayesian updating can occur either because subjects tend to ignore the prior and overreact to the data, or alternatively because of excessive learning rate on prior beliefs and underreact to new measurement/observations.

The complexities of Bayesian procedures may make Bayesian updating rules excessively costly to implement in many practical applications. So, even players who would prefer to use Bayes' rule often rely on simpler, non-Bayesian heuristics for updating beliefs.

> Section review

In this section, we presented various partially distributed strategy-learning schemes for finite games as well as for stochastic games. All these learning algorithms are illustrated in the wireless contexts including power allocation, spectrum access in cognitive radio networks, access control, resource selection in wireless networks, rate bargaining in additive white Gaussian noise channel etc. Partially distributed strategic learning applies to Nash points, bargaining solutions. We have seen that the cost of learning can be introduced in the models to capture realistic behavior of players in changing their actions. However, we have seen the limitations of these learning algorithms in practical scenarios. The perfect monitoring assumptions reduce considerably the applicability of these schemes in autonomous systems.

2.3   Fully Distributed Strategy-Learning

In this section, we present fully distributed strategy-learning algorithms for dynamic games. *Fully distributed learning algorithms* use only own-actions and perceived own-payoffs (completely distributed). We construct learning schemes such that the updating schemes require only own-actions and own-perceived payoffs. Such learning algorithms are very important in a stochastic environment where many parameters are unavailable or unknown. In communication scenarios, this means that the payoff functions are not known by the players (users/mobile devices), only an estimated and noisy numerical value is available.

Information is crucial in the process of decision making.

This section provides a class of learning schemes in which player use less information about the other players, less memory on the history and the players do not need to use the same learning scheme.

A generic scheme for fully distributed strategic learning dynamics is given by

$$x_{j,t+1}(s_j) = F_{j,s_j,t}\left(x_{j,t'}, r_{j,t'}, \hat{\mathbf{r}}_{j,t'}, \ 1 \leq t' \leq t\right)$$

where F_j is the mapping that specifies the learning pattern of player j at time t, s_j is a generic action of player j, and $\hat{\mathbf{r}}_{j,t'}$ is an estimated payoff obtained from the measurements up to $t' - 1$.

Different learning and adaptive algorithms have been studied in the literature [191, 106, 147]. In all these references, the players have to follow the same rule of learning, they have to learn in the same way. We will investigate in detail the following question:

What happens if the players have different learning schemes?

Section 2.3.5 provides an answer for a large class of learning patterns leading to evolutionary game dynamics.

Fully Distributed Strategy-Learning in Wireless Networks

In the wireless literature in the broad sense there are extensive works exploiting learning in games (see [106, 191] and the references therein). The authors in [191] proposed to apply the reinforcement learning algorithm initially introduced by Bush and Mosteller (1955,[42]). Their scheme is a distributed power control policy for networks with a finite number of transmitters and power levels. Based on the sole knowledge of the value of his own payoff at each step, the corresponding algorithm consists, for each transmitter, in updating his strategy namely his probability distribution over the possible power levels. The strategy changes stochastically and leads to a stochastic learning algorithm. Distributed strategic learning for n-player games have been applied to wireless network selection [89], IP TV services [92], dynamic routing [18], power control [106, 191] based on the energy efficiency function, spectrum efficiency [174, 175], wireless network security [199, 197, 173, 198], evolutionary coalitional games for user-centric network selection [90, 91] etc.

2.3.1   Learning by Experimentation

We present a completely uncoupled learning rule such that, when used by all players in a finite game, slot-by-slot play comes close to pure (Nash) equilibrium play a high proportion of the time, provided that the game is generic (interdependent) and has such a pure equilibrium. The interactive learning by experimentation is in discrete time, $t \in \mathbb{Z}_+$. At each time t, a player j has his or her own state $S_{j,t}$ which contains her own decision $a_{j,t}$ and her own perceived payoff $r_{j,t}$ i.e $S_{j,t} = (a_{j,t}, r_{j,t})$. At time $t+1$, each player performs some experiment with some probability $\epsilon_j \in (0,1)$. The player keeps the actions $a_{j,t}$ at time $t+1$ when he does not experiment, otherwise he or she plays $a_j \in \mathcal{A}_j$ drawn uniformly at random. If the received payoff after experimentation $r_{j,t+1}$ is strictly greater than the old payoff $r_{j,t}$, then player j adopts the new state $S_{j,t+1} = (a_{j,t+1}, r_{j,t+1})$, otherwise the player j comes back to its old state $S_{j,t}$.

The next result says that if the finite game is a weakly acyclic game then the resulting Markov chain will visit a pure Nash equilibrium with high proportion of time.

Theorem 2.3.1.1 ([109]). *Given a finite $n-$player weakly acyclic game $\mathcal{G}$, if all players use simple experimentation with experimentation rate $\epsilon > 0$, then for all sufficiently small ϵ a Nash equilibrium is played at least $1 - \epsilon$ of the time.*

⚠ Note that the notion of proportion of visits is different from the previous results of convergence.

Let $\mathcal{G}_\mathcal{A}$ be the set of all $n-$player games with finite action profile space $\mathcal{A}$ that are interactive and that possess at least one pure Nash equilibrium. We now present a result of frequency of visits of *Interactive Trial and Error Learning* proved by [194, 132, 192] for the class $\mathcal{G}_\mathcal{A}$. The result generalizes Theorem 2.3.1.1.

Theorem 2.3.1.2 ([194]). *If all players use interactive trial and error learning and the experimentation rate $\epsilon > 0$ is sufficiently small, then for almost all games in $\mathcal{G}_\mathcal{A}$ a Nash equilibrium is played at least $1 - \epsilon$ of the time.*

Next, we apply this result to discrete power allocation problems without information exchanging on total interference and without knowing the payoff functions.

Example 2.3.1.3 (Learning in discrete power allocation games). *Consider a finite set of transmitters sending information to a finite set of receivers using orthogonal frequencies. Each transmitter (player) is able to transmit to all the channels. The channel gain between a transmitter j and channel l is denoted by $w_{jj'l}$. We assume channel gains change slower than time scaling of the* transmission. Each transmitter j selects of a vector of powers in a finite set $\mathcal{P}_j$. An action for transmitter j is therefore an element of $a_j \in \mathcal{P}_j$. The SINR of transmitter j to channel l is then given by

$$SINR_{j,l}(\mathbf{a}) = \frac{a_{jl} w_{jjl}}{N_{0,l} + \sum_{j' \neq j} a_{j'l} w_{j'jl}},$$

where $N_{0,l}$ is the noise at channel l, $w_{j'jl}$ is the channel gain at link l. The payoff of player j its total sum-rate, i.e.,

$$\tilde{R}_j(\mathbf{a}) = \sum_l b_l \log_2 \left(1 + SINR_{j,l}(\mathbf{a})\right).$$

This strategic power allocation game has been well-studied and is known to have a pure Nash equilibrium, and it is in the specific class of weakly acyclic games. We therefore can apply the learning by experimentation and Theorem 2.3.1.2 says that a pure Nash equilibrium will be visited at least $1 - \epsilon$ fraction of the times for any $\epsilon > 0$. Note that the theorem does not tell how many iterations are needed to be arbitrary close to a Nash equilibrium payoff. Intuitively, the number of iterations may depend on the precision ϵ.

Example 2.3.1.4 (Learning in medium access control games). *Consider the following multiple access game between two mobiles (players). Each mobile can transmit with a power level $p_0 < p_1 < p_2$. Mobile 1 chooses a row, and mobile 2 chooses a column. The action p_0 corresponds to zero power which means does not transmit. The actions p_1, p_2 with $p_1 < p_2$ are strictly positive power levels. Assume that there is a power consumption cost represented by*

	p_2	p_1	p_0
p_2	$(-c(p_2), -c(p_2))$	$(1 - c(p_2), -c(p_1))$	$(1 - c(p_2), 0)$
p_1	$(-c(p_1), 1 - c(p_2))$	$(-c(p_1), -c(p_1))$	$(1 - c(p_1), 0)$
p_0	$(0, 1 - c(p_2))$	$(0, 1 - c(p_1))$	$(0, 0)$

FIGURE 2.6
Multiple access game between two mobiles.

a mapping $c(.)$. If $c \equiv 0$, by applying the learning by trial and error, one finds that an equilibrium configuration with payoffs $(0,0), (1,0)$, or $(0,1)$ will be visited at $1 - \epsilon$ fraction of the time. Assume now that $c(.)$ is non-decreasing:

$$c(p_0) = 0 < c(p_1) < c(p_2) < 1.$$

It is easy to see that the pure Nash equilibria disappears (see Figure 2.6). The game has a unique mixed Nash equilibrium. The outcome of the learning by trial and error algorithm starting from the collision status (p_2, p_2) is a non-equilibrium. We will examine in the next section learning schemes for mixed *equilibria. Another issue that we will address in the last chapter is a learning scheme for global optimum of this game. The profiles (p_1, p_0) and (p_0, p_1) are both unfair, but they are both energy-efficient and successful configurations.*

Example 2.3.1.5 (Learning in Cognitive MAC Game). *The term Cognitive Radio (CR), originally coined in the late 1990s, envisaged a radio that is aware of its operational environment so that it can dynamically and autonomously adjust its radio operating parameters accordingly to adapt to different situations. Cognition is achieved through the so-called cognitive cycle, consisting of the observation of the environment, the orientation and planning that leads to making the appropriate decisions pursuing specific operation goals, and finally the updates regarding the environment. Decisions on the other hand can be reinforced by learning procedures based on the analysis of prior observations and on the corresponding results of prior updates. More than a decade after the cognitive radio concept was born researchers all over the world have devoted significant efforts addressing different technical challenges of cognitive radio networks, mainly covering fundamental problems associated with the cognitive procedures as well as technology enablers of cognitive radio concepts. Another potential offered by cognitive radio networks for bringing Dynamic Spectrum Access (DSA) to reality, thanks to the ability to identify spatial and temporal spectrum gaps not occupied by primary users (white spaces, spectrum holes), and to place secondary/unlicensed transmissions within such spaces.*

Consider a secondary network that coexists with a primary network where each Primary User (PU) is licensed to transmit whenever she wishes for most of time except for the case when the channel is occupied by another PU. Two cognitive users or Secondary Users (SUs) are able to detect the spectral white spaces and may opportunistically transmit their data over the primary channel.

When accessing the channel, it is clear that an interactive situation arises among SUs since they all attempt to access the common channel. Each SU needs to sense the channel before access. This sensing and transmission scheme can be modeled as a two-stage strategic decisions. We focus on the one-shot and two-steps game. At the first step, each SU has two sub-actions : Either to sense the channel (S) or to wait (W), i.e., not to sense the channel. Each sensing operation of SU j induces a sensing cost c_s. Next and once the channel sensing is complete, each SU may attempt to transmit (T) its data if the channel was sensed idle or decide to wait (W), i.e., do not transmit (either because the channel was detected busy or simply the SU prefers to stay quiet). In order to not interfere with the PU network, we note that SU that have not sensed the channel are not allowed to attempt transmission. The actions are then $(S, T), (S, W)$ and (W, W) denoted by ST, SW and WW respectively. The cost of transmission is c_t. Figure 2.7 represents the matrix game.

For any $c_s > 0$, the action SW is weakly dominated by WW. Then, the matrix game above can be reduced to another game with only two actions per SU given in Figure 2.8.

	ST	SW	WW
ST	$(-c_t - c_s, -c_t - c_s)$	$(1 - c_t - c_s, -c_s)$	$(1 - c_t - c_s, 0)$
SW	$(-c_s, 1 - c_t - c_s)$	$(-c_s, -c_s)$	$(-c_s, 0)$
WW	$(0, 1 - c_t - c_s)$	$(0, -c_s)$	$(0, 0)$

FIGURE 2.7
Cognitive MAC Game.

	ST	WW
ST	$(-c_t - c_s, -c_t - c_s)$	$(1 - c_t - c_s, 0)$
WW	$(0, 1 - c_t - c_s)$	$(0, 0)$

FIGURE 2.8
Reduced Cognitive MAC Game.

2.3.2 Reinforcement Learning

Reinforcement learning (RL) is a term that is interpreted differently in different disciplines. The idea of RL in this chapter is the following:

> Reinforcement learners interact with their environment and use their experience to choose or avoid certain actions based on their consequences (past measurements). Actions that led to high payoffs in a certain situation tend to be repeated whenever the same situation recurs, whereas choices that led to comparatively lower payoffs tend to be avoided.

The empirical study of reinforcement learning as a crucial facet of (human and non-human) animal behavior finds its roots in Thorndike's experiments on instrumental learning (Thorndike, 1898, [180]). The results of these experiments were formalized in the "law of effect", one of the most robust properties of learning in the experimental psychology literature.

Reinforcement Learning has originally been studied in the context of the single-player environment. A player receives a numerical payoff signal, which it seeks to maximize. The environment provides this signal as a feedback on the sequence of actions that has been executed by the player. Learners relate the payoff signal to previously executed actions to learn a strategy that maximizes the expected future payoff/reward.

In a game-theoretic setting with multiple players, the payoff function at a given game stage depends on the actions played by the different players. By denoting $a_{j,t}$ the action played by j at time slot t, the payoff for j is written as $\tilde{R}_{j,t}(a_{1,t}, ..., a_{n,t})$. By notational abuse but for the sake of clarity, the perceived payoff (which could be delayed and noisy) of player j at time slot t will be denoted by $r_{j,t}$. We have that $x_{j,t}(s_j) = \Pr[a_{j,t} = s_j]$, $s_j \in \mathcal{A}_j$. The reinforcement learning of [42, 32, 11] consists in updating the probability distribution over the possible actions as follows: $\forall\ j \in \mathcal{N}, \forall s_j \in \mathcal{A}_j$,

$$x_{j,t+1}(s_j) = x_{j,t}(s_j) + \lambda_{j,t} r_{j,t} \left(\mathbb{1}_{\{a_{j,t}=s_j\}} - x_{j,t}(s_j) \right) \qquad (2.27)$$

where $\mathbb{1}_{\{.\}}$ is the indicator function and $\lambda_{j,t}$ is the learning rate of player j at time slot t. The parameter $\lambda_{j,t}$ and the payoff $r_{j,t}$ can be normalized such that $r_{j,t} > 0$, $0 < \lambda_{j,t} r_{j,t} < 1$. A simple interpretation is the following: the action that was played at the last time slot, namely $a_{j,t}$, sees its probability increased (since $\left(\mathbb{1}_{\{a_{j,t}=s_j\}} - x_{j,t}(s_j) \right) \geq 0$) while the other actions see their probability decreased (since $0 - x_{j,t}(s_k) \leq 0$). The key point here is that the increment in the probability of each action s_j depends on the corresponding observed or measured payoff and its learning rate. More importantly, note that in (2.27), for each player, only the value of his individual payoff function at time slot t is required. Therefore, the knowledge of the payoff function $R_j(.)$ is not assumed for implementing the algorithm. In addition the random state $\mathbf{w}_t$ is unknown to the player. This is one of the reasons why deterministic gradient-like techniques are not directly applicable. The algorithm is stochastic and it is a non-model algorithm. In this context, the stochastic gradients and estimated payoffs are similar objects. The generic updating scheme has the following form:

$$\text{Newestimate} \longleftarrow \text{Oldestimate+Stepsize (Target-Oldestimate)} \qquad (2.28)$$

where the target plays the role of the current strategy. The expression [Target - Oldestimate] is an error in the estimation. It is reduced by taking a step size toward the target. The target is presumed to indicate a desirable direction in which to move.

A simple pseudocode for the simultaneous-act games is as follows:

```
for i = 1 to max_of_iterations
        for j = 1 to number_of_players
            action_profile.append(select_action(player-j))
        endfor
        for j = 1 to number_of_players
            playerj.payoff = assign_payoff(action_profile)
        endfor
        for j = 1 to number_of_players
            playerj.update_strategies
        endfor
end
```

Figure 2.9 represents a generic reinforcement algorithm.

It is important to mention that most often game-theoretic modeling assume rational players and that rationality is common knowledge. In the fully

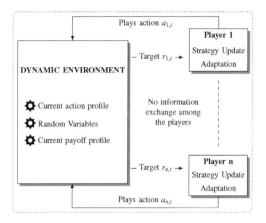

FIGURE 2.9
A generic RL algorithm.

distributed learning schemes described in this chapter, each player is assumed to follow a set of hybrid learning schemes but does not need to know the existence of a game; they do not need whether the other players are rational or not.

Next, we present different variants of reinforcement learning patterns. In the multiplayer setting, we define the environment by a game in normal form and the reinforcement learner updates the strategy.

Bush and Mosteller Reinforcement Learning Algorithms

Reinforcement learners use their experience to choose or avoid certain actions based on their consequences. Actions that led to satisfactory outcomes (i.e. outcomes that met or exceeded aspirations) will tend to be repeated in the future, whereas choices that led to unsatisfactory experiences are avoided. We overview the dynamics of a Bush and Mosteller (1955) stochastic model of reinforcement learning for two-by-two (i.e., two-player two-action) games. In this model, players decide what action to select stochastically: each player's strategy is defined by the probability of undertaking each of the two actions available to them. After every player has selected an action according to their probabilities, every player receives the corresponding payoff and revises her strategy. The revision of strategies takes place following a reinforcement learning approach: players increase their probability of undertaking a certain action if it led to payoffs above their aspiration level, and decrease this probability otherwise. In the Bush and Mosteller (BM) model, the players use only information concerning their own past choices and perceived payoffs, and ignore all the information regarding the payoffs and choices of the other players, i.e., a *fully distributed learning algorithm.*

We now describe the details of the scheme in a finite game in strategic form $\mathcal{G} = (\mathcal{N}, (\mathcal{A}_j), (\tilde{R}_j)_j)$. Denote by $\mathcal{A} = \prod_{j \in \mathcal{N}} \mathcal{A}_j$ the set of strategy profiles, and by $x_j(s_j)$ player j's probability of undertaking action $s_j \in \mathcal{A}_j$. At each time t, each player j chooses an action $a_{j,t}$ and computes her stimulus $\bar{s}_{j,t}$ for the action just chosen $a_{j,t}$ according to the formula

$$\bar{s}_{j,t} = \frac{r_{j,t} - M_j}{\sup_{\mathbf{a}} |R_j(\mathbf{a}) - M_j|},$$

where $r_{j,t}$ denotes the perceived payoff at time t by player j, and M_j is an aspiration level of player j. Hence, the stimulus is always a number in the interval $[-1, 1]$. Note that player j is assumed to know the denominator $\sup_s |r_j(s) - M_j|$. Second, having calculated their stimulus $\bar{s}_{j,t}$ after the outcome a_t, each player j updates her probability $x_j(s_j)$, of undertaking the selected action a_j as follows:

$$x_{j,t+1}(s_j) = \begin{cases} x_{j,t}(s_j) + \lambda_j \bar{s}_{j,t}(1 - x_{j,t}(s_j)) & if \ \bar{s}_{j,t} \geq 0 \\ x_{j,t}(s_j) + \lambda_j \bar{s}_{j,t} x_{j,t}(s_j) & if \ \bar{s}_{j,t} < 0, \end{cases}$$

where λ_j is player j's learning rate ($0 < \lambda_j < 1$). Thus, the higher the stimulus magnitude or the learning rate, the larger the change in probability. The updated probability for the action not selected derives from the constraint that probabilities must add up to one.

Below we summarize the main assumptions made in the Bush & Mosteller (BM) model in terms of the nature of the payoffs, the information players require and the computational capabilities that they have.

- Payoffs: Payoffs and aspiration thresholds are not interpreted as von Neumann-Morgenstern utilities, but as a set of variables measured on an interval scale that is used to calculate stimuli.

- Information: Each player j is assumed to know its action set $\mathcal{A}_j$, and the maximum absolute difference between any payoff she might receive and her aspiration threshold. Players do not use any information regarding the other players.

- Computational capabilities and memory: Players are assumed to know their own mixed strategy $\mathbf{x}$ at any given time. Each player is assumed to be able to conduct the arithmetic operations.

Let X_t be the state of a system in time slot t. X_t is a random variable and $\mathbf{x}$ is a particular value of that variable; the sequence of random variables $\{X_t\}_{t \geq 1}$ constitutes a discrete-time Markov process. We say that the Markov process X_t is ergodic if the sequence of stochastic kernels defined by the t−step transition probabilities of the Markov process converges uniformly to a unique limiting kernel independent of the initial state. The expected motion (also called drift) of the system in state x is given by a function vector $F(\mathbf{x})$ whose components are, for each player, the expected changes in the probabilities of undertaking each of the possible actions, i.e.,

$$F(\mathbf{x}) = \mathbb{E}(X_{t+1} - X_t \mid X_t = \mathbf{x}).$$

This leads to the deterministic dynamical system described by the ordinary differential equation

$$\dot{\mathbf{x}} = F(\mathbf{x}).$$

We define a *self-reinforcing rest point* (S2RP) as an absorbing state of the system where both players receive a positive stimulus. An S2RP corresponds to a pair of pure strategies ($x_j(s_j)$ is either 0 or 1) such that its certain associated outcome gives a strictly positive stimulus to both players.

We now present theoretical results on the asymptotic behavior of the BM learning algorithm. Note that with low learning rates the system may take an extraordinarily long time to reach its long-run behavior.

Proposition 2.3.2.1 ([81]). *In any two-by-two game, assuming players' aspirations are different from their respective payoffs ($\tilde{R}_j(\mathbf{a}) \neq M_j$ for all j and $\mathbf{a}$) and below their respective maximin, the process X_t converges to an S2RP with probability 1 (the set formed by all S2RPs is asymptotically reached with probability 1). If the initial state is completely mixed, then every S2RP can be asymptotically reached with positive probability.*

Proposition 2.3.2.2 ([81]). *In any two-by-two game, assuming players' aspirations are different from their respective payoffs and above their respective maximin:*

- *If there is any S2RP, then the process X_t converges to an S2RP with probability 1 (the set formed by all S2RPs is asymptotically reached with probability 1). If the initial state is completely mixed, then every S2RP can be asymptotically reached with positive probability.*

- *If there is no S2RP then the BM learning process X_t is ergodic with no absorbing state.*

- *S2RPs whose associated outcome is non-Nash equilibrium are unstable for the ODE.*

- *All S2RPs whose associated outcome is a strict Nash equilibrium where at least one unilateral deviation leads to a satisfactory outcome for the nondeviating player are asymptotically stable.*

Example 2.3.2.3. *Consider a discrete noncooperative uplink power control game between mobile users (players). The measure of satisfaction of the mobile user is related to the amount of information that a user can transfer in the lifetime of its battery. Therefore, we choose the throughput per battery life or a mapping of Shannon rate and energy consumption as the payoff function. The payoff of a mobile user has one of the following forms*

-

$$\tilde{R}_j(\mathbf{a}) = \frac{LR}{M} \frac{F_j(SINR_j(\mathbf{a}))}{c_j(a_j)}$$

where a_j is in a finite set of powers $\mathcal{A}_j$, L is the information bits in frame (packets) of $M > L$ bits at a rate R b/s using a_j Watts of power. $c_j(a_j)$ is the energy consumption cost, the SINR of user j is

$$SINR_j(\mathbf{a}) = \frac{\bar{w}}{R} \frac{a_j |h_j|^2}{\sum_{i \neq j} a_i |h_i|^2 + N_0},$$

$\bar{w} > 0$ is the available spread-spectrum bandwidth (in Hz), N_0 is the additive white Gaussian noise (AWGN) power at the receiver (in Watts), and $\{|h_j|^2\}$ is the set of path gains from the mobile to the base station. If the cost function c is the identity function then and the functions $\{F_j\}_j$ are identical to F then one gets the well known energy efficiency function. The channel states are assumed to be nonzero, otherwise the SINR will be zero for any action.

-

$$\tilde{R}_j(\mathbf{a}) = G_j(SINR_j(\mathbf{a})) - c_j(a_j).$$

This payoff is sometimes referred to as the pricing-based payoff function. The case where the mapping

$$G_j = \bar{w} \log\left(1 + SINR_j(\mathbf{a})\right),$$

is known as Shannon information rate.

We know from finite normal form games that, every finite strategic-form game has a mixed strategy Nash equilibrium. That is, with payoff functions, the resulting games have at least one equilibrium in mixed strategies. We now apply Bush and Mosteller's learning algorithms in these power control games. The algorithm works as follows:

BM1: *Set the initial probability vector $\mathbf{x}_{j,0}$, for each j.*

BM2: *At every time slot t, each mobile chooses a power according to its action probability vector $\mathbf{x}_{j,t}$. Thus, the j-th mobile chooses the power s_j at instant t, based on the probability distribution $x_{j,t}(s_j)$. Each mobile obtains a numerical value of its payoff. The payoff to mobile j is $r_{j,t}$. Each mobile updates its action probability according to the Bush and Mosteller learning rule.*

BM3: If the stopping criterion is met then stop, else goto step BM2.

Note that, the above example of SINR can be extended to the spatial location of the mobiles and path loss can be considered. The SINR from transmitter j located at x_j to a receiver located at position y_j is given

$$SINR_j(x, y, h, \mathbf{a}) = \frac{\bar{w}}{R} \frac{\frac{a_j |h_j|^2}{l(\|x_j - y_j\|)}}{N_0(y_j) + \sum_{i \neq j} \frac{a_i |h_i|^2}{l(\|x_i - y_j\|)}},$$

where $\frac{\bar{w}}{R} > 0, N_0(y_j) > 0,$

$$l(d) = c'(d^2 + \epsilon^2)^{\frac{\alpha}{2}}, \ \alpha \geq 2, c' > 0, \epsilon > 0,$$

x denotes $(x_j)_{j \in \mathcal{N}}$, $y = (y_j)_{j \in \mathcal{N}}$, and $h = (h_j)_{j \in \mathcal{N}}$. This defines clearly a robust game because of the randomness on location and the randomness of channel.

A particular case of the Bush and Mosteller model has been applied to discrete power control [191] in wireless networks, for $M_j = \min \tilde{R}_j, \lambda_j = b$ and positive payoffs.

Cross Learning Algorithm (Borgers and Sarin (1993,[32]))

A very simple strategy iteration is the Cross Learning algorithm, a specific type of learning automata. When action a is selected and a normalized payoff $r_{j,t} \in [0, 1]$ is received at time t, then strategy $\mathbf{x}$ is updated according to the following equation:

$$x_{j,t+1} = (1 - r_{j,t})x_{j,t} + r_{j,t}, \ x_{i,t+1} = (1 - r_{it})x_{i,t}, \ \text{for } j \neq i. \qquad (2.29)$$

The information needed to update this algorithm is own-payoff and the own-strategy. So, this learning algorithm is fully distributed. Borgers and Sarin (1993) have shown in [32] that the policy changes induced by this learner converge under infinitesimal time steps to the replicator dynamics [151]. The proof uses basic stochastic approximation techniques (see Appendix). Since the replicator may not converge, the cross learning algorithm does not converge in general. An example of nonconvergence is provided in Chapter 7.

Arthur's Model of Reinforcement Learning

Stochastic approximation examines the behavior of a learning process by investigation of an differential equation derived from the expected motion of the learning process. One classic result is that if the ODE has a global attractor, the learning process converges with probability one to that point. However, most of the ODE approximations of reinforcement learning do not have any global attractors. Those are the evolutionary replicator dynamics, adaptive dynamics, best response dynamics, projection, mean field game dynamics, etc. The models of reinforcement learning we study here are based on those studied by Arthur (1993), and Borgers and Sarin (1993). Consider a finite game

$$\mathcal{G} = \left(\mathcal{N}, (\mathcal{A}_j)_{j \in \mathcal{N}}, (\tilde{R}_j)_{j \in \mathcal{N}} \right),$$

in strategic form: $\mathcal{N}$ is the set of players, $\mathcal{A}_j$ is the set of strategies of player j, and $\tilde{R}_j : \prod_j \mathcal{A}_j \longrightarrow \mathbb{R}$ is a random variable with expectation $\tilde{r}_j(a) = \mathbb{E}(\tilde{R}_j(a))$ that represents player j's payoff function. We assume that the random variable $\tilde{R}_j$ are almost surely bounded and strictly positive. Note that it is usually assumed that payoffs are deterministic functions of strategy profiles, but we relax this assumption in this section. We assume that the players are repeatedly playing the same game $\mathcal{G}$ over a number of time periods. At each time t, under *reinforcement learning*, each player j is assumed each time to have a tendency $\alpha_{j,t}(s_j)$ for each action $s_j \in \mathcal{A}_j$. Let $x_{j,t}(s_j)$ be the probability placed by player j on action s_j at time t. In the models of reinforcement learning, we consider that these probabilities are determined by the following choice rule:

$$x_{j,t}(s_j) = g_{j,s_j,t}(a_{j,t}, r_{j,t}, \mathbf{x}_{j,t-1}).$$

Here, we examine the case where the mapping g_j can be written as

$$\frac{\alpha_{j,t}(s_j)^\gamma}{\sum_{s'_j} \alpha_{j,t}(s'_j)^\gamma}, \quad \gamma \geq 1. \tag{2.30}$$

To complete the reinforcement learning model, we need to define how to update the tendencies $\mathbf{x}_t.$. In this simple model, it takes the form that if player j takes action s_j in period t, then his tendency for s_j is increased by an increment equal to his realized payoff. All other tendencies are unchanged. Let $r_{j,t}$ denotes the random payoff obtained by player j in period t. Thus, we can write the updating rule

$$x_{j,t+1}(s_j) = x_{j,t}(s_j) + r_{j,t+1} \mathbb{1}_{\{a_{j,t+1}=s_j\}} \tag{2.31}$$

This iterative formulation is normalized by dividing by $\frac{C(n+1)}{nC+r_{j,t}}$ for a factor $C > 0$. The reinforcement learning becomes

$$x_{j,t+1}(s_j) = \frac{C(n+1)}{nC + r_{j,t}} \left[x_{j,t}(s_j) + r_{j,t+1} \mathbb{1}_{\{a_{j,t+1}=s_j\}} \right] \tag{2.32}$$

The normalized Arthur model writes:

$$x_{j,t+1}(s_j) = x_{j,t}(s_j) + \lambda_t r_{j,t+1} \left(\mathbb{1}_{\{a_{j,t+1}=s_j\}} - x_{j,t}(s_j) \right), \tag{2.33}$$

$$\forall j, \ \forall s_j \in \mathcal{A}_j, \forall t, \ x_{j,t} \in \Delta(\mathcal{A}_j). \tag{2.34}$$

In order to preserve the simplex of deterministic trajectory, it suffices to have the condition $\lambda_t r_{j,t} \in [0,1]$. We compute the drift (the expected changes in one-step).

$$
\begin{aligned}
F_{j,s_j}(x) &:= \lim_{\lambda_t \to 0} \frac{1}{\lambda_t} \mathbb{E}\left(x_{j,t+1}(s_j) - x_{j,t}(s_j) \mid x_t = x\right) \\
&= \lim_{\lambda_t \to 0} \frac{1}{\lambda_t} \sum_{s'_j \in \mathcal{A}_j} \mathbb{E}\left(x_{j,t+1}(s_j) - x_{j,t}(s_j) \mid \right. \\
&\qquad \left. x_t = x, a_{j,t} = s'_j\right) x_{j,t}(s'_j) \\
&= \sum_{s'_j \neq s_j} x_{j,t}(s'_j) \mathbb{E}\left(-r_{j,t+1} \mid x_t = x, a_{j,t} = s'_j\right) x_{j,t}(s_j) \\
&\qquad + x_{j,t}(s_j) \mathbb{E}\left(r_{j,t+1} \mid x_t = x, a_{j,t} = s_j\right) (1 - x_{j,t}(s_j)) \\
&= x_{j,t}(s_j) \left[\mathbb{E}\left(r_{j,t+1} \mid x_t = x, a_{j,t} = s_j\right) \right. \\
&\qquad \left. + \sum_{s'_j \in \mathcal{A}_j} x_{j,t}(s'_j) \mathbb{E}\left(-r_{j,t+1} \mid x_t = x, a_{j,t} = s'_j\right) \right] \\
&= x_{j,t}(s_j) \left[r_j(e_{s_j}, x_{-j,t}) - \sum_{s'_j \in \mathcal{A}_j} x_{j,t}(s'_j) r_j(e_{s'_j}, x_{-j,t}) \right]
\end{aligned}
$$

The autonomous ordinary differential equation

$$
\dot{x}_{j,t}(s_j) = F_{j,s_j}(x_t),
$$

is a (multitype) replicator equation.

We now use standard techniques from dynamical system viewpoint of stochastic approximation theory [105, 96, 34, 26, 28]. The theory of stochastic approximation largely works by predicting the behavior of stochastic processes by using an ordinary differential equation (ODE) or inclusion (ODI) formed by taking the expected motion of the stochastic process. For example, consider a stochastic process of the form

$$
\mathbf{x}_{t+1} = \mathbf{x}_t + \lambda_t^1 [f(\mathbf{x}_t) + \lambda_t^2(\mathbf{x}_t) + O((\lambda_t^1)^2)], \tag{2.35}
$$

where λ_t^1 is the step size of the process, λ_t^2 is the random component of the process with $\mathbb{E}(\lambda_t^2 \mid \mathbf{x}_t) = 0$.

Remark 2.3.2.4. *Particular case of (2.35) is the model of Arthur (1993) and Borgers & Sarin (1993). The authors in [179] used this stochastic learning procedure and proved the weak convergence to differential equation in finite games. However, the non-Nash rest points have not classified in their work. Their reinforcement learning algorithm is a stochastic approximation version of the replicator dynamics for which is known that, it may not lead to equilibria. See also [74, 140, 190] for limit cycle trajectories of the replicator dynamics.*

Why do we need to develop classes of learning algorithms with pseudo-trajectories in the class of replicator dynamics that does not necessarily converge?

The answer to the above question is not obvious. One of the reasons is that even if the learning scheme does not converge, a nonconvergence result could be itself an interesting result of the behavior of the system and also for the system performance.

Also, the rest points of the replicator dynamics are not necessarily (Nash) equilibria. For example, for finite games, the faces of simplex are invariant under replicator dynamics but the extreme points need not be equilibria.

In the receiver selection game described in Example 2.2.1.6, the vector $(1,0)$ that corresponds to the action R1 is a rest point of the replicator dynamics but it is not equilibrium. Application of reinforcement learning in network security can be found in [197] and in [127] the question of how ignorant but patient cognitive user learn their payoff and optimal strategies is addressed.

The mean or averaged ordinary differential equations (ODEs) derived from the stochastic process above would be a variant of the replicator dynamics

$$\dot{x}_j(s_j) = x_j(s_j) \left[f_{j,s_j}(x) - \sum_{s_j' \in \mathcal{A}_j} x_j(s_j') f_{j,s_j'}(x) \right]$$

which can be written as $\dot{x}_j = G(x_j) r_j(., x_{-j})$, where $G(x_j)$ is a symmetric matrix given by

$$a_j \neq s_j, \ G(x_j)_{a_j s_j} = -x_j(a_j) x_j(s_j),$$

$$G(x_j)_{s_j s_j} = x_j(s_j)(1 - x_j(s_j)).$$

We associate a variant known as the adjusted replicator dynamics that has the following form

$$\dot{x}_{j,s_j} = \frac{F_{j,s_j}(x)}{\sum_{s_j'} x_j(s_j') r_j(e_{s_j'}, x_{-j})} =: K_{j,s_j}(x). \tag{2.36}$$

But despite having ODEs with the same local stability properties, the two processes potentially have different asymptotic behavior. This crucially shows the limits to predicting the behavior of a stochastic process using only the ordinary differential equation. The following result provides a nonconvergence to linearly unstable equilibria with probability zero. See also [27, 179, 178].

Theorem 2.3.2.5 ([76]). *Let x^* be a rest point for the replicator dynamics, or equivalently the adjusted replicator dynamics and $\gamma = 1$. If either*

- x^* *is not a Nash equilibrium,*

- x^* is a Nash equilibrium linearly unstable under the adjusted replicator dynamics, or

- There are two actions and x^* is a Nash equilibrium linearly unstable under the replicator dynamics

then, for the reinforcement learning process defined by the updating strategy (2.30) and updating tendency (2.31), from any initial condition with all tendencies positive, that is, $x_{j,0}(s_j) > 0$ for each player j and all strategies s_j, $P(\lim_t x_t = x^*) = 0$.

2.3.3   Learning Correlated Equilibria

In this section, we present a procedure for converging to the Hannan set and set of correlated equilibria based on works [58, 70] on no-regret dynamics. Compared to the previous fully distributed algorithms, this algorithm will be based on frequencies of play (joint actions).

The notion of *correlated equilibrium* (CE), introduced by Aumann 1974, can be described as follows:

Assume that, before the game is played, each player receives a private signal that does not affect the payoff functions. The player may then choose his action in the game depending on this signal.

A correlated equilibrium of the original game is a Nash equilibrium of the game with the signals.

Considering all possible signal structures generate all correlated equilibria.

If the signals are stochastically independent across the players, it is a Nash equilibrium in mixed or pure strategies of the original game. But the signals could well be correlated, in which case new equilibria can be obtained. Equivalently, a correlated equilibrium is a probability distribution on actions profiles, which can be interpreted as the distribution of play instructions given to the players by some "device" or "referee". Each player is given privately recommendations for his own play only; the joint distribution is known to all of them. Also, for every possible recommendation that a player receives, the player realizes that the instruction provides a best response to the random estimated play of the other players assuming they all follow their recommendations. In [58], Foster and Vohra (1997) have provided a procedure converging to the set of correlated equilibria. The author in [70] gives a simple procedure called *no-regret dynamics* that generates trajectories of play that almost surely converge to the set of correlated equilibria. The result has been extended in [149] to convex and compact set in Hausdorff space (separated space)[1].

Below we present the key element of a *less-regret learning algorithm*.

Definition 2.3.3.1. *Consider a finite game $\mathcal{G}$ in strategic form.*

A probability distribution μ on $\prod_j A_j$ is a correlated equilibrium of if, for every $j \in \mathcal{N}$, every $a_j \in A_j$ and every $s_j \in A_j$ we have

$$\sum_{\mathbf{a}_{-j}} \mu(a) \left[\tilde{R}_j(a_j, \mathbf{a}_{-j}) - \tilde{R}_j(s_j, \mathbf{a}_{-j}) \right] \geq 0.$$

The set of correlated equilibria is convex.

Using the following two properties

- The closure and the interior of a convex set are convex

- The intersection of two convex sets is a convex set

we deduce that the set of correlated equilibria is convex. Note that the empty set is by convention considered to be convex.

A Nash equilibrium is a particular case of correlated equilibrium with independent induced signals.

Note that every Nash equilibrium is a correlated equilibrium. Indeed, Nash equilibria correspond to the special case where μ is a product measure, that is, the play of the different players is independent. Also, the set of correlated equilibria of finite games is nonempty, closed, and convex.

Suppose that the game $\mathcal{G}$ is played many times. Fix an initial distribution $\mathbf{x}_0$ that is chosen arbitrarily at the first time slot. At time $t + 1$, given a history of play $h_t \in (\prod_j A_j)^t$, each player j chooses $a_{j,t+1})$ according to the probability distribution $\mathbf{x}_{j,t+1} \in \mathcal{X}_j$. Given two different actions a_j^* and s_j, define the difference in average up to t as

$$D_j(t+1, a_j^*, s_j) = \frac{1}{t} \sum_{t'=1,\ a_{j,t'}=a_j^*}^{t} \left[\tilde{R}_j(s_j, \mathbf{a}_{-j,t}^*) - \tilde{R}_j(\mathbf{a}_t^*) \right].$$

Define the average regret at period t for not having played, every time that

---

[1]Note that this is different from separable space, which means that it contains a countable dense subset

a_j^* was played in the past, a different strategy s_j as

$$\tilde{R}_j(t+1, a_j^*, s_j) = \max(0, D_j(t+1, a_j^*, s_j)).$$

Fix $\alpha > 0$ to be large enough. Let a_j^* be the last action (of reference) chosen by player j, i.e., $a_j^* = a_{j,t}$. Then the probability distribution $\mathbf{x}_{j,t+1}$ used by j at time $t+1$ is defined by the following:

$$x_{j,t+1}(s_j) := \left\{ \begin{array}{ll} \frac{1}{\alpha}\tilde{R}_j(t+1, a_j^*, s_j) & \forall \, s_j \neq a_j^* \\ 1 - \sum_{\substack{s_j \in \mathcal{A}_j, \\ s_j \neq a_j^*}} x_{j,t+1}(s_j) & \end{array} \right.$$

Note that if α is large enough, this defines a probability distribution. The probabilities of switching to different strategies are proportional to their regrets relative to the current strategy. In particular, if the regrets are small, then the probability of switching from current play is also small.

For every time slot, let $f_t(.)$ be the empirical distribution of the action profiles played up to time t,

$$f_t(\mathbf{a}) = \frac{1}{t} \sum_{t'=1}^{t} \mathbb{1}_{\{\mathbf{a}_{t'}=\mathbf{a}\}}.$$

Theorem 2.3.3.2 ([70]). *If every player plays according to the no-regret dynamics, then the empirical distribution of play $f_t(.) \in \Delta(\prod_j \mathcal{A}_j)$ converges almost surely as t goes to infinity to the set of correlated equilibrium distributions of the one-shot game $\mathcal{G}$.*

Note that this theorem does not tell that the sequence of frequencies $f_t(.)$ converges to a point.

Example 2.3.3.3 (Learning CE in cognitive radio networks; see also Section 7.6). *We consider a cognitive radio network with free spectrum bands. In such a context, the secondary users (unlicensed) have the choice between accessing the licensed spectrum bands owned by primary users (PUs) which is charged as a function of the enjoyed signal-to-interference-plus-noise ratio (SINR) or received power at primary users (PUs), and switching to the unlicensed spectrum bands, which are free of charge but become more crowded when more secondary users (SUs) operate in these spectrum bands. Consequently, the SUs should balance between accessing the free spectrum bands with probably more interference and paying for communication gains by staying with the licensed bands. In this example, we illustrate the case of one free spectrum band available for all SUs and $n'-1$ licensed bands processed by primary users. We denote the set of bands by $\{1,\ldots,n'\}$. A secondary user SU has an additional choice of switching to band n, and the corresponding payoff is $R_j(SINR_j)$. A secondary user j faces the choice of accessing the licensed band with minimum effective price and the free band n'. The primary channel selection game has at least one equilibrium (in pure or mixed strategies). We then address the* *problem of how to reach an equilibrium. We first notice that when using a myopic best response update in the PU selection game, one may have a cycling situation. In fact, during the course of unused primary channel selection, the secondary users may notice that the payoff of accessing a licensed spectrum is higher than staying in the free spectrum, and thus switch to the licensed spectrum accordingly. Since the SUs do this simultaneously, the free spectrum becomes under-loaded and the SUs will switch back to the free spectrum in the next iteration. This phenomenon, in which a player keeps switching between two actions, is known as cycling effect as mentioned above. To eliminate this cycling effect, we use no-regret learning algorithm, which generates empirical distributions that are close to the set of correlated equilibria.*

Exercise 2.5 (Stochastic approximation for CE). *The goal of this exercise is to find the asymptotic pseudo-trajectory of the above no-regret learning scheme by a ordinary differential inclusion (ODI instead of ODE).*

- *Write the no-regret learning in the form of Robbins-Monro inclusion*

- *Use the Blackwell approachability approach*

- *Show that the asymptotic pseudo-trajectory can be written into the form*

$$\dot{y}_t \in \tilde{N}(y_t) - y_t$$

where $\tilde{N}(.)$ is an appropriate set.

- *Conduct a stability analysis by proposing a Lyapunov function for the set of correlated equilibria (CE).*

- *Conclude.*

2.3.4   Boltzmann-Gibbs Learning Algorithms

Boltzmann-Gibbs Distribution

We are particularly interested in Boltzmann-Gibbs-based strategy (called also *softmax*) because its gives a nice approximation of maxima and equilibria in small number of iterations. If the payoff estimations are good enough, then approximated equilibria are obtained. The Boltzmann-Gibbs distribution

$$\mathbb{R}^{|\mathcal{A}_j|} \longrightarrow \Delta(\mathcal{A}_j),$$

$$\tilde{\beta}_{j,\epsilon}(\hat{\mathbf{r}}_{j,t})(s_j) = \frac{e^{\frac{1}{\epsilon_j}\hat{r}_{j,t}(s_j)}}{\sum_{s'_j \in \mathcal{A}_j} e^{\frac{1}{\epsilon_j}\hat{r}_{j,t}(s'_j)}}, \ s_j \in \mathcal{A}_j, \ j \in \mathcal{N}, \qquad (2.37)$$

is known to be close to equilibria for a big parameter $\frac{1}{\epsilon_j}$, which can be interpreted as the rationality level of the player j. The case where the ϵ_j are identically equal to ϵ, has been introduced in [62] and in that context, the factor $\frac{1}{\epsilon}$

is interpreted as a temperature parameter. The Boltzmann-Gibbs distribution is an equilibrium of the *perturbed game* with payoffs: $r_j(\mathbf{x}_j, \mathbf{x}_{-j}) + \epsilon_j H(\mathbf{x}_j)$ where

$$H_j(\mathbf{x}_j) = - \sum_{s_j \in \mathcal{A}_j} x_j(s_j) \log(x_j(s_j)),$$

is the entropy function. The corresponding Nash equilibria of this perturbed game are called *logit equilibria* or *Boltzmann-Gibbs equilibria*.

This rule can be replaced by

$$\frac{\phi_\epsilon(\hat{r}_{j,t}(s_j))}{\sum_{s'_j} \phi_\epsilon(\hat{r}_{j,t}(s'_j))},$$

for a well-chosen ϕ_ϵ.

Boltzmann-Gibbs: the setting

We assume that each player has an estimate of the average payoff of the alternative actions and makes a decision based on this rough information by using a randomized decision rule to revise its strategy. The payoffs of the chosen alternative are then observed and are used to update the strategy for that particular action. Each player experiments only the payoffs of the selected action on that time slot, and uses this information to adjust his or her strategy.

This scheme is repeated several times, generating a discrete time stochastic learning process. Basically, we have three parameters: the current information state, the rule for revision of strategies from the state to randomized actions, and a rule to update the new state. Although players observe only the payoffs of the specific action chosen on the given time slot, the observed values depend on some specific dependency parameters determined by the other players choices revealing implicit information about the system.

> *A natural question is whether such a learning algorithm based on a minimal piece of information may be sufficient to induce coordination and make the system stabilize to an equilibrium.*

The answer to this question is positive for dynamic congestion games, routing games with parallel links, and access point selection games under the *Boltzmann-Gibbs learning scheme*.

Learning in congestion games with vector of payoffs

We consider a finite set of players and finite set of common action space. Denote by $r_{\mathbf{a}}^w(k, t)$ the average $w-$weighted payoff for the action a when k

players have chosen this action at time t. The more an action is congested the lower weighted payoff is, i.e.,

$$\forall t, \ r_a^w(k,t) \geq r_a^w(k+1,t), \ k \leq n,$$

where n is the total number of players. An estimation payoff of player j is a vector $\hat{\mathbf{r}}_j = (\hat{r}_{j,a}^{i'})_{i',\mathbf{a}}$, where $\hat{\mathbf{r}}_j^{i'}$ represents player j's estimate of the average payoff of the action a for the objective i'. Player j can then compute the weighted average estimated payoff $\hat{\mathbf{r}}_j^w = (\hat{r}_{j,a}^w)_a$. Player j can then update its strategy by using the Boltzmann-Gibbs scheme: use the action a with probability $x_{j,a}(\hat{r}) := \dfrac{e^{\frac{1}{\epsilon_j}\hat{r}_{j,a}^w}}{\sum_{a'} e^{\frac{1}{\epsilon_j}\hat{r}_{j,a'}^w}}$, where ϵ_j is a positive parameter. Let $\lambda_t \geq 0$ be a step size satisfying

$$\sum_t \lambda_t = +\infty, \ \sum_t \lambda_t^2 < \infty.$$

We now describe the learning algorithm:

Algorithm 1: Stochastic Learning Algorithm Based on Boltzmann-Gibbs Dynamics

---

forall the *Players* **do**
  Initial Boltzmann distribution $\mathbf{x}_{j,0}$;
  Initialize to some estimations $\hat{r}_{j,a}^{i'}(k_0,0)$;

for *t=1* **to** *max* **do** ;
foreach *Player j* **do**
  Observe its current payoff;
  Update via Boltzmann-Gibbs dynamics $\mathbf{x}_{j,t+1}$;
  Compute the distribution over $a_{j,t+1}$ and $k_{a_{j,t+1}}$ from $\mathbf{x}_{t+1}$;
  Update its estimation via $\hat{r}_{j,a,t+1} = \hat{r}_{j,a,t} + \nu_{t+1}\left(W_{j,a,t+1} - \hat{r}_{j,a,t}\right)$;
  Estimate the random payoffs $\hat{r}_{j,a_{j,t+1}}(k_{a_{j,t+1}}, t+1)$;

The learning process may then be described as follows. At stage $t+1$ the past estimation $\hat{r}_{j,a}^w(k_{a,t},t)$ determines the transition probabilities $\mathbf{x}_{j,t}(a) = x_{j,a}(\hat{r}_{j,a}^w(k_{a,t},t))$, which are used by player j to experiment a random action $a_{j,t+1}$. The action profile of all the players determines a total random number $k_{a,t+1}$ of players j such that $a_{j,t+1} = a$. The weighted payoffs of an a are then $\hat{r}_{j,t+1}^w(k_{a_{j,t+1},t+1}) = r_{j,a}^w(k,t+1)$ if $a_{j,t+1} = a, k_{a,t+1} = k$. Finally, each player j observes only the payoff of the chosen alternative $a_{j,t+1}$ and updates his, or her estimations by averaging

$$\hat{r}_{j,a,t+1}^{i'} = (1 - \nu_{t+1})\hat{r}_{j,a,t}^{i'} + \nu_{t+1}r_{j,a,t+1}^{i'} \text{ if } a_{j,t+1} = a.$$

Otherwise the estimation is unchanged: $\hat{r}_{j,a,t+1}^{i'} = \hat{r}_{j,a,t}^{i'}$. Define

$$W_{j,a,t+1} = \begin{cases} r_{j,a,t+1} & \textit{if } a_{j,t+1} = a \\ \hat{r}_{j,a,t} & \text{otherwise} \end{cases}$$

The learning algorithm can be rewritten as

$$\hat{r}_{t+1} - \hat{r}_t = \nu_{t+1} \left(W_{t+1} - \hat{r}_t \right).$$

This process has the form of a stochastic learning algorithm with the distribution of the random vector W_t being determined by the individual updating rules that depend upon the estimations. Assuming that the payoff functions are bounded, the sequences generated by the learning algorithm is also bounded. Hence, the asymptotic behavior of our learning algorithm can be studied by analyzing the continuous adaptive dynamics of the drift $\mathbb{E}(W_{t+1} \mid \hat{r}_t)$. It is easy to see that the expectation of W given $\hat{r}$ is

$$\mathbb{E}(W_{j,a} \mid \hat{r}) = \mathbf{x}_{j,a}(\hat{r})\bar{r}_{j,a} + (1 - \mathbf{x}_{j,a}(\hat{r}))\hat{r}_{j,a}.$$

Hence, the expected change in one time slot can be written as

$$\mathbb{E}\left(W_{j,a} - \hat{r}_{j,a} \right) = \mathbf{x}_{j,a}(\hat{r})\bar{r}_{j,a} + (1 - \mathbf{x}_{j,a}(\hat{r}))\hat{r}_{j,a} - \hat{r}_{j,a}$$

$$= \mathbf{x}_{j,a}(\hat{r})[\bar{r}_{j,a} - \hat{r}_{j,a}].$$

We deduce the following lemma:

Lemma 2.3.4.1. *The stochastic learning algorithm generates the continuous time dynamics given by*

$$\frac{d}{dt}\hat{r}_{j,a,t} = \mathbf{x}_{j,a,t}(\hat{r}) \left(\bar{r}_{j,a}(t) - \hat{r}_{j,a,t} \right) \tag{2.38}$$

where

$$\bar{r}_{j,a}(t) = \mathbb{E}\left(r_{j,a}(\sum_{j'} Ber^{j'}_{a,t}, t) \mid Ber^{j}_{a,t} = 1 \right),$$

represents the average payoff observed by player j when he or she chooses the action a and the other players choose it with probabilities $\mathbf{x}_{j',a}$. The random variable $Ber^{j}_{a,t}$ denotes a Bernoulli random variable with the parameter $\mathbb{P}(Ber^{j}_{a,t} = 1) = \mathbf{x}_{j,a}$.

Proposition 2.3.4.2. *The Boltzmann-Gibbs-based stochastic learning algorithm converges almost surely to Nash equilibria.*

The proof of Proposition 2.3.4.2 follows the same line as in [28, 26] or Theorem 8 in [47]. Using Proposition 4.1 and 4.2 of [26] with the limit set Theorem 5.7 it follows that, the symmetric Nash equilibrium is a global attractor and the function $y \longmapsto \| y - x \|_\infty$ is a Lyapunov function for the dynamics (2.38).

Transition from Micro to Macro in Congestion Games

Define the scaled payoff functions as $r_a^{n,i'}\left(\frac{k_a}{n}, t\right)$ and the mean field profile

$$X_t^n(a) = \frac{1}{n} \sum_{j=1}^{n} \mathbb{1}_{\{a_{j,t}=a\}}.$$

Assuming that the second moment of the number of players that use the same action is finite[2], and the mean process converges weakly to a mean field limit[163], and the solution of a system of ordinary differential equations given by the drift $\frac{1}{\Delta_n} f^n(x(t))$ where

$$f^n(x(t)) = \mathbb{E}\left(X^n(t+\Delta_n) - X^n(t) \mid X^n(t) = x(t)\right),$$

is the expected change in the system in one-time slot with duration Δ_n. For example the Boltzmann-Gibbs dynamics (also called logit dynamics or smooth best response dynamics) is given by

$$\dot{x}_{j,a}(t) = \sum_{a'} x_{j,a'}(t) \rho_{a'a}^j(x(t), t) - x_{j,a}(t),$$

where

$$\rho_{a'a}^j(x(t)) = \frac{e^{\frac{r_{j,a}(x(t),t)}{\epsilon_j}}}{\sum_{\bar{a}} e^{\frac{r_{j,\bar{a}}(x(t),t)}{\epsilon_j}}}.$$

Players from class j can learn $x(t)$ via the ordinary differential equation (ODE) and can use an action a with probability $\mathbf{x}_{j,a}$. Note that our study can be extended to the case where one has both atomic and nonatomic players by considering the weighted population profile:

$$\tilde{X}^n(t)_{(j,a)} = \frac{1}{\sum_j \gamma_j^n} \sum_j \gamma_j^n \delta_{a_{j,t}=a},$$

where γ_j^n is the weight ("power") of j in the population of size n.

2.3.5 Hybrid Learning Scheme

In this section we present how to combine various learning schemes based on mean field game dynamics ([92, 164, 153, 140, 89], [176]). Different learning and adaptive algorithms have been studied in the literature [191, 106, 147]. In all these references, the players have to follow the same rule of learning, they have to learn in the same way. We now ask the following question:

> What happens if the players have different learning schemes?

---

[2] Notice that the number of players that can interact can be very large.

Our aim here is to propose a class of learning schemes in which players use less information about the other players, less memory on the history, and need not use the same learning scheme. The players will be able to change learning scheme during the interaction leading to the evolutionary game dynamics with migration [172, 153, 170]. Discrete-time evolutionary game dynamics are provided in Section 4.12.

Below we list some mean field limits that we borrow from evolutionary game dynamics [196, 195]. They are in the form

$$\dot{x}_a^\theta(t) = \sum_{\bar{a}} x_{\bar{a}}^\theta(t) \rho_{\bar{a},a}^\theta(x(t)) - x_a^\theta(t) \sum_{\bar{a}} \rho_{a,\bar{a}}^\theta(x(t)),$$

where θ denotes the class of users and ρ is the migration rate between strategies.

- *Excess payoff dynamics:* Brown-von Neumann-Nash dynamics is one of well-known excess payoff dynamics. *Brown-von Neumann-Nash dynamics* (BNN) is obtained for

$$\rho_{a',a}^{1,\theta}(x(t)) = \max(0, r_a^\theta(x(t)) - \sum_{\bar{a}} x_{\bar{a}}^\theta(t) r_{\bar{a}}^\theta(x(t)).$$

The set of rest points of the BNN dynamics is exactly the set of (Nash) equilibria.

- *Imitation of neighbors:* The *replicator dynamics* is obtained for

$$\rho_{a',a}^{2,\theta}(x(t)) = x_a^\theta(t) \max(0, r_a^\theta(x(t)) - r_{a'}^\theta(x(t))).$$

The set of rest points of the replicator dynamics contains the set of equilibrium states but it can be much bigger since it is known that the replicator dynamics may not lead to equilibria. Typically the corners are rest points and the faces of the simplex are invariant but may not be an equilibrium.

- *Boltzman-Gibbs dynamics* also called *smooth best response* dynamics or logit dynamics is obtained for

$$\rho_{a',a}^{4,\theta}(x(t)) = \frac{e^{\frac{r_a^\theta(x(t))}{\epsilon}}}{\sum_{\bar{a}} e^{\frac{r_{\bar{a}}^\theta(x(t))}{\epsilon}}}, \quad \epsilon > 0.$$

Using the logit map, it can be shown that the time average of the replicator dynamics is a perturbed solution of the best reply dynamics. The rest points of the Smith dynamics are approximated equilibria. More details on logit learning can be found in [60].

- *Pairwise comparison dynamics:* For example, the generalized Smith dynamics is obtained for

$$\rho_{a',a}^{5,\theta} = \max(0, r_a^\theta(x(t)) - r_{a'}^\theta(x(t)))^\gamma, \quad \gamma \geq 1.$$

The set of rest points of the generalized Smith dynamics is exactly the set of equilibria. Sandholm [140] proved that under this class of dynamics, global convergence holds in stable games and in potential games etc. Extension to evolutionary game dynamics with *migration between location of players* and application to hybrid power control in wireless communications can be found in [172].

- *Best response dynamics* is obtained for ρ^6 equal to the best reply to $x(t)$. The set of rest points of the best response dynamics is exactly the set of equilibria. More details on best response dynamics can be found in [63].

- *Ray-projection dynamics* is a myopic adaptive dynamic in which a subpopulation grows when its expected payoff is less than the ray-projection payoff of all the other classes. It is obtained $\rho^{3,\theta}_{a',a}(x(t)) = \Lambda^\theta_a(x(t))$ where Λ is a continuous map.

Notice that the replicator dynamics, best response dynamics and logit dynamics can be obtained as a particular case of the ray-projection dynamics.

Consider a population in which the players can adopt different learning schemes in $\{\rho^1, \rho^2, \rho^3, \ldots\}$ (finite, say with size κ). Then, based on the composition of population and the use of each learning scheme we build an hybrid game dynamics. The intra-incoming and the intra-outgoing flow as well as the inter-neighborhood flow are expressed in term of the weighted combination of different learning schemes picked from the set $\{\rho^1, \rho^2, \rho^3, \ldots\}$.

Define the property of *weighted equilibrium stationarity* (WES,[153, 160]) as follows:

(WES)   *Every rest point of the hybrid game dynamics generated by the weighted payoff is a weighted equilibrium and every constrained weighted equilibrium is a rest point of the dynamics.*

Note that this property is not satisfied by the well-known replicator dynamics as it is known that the replicator dynamics may not lead to Nash equilibria. We have the following result:

Proposition 2.3.5.1. *Let $\tilde{\lambda}^j$ be the proportion the players that adopt the learning scheme ρ^j. If all the learning schemes contained in the support of $\tilde{\lambda} = (\tilde{\lambda}^1, \ldots, \tilde{\lambda}^\kappa) \in \mathbb{R}^\kappa_+$ satisfy the property (WES), then the hybrid mean field limit game dynamics generated by these learning schemes satisfies also the weighted equilibrium stationarity property.*

Proof. If x is an equilibrium, then x is a rest point of all the dynamics in the support of $\tilde{\lambda}$. We now that prove that any rest point of the combined dynamics in the support of $\tilde{\lambda}$ is an equilibrium. Suppose that it is not the case. Then, there exists at least one j such that x is not rest point of the dynamics generated by the learning scheme ρ^j which satisfies (WES). This means that x is not an equilibrium. We conclude that any rest point of combined dynamics is an equilibrium. This completes the proof. □

Remark 2.3.5.2 (How to eliminate the rest points which are not equilibria?). *Consider the family of learning schemes generated by the revision protocol:*

$$\rho_{a',a}^{\gamma,\theta} = \max(0, r_a^\theta(x(t)) - r_{a'}^\theta(x(t)))^\gamma, \ \gamma \geq 1.$$

It is easy to see that this family of learning pattern satisfies the WES property. We deduce that if t 99% of the players use a learning scheme via ρ^γ and 1% of the population use a replicator-based learning scheme then the resulting combined dynamics satisfies the WES property. We conclude that every rest point of the replicator dynamics which is a nonNash equilibrium will be eliminated using this new combined dynamics. This says that players can learn in a bad way but if the fraction of good learners is nonzero, then the rest points of resulting combined dynamics will be equilibria.

2.3.6   Fast Convergence of Evolutionary Dynamics

Fix a population game with size n and an hybrid evolutionary dynamics. Let $X_t^n(a)$ be the proportion of players with action $a \in \mathcal{A}$ at time t. Let $E^* \subseteq \mathcal{A}$ be the set of Nash equilibria of the population game. Let

$$\mathcal{N}_{E^*,\epsilon} = \{x \mid d(x, E^*) < \epsilon\}$$

where $d(x, E^*)$ denotes $\inf_{e^* \in E^*} d(x, e^*)$.

Let $\bar{f}_{t,\epsilon}^n = \frac{1}{t} \int_0^t \mathbb{1}_{\{X_s^n \in \mathcal{N}_{E^*,\epsilon}\}} \, ds$

This is the proportion of time that the process spends within ϵ of one or more Nash equilibria up to time t.

Given ϵ, we say that the mean field process X_s^n is close to x^* with $1 - \epsilon$ proportion of time if for all initial conditions x_0,

$$\mathbb{P}\left(\liminf_{t \to +\infty} \bar{f}_{t,\epsilon}^n \geq 1 - \epsilon\right) \geq 1 - \epsilon.$$

Fix a small $\epsilon > 0$. If for a finite T, the frequency

$$\bar{f}_{t,T,\epsilon}^n = \frac{1}{T} \int_t^{T+t} \mathbb{1}_{\{X_s^n \in \mathcal{N}_{E^*,\epsilon}\}} \, ds$$

satisfies

$$\mathbb{P}\left(\bar{f}_{t,T,\epsilon}^n \geq 1 - \epsilon\right) \geq 1 - \epsilon,$$

such that for all sufficiently large n, all initial conditions x_0 and t, then we refer to as fast convergence.

Generically, (finite) potential population games and stable population games provide very fast convergence within a range ϵ under various hybrid evolutionary game dynamics.

2.3.7   Convergence in Finite Number of Steps

In this section we briefly discuss discrete time mean field game dynamics that require only a few number of iterations to be close to their stationary points. The difference equation that we provide below converges in a finite number of iterations in the class of robust population games studied in Chapters 1 and 9. Note that this property does not hold for the replicator equation, where the convergence time is unbounded if not started from a rest point.

Consider a large population of players. Each member of the population can choose an action in the set $\mathcal{A}$, which we assume to be finite (and nonempty). Let $\mathcal{X} = \Delta(\mathcal{A})$ be the $(\mid \mathcal{A} \mid -1)$-dimensional simplex of $\mathbb{R}^{\mid \mathcal{A} \mid}$. Let $\mathcal{W}$ be the state space. The payoff function of a generic player with the action a is

$$r_a : \mathcal{X} \longrightarrow \mathbb{R},$$

defined by

$$r_a(x) = \mathbb{E}_{\mathbf{w}} \left(\tilde{R}_a(\mathbf{w}, x) \right), \ a \in \mathcal{A}.$$

The collection $(\mathcal{W}, \mathcal{A}, \tilde{R}(.,.))$ defines a robust population game. Let $\{x_t\}$ be the sequence of population states generated by

$$x_{t+1} = x_t + \bar{\epsilon}_{H(x_t)}[S(x_t) - x_t],$$

where

- $H(x_t) = \begin{cases} y_t \in H_2(x_t) & \text{if } H_2(x_t) \neq \emptyset \\ x_t & \text{if } H_2(x_t) = \emptyset \end{cases}$ where

$$H_2(x_t) = \{z \in \mathcal{X} \mid \langle z - x, r(x_t) \rangle > 0\}$$

-

$$\bar{\epsilon}_y(x) = \inf\{\epsilon \in (0,1) \mid \langle y - x, r((1 - \epsilon)x + \epsilon y) \rangle \leq 0\}$$

- $\bar{\epsilon}_{H(x_t)} \in [0,1]$ denotes $\bar{\epsilon}_y(x_t)$ for the selected strategy $y = H(x_t)$.

For a pure strategy selection process, one can replace H by

$$H_{boundary}(x_t)$$

$$= \begin{cases} e_a & \text{if } r_a(x_t) > \langle x_t, r(x_t) \rangle \\ \left(1 + \frac{x_t(a)}{1 - x_t(a)}\right) x_t - \left(\frac{x_t(a)}{1 - x_t(a)}\right) e_a & \text{if } x_t(a) > 0 \text{ and } r_a(x_t) < \langle x_t, r(x_t) \rangle \\ x_t & \text{otherwise} \end{cases}$$

This selection process converges in a finite number of steps in the robust anticoordination games studied in Chapters 1 and 9.

2.3.8 Convergence Time of Boltzmann-Gibbs Learning

Most learning algorithms proposed in the literature fail to converge in general. Even when convergence occurs in some specific class of games, the speed of convergence can be arbitrary slow. In this subsection, we present the convergence time of the Boltzmann-Gibbs learning algorithm for potential games with bounded rationality. We study novel Boltzmann-Gibbs learning algorithms by adjusting the way to update a strategy and prove that the convergence to stationary distributions can be reduced from exponential time to linear time in the number of players.

Information Theoretical Interpretation

The information theoretical approach takes into account the *cost of computation* of the optimal strategy distribution. Information theory provides a principled way to modify strategic form games to accommodate bounded rationality. This is done by following information theory's description that, given only partial knowledge concerning the distributions the players are using, we should use the minimum information principle to infer those distributions. Doing so results in the principle that the outcome under bounded rationality is the minimizer of a certain set of coupled Lagrangian functions of the joint distribution.

In the standard Boltzmann-Gibbs learning rule or logit rule, each player j chooses its action s_j with probability

$$\tilde{\beta}_{j,\epsilon}(r)(s_j) = \frac{e^{\frac{1}{\epsilon_j} r_j(\mathbf{e}_{s_j}, x_{-j})}}{\sum_{s'_j} e^{\frac{1}{\epsilon_j} r_j(\mathbf{e}_{s'_j}, x_{-j})}},$$

where the $\{\epsilon_j\}_j$ are some nonnegative constants. As $\bar{\epsilon} = \max_j \epsilon_j$ becomes smaller, the player chooses strategy with the highest payoff with higher probability. The parameters $\{\frac{1}{\epsilon_j}\}_j$ can then be interpreted as the level of rationality of the players. Below we omit the dependency in r and $\tilde{\beta}_{j,\epsilon}(r)(s_j)$ will be simply denoted by $\tilde{\beta}_{j,\epsilon}(s_j)$.

Lemma 2.3.8.1. *The distribution $\tilde{\beta}_{j,\epsilon}$ is the maximizer of the modified game with payoffs $\tilde{v}_j(x) = r_j(x) + \epsilon_j H(x)$ where H is the entropy function.*

Lemma 2.3.8.2. *For exact potential games with potential function V^n, one has*

$$\tilde{\beta}_{j,\epsilon}(a_j) = \frac{e^{\frac{1}{\epsilon_j} r_j(\mathbf{e}_{a_j}, x_{-j})}}{\sum_{s_j} e^{\frac{1}{\epsilon_j} r_j(\mathbf{e}_{s_j}, x_{-j})}} \tag{2.39}$$

$$= \frac{e^{\frac{1}{\epsilon_j} V^n(\mathbf{e}_{a_j}, x_{-j})}}{\sum_{s_j} e^{\frac{1}{\epsilon_j} V^n(\mathbf{e}_{s_j}, x_{-j})}}, \tag{2.40}$$

Proof.

$$\tilde{\beta}_{j,\epsilon}(s_j) = \frac{e^{\frac{1}{\epsilon_j} r_j(\mathbf{e}_{s_j}, x_{-j})}}{\sum_{s'_j} e^{\frac{1}{\epsilon_j} r_j(\mathbf{e}_{s'_j}, x_{-j})}} = \frac{e^{\frac{1}{\epsilon_j}(r_j(\mathbf{e}_{s_j}, x_{-j}) - r_j(\mathbf{e}_{a_j^*}, x_{-j}))}}{\sum_{s'_j} e^{\frac{1}{\epsilon_j}[r_j(\mathbf{e}_{s'_j}, x_{-j}) - r_j(\mathbf{e}_{a_j^*}, x_{-j})]}}$$

$$= \frac{e^{\frac{1}{\epsilon_j}[V^n(\mathbf{e}_{s_j}, x_{-j}) - V^n(\mathbf{e}_{a_j^*}, x_{-j})]}}{\sum_{s'_j} e^{\frac{1}{\epsilon_j}[V^n(\mathbf{e}_{s'_j}, x_{-j}) - V^n(a_j^*, x_{-j})]}} = \frac{e^{\frac{1}{\epsilon_j} V^n(\mathbf{e}_{s_j}, x_{-j})}}{\sum_{s'_j} e^{\frac{1}{\epsilon_j} V^n(\mathbf{e}_{s'_j}, x_{-j})}},$$

$\square$

To establish the results on the speed of convergence of Boltzmann-Gibbs learning algorithms, we will need some reasonable bounds on the time it takes for the process to reach close to its stationary distribution, and then we will quantify the distance between the stationary distribution and an ϵ−equilibrium. We first need some definitions of stationary distributions of Markov processes.

The full support of the probability of choosing an action given the Boltzmann-Gibbs rule ensures that at each updating time, every strategy in $\mathcal{A}$ has a positive probability of being chosen by the revising player. Therefore, there is a positive probability that the Markov chain X_t^n will transit from any given current state $\mathbf{x}$ to any other given state y within a finite number of steps. A Markov process with this property is said to be *irreducible*. The constant jump rate Markov chain X_t^n is said to be *reversible* if it admits a reversible distribution: a probability distribution $\bar{\omega}^n$ on S satisfies the balance conditions:

$$\bar{\omega}_x^n P_{xy}^n = \bar{\omega}_y^n P_{yx}^n, \quad \forall (x,y) \in S^n \times S^n. \tag{2.41}$$

By taking the sum of this equation over $x \in S^n$, we obtain that a reversible distribution is also a stationary distribution.

The continuous-time Markov process over a finite state space S^n can be characterized using a discrete-time Markov chain with transition matrix P^n. The matrix $H_t^n = e^{t(P^n - I)}$ is the transition matrix of the continuous process. I is the identity matrix of size $|S^n|$.

Next, we provide conditions under which the time t distributions of a Markov chain or process converge to its stationary distribution i.e $H_t^{n,x} \longrightarrow$

$\bar{\omega}^n$. In the continuous-time setting, irreducibility is sufficient for convergence in distribution; in the discrete-time setting, aperiodicity is also required.

Lemma 2.3.8.3. *Suppose that X_t^n is either an irreducible aperiodic Markov chain or an irreducible Markov process, and that its stationary distribution is $\bar{\omega}^n$. Then for any initial distribution μ_0, we have that*

$$\lim_{t \longrightarrow \infty} \mathbb{P}_{\mu_0}(X_t^n = x) = \bar{\omega}_x^n,$$

$$\lim_{t \longrightarrow \infty} H_t^{n,x} = \bar{\omega}^n.$$

We use logarithmic Sobolev inequalities for finite Markov chains established by Diaconis and Saloff-Coste (1995). The inequality provides rate bounds for the convergence of Markov chains on finite state spaces to their stationary distributions. Suppose that players in the game G have the same action set $\mathcal{A}_j = \mathcal{A}$ and each player's payoffs are defined by the same function of the player's own action and the overall distribution of strategies $\xi(a) \in S^n$, defined by

$$\xi_b(a_1, \ldots, a_n) = \frac{1}{n} \sum_{j=1}^{n} \mathbb{1}_{\{a_j = b\}}.$$

There is a population game

$$F^n : S^n \longrightarrow \mathbb{R}^n,$$

that represents G in the sense that $R_j(a) = F_{j,a_j}^n(\xi(a))$. Suppose that $\mathcal{G}$ is a normal form potential game with same action set and with potential function V^n. If $\xi(a) = \xi(a')$, then V^n satisfies $V^n(a) = V^n(a')$. Furthermore, the population game defined by F^n is a potential game, and that its potential function is given by $W^n(x) = V^n(a)$, where

$$a \in \xi^{-1}(x) = \{a', \ \xi(a') = x\}.$$

The Boltzmann-Gibbs learning rule induces an irreducible and reversible Markov chain X_t on the finite state space

$$S^n = \{(\frac{x_1}{n}, \ldots, \frac{x_m}{n}) \mid x_i \in \mathbb{Z}_+, \ \sum_a x_a = n\}.$$

The cardinality of S^n is $\binom{n+m-1}{m-1}$. If the players use a Boltzmann-Gibbs learning rule, the Markov process X_t^n is reversible. The stationary distribution weight $\bar{\omega}_x^n$ is proportional to the product of two terms. The first term is a multinomial coefficient, and represents the number of ways of assigning the n agents to strategies in $\mathcal{A}$ so that each strategy a is chosen by precisely nx_a players. The second term is an exponential function of the value of the potential at state x. Thus, the value of $\bar{\omega}_x^n$ balances the value of the potential at state x with the likelihood that profile x would arise were players assigned to strategies at random.

Proposition 2.3.8.4. *Let F^n be a finite potential game with potential function V^n, and suppose that the players and follow a Boltzmann-Gibbs learning rule with the same rationality level $\frac{1}{\epsilon}$ and revise their strategies by taking into the strategy without their own-effect i.e x is replaced by $\frac{nx - \mathbf{e}_a}{n-1}$; then the Markov process X_t^n is reversible with stationary distribution*

$$\bar{\omega}_x^n = \frac{1}{K^n} \binom{n}{nx_1, nx_2, \ldots, nx_m} e^{\frac{1}{\epsilon} V^n(x)},$$

for $x \in S^n$, the number K^n is chosen such that $\sum_{x \in S^n} \bar{\omega}_x^n = 1$.

For the proof, we need to verify the reversibility condition, it is enough to check that the equality

$$\bar{\omega}_x^n P_{xy}^n = \bar{\omega}_y^n P_{yx}^n, \quad \forall (x,y) \in S^n \times S^n.$$

holds for pairs of states x, y such that $y = x + \frac{1}{n}(\mathbf{e}_b - \mathbf{e}_a)$,

$$\bar{\omega}_x^n P_{xy}^n = \bar{\omega}_y^n P_{yx}^n.$$

Proof. For pairs of states x, y such that $y = x + \frac{1}{n}(\mathbf{e}_b - \mathbf{e}_a)$, denote by $z = y - \frac{1}{n}\mathbf{e}_a = x - \frac{1}{n}\mathbf{e}_a$. z represents both the distribution of action of the other $a-$player when the Markov state is x, and the distribution of opponents of a $b-$player at state y. Thus, in both cases, a player who is revising strategy its by omitting it-self will consider the payoff vector $\bar{F}_a^n(z) = F_a^n(z + \frac{1}{n}\mathbf{e}_a)$ where z satisfies $\sum_a z_a = \frac{n-1}{n}$. Thus,

$$V^n(x) - V^n(y) = F_a^n(x) - F_b^n(x) = F_a^n(z + \frac{1}{n}\mathbf{e}_a) - F_b^n(z + \frac{1}{n}\mathbf{e}_b).$$

Using the above relations, we verify that $\bar{\omega}_x^n P_{xy}^n = \bar{\omega}_y^n P_{yx}^n$. $\qquad\square$

Using Boltzmann-Gibbs learning combined with imitation dynamics with at least one player on each action, [140] shows that the stationary distribution is of order

$$\frac{e^{V^n(x)/\epsilon}}{\sum_{x \in S^n} e^{V^n(x)/\epsilon}}.$$

Proposition 2.3.8.5. *The convergence time to an $\eta-$equilibrium under Boltzmann-Gibbs learning is in order of $n \log \log(n) + \log(\frac{1}{\eta})$ if the rationality level is sufficiently large.*

Proof. The proof uses logarithmic Sobolev inequalities for Markov chains on finite spaces which show rates of convergence of Markov chains on finite state spaces to their stationary distributions and we combine this result with the fact that the stationary distribution is an $\epsilon-$best response. Following Diaconis and Saloff-Coste (1996), the total variation of

$$\| H_t^n - \bar{\omega}^n \|_{TV} = \sum_{A \subseteq S^n} |H_t^n(A) - \bar{\omega}^n(A)| = \frac{1}{2} \sum_{x \in S^n} |H_t^n(x) - \bar{\omega}_x^n|$$

$$\| H_t^n - \bar{\omega}^n \|_{TV} \leq e^{-c},$$

for $t \geq \frac{1}{\lambda}(\log(\log(\frac{1}{\min_x \bar{\omega}_x^n})) + c)$ where $1 - \lambda$ is the largest eigenvalue of the operator $\frac{P^n + {}^t(P^n)}{2}$ which is greater than $\frac{c_1}{n} e^{-\frac{1}{\epsilon}}$. The stationary distribution $\bar{\omega}^n$ generates an $\frac{\eta}{2}$-equilibrium for ϵ sufficiently small. It suffices to take

$$\epsilon \leq \frac{\eta}{2m \log m (M+1)}, \quad M = \sup_x V(x).$$

This completes the proof.

□

Example 2.3.8.6. *An n-player congestion game is a game in which each player's strategy consists of a set of resources, and the cost of the set of strategy, say, depends only on the number of players using each resource, i.e., the cost takes the form $c_e(\xi_e(a))$, where $\xi_e(a)$ is the number of players using resource e and c_e is a nonnegative increasing function. A standard example is a network congestion game on a directed graph (e.g. road or transportation network [21]), in which each player must select a path from some source to some destination, and each edge has an associated delay function that increases with the number of players using the edge. It is well known that this game admits potential function,*

$$V^n(\xi) = -\sum_e \int_0^{\xi_e(a)} c_e(y) \, dy.$$

Applying the Boltzmann-Gibbs learning algorithms, one can use the result 2.3.8.5. The convergence time to be η-close to potential maximizer is in order of $n \log \log n + \log(\frac{1}{\eta})$ for any initial condition for level of rationality sufficiently large.

2.3.9 Learning Satisfactory Solutions

In this section, we provide a fast and fully distributed learning algorithm for constrained games. The algorithm does not require knowledge of past actions used by other players. It is completely based on the perceived payoffs. We show that the algorithm converges to a satisfactory solution and explicitly give the convergence time for a specific class of payoff functions.

We consider l resources. The set of actions is denoted by $\mathcal{A} = \{1, 2, \ldots, l\}$. Each resource j is associated to a payoff $f_j^n : \mathbb{R}^l \longrightarrow \mathbb{R}$. We consider the finite number of players as well as the asymptotic case (the mean field limit where the number of players tends to infinity). The fraction of users assigned to resource j is denoted by x_j^n. The vector $\mathbf{x}^n = (x_j^n)_j$ satisfies $\forall j, x_j^n \geq 0$, $\sum_j x_j^n = 1$. The state of the system at time t is a vector $\mathbf{x}^n(t) = (x_j^n(t))_j$. The induced payoff by the players on resource j is $f_j^n(\mathbf{x}^n)$. For the asymptotic case we will examine the case where $f_j(\mathbf{x}) = f_j(x_j)$, i.e., the payoff at resource

j depends only on the load x_j. We assume the existence of a minimum payoff requirement of each user. Each user is satisfied if his or her payoff is at least U. If we refer to the throughput, this gives a minimum throughput requirement to each user: $f_j(\mathbf{x}) \geq U,\ \forall j$. This gives a constraint satisfaction problem. A system state x that satisfies $\forall j \in \mathcal{A},\ f_j(\mathbf{x}) \geq U$ is a *satisfactory solution*. Since the throughput decreases with the load, there is a threshold x_j^* such that if $f_j(x) < U$ then $x_j > x_j^*$ (the QoS constraint is violated). The maximum load of channel j such that the constraint is satisfied is then

$$M_j = \sup\{x_j,\ f_j(x_j) \geq U\},$$

which is typically $f_j^{-1}(U)$ for bijective payoffs. We now present the fully distributed learning algorithm for satisfaction solution.

Algorithm 2: Distributed learning algorithm

---

forall the *users* **do**
  | Initialize to some assignment;

for t=1 **to** *max* **do** do;
foreach *user* i **do**
  | Denote by $j_t = j$ the resource where user i is connected to ;
  | Observe the payoff $f_j(\mathbf{x})$;
  | If $f_j(\mathbf{x}) < U$ then with probability $(U - f_j)/Af_j$ choose randomly
  | j', and set $j_{t+1} = j'$;

Assume that the constraint set

$$\{\mathbf{x} \mid \forall j,\ f_j(\mathbf{x}) \geq U,\ x_j \geq 0,\ \sum_j x_j = 1\},$$

is nonempty. We show that the learning algorithm converges to a system state where every user has at least the payoff U in the long run. To that end, we introduce a Lyapunov function that is zero only at a satisfactory solution and show that the Lyapunov function decreases towards zero. Consider the mapping $L(\mathbf{x}) = \sum_{j \in \mathcal{A}} \max(0, x_j - M_j)$. The following holds:

- $L(\mathbf{x}) \geq 0$

- $[L(\mathbf{x}) = 0,\ \mathbf{x} \in \Delta(\mathcal{A})] \Longleftrightarrow \mathbf{x}$ is a satisfactory solution.

- Denote by SSS be the set of satisfactory solution. If $\mathbf{x}_t$ is the current system (not in SSS) and $\mathbf{x}_{t+1}$ denotes the next state generated by the learning algorithm, then $L(\mathbf{x}_{t+1}) - L(\mathbf{x}_t) < 0$.

We determine explicitly the convergence of the learning algorithm for any initial condition. We first define an approximate η−satisfactory solution: Let

$$B_{\eta'}(\mathbf{x}) = \{j,\ f_j(\mathbf{x}) < (1 - \eta')U\},$$

The system x is η−satisfactory solution for η'−sufficiently small, $\sum_{j \in B_{\eta'}(\mathbf{x})} x_j \leq \eta$.

Proposition 2.3.9.1. *For any initial system state* $\mathbf{x}_0$*, the time to reach an approximate* $\eta-$*satisfactory solution is on the order of* $\frac{|A|}{\eta^2}$*.*

Proof. Consider a state x that is not an $\epsilon-$satisfactory solution, i.e., $\sum_{j \in B_\epsilon(\mathbf{x})} x_j \geq \epsilon$.

It is sufficient to establish the result for the channel j such that $f_j(\mathbf{x}) < (1 - \frac{\epsilon}{M+1})U$ to the Lyapunov function. The fraction of users leaving these actions is

$$w_* = \sum_{j \in B_\epsilon(\mathbf{x})} x_j \frac{U - f_j(\mathbf{x})}{Af_j(\mathbf{x})} \geq \epsilon \frac{\epsilon U}{A(1 - \frac{\epsilon}{M+1})U}.$$

$V(\mathbf{x}) \geq \epsilon$ implies that there exists j such that $M_j - x_j \geq \frac{\epsilon}{m}$ and $\frac{w_*}{l}$ of the users migrate to j. The Lyapunov function is reduced at least by $D = \frac{\epsilon^2}{lA(1 - \frac{\epsilon}{M+1})}$. Thus,

$$V(\mathbf{x}_T) = \sum_{t=1}^{T}(V(\mathbf{x}_t) - V(\mathbf{x}_{t-1})) + V(\mathbf{x}_0) \leq -TD + V(x_0).$$

Hence, $V(\mathbf{x}_T) \leq \epsilon$ when

$$T \geq (V(\mathbf{x}_0) - \epsilon)\frac{1}{D} = (V(\mathbf{x}_0) - \epsilon)\frac{lA(1 - \frac{\epsilon}{M+1})U}{\epsilon^2 U}$$

which is on the order of $V(\mathbf{x}_0)\frac{lA}{\epsilon^2}$. $\square$

Example 2.3.9.2 (QoS - based dynamic spectrum access). *Consider a dynamic spectrum access under quality of service requirement. Denote by l the total number of channels in the system. x_j denotes the load at channel j. Each user satisfaction level is at U (which can represent a throughput). Assume a large number of users spatially distributed in the system. Assume that the throughput at channel j, f_j depends only on the load x_j. Assuming that the throughput U is feasible, we apply the learning algorithm and an $\eta-$satisfactory solution is obtained after at most $M\frac{l}{\eta^2}$ iterations.*

We summarize some comparative properties of the different learning schemes in Table 2.2. where

- LC: Limit cycle

- PD: Partially distributed, FD: fully distributed,

- RL: Reinforcement learning,

- NE: Nash equilibrium, CE: Correlated equilibrium,

- SS: Satisfactory solution,

- $O(\frac{1}{\eta})$: The time the algorithm takes to be close to the solution with a small error tolerance η,

TABLE 2.2
Comparative properties of the different learning schemes

	PD	FD	Outcome
Fictitious play	yes	no	NE
Best response	yes	no	NE
Smooth best response	yes	no	NE
Cost-to-Learn	yes		NE
RL: Arthur's model	no	yes	NE or boundary, LC
RL: Borger & Sarin	no	yes	NE or boundary, LC
Boltzmann-Gibbs	no	yes	NE, LC
No-regret	no	yes	CE
Trial and error	no	yes	NE
Bargaining	yes	no	Pareto optimal
Satisfactory learning	no	yes	SS

- Boundary: the boundary of the simplex.

Note that the convergence time of best reply dynamics and satisfactory learning is finite in some classes of games (potential games, supermodular games, dominance solvable games) if one player moves at a time. The convergence time is exponential for log-linear dynamics. For potential games, it can be reduced to $n \log \log n + \log(\frac{1}{\eta})$.

2.4 Stochastic Approximations

We will focus mainly on the following stochastic iterative processes in Euclidean space:

$$x_{k+1} = x_k + \lambda_k f(x_k), \qquad (2.42)$$

$$x_{k+1} = x_k + \lambda_k \left[f(x_k) + M_{k+1} \right] \qquad (2.43)$$

$$x_{k+1} = x_k + \lambda_k \left[f(x_k) + M_{k+1} + \epsilon_{k+1} \right] \qquad (2.44)$$

$$x_{k+1} = x_k + \lambda_k \left[f(w_k, x_k) + M_{k+1} + \epsilon_{k+1} \right] \qquad (2.45)$$

$$x_{k+1} - x_k - \lambda_k (M_{k+1} + \epsilon_{k+1}) \in \lambda_k F(x_k), \qquad (2.46)$$

where M_k is a martingale difference, ϵ_k is vanishing sequence, f is vectorial function, and w_k is Markovian with transition $q_{wxw'}$ with unique stationary distribution π_x. We distinguish two cases: constant learning rate $\lambda > 0$ and vanishing learning rate λ_k such that $\sum_{k \geq 1} \lambda_k = +\infty$, F is correspondence in appropriate dimension.

A direct analysis to the above stochastic processes is, in general, difficult. Under suitable conditions, nice properties of the stochastic processes can be obtained from deterministic trajectories. This is the *dynamical system view point of stochastic approximations*. There is also a probabilistic view of stochastic approximations. See the appendix for more details.

2.5 Chapter Review

Exercise 2.6 (Solution concepts). *You can think of this question as "true or false and explain why" discussion of the italicized statements, although parts of it are somewhat more specific.*

(a) *Cooperative game theory guarantees Pareto optimality but noncooperative game theory does not.*

  − *Maxmin fair allocations are Pareto optimal.*

  − *Bargaining solutions are Pareto optimal.*

(b) *Correlated Nash equilibrium improves the performance of the system compared to Nash equilibrium.*

(c) *Hierarchical solutions are better than Nash equilibria.*

(d) *Hierarchical solutions are more fair than Nash equilibria.*

(e) *In zero-sum games, saddle points are interchangeable.*

(f) *Conjectural variation solution may coincide with Nash equilibrium.*

(g) *Conjectural variation may lead to Pareto optimal solutions.*

(h) *Conjectural variation may lead to hierarchical solutions.*

(i) *Conjectural variation may lead to global optima.*

Exercise 2.7 (Learning). • *Give an example of a networked game in which the fictitious play converges. Propose a nonconvergent game under fictitious play.*

• *Give an example of a networked game in which the Cournot tâtonnement converges. Give an example of a networked game in which the Cournot tâtonnement does not lead to equilibria.*

• *Give sufficient conditions for convergence of iterative best-response scheme. Give an example of non-convergence of iterative best-response scheme.*

- *Give an example of reinforcement learning scheme that converges almost surely to equilibria in generic two-by-two games. Give an example of reinforcement learning that converges almost surely to equilibria in pseudo-potential games.*

- *Propose a two-by-two game and learning algorithm under which the convergence time is finite. Propose a two-by-two game and a learning algorithm under which the convergence time is infinite.*

- *Give an example of a networked game that is not a potential game but a reinforcement learning converges to equilibria.*

- *Propose a Q-learning algorithm that leads to the discounted value in two-player zero-sum stochastic games.*

- *Propose a learning algorithm for conjectural variation solutions.*

Exercise 2.8 (Dynamics). *Recall the Nash stationary property: The rest points of the dynamics are exactly the Nash equilibria of the game.*

- *Propose an iterative version of the Brown-Neumann-Nash dynamics that satisfies the Nash stationarity*

- *Propose two examples of game dynamics in continuous time that satisfy the Nash stationarity property*

- *Propose an example of dynamics that does not satisfy the Nash stationarity property*

- *Show that the replicator dynamics does not satisfy the Nash stationary property.*

- *Propose an evolutionary game dynamics that converges to equilibria in any finite games with monotone payoffs.*

2.6 Discussions and Open Issues

We have presented various learning algorithms for games with different information levels: partially distributed learning algorithms and fully distributed learning algorithms. Convergence, non-convergence, stability and applicability to engineering problems have been discussed. We have also studied convergence time of some specific learning algorithms. However some interesting questions remain open:

For what classes of wireless games can one design *distributed fast algorithms* that get close to equilibrium?

This is a very challenging open question. Partial answers will be given in the next chapters.

If convergence occurs, what are the theoretical bounds to the rate at which convergence to equilibrium can occur?

We have illustrated examples of finite games in Chapter 2 in which the iterative best response algorithm converges in a finite number of steps (using improvement path techniques), but in general the convergence time of learning algorithms (when it converges) can be arbitrary large.

What are the speeds of convergence of the learning schemes (when convergence occurs)?

The learning algorithms may not converge but in some special cases in which convergence occurs, we have observed different speeds of convergence to rest point when different learning rates are used by the players. We have observed that increasing the learning rates can accelerate the convergence to rest points. However, in some learning algorithms such as logit response learning, the time it takes to converge (close) to the stationary distribution is exponential in the number of players/actions which is not a positive result in the sense of algorithmic theory.

What type of learning and adaptive behavior do players use in large interactive systems?

We have presented above strategy learning frameworks in which players use different learning schemes and learn in different ways. This new approach has the advantages capturing many observed scenarios. We gave qualitative properties the combined dynamics obtained when the players learn in different way. In particular, in the presence of good learners, bad learners, slow learners or quick learners, our approach can be useful if the fraction of players that are good learners is not too small. However, the analysis under these mixtures of different learners seems more complicated in terms of stability/instability analysis.

> Are stationary-equilibrium approaches the right concepts for predicting long-term behavior in games?

Since most (stationary) equilibrium concepts are known to be inefficient in notable scenarios (power allocation, access control, spectrum management, routing, etc.) and most of the proposed learning algorithms fail to converge; or the outcome will not be played. It is reasonable to ask if the proposed solution concepts are well adapted. This aspect is an active area, and many researchers are currently focusing on set-valued solution concepts instead of point-solution concepts, robust games, game situations, adding both endogenous and exogenous context of the classical formulation, non-equilibrium approach etc. This question remains an open issue.

> How to design learning procedures to be close to the set of global optima?

An interesting direction is the development of fully distributed learning algorithms leading to *global optima* (or approximating global optima) in simple classes of dynamic engineering games (and beyond). In Chapter 8, we will present specific games in which global optima can be reached using fully distributed learning schemes [92, 162].

> Difference between the learning schemes

Analyzing the speed of learning and the convergence time of the various distributed strategic learning algorithms is important in two ways. First of all, the learning rates need to be tuned in such a way that the learners behave reliably, without slowing down the learning processes more than necessary. Secondly, when different learners are engaged in a long-run game it might be that they do not learn equally fast. This can lead to artifacts caused by the main difference in learning speed rather then a true differentiation in qualitative learning dynamics. Therefore, it is necessary to analyze how the different step size parameters (λ, ν, μ) relate in order to elicit the underlying qualitative differences.

⚠ Make sure that two learning schemes can be compared before comparing their convergence time!

There are many learning algorithms developed in the engineering literature that are relevant, it is nice to get comparative results. Then natural questions are

- What are the differences between these learning algorithms?

- Can we compare of them?

- What are the appropriate comparison metrics?

> Given several stochastic learning schemes, how can the convergence time be compared to a target set?

Below we explain that it is not obvious to compare two learning algorithms in terms of convergence (when they are converging). It is not always possible to say that one algorithm is better than another one. To illustrate this statement, consider two distributed learning algorithms L_1 and L_2.

Suppose that learning pattern L_1 generates a unique deterministic sequence of actions during the play leading to a deterministic trajectory given any starting point. Assume that the learning algorithm L_2 defines a stochastic trajectory (over the distribution of actions). Then, are we able to compare L_1 and L_2 graphically or by simulation?

Generally, the answer to this question is NO, and below we provide the main reasons why it is not fair to compare them in some sense:

- L_1 is deterministic (sure) and L_2 is stochastic, i.e., the outcome of the plays is probabilistic, and thus the starting point may play a crucial role. It is not necessarily fair to choose the same starting point.

- The two algorithms are not in the same category. Therefore, a comparison needs to be clearly defined. To see why, consider the following two objects:

$$T_1 = \inf\{t > 0 \mid x_t^{(L_1)} \in O\},$$

where O is a target set. It could be the set of equilibria, a singleton, set of bargaining solutions, set of Pareto optimal solutions, or evolutionary stable sets. Similarly,

$$T_2 = \inf\{t > 0 \mid x_t^{(L_2)} \in O\}.$$

Note that T_1 is a deterministic positive real number and T_2 is a positive random variable. Hence comparing T_1 and T_2 does not make sense. Suppose that we run the two learning algorithms. One run of trajectory gives, for example,

$$T_1 > T_2(run_1)$$

where $T_2(run_1)$ is the realized value of the random variable T_2 at the first run (run_1). This does not mean that the algorithm L_1 is better than the algorithm L_2 because at the second run, one may get

$$T_1 < T_2(run_2).$$

Proposed metrics

- The probability that T_2 is less than T_1 can be defined and it is well-posed. Let
$$\mu_{12} := \mathcal{P}(\{\omega, \ T_1 \leq T_2(\omega)\}).$$

If $\mu_{12} = 1$, we say that the convergence time of L_1 is almost surely greater than the convergence time of L_2. If $\mu_{12} = 1 - \epsilon, \epsilon \in (0,1)$, we say that the convergence time of L_1 is less than the convergence time of L_2 in ϵ per cent of the cases.

- T_1 can be compared with the expected value of T_2,
$$\mathbb{E}(T_2) = \sum_{k>0} \mathbb{P}(T_2 = k)k.$$

In terms of simulations, the value of $\mathbb{E}(T_2)$ can be computed (approximately) by running L_2 several times and then taking the average. However this approach can be time-consuming.

- Find a deterministic pseudo-trajectory for L_2 like an ODE representation or an ODI (ordinary differential inclusion) representation. Then, $x_t^{L_1}$ can be compared with a scaled version $\tilde{x}_{\lambda t}^{(L_2)}$. However, $\tilde{x}_{\lambda t}^{(L_2)}$ and $x_{\lambda t}^{(L_2)}$ can be far away at the beginning of the simulation.

- Compare the rate of convergence of the two algorithms if possible

- Adjustment of the growth rate of the ODEs if possible. See the chapter on risk-sensitivity.

Proposed metrics for deterministic algorithms

Below we introduce the notion of fast and slow deterministic algorithms. Let x_t and y_t be two convergent sequences generated by deterministic iterates. Assume that $\lim_t x_t = \lim_t y_t = x^*$ and the ratio $\frac{\|x_t - x^*\|}{\|y_t - y^*\|}$ has a limit denoted by δ (finite). We say that x is faster than y if $\delta = 0$. If $\delta > 0$ then we say that the two sequences have similar rate. If δ is the opposite argument to $\delta = 0$ is used.

In Chapter 3, we study payoff learning which complements the strategy dynamics developed in this chapter.

3

Payoff Learning and Dynamics

3.1 Introduction

A central learning problem in dynamic environments is balancing exploration of untested actions against exploitation of actions that are known to be good. The benefit of exploration can be estimated using the notion *value of information* (VoI), i.e., the expected improvement in future decision quality that might arise from the information acquired by exploration.

In this chapter we study games with (numerical) noisy payoffs. The payoff-learning is sometimes referred to as Q-learning [188, 189]. Here we focus on specific classes of stochastic games with incomplete information in which the state transitions are action-independent. We develop fully distributed iterative schemes to learn expected payoff functions and near-equilibrium payoffs. Based on stochastic approximation techniques, we show that the payoff estimations can be approximated by differential equations. This leads to *payoff dynamics*.

The basic setting

We consider an interactive system with n potential players, $1 \leq n < +\infty$. We denote the set of players by $\mathcal{N}$. Each player $j \in \mathcal{N}$ has a finite number of actions. The action space of player j is denoted by $\mathcal{A}_j$. The payoff of player j is a random variable $R_{j,t}$ with realized value denoted by $r_{j,t}$ at time t. The random variable $R_{j,t}$ is indexed by a state $\mathbf{w}$ in some set $\mathcal{W}$ and the chosen strategies of all the players.

- We assume that the players do not observe the past actions/strategies played by the others (imperfect monitoring).

- We assume that the players do not know the state of the game i.e the value of $\mathbf{w}_t$ at time t is unknown.

- We assume that the players do not know the expression of their payoff functions.

Information	Known	Unknown
Action spaces of the others	no	yes
Last action of the others	no	yes
Payoff functions of the others	no	yes
Own-payoff function	no	yes
Number of players	no	yes
Measurement own-payoff	yes	no
Last own-action	yes	no
Capability	arithmetic operations	

TABLE 3.1
Information assumptions

At each time t, each player chooses an action and receives a numerical noisy value of its payoff (*perceived payoff*) at that time. We assume that the transition probabilities between the states $\mathbf{w}$ are independent and unknown to the players.

The long-run game is described as follows.

- At time slot $t = 0$, each player $j \in \mathcal{N}$ chooses an action $a_{j,0} \in \mathcal{A}_j$ and perceives a numerical value of its payoff which corresponds to a realization of the random variables depending on the actions of the other players and the state of the nature etc. He initializes its estimation to $\hat{\mathbf{r}}_{j,0}$.

- At time slot t, each player j estimates of his payoffs, chooses an action based its own-experiences and experiments a new strategy. Each player j receives an output $r_{j,t} \in \mathbb{R}$. Based on this target $r_{j,t}$ the player j updates his estimation vector $\hat{\mathbf{r}}_{j,t}$ and builds the strategy $\mathbf{x}_{j,t+1}$ for the next time slot. The strategy $\mathbf{x}_{j,t+1}$ is a function only of $\mathbf{x}_{j,t}, \hat{\mathbf{r}}_{j,t}$ and the target value. Note that the exact value of the state of the nature $\mathbf{w}_t$ at time t and the past strategies $\mathbf{x}_{-j,t-1} := (\mathbf{x}_{k,t-1})_{k \neq j}$ of the other players and their past payoffs $\mathbf{r}_{-j,t-1} := (r_{k,t-1})_{k \neq j}$ are unknown to player j at time t.

- The game moves to $t + 1$.

Table 3.1 summarizes the information available to the players.

In order to construct the strategies of the dynamic game, we first need to define what are the histories, i.e., collection of the observations, measurements and experiments made by each player up to the current time.

Histories

A player's private information consists of his past own-actions and perceived own-payoffs.

A private history of length t for player j is a collection

$$h_{j,t} = (a_{j,0}, r_{j,0}, a_{j,1}, r_{j,1}, \ldots, a_{j,t-1}, r_{j,t-1}) \in H_{j,t} := (\mathcal{A}_j \times \mathbb{R})^t.$$

Behavioral Strategy

From a private history, each player built his or her strategy at the current time. These strategies are called behavioral strategies.

A behavioral strategy for player j at time t is a mapping

$$\tilde{\tau}_{j,t} : H_{j,t} \longrightarrow \mathcal{X}_j.$$

The collection $\tilde{\tau}_j = (\tilde{\tau}_{j,t})_t$ is a behavioral strategy. The set of complete histories of the dynamic game after t stages is

$$H_t = (\mathcal{W} \times \prod_j \mathcal{A}_j \times \mathbb{R}^n)^t,$$

and it describes the states, the chosen actions and the received payoffs for all the players at all past stages before t. A strategy profile $\tilde{\tau} = (\tilde{\tau}_j)_{j \in \mathcal{N}}$ and an initial state $\mathbf{w}$ induce a probability distribution $P_{\mathbf{w}, \tilde{\tau}}$ on the set of plays

$$H_\infty = (\mathcal{W} \times \prod_j \mathcal{A}_j \times \mathbb{R}^n)^{\mathbb{N}}.$$

Long-term payoff

We focus on the average payoff i.e $F_{j,T} = \frac{1}{T} \sum_{t=1}^{T} r_{j,t}$. Given a initial state $\mathbf{w}$ and a strategy profile $\tilde{\tau}$, The payoff of player j is the superior limiting of the Cesaro-mean payoff $\mathbb{E}_{\mathbf{w}, \tilde{\tau}} F_{j,T}$. We assume that $\mathbb{E}_{\mathbf{w}, \tilde{\tau}} F_{j,T}$ has a limit. This allows us to rely the stationary payoff of the dynamic game with the payoff of a static game called expected game.

Exercise 3.1. *Explain why the number of state-independent Nash equilibria in robust games is generically finite and odd.*

Let K be a real closed field.

A $K-$polynomial system is a finite number of polynomial equations and inequalities $(P < 0, P > 0, P = 0)$ with coefficients in K. A semi-algebraic set in K^d is a finite (disjoint) union of sets defined by a $K-$polynomial system. A semi-algebraic function (correspondence) is one whose graph is semi-algebraic.

Team problem with uncertainty

A team problem with uncertainty is a robust game with finite strategy spaces where all players have the same uncertain pay-off function.

Theorem 3.1.0.3. *In a team problem with uncertainty and finite action profiles, one has finitely many equilibrium pay-offs.*

A main tool for its proof is Morse-Sard's Theorem [117, 141], which also plays an essential role in the fact that generically the number of equilibrium is odd in finite games.

Theorem 3.1.0.4 (Morse-Sard's Theorem, Morse 1939[117], Sard 1942,[141]). *Let $\mathcal{D}$ and $\mathcal{A}$ be two smooth manifolds Let $g : \mathcal{D}^m \longrightarrow \mathcal{A}^k$ be a smooth map from a $m-$dimensional manifold $\mathcal{D}$ to an $k-$dimensional manifold $\mathcal{A}$. Assume that the rank of the Jacobian matrix is (strictly) less than k. Then, the set of critical values of g is of Lebesgue measure 0.*

The Morse-Sard's theorem asserts that the image of the set of critical points of a smooth function g from one manifold to another has Lebesgue measure zero.

Proof. Consider a finite team game with uncertain payoff. Generically, the payoff are nontrivial. To show that such situation has finitely many equilibrium pay-offs, it suffices to prove that there are finitely many pay-offs to completely mixed equilibria because the number of pure action profiles is finite (and hence finite number of equilibrium payoffs in this class). In a strictly mixed equilibrium of the team game, the gradient of the payoff function is zero. Now, we use Sard's theorem (Theorem 3.1.0.3) to deduce that the set of completely mixed equilibrium payoffs has Lebesgue measure zero.

Use the classical property of semi-algebraic sets: the range and inverse ranges of semi-algebraic sets by semi-algebraic functions are semi-algebraic. Thus, generically, the set of equilibrium payoffs is semi-algebraic. The semi-algebraicity implies that it is a finite union of intervals. Hence, generically the stated result holds. □

3.2 Learning Equilibrium Payoffs

In this subsection, we will develop a fully distributed learning procedure to learn the expected payoffs. Note that, since the state is unknown, the players do not know which game they are playing. There is a sequence of games generated by the value of the random variable $\mathcal{W}$ that determines the states of which the number interacting players varies over time. In the case where only a single state is available (the Dirac distribution case), the exact payoff is learned.

We consider the payoff-RL in the following form: (Payoff-RL)

$$
\begin{cases}
\mathbf{x}_{j,t+1} &= f_j(a_{j,t}, \hat{\mathbf{r}}_{j,t}, \mathbf{x}_{j,t}) \\
\hat{\mathbf{r}}_{j,t+1} &= g_j(\nu_{j,t}, a_{j,t+1}, r_{j,t+1}, \mathbf{x}_{j,t}, \hat{\mathbf{r}}_{j,t}) \\
& \quad j \in \mathcal{N}, t \geq 0, a_{j,t} \in \mathcal{A}_j \\
& \quad \mathbf{x}_0 \in \prod_{j \in \mathcal{N}} \mathcal{X}_j, \\
& \quad \hat{\mathbf{r}}_0 \in \mathbb{R}^{\sum_{j \in \mathcal{N}} |\mathcal{A}_j|}.
\end{cases}
$$

where

- The functions f_j, $j \in \mathcal{N}$ are based on estimated payoff and perceived measured payoff, such that the invariance of simplex is preserved. The function f_j defines the strategy pattern of player j at the current time.

- The functions g, and ν are well-chosen in order to have a good estimation of the payoffs. Most of the time we will assume that $\nu \in l^2 \backslash l^1$, which is a standard assumption in Robbins-Monro-like procedures.

Below we provide different examples of payoff-learning procedures.

Boltzmann-Gibbs-based Payoff-RL

The Boltzmann-Gibbs-based Payoff-RL consists of constructing the estimates of the payoff based Boltzmann-Gibbs rule, i.e., the Boltzmann-Gibbs strategy defined in chapter 2 will be used to learn the expected payoffs.

Let $\tilde{\beta}_{j,\epsilon_j} : \mathbb{R}^{\sharp \mathcal{A}_j} \longrightarrow \mathcal{X}_j$ be the Boltzmann-Gibbs strategy defined by

$$
\tilde{\beta}_{j,\epsilon_j}(\hat{\mathbf{r}}_{j,t})(s_j) = \frac{e^{\frac{\hat{r}_{j,t}(s_j)}{\epsilon_j}}}{\sum_{s_j' \in \mathcal{A}_j} e^{\frac{\hat{r}_{j,t}(s_j')}{\epsilon_j}}}.
$$

By choosing

$$
f_j = \tilde{\beta}_{j,\epsilon_j}(\hat{\mathbf{r}}_{j,t}),
$$

$$
g_j = \hat{\mathbf{r}}_{j,t}(s_j) + \nu_{j,t} \mathbb{1}_{\{a_{j,t+1}=s_j\}}(r_{j,t+1} - \hat{\mathbf{r}}_{j,t}(s_j)),
$$

one gets **(BG-based Payoff-RL)**

$$
\begin{cases}
\mathbf{x}_{j,t+1} &= \tilde{\beta}_{j,\epsilon_j}(\hat{\mathbf{r}}_{j,t}) \\
\hat{\mathbf{r}}_{j,t+1}(s_j) &= \hat{\mathbf{r}}_{j,t}(s_j) + \nu_{j,t} \mathbb{1}_{\{a_{j,t+1}=s_j\}}(r_{j,t+1} - \hat{\mathbf{r}}_{j,t}(s_j)) \\
& \quad j \in \mathcal{N}, \ s_j \in \mathcal{A}_j
\end{cases}
$$

Different variations of this scheme are possible:

- The term

$$
\nu_{j,t}(s_j) = \frac{1}{1 + \sum_{t'=1}^{t} \mathbb{1}_{\{a_{j,t'}=s_j\}}},
$$

i.e., the learning rate takes into account how many times the actions s_j have been chosen.

- The parameter ϵ_j, which can be seen as a temperature parameter or perturbation parameter, can be time-varying, i.e., $\epsilon_{j,t}$.

Algorithm 3: BG-Payoff-RL

---

forall the *Players* **do**
     Initialize to some assignment;

for *t=1* **to** *max* **do** do;
foreach *Player j* **do**
     Choose $a_{j,t} = s_j$ the action picked by Player j according to $x_{j,t}$;

foreach *Player j* **do**
     Observe the payoff $r_{j,t}$;
     Update the estimates $\hat{r}_{j,t+1}$;
     Construct the new Boltzmann-Gibbs strategy based $\hat{r}_{j,t+1}$;

Example 3.2.1 (Boltzmann-Gibbs Payoff-RL with multiplicative weights). *The multiplicative Boltzmann-Gibbs Payoff-RL is given by* **(Payoff-M-BG)**

$$
\begin{cases}
\mathbf{x}_{j,t+1}(s_j) &= \dfrac{(1+\frac{1}{\epsilon_j})^{\hat{r}_{j,t}(s_j)}}{\sum_{s'_j \in \mathcal{A}_j} (1+\frac{1}{\epsilon_j})^{\hat{r}_{j,t}(s'_j)}} \\
\hat{\mathbf{r}}_{j,t+1}(s_j) &= \hat{\mathbf{r}}_{j,t}(s_j) + \nu_{j,t}\mathbb{1}_{\{a_{j,t+1}=s_j\}}\left(r_{j,t+1} - \hat{\mathbf{r}}_{j,t}(s_j)\right), \\
& \qquad\qquad j \in \mathcal{N}, \ s_j \in \mathcal{A}_j
\end{cases}
$$

Imitative Boltzmann-Gibbs based Payoff-RL

Imitation has been defined as a significant increase in the probability of an action as a result of observing the measurement or a result of observing another player performing that action. Informally, it is a learning that involves doing or saying something by experimenting/watching it done by another, usually several times. Here we provide a very specific class of imitative learning.

Let

$$
\sigma_{j,\epsilon} : \ \mathbb{R}^{\sharp \mathcal{A}_j} \longrightarrow \mathcal{X}_j,
$$

be the imitative Boltzmann-Gibbs strategy defined by

$$
\sigma_{j,\epsilon}(\hat{\mathbf{r}}_{\mathbf{j},\mathbf{t}})(s_j) = \frac{x_{j,t}(s_j)e^{\frac{\hat{r}_{j,t}(s_j)}{\epsilon_j}}}{\sum_{s'_j \in \mathcal{A}_j} x_{j,t}(s'_j)e^{\frac{\hat{r}_{j,t}(s'_j)}{\epsilon_j}}}.
$$

The imitative Boltzmann-Gibbs (IBG) based Payoff-RL is given by **(Payoff-IBG):**

$$
\begin{cases}
\mathbf{x}_{j,t+1} &= \sigma_{j,\epsilon}(\hat{\mathbf{r}}_{\mathbf{j},\mathbf{t}}) \\
\hat{\mathbf{r}}_{j,t+1}(s_j) &= \hat{\mathbf{r}}_{j,t}(s_j) + \nu_{j,t}\mathbb{1}_{\{a_{j,t+1}=s_j\}}\left(r_{j,t+1} - \hat{\mathbf{r}}_{j,t}(s_j)\right) \\
& \qquad\qquad j \in \mathcal{N}, \ s_j \in \mathcal{A}_j
\end{cases}
$$

Imitative Boltzmann-Gibbs with multiplicative weights payoff-RL

The multiplicative weight imitative BG based payoff-RL is given by (**Payoff-M-IBG**)

$$
\begin{cases}
\mathbf{x}_{j,t+1}(s_j) &= \dfrac{x_{j,t}(s_j)(1+\epsilon_j)^{\hat{r}_{j,t}(s_j)}}{\sum_{s'_j \in \mathcal{A}_j} x_{j,t}(s'_j)(1+\epsilon_j)^{\hat{r}_{j,t}(s'_j)}} \\
\hat{\mathbf{r}}_{j,t+1}(s_j) &= \hat{r}_{j,t}(s_j) + \nu_{j,t}\mathbb{1}_{\{a_{j,t+1}=s_j\}}\left(r_{j,t+1} - \hat{r}_{j,t}(s_j)\right) \\
& \qquad\qquad j \in \mathcal{N}, \ s_j \in \mathcal{A}_j
\end{cases}
$$

No-regret based CODIPAS-RL

$$
x_{j,t+1}(s_j) - x_{j,t}(s_j) = \mathbb{1}_{\{j \in \mathcal{B}^n(t)\}}\left(\underline{R}_t(s_j) - x_{j,t}(s_j)\right), \tag{3.1}
$$

$$
\hat{r}_{j,t+1}(s_j) - \hat{r}_{j,t}(s_j) = \mathbb{1}_{\{a_{j,t+1}=s_j, j \in \mathcal{B}^n(t+1)\}}\left(r_{j,t+1} - \hat{r}_{j,t}(s_j)\right) \tag{3.2}
$$

$$
\theta_j(t+1) = \theta_j(t) + \mathbb{1}_{\{j \in \mathcal{B}^n(t)\}} \tag{3.3}
$$

$$
\underline{R}_t(s_j) = \frac{\phi([\hat{r}_{j,t}(s_j) - r_{j,t}]_+)}{\sum_{s'_j}\phi([\hat{r}_{j,t}(s'_j) - r_{j,t}]_+)} \tag{3.4}
$$

where $\mathcal{B}^n(t)$ denotes the set of active players (out of n) at time t. The set $\mathcal{B}^n(t)$ is a random set. Thus, one gets a dynamic game with random set of interacting players. The novelty here is the term $\theta_j(t)$ which indicates how many times player j has been active up to t. Each player updates only when he or she is active. Thus, one has asynchronous clocks. The detailed analysis of such a system will be given in the *Random Updates* Chapter.

The strategy learning based on the non-regret rule is known to convergent to the set of correlated equilibria. Here the non-regret is based on the estimations and random updates.

Imitative no-regret based CODIPAS-RL

$$
x_{j,t+1}(s_j) - x_{j,t}(s_j) = \lambda_{\theta_j(t)}\mathbb{1}_{\{j \in \mathcal{B}^n(t)\}}\left(IR_t(s_j) - x_{j,t}(s_j)\right), \tag{3.5}
$$

$$
\hat{r}_{j,t+1}(s_j) - \hat{r}_{j,t}(s_j) = \nu_{\theta_j(t)}\mathbb{1}_{\{a_{j,t+1}=s_j, j \in \mathcal{B}^n(t+1)\}}\left(r_{j,t+1} - \hat{r}_{j,t}(s_j)\right), \tag{3.6}
$$

$$
\theta_j(t+1) = \theta_j(t) + \mathbb{1}_{\{j \in \mathcal{B}^n(t+1)\}} \tag{3.7}
$$

$$
IR_t(s_j) = \frac{x_{j,t}(s_j)\phi([\hat{r}_{j,t}(s_j) - r_{j,t}]_+)}{\sum_{s'_j} x_{j,t}(s'_j)\phi([\hat{r}_{j,t}(s'_j) - r_{j,t}]_+)} \tag{3.8}
$$

where $[y]_+ = \max(0, y)$, ϕ is a positive and increasing function.

3.3   Payoff Dynamics

We examine the connection between payoff-RL and differential equations. We consider the payoff reinforcement learning

$$\hat{\mathbf{r}}_{j,t+1}(s_j) = \hat{\mathbf{r}}_{j,t}(s_j) + \nu_{j,t} \mathbb{1}_{\{a_{j,t+1}=s_j\}} \left(r_{j,t+1} - \hat{\mathbf{r}}_{j,t}(s_j)\right).$$

By taking the expected changes in one-time slot,

$$\mathbb{E}\left(\frac{\hat{\mathbf{r}}_{j,t+1}(s_j) - \hat{\mathbf{r}}_{j,t}(s_j)}{\nu_{j,t}} \mid \hat{\mathbf{r}}_{j,t+1}, \mathbf{x}_t = \mathbf{x} \right) = \mathbf{x}_j(s_j) \left[\mathbb{E}_{\mathbf{w}} R_j(\mathbf{w}, e_{s_j}, \mathbf{x}_{-j,t}) - \hat{\mathbf{r}}_{j,t}(s_j) \right],$$

Asymptotically, the limit set of the process $\hat{\mathbf{r}}_{j,t}$ can be relied to the ordinary differential equation (ODE) given by

$$\frac{d}{dt}\hat{\mathbf{r}}_{j,t}(s_j) = \mathbf{x}_{j,t}(s_j) \left[\mathbb{E}_{\mathbf{w}} R_j(\mathbf{w}, e_{s_j}, \mathbf{x}_{-j,t}) - \hat{\mathbf{r}}_{j,t}(s_j) \right], s_j \in \mathcal{A}_j, \ j \in \mathcal{N} \quad (3.9)$$

Moreover, if $\mathbf{x}_t \longrightarrow \mathbf{x}^*$, the point $\mathbb{E}_{\mathbf{w}} R_j(\mathbf{w}, \mathbf{x}^*)$ is globally asymptotically stable under (3.9). Using the above approach in our setting, one has that, provided each action is chosen infinitely often, the Q-values in this algorithm will converge almost surely to the expected action values, that is $\hat{\mathbf{r}}_{j,t} \longrightarrow \mathbb{E}_{\mathbf{w}} R_j(\mathbf{w}, \mathbf{x}^*)$ as $t \longrightarrow \infty$.

3.4   Routing Games with Parallel Links

Routing in discrete time and a finite number of players is originally discussed in transportation problems. Such settings also show up in context of networks where the flow of a user is un-splittable. In this section, we assume that the cost function is not known to players. Moreover, the variation in payoff reflects the changes in the dynamic environment. Therefore, this justifies that it is difficult to know the cost in advance rather the players need to learn the cost during the game, and hence their strategies.

We discuss a suitable learning mechanism for equilibrium cost in dynamic routing games. We assume that players can:

- estimate the average cost and

- delay time of the alternative routes. Players make a decision based on this rough information by using a randomized decision rule to revise their strategy.

Routing	Game setting
Route	Action
Set of routes	Pure action spaces
Delay	Cost
Average Delay	Average Cost

TABLE 3.2
Routing versus game theoretic parameters

The costs and time delays of the chosen alternative are then observed and is used to update the strategy for that particular route. Each player experiments with only the costs and time delays of the selected route on that time slot, and uses this information to adjust his strategy for that particular route. This scheme is repeated every slot, generating a discrete time stochastic learning process. Basically we have three parameters:

- the state;

- the rule for revision of strategies from the state to randomized actions; and,

- a rule to update the new state.

Although players observe only the costs of the specific route chosen on the given time slot, the observed values depend on the congestion levels determined by the others players choices revealing implicit information on the entire system. The natural question is whether such a learning algorithm based on a minimal piece of information may be sufficient to induce coordination [1] and make the system stabilize to an equilibrium. The answer to this question is positive for dynamic routing games on:

- Parallel links and a *Boltzmann-Gibbs dynamics* for route selection.

- General topology with monotone cost functions (in the vectorial sense).

We consider a finite set of players and finite set of routes. Denote by $C_{\underline{r}}^w(k,t)$ the average $w-$weighted cost for the path $\underline{r}$ when k players chose this path at time t. The weight w simply depicts that the effective cost is the weighted sum of several costs depending on certain objective. For example, the objective can be a delay cost and memory cost to be minimized. These can be combined together with a certain weight w. Again, the weight w could also be different for different players due their objectives. Henceforth, we omit w and work with generic cost $C_{\underline{r}}(k,t)$ for simplicity of notation.

Table 3.2 summarizes the details of the game.

---

[1] Note that we do not assume coordination between players; there is not central authority and there is no signaling scheme to the players.

An estimation of player j is a vector $C_{j,t} = (C_{j,t}(\underline{r}))_{\underline{r}}$ where $C_{j,t}(\underline{r})$ represents player j's estimate of the average cost of route $\underline{r}$ (the weighted cost composition).

Player j updates his or her randomized actions using the Boltzmann-Gibbs scheme which consists of choosing the route $\underline{r}$ with probability

$$x_j(C)(\underline{r}) := \frac{e^{-\frac{1}{\epsilon_j}C_{j,t}(\underline{r})}}{\sum_{r'} e^{-\frac{1}{\epsilon_j}C_{j,t}(r')}}, \ \epsilon_j > 0.$$

Congestion is captured by the inequality

$$\forall t, \ C_{\underline{r}}(k,t) \le C_{\underline{r}}(k+1,t), \ k \le n,$$

where n is the total number of players. A detailed analysis of routing games without knowledge of payoff functions can be found in [18].

3.5   Numerical Values of Payoffs Are Not Observed

What happens if the players do not even observe numerical values of their own-payoffs?

Most of the learning schemes developed are based on numerical measurement of the payoffs. However, the numerical value may not be known in many practical scenarios. The payoffs are, in general, real numbers. In terms of bits, one needs to convert the truncated version of those real numbers into binary codes and feed back it. This means that it may require many bits if one wants good approximations. For low complexity, one can feedback only a payoff-based signal. Each private signal will indicate if the payoff of the corresponding player belongs to some set or not. Below we provide a generic scheme to learn the outage probability, the error probability, and the success probability. The key idea behind the approach is to construct a probabilistic payoff function based on the expected value of the private signals.

Let O_j be the set of targets of player j. Suppose that if the payoff of player j belongs to O_j, then the player will receive a signal (which can be noisy) $\tilde{s}_{j,t}$ that says yes or no, ACK/NACK, 0 or 1. Then, one can construct a stochastic approximation scheme for the frequency of visit of the signal to the set $\{1\}$ and learn the probability $P(\tilde{s}_j \in O_j)$. Then, the novel objective of the player j could be to

- Minimize the outage probability $P(\tilde{s}_j \in O_j^c)$ where O_j^c denotes the complementary of the set of O_j.

- Maximize the success probability $P(\tilde{s}_j \in O_j)$

Remark 3.5.1. *The approach of private signals based on payoffs could be used to learn satisfactory solutions. This will be discussed in Chapter 6.*

Open issues

The payoff reinforcement learning (Payoff-RL) is an attractive method of learning because of the simplicity of the computational demands per iteration, and also because of this proof of convergence to a global payoff function under stationary strategies. One catch, though, is that the system has to be an interactive Markov decision process and the strategies are restricted. As it is well-known, even interesting toy examples are liable to break the Markov model. The extension of the payoff-RL to a non-Markovian setting is an active research area.

In Chapter 4, we present combined fully distributed payoff and strategy reinforcement learning (CODIPAS-RL) The chapter uses the second part of Chapter 2 as well as the payoff dynamics of Chapter 3.

4

Combined Learning

4.1 Introduction

In Chapter 2, we have seen various classes of strategy learning schemes. In Chapter 3 we have provided payoff-learning schemes. In this chapter, we will go one step further by combining the two approaches.

> Distributed schemes may have local information but may also have uncertain knowledge of the system parameters. The challenge is to design fast, convergent, fully distributed learning algorithms that perform well even in the presence of errors (noises) about the system measurements. As we will see, information is crucial in the process of decision making.

In many dynamic interactions, it is convenient to have a learning and adaptive procedure that works with minimal information (does not require any explicit information about the other players actions or payoffs) and less memory (small number of parameters in terms of past own-actions and past own-payoffs) as possible. Such a rule is said to be model-free (nonmodel) or uncoupled or *fully distributed*. However, we have seen in the previous chapters that for a large class of games, no such general algorithm causes the players' period-by-period behavior to converge to Nash equilibrium. Hence, there is no guarantee for fully distributed learning algorithms and dynamics that behaviors will come close to (Nash) equilibrium most of the time.

However, for weaker notions of convergence or other solution concepts, there are many promising results:

- In terms of frequency of plays, it can be shown that *regret minimizing procedures* can cause the empirical frequency distribution of play to converge to the set of correlated (Nash) equilibria. Note that the set of correlated equilibria is convex and includes the convex hull of the set of Nash equilibria.

- In the same line of research, one can define the frequency of visits of the interactive Markov chain in a specific set. This is captured by the so-called *interactive trial and error learning*. As we have seen in Chapter 2, the interactive trial and error learning implements Nash equilibrium behavior in any game with generic payoffs (interdependent games) and that has at least one pure Nash equilibrium. One can modify the reactions of the players to make high frequency visit to Pareto optimal solutions, which exist for finite games.

Most of these studies focus on *strategy learning* in the sense that the players adapt their strategies in order to learn an optimal strategy in the long-term.

The goal of this chapter to study combined learning in dynamic games [20] with incomplete information and imperfect payoffs. We develop fully distributed iterative schemes to learn both payoff functions and optimal strategies. Based on stochastic approximation techniques, we show that the iterative schemes can be approximated by differential equations or inclusions (deterministic or stochastic). This leads to novel game dynamics called *combined dynamics*. The game dynamics are closely related to *evolutionary game dynamics with migration* [172, 153]. Considering the class of aggregative robust games and robust pseudo-potential games, we apply the heterogeneous and the hybrid learning schemes to wireless networking and communications.

Let $\mathcal{G}$ be a game where each player has a finite number of actions. If the player knows the payoff function and the past actions played by the others one can use the fictitious play algorithms, the best response dynamics, the gradient-based algorithms etc. The same techniques can be applied to continuous-kernel games. Even under the observation of past actions and payoffs of the others players, these algorithms may not converge [158, 170].

In many situations the payoff functions of players are very complex in terms of their preferences. Simple machines may not be able to know these complex mathematical expressions due to their computational limitations.

> A natural question in this context is, *how will the players learn their payoff functions jointly with associated optimal strategies?*

This question will be examined in detail in the next sections. In terms of information, additional learning difficulties arise when the players do not know the payoff function nor the last actions of the other players. Now, each player needs to interact with the dynamic environment to find out his or her optimal payoff and his or her optimal strategy.

The players build their strategies and update them by using only their own-actions and estimated own-payoffs.

As discussed in Chapter 1, we develop various COmbined fully DIstributed PAyoff and Strategy-Reinforcement Learning algorithms (CODIPAS-RL) for dynamic robust games. Under CODIPAS-RL, the players can learn their expected payoffs and their optimal strategies by using some simple iterative techniques based on own-experiences. The actions that give good performance are reinforced and new actions are explored. Such an approach is called model of *reinforcement learning*. When the player can also directly learn about his or her optimal strategy without knowing the payoff function, the approach is called *model-free reinforcement learning* (see [42] and the references therein). The proposed CODIPAS-RL scheme is in the class of model-free reinforcement learning for both strategies and payoffs learning (also called Q-value learning).

Compared to the standard reinforcement learning algorithms [11, 32, 139] in which the expected payoff function is not learned, the CODIPAS-RL has the advantage of estimating the optimal expected payoffs in parallel. The estimated payoffs give the expected payoff function if all the actions have been sufficiently explored. CODIPAS-RL is a good learning scheme for games under uncertainty or robust games [1, 162]. In particular, in some class of matrix games with random entries, the CODIPAS-RL applies as well.

Inspired by stochastic approximation techniques (see Kushner and Clark (1978,[96]), Benaim (1999, [26]), Borkar (1997, [34]), Collins and Leslie (2003, [103])), we construct a class of combined learning framework for both payoff and strategies (joint learning) and show that, under suitable conditions, CODIPAS-RL with different learning rates can be studied by their deterministic ordinary differential equation (ODE) counterparts. The resulting ordinary differential equations may differ from the classical imitation dynamics, the replicator dynamics, the (myopic) best response (BR) dynamics, the logit dynamics, and fictitious play, etc.

In a dynamic unknown environment, fully distributed learning schemes are essential for the applications of game theory. In decentralized networks where the traffic, the topology, and the channel states may vary over time and the communications between players (users, nodes, mobile devices) are difficult, it is important to investigate games of incomplete information and fully distributed learning algorithms which demand a minimal amount of information from the other players. We introduce robust aggregative games and derive basic properties of the learning scheme for this class of robust games. We apply this learning scheme to resource selection/allocation problems in wireless networking.

> Each player adapts jointly his or her strategies and payoff estimations over the interactions.

The chapter tackles three fundamental issues in the study of dynamic interactive systems:

- Providing solution concepts for dynamic interactions with uncertainty and incomplete information, where the payoff functions, the number of players, and the current state information are unavailable.

- Incorporating uncertainty about the number of active players, and learning the payoff functions in parallel with the strategy learning in the context of dynamic robust games is one of the most fundamental settings in engineering systems.

- Providing decentralized, fully distributed payoff and strategy learning for both discrete and continuous action spaces.

The remainder of this chapter is structured as follows. Section 4.2 presents the model and notations. Then, we propose combined fully distributed payoff and strategy learning based on Boltzmann-Gibbs schemes. Section 4.4 analyzes the convergence to novel game dynamics based on heterogeneous learning. Section 4.4.3 gives an illustrative example of robust games and learning in wireless networking.

4.2   Model and Notations

In this section, we describe the notations and the assumptions required for the CODIPAS-RL scheme to be applied to a given dynamic interactive scenario [162, 89].

We consider a system with a maximum of n potential players. We denote the set of players by $\mathcal{N}$. Each player $j \in \mathcal{N}$ has a finite number of actions (pure strategies), denoted by $\mathcal{A}_j$. The payoff of the player j is a random variable $R_{j,t}$ with realized value denoted by $r_{j,t}$ at time t. The random variable $R_{j,t}$ is indexed by $\mathbf{w}$ in some set $\mathcal{W}$ and the chosen strategies of all the players. We assume that the players do not observe the past strategies played by the others (imperfect monitoring). We assume that the players do not know the state of the game, i.e., the value $\mathbf{w}_t$ at time t is unknown. We assume that the players do not know the expression of their payoff functions. At each time t, each player chooses an action and receives a numerical noisy value of its payoff (*perceived payoff*) at that time. The perceived payoff can be interpreted as a measured metric in the random environment. We assume that the transition probabilities between the states $\mathbf{w}$ are independent and unknown to the players. The stochastic game is described by

$$\mathcal{G} = (\mathcal{N}, (\mathcal{A}_j)_{j \in \mathcal{N}}, \mathcal{W}, (R_j(\mathbf{w}, .))_{j \in \mathcal{N}, \mathbf{w} \in \mathcal{W}})$$

where

- $\mathcal{N} = \{1, 2, \ldots, n\}$ is unknown to the players.

- $j \in \mathcal{N}$, $\mathcal{A}_j$, is the set of actions available to player j. In each state, the same action spaces are available. Player j does not know the action spaces of the other players.

- *No information on the current state:* $\mathcal{W}$ is the set of possible states, a compact subset of $\mathbb{R}^k$ for some integer k. We assume that the state space $\mathcal{W}$ and the probability transition on the states are both *unknown* to the players. The state is represented by an independent random variable (the transitions between the states are independent of the chosen actions). We assume that the current state w_t is unknown to the players.

- *Payoff functions are unknown:* $R_j : \mathcal{W} \times \prod_{j'=1}^{n} \mathcal{X}_{j'} \longrightarrow \mathbb{R}$ where $\mathcal{X}_j := \Delta(\mathcal{A}_j)$ is the set of probability distributions over $\mathcal{A}_j$, i.e.,

$$\mathcal{X}_j := \left\{ \mathbf{x}_j \mid x_j(s_j) \in [0,1], \sum_{s_j \in \mathcal{A}_j} x_j(s_j) = 1 \right\}; \qquad (4.1)$$

We denote by $\tilde{R}_j$ the payoff function restricted by $\mathcal{W} \times \prod_{j'=1}^{n} \mathcal{A}_{j'}$.

A stage game at time t in some state w_t is a collection

$$\mathcal{G}(\mathbf{w}_t) = \left(\mathcal{N}, (\mathcal{A}_j)_{j \in \mathcal{N}}, (R_j(\mathbf{w}_t, .))_{j \in \mathcal{N}} \right).$$

Since the players do not observe the past actions of the other players, we consider strategies used by both players to be only dependent on their current perceived own-payoffs. Denote by $x_{j,t}(s_j)$ the probabilities of player j choosing s_j at time t, and let $\mathbf{x}_{j,t} = [x_{j,t}(s_j)]_{s_j \in \mathcal{A}_j} \in \mathcal{X}_j$ be the mixed (state-independent) strategy of player j.

4.2.1 Description of the Dynamic Game

The payoffs $R_{j,t}$ are real-valued random variables. The payoff functions are unknown to the players. We assume that the distribution or the law of the possible payoffs is also unknown (This is in contrast to standard approaches of learning with noise based on stochastic evolutionary dynamics in which the law of evolution of the payoffs is assumed to be known by the players). The basic idea of CODIPAS is to learn the expected payoffs simultaneously with the optimal strategies during a long-run interaction, without Bayesian assumptions on the initial beliefs formed over the possible states. The long-run game is described as follows.

- At time slot $t = 0$, each player $j \in \mathcal{N}$ chooses an action $a_{j,0} \in \mathcal{A}_j$ and perceives a numerical value of its payoff that corresponds to a realization of the random variables depending on the actions of the other players and the state of the nature, etc. He initializes its estimation to $\hat{r}_{j,0} \in \mathbb{R}^{|\mathcal{A}_j|}$.

- At time slot t, each player j has an estimation of his or her payoffs, chooses an action based his or her own-experiences and experiment a new strategy. Each player j receives an output $r_{j,t}$. Based on this target $r_{j,t}$, the player j updates its estimation vector $\hat{\mathbf{r}}_{j,t}$ and builds a strategy $\mathbf{x}_{j,t+1}$ for the next time slot. The strategy $\mathbf{x}_{j,t+1}$ is a function only of $\mathbf{x}_{j,t}, \hat{\mathbf{r}}_{j,t}$ and the target value. Note that the exact value of the state of the nature $\mathbf{w}_t$ at time t and the past strategies $\mathbf{x}_{-j,t-1} := (\mathbf{x}_{k,t-1})_{k \neq j}$ of the other players and their past payoffs $\mathbf{r}_{-j,t-1} := (r_{k,t-1})_{k \neq j}$ are unknown to player j at time t.

- The game moves to $t + 1$.

> ⚠ You don't have to know you are playing game to be in a game!

Histories

A player's information consists of his past own-actions and perceived own-payoffs. A private history of length t for player j is a collection

$$h_{j,t} = (a_{j,0}, r_{j,0}, a_{j,1}, r_{j,1}, \ldots, a_{j,t-1}, r_{j,t-1}) \in H_{j,t} := (\mathcal{A}_j \times \mathbb{R})^t.$$

Behavioral Strategy

A behavioral strategy for player j at time slot t is a mapping

$$\tilde{\tau}_{j,t} : H_{j,t} \longrightarrow \mathcal{X}_j.$$

Let $\tilde{\tau}_j := (\tilde{\tau}_{j,t})_t$. The set of complete histories of the dynamic game after t stages is $H_t = (\mathcal{W} \times \prod_j \mathcal{A}_j \times \mathbb{R}^n)^t$; it describes the states, the chosen actions and the received payoffs for all the players at all past stages before t. A strategy profile $\tilde{\tau} = (\tilde{\tau}_j)_{j \in \mathcal{N}}$ and a initial state $\mathbf{w}$ induce a probability distribution $P_{\mathbf{w},\tilde{\tau}}$ on the set of plays

$$H_\infty = (\mathcal{W} \times \prod_j \mathcal{A}_j \times \mathbb{R}^n)^\infty.$$

We focus on the limiting of the average payoff, i.e., $F_{j,T} = \frac{1}{T} \sum_{t=1}^T r_{j,t}$. Given an initial state $\mathbf{w}$ and a strategy profile $\tilde{\tau}$, the payoff of player j is the superior limiting of the Cesaro-mean payoff $\mathbb{E}_{\mathbf{w},\tilde{\tau}} F_{j,T}$. We assume that $\mathbb{E}_{\mathbf{w},\tilde{\tau}} F_{j,T}$ has a limit.

Definition 4.2.1.1 (Expected robust game). *We define the expected robust game as the collection*

$$\left(\mathcal{N}, (\mathcal{X}_j)_{j \in \mathcal{N}}, \mathbb{E}_{\mathbf{w}} R_j(\mathbf{w}, .) \right).$$

Definition 4.2.1.2 (Equilibrium of the expected robust game). *A strategy profile* $(\mathbf{x}_j)_{j \in \mathcal{N}} \in \prod_{j=1}^n \mathcal{X}_j$ *is a (mixed) stationary equilibrium for the expected robust game if and only if*

$$\forall j \in \mathcal{N}, \quad \mathbb{E}_{\mathbf{w}} R_j(\mathbf{w}, \mathbf{y}_j, \mathbf{x}_{-j}) \leq \mathbb{E}_{\mathbf{w}} R_j(\mathbf{w}, \mathbf{x}_j, \mathbf{x}_{-j}), \; \forall \mathbf{y}_j \in \mathcal{X}_j. \tag{4.2}$$

where $\mathbf{x}_{-j}$ *denotes* $(\mathbf{x}_k)_{k \neq j}$.

The existence of solution of Equation (4.2) is equivalent to the existence of solution of the following *variational inequality problem*: find $\mathbf{x}$ such that

$$\langle \mathbf{x} - \mathbf{y}, V(\mathbf{x}) \rangle \geq 0, \; \forall \mathbf{y} \in \prod_j \mathcal{X}_j,$$

where $\langle .,. \rangle$ is the inner product, $V(\mathbf{x}) = [V_1(\mathbf{x}), \ldots, V_n(\mathbf{x})]$,

$$V_j(\mathbf{x}) = [\mathbb{E}_{\mathbf{w}} R_j(\mathbf{w}, \mathbf{e}_{s_j}, \mathbf{x}_{-j})]_{s_j \in \mathcal{A}_j},$$

$\mathbf{e}_{s_j}$ is the unit vector with 1 at the position of s_j and 0 otherwise.

Note that an equilibrium of the expected robust game may not be an equilibrium at each time slot. This is because $\mathbf{x}$ being an equilibrium for expected robust game does not imply that $\mathbf{x}$ is an equilibrium of the game $\mathcal{G}(\mathbf{w})$ for some state $\mathbf{w} \in \mathcal{W}$. However under suitable conditions, the dynamic game with limiting of Cesaro-mean payoff corresponds to the expected game, which has at least one stationary equilibrium. The existence of such equilibrium points are guaranteed since the mappings

$$r_j : (\mathbf{x}_j, \mathbf{x}_{-j}) \longmapsto \mathbb{E}_{\mathbf{w}} R_j(\mathbf{w}, \mathbf{x}_j, \mathbf{x}_{-j}),$$

are jointly continuous, quasi-concave in $\mathbf{x}_j$, the spaces $\mathcal{X}_j$, are nonempty, convex and compact sets. The result follows by using Kakutani fixed point theorems [85].

In Section 4.6, we provide an algorithm to solve *variational inequalities* for continuous action spaces.

4.2.2 Combined Payoff and Strategy Learning

In this subsection, we will develop a fully distributed learning procedure to learn the expected payoffs as well as the optimal strategies associated with the learned payoffs. Note that, since the state of the game is unknown, the players do not know which game they are playing. We have a sequence of games generated by the value of random variable $\mathcal{W}$ that determines the states in which the number interacting players vary over time. In the case where only a single state is available (the Dirac distribution case), the exact payoff is learned.

Table 4.1 summarizes the information available to the players. We consider CODIPAS in the following form:

$$(CODIPAS - RL) \begin{cases} \mathbf{x}_{j,t+1} = f_j(\lambda_{j,t}, a_{j,t}, r_{j,t}, \hat{\mathbf{r}}_{j,t}, \mathbf{x}_{j,t}) \\ \hat{\mathbf{r}}_{j,t+1} = g_j(\nu_{j,t}, a_{j,t}, r_{j,t}, \mathbf{x}_{j,t}, \hat{\mathbf{r}}_{j,t}) \\ \qquad j \in \mathcal{N}, t \geq 0, a_{j,t} \in \mathcal{A}_j. \end{cases}$$

where

- The functions f and λ are based on estimated payoff and perceived measured payoff (possibly with small delays) such that the invariance of simplex is preserved almost surely. The function f_j defines the strategy learning pattern of player j, and $\lambda_{j,t}$ is its strategy learning rate. If at least two of the functions f_j are different, then we refer to it as *heterogeneous learning* in the sense that the learning schemes of the players are different. Let l^2 be the space of sequences $\{\kappa_t\}_{t \geq 0}$ such that $\sum_t |\kappa_t|^2 < +\infty$. and

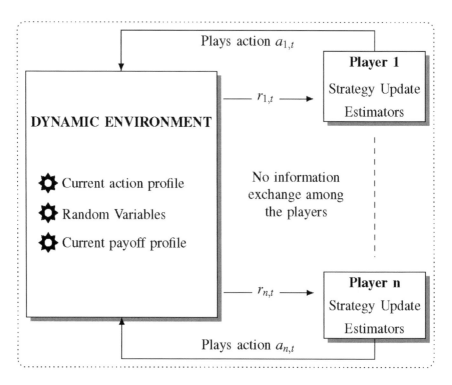

FIGURE 4.1
A generic CODIPAS-RL algorithm.

l^1 be the space of sequences $\{\kappa_t\}_{t\geq 0}$ such that $\sum_t |\kappa_t| < +\infty$. We assume that

$$\lambda_{j,t} \geq 0, \ \sum_t \lambda_{j,t} = \infty, \ \sum_t \lambda_{j,t}^2 < \infty.$$

That is, $\lambda_j \in l^2 \backslash l^1$. If all the functions f_j are identical but the learning rates λ_j are different, we refer to *learning with different speed*: slow learners, medium learners, and fast learners, etc.

- The functions g, and ν are wellchosen in order to have a good estimation of the payoffs. We assume that $\nu \in l^2 \backslash l^1$.

Why should the learning rate $\lambda_{j,t}$ sum to infinity?

Information	Known	Unknown
Existence of being in a game	No	
Existence of opponents	No	
Current game data	No	Yes
Current state	Not observed	
Nature of the noise	No	Yes
Action spaces of the others	No	Yes
Last action of the others	No	Yes
Payoff functions of the others	No	Yes
Own-payoff function	No	Yes
Number of players	No	Yes
Measurement own-payoff	Yes	No
Last own-action	Yes	No
Capability	Arithmetic operations	

TABLE 4.1
Basic assumptions for CODIPAS.

The learning rate $\lambda_{j,t}$ should be a positive real value and can be seen as a step-size. The horizon of game should be unbounded in order to have enough measurements. Therefore, the sequence $\sum_{t=1}^{T} \lambda_{j,t}$ should go to infinity when the horizon goes to infinity. However, the second condition $\lambda_j \in l^2$ is not required. As we will see, the important term is $\sum_t \lambda_{j,t} M_{j,t}$, where $M_{j,t}$ is a noise term that appears when we write the learning scheme in the form of Robbins-Monro.

Informally, each player that employs the CODIPAS-RL process acts according to the following three steps:

- Step 1: When in some situation,

- Step 2: Do some behavior (randomized actions dictates by the strategy),

- Step 3: Get some consequence (perceived payoffs) and hence update the behavior.

Lemma 4.2.2.1. *The above strategies generated by the learning process are in the class of behavioral strategies.*

Proof. Let us start with an initial state w_0 with some probability $\mathbb{P}_{w_0}$. Assume that the initial state of the algorithm x_0, is drawn with some probability measure $\mathbb{P}_{x_0}$. It is clear that the estimated payoffs are generated by these random variables (assuming that the learning rates are deterministic). Then, x_1 depends on (x_0, w_0, M_1); hence, it is a mapping from H_1 to the set $\prod_{j \in \mathcal{N}} \mathcal{X}_j$. Using the iterative stochastic equation, one has that $x_{j,t}$ is adapted to the filtration generated by the previous variables up to $t-1$ and the martingale $M_{t'}$, $t' \leq t$. This means that

$$x_{j,t} \in \sigma((x_{t'})_{t' \leq t-1}, \ (w_{t'})_{t' \leq t}, (M_{t'})_{t' \leq t}).$$

Hence, CODPAS-RL generates a behavioral strategy.

□

Cauchy sequence

The Cauchy sequence, named after Augustin-Louis Cauchy, is a sequence whose elements become arbitrarily close to each other as the sequence progresses. To be more precise, given any positive number, we can always drop some terms from the start of the sequence, so that the maximum of the distances between any two of the remaining elements is smaller than that number.

Formally, given a metric space $(\mathcal{X}, \tilde{d})$, a sequence $(\lambda_t)_t$ is Cauchy, if for every positive real number $\epsilon > 0$ there is a positive integer t_0 such that for all natural numbers $k, k' > t_0$, the distance $\tilde{d}(\lambda_k, \lambda_{k'})$ is less than ϵ.

Hilbert space

Hilbert space, named after David Hilbert, generalizes the notion of Euclidean space. It extends the methods of vector algebra and calculus from the two-dimensional Euclidean plane and three-dimensional space to spaces with any finite or infinite number of dimensions. A Hilbert space is an abstract vector space possessing the structure of an inner product that allows length and angle to be measured. Furthermore, Hilbert spaces are required to be *complete* (if every Cauchy sequence of points in $\mathcal{X}$ has a limit that is also in $\mathcal{X}$), a property that stipulates the existence of enough limits in the space to allow the techniques of calculus to be used.

Exercise 4.1. *Show that the space l^2 is Hilbert space.*

The information needed to update the strategy and the payoff estimation is given in Figure 4.1. A simple pseudocode for the simultaneous-act is as follows:

```
for i = 1 to max_of_iterations
        for j = 1 to number_of_players
            action_profile.append(select_action(player-j))
        endfor
        state=generate_state_vector
        for j = 1 to number_of_players
            playerj.payoff = assign_payoff(state,action_profile)
        endfor
        for j = 1 to number_of_players
            playerj.update_strategies
```

```
               playerj.update_payoff_estimations
         endfor
end
```

Example 4.2.3 (CODIPAS-RL based on [11, 32]). *Let $s_j \in \mathcal{A}_j$, and f_{j,s_j} be the function*

$$f_{j,s_j}(\lambda_{j,t}, a_{j,t}, r_{j,t}, \hat{\mathbf{r}}_{j,t}, \mathbf{x}_{j,t}) = x_{j,t}(s_j) + \lambda_{j,t} r_{j,t} \left(\mathbb{1}_{\{a_{j,t}=s_j\}} - x_{j,t}(s_j) \right),$$

and let g_{j,s_j} be

$$g_{j,s_j}(\nu_{j,t}, a_{j,t}, r_{j,t}, \mathbf{x}_{j,t}, \hat{\mathbf{r}}_{j,t}) = \hat{\mathbf{r}}_{j,t}(s_j) + \frac{\nu_{j,t}}{x_{j,t}(s_j)} \mathbb{1}_{\{a_{j,t}=s_j\}} \left(r_{j,t} - \hat{\mathbf{r}}_{j,t}(s_j) \right),$$

then one gets:

$$\textbf{(A)} \quad \begin{cases} \mathbf{x}_{j,t+1} & = \quad \mathbf{x}_{j,t} + \lambda_{j,t} r_{j,t+1} \left(\mathbb{1}_{\{a_{j,t+1}=s_j\}} - \mathbf{x}_{j,t} \right) \\ \hat{\mathbf{r}}_{j,t+1}(s_j) & = \quad \hat{\mathbf{r}}_{j,t}(s_j) + \frac{\nu_{j,t}}{x_{j,t+1}(s_j)} \mathbb{1}_{\{a_{j,t+1}=s_j\}} \left(r_{j,t+1} - \hat{\mathbf{r}}_{j,t}(s_j) \right) \\ & \qquad \qquad j \in \mathcal{N}, \ s_j \in \mathcal{A}_j \end{cases}$$

This is a CODIPAS-RL algorithm. It has been shown in [162, 89] that this algorithm gives nice properties of convergence in a wide range of applications that include robust potential games and anticoordination games under independent and identically distributed (i.i.d) noise. However, the learning rate and the initial conditions have to be well-chosen. The learning rate can be constant or time-varying, decreasing or not depending on the scenario under consideration. Since $\mathbb{1}_{\{a_{j,t}=s_j\}} = 0$, if the action s_j has not been played by player j at time t, the payoff-learning does not update all the components of estimates at the same time. Only the component corresponding to the action that has been played needs to be updated. This reduces the computational complexity of CODIPAS-RL. The term $\frac{1}{x_{j,t}(s_j)}$ in the payoff learning can be kept only if all the components of $x_{j,t}$ are strictly positive. If this is not the case, the term is replaced by 1. Notice that the factor $\frac{1}{x_{j,t}(s_j)}$ plays an important role in the speed of convergence of the payoff-learning and hence in the strategy-learning.

The above scheme can be generalized as follows. The numerical value of perceived payoffs are normalized such that $\lambda_{j,t} r_{j,t} \in [0,1]$.

$$\begin{cases} x_{j,t+1}(s_j) = x_{j,t}(s_j) + \begin{cases} A_{j,t}^1 \text{ if } a_{j,t} = s_j \\ A_{j,t}^2 \text{ if } a_{j,t} \neq s_j \end{cases} \\ A_{j,t}^1 := \lambda_{j,t} \sum_{s'_j \neq s_j} \Psi_{j,s_j s'_j}(\mathbf{x}_{j,t}) x_{j,t}(s'_j) - \lambda_{j,t}(1 - r_{j,t}) \sum_{s'_j \neq s_j} \Phi_{j,s_j s'_j}(\mathbf{x}_{j,t}) \\ A_{j,t}^2 := -\lambda_{j,t} \Psi_{j,s_j s'_j}(\mathbf{x}_{j,t}) x_{j,t}(s'_j) + \lambda_{j,t}(1 - r_{j,t}) \Phi_{j,s_j s'_j}(\mathbf{x}_{j,t}) \\ \hat{r}_{j,t+1}(s_j) = \hat{r}_{j,t}(s_j) + \nu_t \mathbb{1}_{\{a_{j,t+1}=s_j\}}(r_{j,t+1} - \hat{r}_{j,t}(s_j)) \\ \mathbf{x}_{j,0} \in \Delta(\mathcal{A}_j) \\ j \in \mathcal{N}, \ s_j \in \mathcal{A}_j, \end{cases}$$

where the functions Φ and Ψ should satisfy the following inequalities:

$$\Psi_{j,s_j s'_j}(\mathbf{x}_{j,t}) \leq \mathbf{x}_{j,t}(s'_j), \ \forall j \in \mathcal{N}, \ s'_j \in \mathcal{A}_j \tag{4.3}$$

$$\sum_{s'_j \neq s_j} \Phi_{j,s_j s'_j}(\mathbf{x}_{j,t}) \leq \mathbf{x}_{j,t}(s_j) \ \forall s_j \tag{4.4}$$

The deterministic differential equation is given by

$$\dot{x}_{j,t}(s_j) = \sum_{s'_j \neq s_j} x_{j,t}(s_j) \left[\Psi_{j,s_j s'_j}(\mathbf{x}_{j,t}) + \Phi_{j,s_j s'_j}(\mathbf{x}_{j,t}) \right] \left(r_j(e_{s_j}, x_{-j,t}) - r_j(e_{s'_j}, x_{-j,t}) \right) ($$

Example 4.2.4 (Boltzmann-Gibbs-based CODIPAS-RL). *We are particularly interested in Boltzmann-Gibbs-based strategy (called also softmax) because its gives a nice approximation of maxima and equilibria in a small number of iterations. If the payoff estimations are good enough, then approximated equilibria are obtained. The Boltzmann-Gibbs distribution*

$$\tilde{\beta}_{j,\epsilon}(\hat{\mathbf{r}}_{j,t})(s_j) = \frac{e^{\frac{1}{\epsilon_j}\hat{r}_{j,t}(s_j)}}{\sum_{s'_j \in \mathcal{A}_j} e^{\frac{1}{\epsilon_j}\hat{r}_{j,t}(s'_j)}}, \ s_j \in \mathcal{A}_j, \ j \in \mathcal{N}$$

are known to be close to equilibria for a big parameter $\frac{1}{\epsilon_j}$, which can be interpreted as the rationality level of the player j. The case where the ϵ_j are identically equal to ϵ has been introduced in [62] and in that context, the factor $\frac{1}{\epsilon}$ is interpreted as a temperature parameter. The Boltzmann-Gibbs distribution is a $0-$equilibrium of the perturbed game with payoffs: $r_j(\mathbf{x}_j, \mathbf{x}_{-j}) + \epsilon_j H(\mathbf{x}_j)$ where

$$H(\mathbf{x}_j) = - \sum_{s_j \in \mathcal{A}_j} x_j(s_j) \log(x_j(s_j)),$$

is the entropy function. The corresponding Nash equilibria of this perturbed game are called logit equilibria or Boltzmann-Gibbs equilibria. By choosing $f_j = \mathbf{x}_{j,t} + \lambda_{j,t} \left[\tilde{\beta}_{j,\epsilon}(\hat{\mathbf{r}}_{j,t}) - \mathbf{x}_{j,t} \right]$, one gets

$$\textbf{(BG)} \ \begin{cases} \mathbf{x}_{j,t+1} &= \quad (1 - \lambda_{j,t})\mathbf{x}_{j,t} + \lambda_{j,t}\tilde{\beta}_{j,\epsilon}(\hat{\mathbf{r}}_{j,t}) \\ \hat{\mathbf{r}}_{j,t+1}(s_j) &= \quad \hat{r}_{j,t}(s_j) + \nu_{j,t}\mathbb{1}_{\{a_{j,t+1}=s_j\}}(r_{j,t+1} - \hat{r}_{j,t}(s_j)) \\ & \qquad\qquad j \in \mathcal{N}, \ s_j \in \mathcal{A}_j \end{cases}$$

> *The Boltzmann-Gibbs based CODPIPAS-RLs outperform the standard RL algorithms.*

Consider the cycling game with three actions $\mathcal{A} = \{1, 2, 3\}$ such that 1 beats 3 which beats 2. After the matching, the winner gets a payoff of μ (the probability of the meeting occurs), otherwise the payoff is 0. This leads to the following two-by-three non-zero-sum matrix game.

$P1 \backslash P2$	1	2	3
1	$0,0$	$\mu,0$	$0,\mu$
2	$0,\mu$	$0,0$	$\mu,0$
3	$\mu,0$	$0,\mu$	$0,0$

This game is equivalent to the Shapley game. The game is not a potential game since it has no equilibrium in pure strategies. The Boltzmann-Gibbs based CODIPAS-RL outperforms the standard reinforcement learning algorithms in this Shapley game. It is known that many algorithms have trouble with Shapley's game. For instance, fictitious play cycles asymptotically. It does look like it is converging, but the number of iterations to get to the next cycle increases exponentially. The replicator dynamics, the logit-dynamics, and the best response dynamics fail also to converge [158, 161] to the equilibrium. But the Boltzmann-Gibbs based CODIPAS-RL with slow learners and fast learners converges to the approximated equilibrium of the game [103, 158, 161]. The result can be extended to more general class of cycling games by using the Poincaré-Bendixson theorem and the Dulac criterion.

⚠ Uniqueness does not imply stability

In the above example, the expected robust game has no equilibrium in pure strategies. The game has a unique equilibrium in mixed strategies. The unique equilibrium is symmetric and is located at the relative interior of the simplex. However, uniqueness does not imply stability. Replicator dynamics and best dynamics applied to the above expected robust game do not converge to the unique equilibrium point. In other words, the unique equilibrium point is unstable with the respect to these dynamics.

Example 4.2.5 (Boltzmann-Gibbs CODIPAS-RL with multiplicative weights). *The multiplicative Boltzmann-Gibbs (M-BG) CODIPAS-RL is given by*

$$
\textbf{(M-BG)} \quad
\begin{cases}
\mathbf{x}_{j,t+1}(s_j) & = \dfrac{(1+\frac{1}{\lambda_{j,t}})^{\hat{r}_{j,t}(s_j)}}{\sum_{s_j'\in\mathcal{A}_j}(1+\frac{1}{\lambda_{j,t}})^{\hat{r}_{j,t}(s_j')}} \\[2ex]
\hat{\mathbf{r}}_{j,t+1}(s_j) & = \hat{\mathbf{r}}_{j,t}(s_j) + \nu_{j,t}\mathbb{1}_{\{a_{j,t+1}=s_j\}}(r_{j,t+1}-\hat{\mathbf{r}}_{j,t}(s_j)) \\
& \qquad j\in\mathcal{N},\ s_j\in\mathcal{A}_j
\end{cases}
$$

Example 4.2.6 (Imitative Boltzmann-Gibbs based CODIPAS-RL). *Let* $\sigma_{j,\epsilon}:$ $\mathbb{R}^{\sharp\mathcal{A}_j}\longrightarrow\mathcal{X}_j$ *be the imitative Boltzmann-Gibbs strategy defined by*

$$
\sigma_{j,\epsilon}(\hat{\mathbf{r}}_{\mathbf{j,t}})(s_j) = \frac{x_{j,t}(s_j)e^{\frac{\hat{r}_{j,t}(s_j)}{\epsilon_j}}}{\sum_{s_j'\in\mathcal{A}_j}x_{j,t}(s_j')e^{\frac{\hat{r}_{j,t}(s_j')}{\epsilon_j}}}.
$$

The imitative Boltzmann-Gibbs (IBG) based CODIPAS-RL is given by

$$\textbf{(IBG)} \begin{cases} \mathbf{x}_{j,t+1} & = & (1-\lambda_{j,t})\mathbf{x}_{j,t} + \lambda_{j,t}\sigma_{j,\epsilon}(\hat{\mathbf{r}}_{\mathbf{j},\mathbf{t}}) \\ \hat{\mathbf{r}}_{j,t+1}(s_j) & = & \hat{\mathbf{r}}_{j,t}(s_j) + \nu_{j,t}\mathbb{1}_{\{a_{j,t+1}=s_j\}}\left(r_{j,t+1} - \hat{\mathbf{r}}_{j,t}(s_j)\right) \\ & & j \in \mathcal{N}, \; s_j \in \mathcal{A}_j \end{cases}$$

Example 4.2.7 (Imitative Boltzmann-Gibbs with multiplicative weights CODIPAS-RL). *The multiplicative weight imitative BG based CODIPAS-RL is given by*

$$\textbf{(M-IBG)} \begin{cases} \mathbf{x}_{j,t+1}(s_j) & = & \dfrac{x_{j,t}(s_j)(1+\lambda_{j,t})^{\hat{r}_{j,t}(s_j)}}{\sum_{s'_j \in \mathcal{A}_j} x_{j,t}(s'_j)(1+\lambda_{j,t})^{\hat{r}_{j,t}(s'_j)}} \\ \hat{\mathbf{r}}_{j,t+1}(s_j) & = & \hat{\mathbf{r}}_{j,t}(s_j) + \nu_{j,t}\mathbb{1}_{\{a_{j,t+1}=s_j\}}\left(r_{j,t+1} - \hat{\mathbf{r}}_{j,t}(s_j)\right) \\ & & j \in \mathcal{N}, \; s_j \in \mathcal{A}_j \end{cases}$$

This scheme is inspired by the classical discrete-time replicator equation given by

$$x_{j,t+1}(s_j) = x_{j,t}(s_j)\frac{\hat{\mathbf{r}}_{j,t}(s_j) + c}{c + \sum_{s'_j} x_{j,t}(s'_j)\hat{\mathbf{r}}_{j,t}(s'_j)},$$

where c is chosen such that the denominator does not vanish and the term $\hat{\mathbf{r}}_{j,t}(s_j) + c \geq 0$. If $\hat{\mathbf{r}}_{j,t}(s_j)$ is the expected payoff obtained at s_j, this discrete equation approaches the adjusted replicator dynamics. Since the expected payoff is not known, we replace it by

$$\frac{1}{\lambda_{j,t}}\left[(1+\lambda_{j,t})^{\hat{r}_{j,t}(s_j)} - 1\right]$$

Exercise 4.2. *Compute the following limit*

$$\lim_{\epsilon \longrightarrow 0^+} \frac{(1+\epsilon)^r - 1}{\epsilon}$$

4.3   Pseudo-Trajectory

Most of the convergence proofs of reinforcement learning algorithms rely on the theory of stochastic approximation. Those use the ordinary differential equation method, which relates the inhomogeneous, random, discrete time process to an autonomous, deterministic, continuous time dynamical system. Using the development of the differential equation or inclusion (ordinary or stochastic) approach to stochastic approximation studied in [33, 96, 34, 26, 29], our learning algorithm can be written in the form of Robbin-Monro iterative schemes

$$x_{t+1} = x_t + \nu_t(f(x_t) + M_{t+1}),$$

approximated by the ODE $\dot{x} = f(x)$ under standard assumptions, where f is a vectorial function, M is a noise, and $\nu \in l^2\backslash l^1$ is a learning rate.

4.3.1 Convergence of the Payoff Reinforcement Learning

We consider the payoff RL

$$\hat{\mathbf{r}}_{j,t+1}(s_j) = \hat{\mathbf{r}}_{j,t}(s_j) + \nu_{j,t}\mathbb{1}_{\{a_{j,t+1}=s_j\}}\left(r_{j,t+1} - \hat{\mathbf{r}}_{j,t}(s_j)\right).$$

By taking the expected changes in one-time slot,

$$\mathbb{E}\left(\frac{\hat{\mathbf{r}}_{j,t+1}(s_j) - \hat{\mathbf{r}}_{j,t}(s_j)}{\nu_{j,t}} \,|\hat{\mathbf{r}}_{j,t}, \mathbf{x}_t = \mathbf{x}\right) = \mathbf{x}_{j,t}(s_j)\left[\mathbb{E}_{\mathbf{w}} R_j(\mathbf{w}, \mathbf{e}_{s_j}, \mathbf{x}_{-j}) - \hat{\mathbf{r}}_{j,t}(s_j)\right],$$

Asymptotically, the limit set of the process $\hat{\mathbf{r}}_{j,t}$ can be approximated with the ODE

$$\frac{d}{dt}\hat{\mathbf{r}}_{j,t}(s_j) = \mathbf{x}_{j,t}(s_j)\left[\mathbb{E}_{\mathbf{w}} R_j(\mathbf{w}, \mathbf{e}_{s_j}, \mathbf{x}_{-j,t}) - \hat{\mathbf{r}}_{j,t}(s_j)\right], \qquad (4.6)$$

Moreover, the point $\mathbb{E}_{\mathbf{w}} R_j(\mathbf{w}, \mathbf{x})$ is globally asymptotically stable under (4.6). Using the above approach, one can conclude:

> Provided that each action is chosen infinitely often, the Q-values (estimated payoffs) in this algorithm will converge almost surely to the expected action values, that is $\hat{\mathbf{r}}_{j,t}(s_j) \longrightarrow \mathbb{E}_{\mathbf{w}} R_j(\mathbf{w}, e_{s_j}, \mathbf{x}_{-j})$ as $t \longrightarrow \infty$.

4.3.2 Folk Theorem

The strategies $\{\mathbf{x}_{j,t}\}_{t\geq 0}$ generated by these learning schemes are in the class of behavioral strategies $\tilde{\tau}$ described above. The asymptotic behaviors of the CODIPAS-RL (A) are related to the multi-type replicator dynamics combined with the payoff dynamics. An important result in the field of *evolutionary game dynamics* is the *folk theorem* (evolutionary version). It states that, under the replicator dynamics of the expected two-player game, the following properties are satisfied:

Theorem 4.3.2.1 (Folk Theorem - evolutionary version).   • *Every Nash equilibrium of the expected game is a rest point.*

- *Every strict (Nash) equilibrium of the expected game is asymptotically stable.*

- *Every stable rest point is a Nash equilibrium of the expected game.*

- *If an interior orbit converges, its limit is a Nash equilibrium of the expected game.*

For a proof of all these statements, we apply the results in [190, 74] to the expected game. We refer to [73, 43] for more recent analysis of the stochastic replicator dynamics. In the two-player case, there are many other interesting properties: If the time average $\frac{1}{T}\int_0^T x_{j,t}(s_j)\,dt$ of an interior orbit converges, the limit is a Nash equilibrium. The limit set of

$$P_{s_j s_k}^T := \frac{1}{T}\int_0^T x_{j,t}(s_j)x_{j,t}(s_k)\,dt,$$

leads to coarse correlated equilibria also called *Hannan set*.

Assume that the learning rates are wellchosen.

Proposition 4.3.2.2. • *In convergent scenarios, the strategy-reinforcement learning of CODIPAS-RL has the same speed of convergence as in [179, 56, 139].*

- *If all the actions are sufficiently used then, the payoff-reinforcement learning of CODIPAS-RL converges to the expected payoffs,*

- *The folk theorem holds for the strategy-reinforcement learning of CODIPAS-RL.*

Outline of Proof. • Choose $\lambda_{j,t} = k_j\lambda_t$ with $\lambda \in l^2\backslash l^1$. We define the drift as the expected change in the strategy in one slot:

$$
\begin{aligned}
drift &= \mathbb{E}\left(\frac{x_{j,t+1}(s_j) - x_{j,t}(s_j)}{\lambda_t} \mid \mathbf{x}_t = \mathbf{x}\right) \qquad (4.7)\\
&= k_j\mathbb{E}\left[r_{j,t+1}(\mathbb{1}_{\{a_{j,t+1}=s_j\}} - x_{j,t}(s_j)) \mid \mathbf{x}_t = \mathbf{x}\right]\\
&= k_j x_j(s_j)\left[\mathbb{E}_{\mathbf{w}}R_j(\mathbf{w}, \mathbf{e}_{s_j}, \mathbf{x}_{-j}) - \sum_{s_j'\in\mathcal{A}_j} x_j(s_j')\mathbb{E}_w R_j(\mathbf{w}, \mathbf{e}_{s_j'}, \mathbf{x}_{-j})\right].
\end{aligned}
$$

when λ_t goes to zero one gets the rescaled replicator dynamics.

- Then, we use [26, 33] to establish the comparison with the works in [179, 56, 139] in term of speed of almost sure convergence. Since the strategy reinforcement-learning leads to an adjusted version of the replicator dynamics, the folk theorem holds.

- Similarly, if $\mu_{j,t} = l_j\mu_t \in l^2\backslash l^1$

$$\mathbb{E}\left(\frac{\hat{\mathbf{r}}_{j,t+1} - \hat{\mathbf{r}}_{j,t}}{\mu_t} \mid \mathbf{x}_t = \mathbf{x}, \hat{\mathbf{r}}_{j,t}\right) \qquad (4.8)$$

The component s_j is $l_j[\mathbb{E}_{\mathbf{w}}R_j(\mathbf{w}, \mathbf{e}_{s_j}, \mathbf{x}_{-j}) - \hat{r}_{j,t}(s_j) \mid \hat{\mathbf{r}}_t, \mathbf{x}_t]$. This gives the differential equation

$$\frac{d}{dt}\hat{r}_{j,t}(s_j) = l_j[\mathbb{E}_{\mathbf{w}}R_j(\mathbf{w}, \mathbf{e}_{s_j}, \mathbf{x}_{-j}) - \hat{r}_{j,t}(s_j)],$$

which is globally asymptotically convergent to the expected payoff $\mathbb{E}_{\mathbf{w}} R_j(\mathbf{w}, \mathbf{e}_{s_j}, \mathbf{x}_{-j})$. This completes the proof.

$\square$

Using the result of Hopkins and Posch (2005, Theorem 4, [76]), the strategy-learning of CODIPAS-RL converges with probability zero to rest points corresponding to state-independent Nash equilibria of the underlying expected robust game unstable under the replicator dynamics or rest points not corresponding to a Nash equilibrium of the underlying expected robust game.

ODE of Boltzmann-Gibbs CODIPAS-RL

The ODE of the Boltzmann-Gibbs-based CODIPAS-RL is given by

$$\textbf{(ODE: BG)} \left\{ \begin{array}{rcl} \dot{\mathbf{x}}_{j,t} & = & \tilde{\beta}_{j,\epsilon}(\hat{\mathbf{r}}_{j,t}) - \mathbf{x}_{j,t} \\ \frac{d}{dt}\hat{\mathbf{r}}_{j,t}(s_j) & = & \mathbf{x}_{j,t}(s_j)\left[\mathbb{E}_{\mathbf{w}} R_j(\mathbf{w}, \mathbf{e}_{s_j}, \mathbf{x}_{-j,t}) - \hat{r}_{j,t}(s_j)\right] \end{array} \right.$$

ODE of Imitative Boltzmann-Gibbs-Based CODIPAS-RL

The ODE of the Imitative Boltzmann-Gibbs-based CODIPAS-RL is given by

$$\textbf{(ODE: IBG)} \left\{ \begin{array}{rcl} \dot{\mathbf{x}}_{j,t} & = & \sigma_{j,\epsilon}(\hat{\mathbf{r}}_{j,t}) - \mathbf{x}_{j,t} \\ \frac{d}{dt}\hat{\mathbf{r}}_{j,t}(s_j) & = & \mathbf{x}_{j,t}(s_j)\left[\mathbb{E}_{\mathbf{w}} R_j(\mathbf{w}, \mathbf{e}_{s_j}, \mathbf{x}_{-j,t}) - \hat{r}_{j,t}(s_j)\right] \end{array} \right.$$

4.3.3 From Imitative Boltzmann-Gibbs CODIPAS-RL to Replicator Dynamics

We compute the drift D which is the expected changes in one-time slot for (M-IBG) is

$$D = \frac{x_{j,t+1}(s_j) - x_{j,t}(s_j)}{\lambda_{j,t}} \tag{4.9}$$

$$= \frac{1}{\lambda_{j,t}} \left[\frac{x_{j,t}(s_j)(1 + \lambda_{j,t})^{\hat{r}_{j,t}(s_j)}}{\sum_{s_j' \in \mathcal{A}_j} x_{j,t}(s_j')(1 + \lambda_{j,t})^{\hat{r}_{j,t}(s_j')}} - x_{j,t}(s_j) \right] \tag{4.10}$$

$$= \frac{x_{j,t}(s_j)}{\sum_{s_j' \in \mathcal{A}_j} x_{j,t}(s_j')(1 + \lambda_{j,t})^{\hat{r}_{j,t}(s_j')}} \times$$

$$\left[\frac{(1 + \lambda_{j,t})^{\hat{r}_{j,t}(s_j)} - 1}{\lambda_{j,t}} - \sum_{s_j' \in \mathcal{A}_j} x_{j,t}(s_j') \frac{(1 + \lambda_{j,t})^{\hat{r}_{j,t}(s_j')} - 1}{\lambda_{j,t}} \right] \tag{4.11}$$

Using the fact that $\lambda_{j,t} \longrightarrow 0$, the term $\frac{(1 + \lambda_{j,t})^{\hat{r}_{j,t}(s_j)} - 1}{\lambda_{j,t}} \longrightarrow \hat{r}_{j,t}(s_j)$, and taking the conditional expectation, we conclude that the drift goes to the multi-type replicator equation:

$$\dot{\mathbf{x}}_{j,t}(s_j) = \mathbf{x}_{j,t}(s_j)\left[\mathbb{E}\hat{r}_{j,t}(s_j) - \sum_{s'_j \in \mathcal{A}_j} \mathbb{E}\hat{r}_{j,t}(s'_j)\mathbf{x}_{j,t}(s'_j)\right].$$

Hence, the ODE of imitative Boltzmann-Gibbs CODIPAS-RL with multiplicative weights is given by

$$\textbf{(ODE: M-IBG)} \begin{cases} \dot{\mathbf{x}}_{j,t}(s_j) & = & \mathbf{x}_{j,t}(s_j)\left[\hat{r}_{j,t}(s_j) - \sum_{s'_j \in \mathcal{A}_j} \hat{r}_{j,t}(s'_j)\mathbf{x}_{j,t}(s'_j)\right] \\ \frac{d}{dt}\hat{\mathbf{r}}_{j,t}(s_j) & = & \mathbf{x}_{j,t}(s_j)\left[\mathbb{E}_{\mathbf{w}} R_j(\mathbf{w}, \mathbf{e}_{s_j}, \mathbf{x}_{-j,t}) - \hat{r}_{j,t}(s_j)\right] \end{cases}$$

4.4 Hybrid and Combined Dynamics

In this section we derive novel game dynamics obtained by combining different learning schemes adopted by the players.

> The two-time scaling technique leads to a novel class of game dynamics called *composed game dynamics*, which have nice convergence properties in many cycling games.

4.4.1 From Boltzmann-Gibbs-Based CODIPAS-RL to Composed Dynamics

We first examine the two player case. We say that Player 1 is a slow learner if $\frac{\lambda_{1,t}}{\lambda_{2,t}} \longrightarrow 0, \frac{\lambda_{2,t}}{\nu_t} \longrightarrow 0$. In that case, the ODE is given by

$$\begin{cases} \dot{x}_{1,t} = \beta_{1,\epsilon}(\beta_{2,\epsilon}(x_{1,t})) - x_{1,t} \\ x_{2,t} \longrightarrow \beta_2(x_1) \end{cases}$$

where $\beta_{j,\epsilon} : \prod_{j' \neq j} \mathcal{X}_{j'} \longrightarrow \mathcal{X}_j$ is a smooth best-response i.e

$$\beta_{j,\epsilon}(\mathbf{x}_{-j}) = \tilde{\beta}_{j,\epsilon}(\mathbb{E}_{\mathbf{w}} R_j(\mathbf{w}, ., \mathbf{x}_{-j})).$$

Similarly, if Player 2 is a slow learner:

$$\frac{\lambda_{2,t}}{\lambda_{1,t}} \longrightarrow 0, \frac{\lambda_{1,t}}{\nu_t} \longrightarrow 0$$

$$\begin{cases} \dot{x}_{2,t} = [\beta_{2,\epsilon}(\beta_{1,\epsilon}(x_{2t}))] - x_{2t}. \\ x_{1,t} \longrightarrow \beta_{1,\epsilon}(x_2) \end{cases}$$

Note that the dynamics is completely uncoupled. Each player can follow its dynamics independently of the strategies of the other player. The dynamic is known to be convergent for the Shapley game which is a modified version of the Rock-Paper-Scissor game. The replicator dynamics, the best response dynamics, the linear-dynamics, and the fictitious play fail to converge for these two classes of games.

Combining the two ODEs together, one gets new game dynamics:

$$\begin{cases} \dot{x}_{1,t}(s_1) = [\beta_{1,\epsilon}(\beta_{2,\epsilon}(x_{1,t}))]_{s_1} - x_{1,t}(s_1), \ s_1 \in \mathcal{A}_1 \\ \dot{x}_{2,t}(s_2) = [\beta_{2,\epsilon}(\beta_{1,\epsilon}(x_{2,t}))]_{s_2} - x_{2,t}(s_2), \ s_2 \in \mathcal{A}_2 \end{cases}$$

4.4.2 From Heterogeneous Learning to Novel Game Dynamics

An example of heterogeneous learning with two players is obtained by combining the CODIPAS-RLs of Examples 4.2.6 and 4.2.7:

$$\text{(HCODIPAS-RL)} \begin{cases} \mathbf{x}_{1,t+1} = (1 - \lambda_{1,t})\mathbf{x}_{1,t} + \lambda_{1,t}\tilde{\beta}_{1,\epsilon}(\hat{\mathbf{r}}_{1,t}) \\ \hat{\mathbf{r}}_{1,t+1}(s_1) = \hat{\mathbf{r}}_{1,t}(s_1) + \nu_{1,t}\mathbb{1}_{\{a_{1,t+1}=s_1\}} (r_{1,t+1} - \hat{\mathbf{r}}_{1,t}(s_1)) \\ \mathbf{x}_{2,t+1} = \dfrac{x_{2,t}(s_2)(1 + \lambda_{2,t})^{\hat{r}_{2,t}(s_2)}}{\sum_{s_2' \in \mathcal{A}_2} x_{2,t}(s_2')(1 + \lambda_{2,t})^{\hat{r}_{2,t}(s_2')}} \\ \hat{\mathbf{r}}_{2,t+1}(s_2) = \hat{\mathbf{r}}_{2,t}(s_2) + \nu_{2,t}\mathbb{1}_{\{a_{2,t+1}=s_2\}} (r_{2,t+1} - \hat{\mathbf{r}}_{2,t}(s_2)) \end{cases}$$

(4.12)

By combining the standard reinforcement learning algorithm with Boltzmann-Gibbs learning for which the rest points are approximated equilibria, we prove the convergence of heterogeneous learning to combined (hybrid) dynamics.

Proposition 4.4.2.1. *The asymptotic pseudo-trajectory of (HCODIPAS-RL) is given by the following system of differential equations:*

$$\begin{cases} \frac{d}{dt}\hat{r}_{1,t}(s_1) &= x_{1,t}(s_1)\left(\mathbb{E}_{\mathbf{w}} R_1(\mathbf{w}, e_{s_1}, \mathbf{x}_{2,t}) - \hat{r}_{1,t}(s_1)\right), \ s_1 \in \mathcal{A}_1 \\ \dot{\mathbf{x}}_{1,t} &= \tilde{\beta}_{1,\epsilon}(\hat{r}_{1,t}) - \mathbf{x}_{1,t} \\ \frac{d}{dt}\hat{r}_{2,t}(s_2) &= x_{2,t}(s_2)\left(\mathbb{E}_{\mathbf{w}} R_2(\mathbf{w}, \mathbf{x}_{1,t}, e_{s_2}) - \hat{r}_{2,t}(s_2)\right), \ s_2 \in \mathcal{A}_2 \\ \dot{x}_{2,t}(s_2) &= x_{2,t}(s_2)\left[\hat{r}_{2,t}(s_2) - \sum_{s_2' \in \mathcal{A}_2} \hat{r}_{2,t}(s_2')x_{2,t}(s_2')\right] \end{cases}$$

(4.13)

Moreover, if $\frac{\lambda_{j,t}}{\nu_{j,t}} \longrightarrow 0$, then the system reduces to

$$\begin{cases} \dot{x}_{1t}(s_1) &= \tilde{\beta}_{1,\epsilon}(\mathbb{E}_{\mathbf{w}} R_1(\mathbf{w}, e_{s_1}, \mathbf{x}_{2t})) - x_{1,t}(s_1) \\ \dot{x}_{2t}(s_2) &= x_{2t}(s_2)\left[\mathbb{E}_{\mathbf{w}} R_2(\mathbf{w}, \mathbf{x}_{1t}, e_{s_2}) - \sum_{s_2' \in \mathcal{A}_2} \mathbb{E}_{\mathbf{w}} R_2(\mathbf{w}, \mathbf{x}_{1t}, e_{s_2'})x_{2t}(s_2')\right] \\ & s_1 \in \mathcal{A}_1, \ s_2 \in \mathcal{A}_2. \end{cases}$$

(4.14)

Proof. We first look at the case of the same learning rate λ_t for the strategies but different from the payoff learning rate ν_t. Assume that the ratio $\frac{\lambda_t}{\nu_t} \longrightarrow 0$. The scheme can be written as

$$\begin{cases} \mathbf{x}_{t+1} &= \mathbf{x}_t + \lambda_t[\tilde{f}(\mathbf{x}_t, \hat{\mathbf{r}}_t) + M_{t+1}^{(1)}] \\ \hat{\mathbf{r}}_{t+1} &= \hat{\mathbf{r}}_t + \nu_t[\tilde{g}(\mathbf{x}_t, \hat{\mathbf{r}}_t) + M_{t+1}^{(2)}] \end{cases}$$

where $M_{t+1}^{(k)}$, $k \in \{1, 2\}$ are noises. By rewriting the first equation as

$$\mathbf{x}_{t+1} = \mathbf{x}_t + \nu_t \frac{\lambda_t}{\nu_t}(\tilde{f}(\mathbf{x}_t, \hat{\mathbf{r}}_t) + M_{t+1}^{(1)}) = \mathbf{x}_t + \nu_t \tilde{M}_{t+1}^{(1)},$$

where $\tilde{M}_{t+1}^{(1)} = \frac{\lambda_t}{\nu_t}(\tilde{f}(\mathbf{x}_t, \hat{\mathbf{r}}_t) + M_{t+1}^{(1)})$. Then, by taking the conditional expectation, we obtain that

$$\begin{cases} \frac{\mathbf{x}_{t+1} - \mathbf{x}_t}{\nu_t} &= \tilde{M}_{t+1}^{(1)} \\ \frac{\hat{\mathbf{r}}_{t+1} - \hat{\mathbf{r}}_t}{\nu_t} &= \tilde{g}(\mathbf{x}_t, \hat{\mathbf{r}}_t) + M_{t+1}^{(2)} \end{cases}$$

For t sufficient large, it is plausible to view the mapping x_t as quasi-constant when analyzing the behavior of $\hat{\mathbf{r}}_t$ i.e., the ODE given by drift (expected change in one time slot) is quasi-constant:

$$\begin{cases} \mathbb{E}\left(\frac{\mathbf{x}_{t+1} - \mathbf{x}_t}{\nu_t} \mid \mathcal{F}_t\right) &= \mathbb{E}\left(\tilde{M}_{t+1}^{(1)} \mid \mathcal{F}_t\right) \longrightarrow 0 \\ \mathbb{E}\left(\frac{\hat{\mathbf{r}}_{t+1} - \hat{\mathbf{r}}_t}{\nu_t} \mid \mathcal{F}_t\right) &= \mathbb{E}\left(\tilde{g}(\mathbf{x}_t, \hat{\mathbf{r}}_t) + M_{t+1}^{(2)} \mid \mathcal{F}_t\right) \longrightarrow \mathbb{E}\tilde{g}(\mathbf{x}_t, \hat{\mathbf{r}}_t) \end{cases}$$

where $\mathcal{F}_t$ is the filtration generated by $\{\mathbf{x}_{t'}, r_{t'}, \mathbf{w}_{t'}, \hat{\mathbf{r}}_{t'}\}_{t' \le t}$.

Equivalently,

$$\begin{cases} \dot{\mathbf{x}}_t &= 0 \\ \frac{d}{dt}\hat{\mathbf{r}}_t &= \mathbb{E}\tilde{g}(\mathbf{x}_t, \hat{\mathbf{r}}_t) \end{cases}$$

Since the component s_j of the function $\tilde{g}$ is $\mathbb{E}_{\mathbf{w}} R_j(\mathbf{w}, \mathbf{e}_{s_j}, \mathbf{x}_{-j,t}) - \hat{\mathbf{r}}_{j,t}$, the second system is globally convergent to $\mathbb{E}_{\mathbf{w}} R_j(\mathbf{w}, \mathbf{e}_{s_j}, \mathbf{x}_{-j})$. Then, one gets the sequences

$$(x_t, \hat{\mathbf{r}}_t)_t \longrightarrow \{(\mathbf{x}, \mathbb{E}_{\mathbf{w}} R_j(\mathbf{w}, \mathbf{e}_{s_j}, \mathbf{x}_{-j})), \; \mathbf{x} \in \prod_{j \in \mathcal{N}} \mathcal{X}_j\}.$$

Now, consider the first equation

$$\mathbf{x}_{t+1} = \mathbf{x}_t + \lambda_t(\tilde{f}(\mathbf{x}_t, \hat{\mathbf{r}}_t) + M_{t+1}^{(1)}),$$

which can be rewritten as

$$\begin{aligned} \mathbf{x}_{j,t+1} &= \mathbf{x}_{j,t} + \lambda_t(\tilde{f}_j(\mathbf{x}_t, \mathbb{E}_{\mathbf{w}} R_j(\mathbf{w}, \mathbf{e}_{s_j}, \mathbf{x}_{-j,t})) \\ &+ \tilde{f}_j(\mathbf{x}_t, \hat{\mathbf{r}}_t) - \tilde{f}_j(\mathbf{x}_t, \mathbb{E}_{\mathbf{w}} R_j(\mathbf{w}, \mathbf{e}_{s_j}, \mathbf{x}_{-j,t})) + M_{t+1}^{(1)}) \end{aligned}$$

By denoting

$$M_{t+1}^{(3)} := \tilde{f}_j(\mathbf{x}_t, \hat{\mathbf{r}}_t) - \tilde{f}_j(\mathbf{x}_t, \mathbb{E}_{\mathbf{w}} R_j(\mathbf{w}, \mathbf{e}_{s_j}, \mathbf{x}_{-j,t})) + M_{t+1}^{(1)}$$

which goes to zero when taking the conditional expectation in $\mathcal{F}_t$. The equation can be approximated asymptotically by

$$\mathbf{x}_{j,t+1} = \mathbf{x}_{j,t} + \lambda_t(\tilde{f}_j(\mathbf{x}_t, \mathbb{E}_\mathbf{w} R_j(\mathbf{w}, ., \mathbf{x}_{-j,t}) + M_{t+1}^{(3)}).$$

This last learning scheme has the same asymptotic pseudo-trajectory as the ODE

$$\dot{\mathbf{x}}_{j,t}(s_j) = \tilde{f}_{j,s_j}(\mathbf{x}_t, \mathbb{E}_\mathbf{w} R_j(\mathbf{w}, \mathbf{e}_{s_j}, \mathbf{x}_{-j,t}))$$

For same rate or proportional learning rates λ and ν, the dynamics are multiplied by the ratio. Hence, the announced results follow: the first equation is for

$$\tilde{f}_1 = \tilde{\beta}_{1,\epsilon}(\hat{r}_{1,t}) - \mathbf{x}_{1,t},$$

the second pair of equations is obtained for the imitative Boltzmann-Gibbs CODIPAS-RL with multiplicative weights. This completes the proof. $\qquad \square$

Following the above techniques, one gets the following result in the presence of slow and fast learners in a heterogeneous setting. Let

$$\tilde{\xi}_j(\mathbf{x}_{-j})(s_j) = \frac{e^{\int_0^t \mathbb{E}_\mathbf{w} R_j(\mathbf{w}, \mathbf{e}_{s_j}, \mathbf{x}_{-j,t'})\, dt'}}{\sum_{s_j' \in \mathcal{A}_j} e^{\int_0^t \mathbb{E}_\mathbf{w} R_j(\mathbf{w}, \mathbf{e}_{s_j'}, \mathbf{x}_{-j,t'})\, dt'}}$$

The function $\tilde{\xi}$ is an explicit solution of the replicator equation. One has that,

$$\tilde{\xi}_j(\{\mathbf{x}_{-j,t}\}_t)(s_j) = \beta_{j,\frac{1}{t}}(M_{R,t})(s_j),$$

where

$$M_{R,t}(s_j) = \frac{1}{t}\int_0^t \mathbb{E}_\mathbf{w} R_j(\mathbf{w}, \mathbf{e}_{s_j}, \mathbf{x}_{-j,t'})\, dt'.$$

For the two-player case, one can use the linearity property to compute $M_{R,t}$.

$$M_{R,t}(s_1) = \frac{1}{t}\int_0^t \mathbb{E}_\mathbf{w} R_1(\mathbf{w}, \mathbf{e}_{s_1}, \mathbf{x}_{2,t'})\, dt' \qquad (4.15)$$

$$= \mathbb{E}_\mathbf{w} \frac{1}{t}\int_0^t R_1(\mathbf{w}, \mathbf{e}_{s_1}, \mathbf{x}_{2,t'})\, dt' \qquad (4.16)$$

$$= \mathbb{E}_\mathbf{w} R_1\left(\mathbf{w}, \mathbf{e}_{s_1}, \frac{1}{t}\int_0^t \mathbf{x}_{2,t'}\, dt'\right) \qquad (4.17)$$

$$= \mathbb{E}_\mathbf{w} R_1(\mathbf{w}, \mathbf{e}_{s_1}, \bar{\mathbf{x}}_{2,t}) \qquad (4.18)$$

$$= r_1(\mathbf{e}_{s_1}, \bar{\mathbf{x}}_{2,t}) \qquad (4.19)$$

where $\bar{\mathbf{x}}_{2,t}(s_2) = \frac{1}{t}\int_0^t \mathbf{x}_{2,t'}(s_2)\, dt'$. This means that in the two-player case,

$$\tilde{\xi}_1(\{\mathbf{x}_{2,t}\}_t)(s_1) = \tilde{\beta}_{1,1}(r_1(., \bar{\mathbf{x}}_2)) = \beta_{1,1}(\bar{\mathbf{x}}_2)$$

i.e., a smooth best response to the time average of the strategies of the other player.

Note that this property does not hold for more than two players. The reason is that the limiting time average of the cross-product differs from the product of the time average of each component. For $\mathbf{x}_{-j,t'}$ constant, we define $\xi_j : \prod_{j' \neq j} \mathcal{X}_{j'} \longrightarrow \mathcal{X}_j$ given by

$$\xi_j(\mathbf{x}_{-j})(s_j) = \frac{e^{t\mathbb{E}_{\mathbf{w}} R_j(\mathbf{w}, \mathbf{e}_{s_j}, \mathbf{x}_{-j})}}{\sum_{s'_j \in \mathcal{A}_j} e^{t\mathbb{E}_{\mathbf{w}} R_j(\mathbf{w}, \mathbf{e}_{s'_j}, \mathbf{x}_{-j})}} = \beta_{j, \frac{1}{t}}(\mathbf{x}_{-j})(s_j).$$

This says that the time-average of the replicator equation by fixing the strategies of the others is closely related to the best response. A similar result can be found in [75, 2, 140] by using the logit map.

Proposition 4.4.2.2 (Slow and fast learners in CODIPAS-RL). • *Assume that Player 1 is a slow learner of (BG), and Player 2 is a fast learner of (BG), then almost surely,* $\|\mathbf{x}_{2t} - \beta_{2,\epsilon}(\mathbf{x}_1)\| \longrightarrow 0$, *as t goes to infinity and,*

$$\dot{x}_{1,t} = \beta_{1,\epsilon}(\beta_{2,\epsilon}(\mathbf{x}_{1,t})) - \mathbf{x}_{1,t},$$

is the asymptotic pseudo-trajectory of $\{\mathbf{x}_{1t}\}_{t \geq 0}$.

• *Assume that Player 1 is a slow learner of (M-IBG) and Player 2 is a fast learner of (M-IBG). Then, almost surely,*

$$\|\mathbf{x}_{2,t} - \xi_2(\mathbf{x}_1)\| \longrightarrow 0,$$

as t goes to infinity and,

$$\dot{x}_{1,t}(s_1) = x_{1,t}(s_1) \left[\mathbb{E}R_1(\mathbf{w}, \mathbf{e}_{s_1}, \xi_2(\mathbf{x}_{1,t})) \right. $$
$$\left. - \sum_{s'_1 \in \mathcal{A}_1} x_{1,t}(s'_1) \mathbb{E}R_1(\mathbf{w}, \mathbf{e}_{s'_1}, \xi_2(\mathbf{x}_{1,t})) \right],$$

is the asymptotic pseudo-trajectory of $\{\mathbf{x}_{1,t}\}_{t \geq 0}$.

• *Assume that Player 1 is a slow learner of (IBG) and Player 2 is a fast learner of (BG). Then, almost surely,*

$$\|\mathbf{x}_{2t} - \beta_{2,\epsilon}(\mathbf{x}_1)\| \longrightarrow 0,$$

as t goes to infinity and,

$$\dot{x}_{1,t} = \sigma_{1,\epsilon}(\beta_{2,\epsilon}(\mathbf{x}_{1,t})) - \mathbf{x}_{1,t},$$

is the asymptotic pseudo-trajectory of $\{\mathbf{x}_{1t}\}_{t \geq 0}$.

• *Assume that Player 1 is a slow learner of (M-IBG) and Player 2 is a fast learner of (BG). Then, almost surely,*

$$\|\mathbf{x}_{2t} - \beta_{2,\epsilon}(\mathbf{x}_1)\| \longrightarrow 0,$$

as t goes to infinity and,

$$\dot{x}_{1,t}(s_1) = x_{1,t}(s_1)\left[\mathbb{E}R_1(\mathbf{w}, \mathbf{e}_{s_1}, \beta_{2,\epsilon}(\mathbf{x}_{1,t})) \qquad (4.20)\right.$$
$$\left. - \sum_{s_1' \in \mathcal{A}_1} x_{1,t}(s_1')\mathbb{E}R_1(\mathbf{w}, \mathbf{e}_{s_1'}, \beta_{2,\epsilon}(\mathbf{x}_{1,t}))\right]$$

is the asymptotic pseudo-trajectory of $\{\mathbf{x}_{1,t}\}_{t\geq 0}$.

• *Assume that Player 1 is a slow learner of (M-IBG) and Player 2 is a fast learner of (IBG). Then almost surely,*

$$\|\mathbf{x}_{2,t} - \sigma_{2,\epsilon}(\mathbf{x}_1)\| \longrightarrow 0,$$

as t goes to infinity and,

$$\dot{x}_{1,t}(s_1) = x_{1,t}(s_1)\left[\mathbb{E}R_1(\mathbf{w}, \mathbf{e}_{s_1}, \sigma_{2,\epsilon}(\mathbf{x}_{1,t})) \qquad (4.21)\right.$$
$$\left. - \sum_{s_1' \in \mathcal{A}_1} x_{1,t}(s_1')\mathbb{E}R_1(\mathbf{w}, \mathbf{e}_{s_1'}, \sigma_{2,\epsilon}(\mathbf{x}_{1,t}))\right], \qquad (4.22)$$

is the asymptotic pseudo-trajectory of $\{\mathbf{x}_{1,t}\}_{t\geq 0}$.

4.4.3 Aggregative Robust Games in Wireless Networks

The focus of our analysis in this subsection is on aggregative games. In an aggregative game, the payoff of each player is a function of the player's own action and of the sum of the actions (or, the weighted sum action) of all players. The class of aggregative games contains many interesting games from economics [53]. A wide class of oligopoly games, including Cournot's original model, models of the joint exploitation of a common resource, and collective actions models are in this class. We provide a specific class of aggregation robust games well-used in wireless networking and communication systems.

Definition 4.4.3.1. *The game* $\mathcal{G}$ *is an aggregative robust game if there exists a mapping* Φ *such that*

$$\tilde{R}_j(\mathbf{w}, \mathbf{a}) = \Phi_j(w_j a_j, \sum_{i=1}^{n} w_i a_i).$$

Examples of Payoffs

• Energy efficiency function as payoff: The payoff function of player j is given by

$$\Phi_j(\alpha_j, I) = \frac{v_j(1 + \frac{\alpha_j}{N_0 + I})}{c_j(\alpha_j)},$$

where $I = \sum_{i \neq j} w_i a_i$ the interference to player j, N_0 is a background noise, v_j is a positive real-valued function, and $c_j(.)$ is a cost for power consumption. The parameter α_j corresponds to the received power from player j, i.e $\alpha_j = w_j a_j$ where w_j is the channel gain (positive random variable).

• Shannon rate as payoff: The Shannon rate is given by

$$\tilde{R}_j(\mathbf{w}, \mathbf{a}) = \log_2 \left(1 + \frac{w_j a_j}{N_0 + \sum_{i \neq j} w_i a_i} \right),$$

• Cost of energy consumption per capacity :

$$\tilde{R}_j(\mathbf{w}, \mathbf{a}) = \frac{c_j(a_j)}{f_j \left(\frac{w_j a_j}{N_0 + \sum_{i \neq j} w_i a_i} \right)}$$

Aggregative games are an important class of games in wireless networks. Aggregative terms are observed in interference control, access control, frequency selection, congestion control, flocking behavior of sensors, consensus problems between different nodes or data centers, etc.

4.4.3.2 Power Allocation as Aggregative Robust Games

The problem of competitive Shannon rate maximization is an important signal-processing problem for power-constrained multiuser systems. It involves solving the power allocation problem for mutually interfering players (transmitters) operating across multiple frequencies. The channel state information is usually estimated at the receiver by using a training sequence, or semiblind estimation methods. Obtaining channel state information at the transmitter requires either a feedback channel from the receiver to the transmitter, or exploiting the channel reciprocity such as in time division duplexing systems. While it is a reasonable approximation to assume perfect channel state information at the receiver, usually channel state information at the transmitter cannot be assumed perfect, due to many factors such as inaccurate channel estimation, erroneous or outdated feedback, or frequency offsets between the reciprocal channels. Therefore, the imperfectness of the channel state from the transmitter-side has to be taken into consideration in any practical communication system.

We consider networks that can be modeled by a parallel multiple access channel, which consists of several orthogonal multiple access channels. The sets of transmitters (players) and channels are respectively denoted by $\mathcal{N}$ and $\mathcal{L}$. The cardinality of $\mathcal{N}$ is n, and the cardinality of $\mathcal{L}$ is L. Now, since transmitters are power-limited [172, 170], we have that

$$\forall j \in \mathcal{N}, \quad \sum_{l=1}^{L} a_{j,l} \leq a_{j,\max}. \tag{4.23}$$

where $a_{j,\max}$ is the maximum power budget available to player j at any time. The discrete power allocation set is therefore chosen such that

$$\mathcal{A}_j \subset \bar{\mathcal{A}}_j = \{(a_{j,1}, \ldots, a_{j,L}) \in \mathbb{R}_+^L \mid \sum_{l \in \mathcal{L}} a_{j,l} \leq a_{j,\max}\}.$$

We assume that player j does not know w_j. The payoff function $\tilde{R}_j$ is not known to player j and has the form of the sum-rate over all the channels. If Φ is decomposable such that

$$\tilde{R}_j(\mathbf{w}, \mathbf{a}) = \sum_{l=1}^{L} \Phi_{j,l}(w_{j,l} a_{j,l}, \sum_{i \neq j}^{n} w_{i,l} a_{i,l})$$

where $\Phi_{j,l}(\alpha, I_l) = \log_2\left(1 + \frac{w_{j,l} a_{j,l}}{N_0 + \sum_{i \neq j}^{n} w_{i,l} a_{i,l}}\right)$, then, the game is a sort of weighted aggregative robust game with unknown channel gain w. Following the work of Monderer and Shapley [116], we define a potential game in the context of randomness as follows:

Definition 4.4.3.3 (State-dependent potential games). *We say that the game* $\mathcal{G}(\mathbf{w})$ *is a potential game if there exists a function* $\phi(\mathbf{w}, \mathbf{a})$ *and functions* $\tilde{B}_j(\mathbf{w}, \mathbf{a}_{-j})$ *such that the payoff function of player j can be written as*

$$\tilde{R}_j(\mathbf{w}, \mathbf{a}) = \phi(\mathbf{w}, \mathbf{a}) + \tilde{B}_j(\mathbf{w}, \mathbf{a}_{-j}).$$

Definition 4.4.3.4 (State-dependent best-response potential games). *We say that the game* $\mathcal{G}(\mathbf{w})$ *is a best-response potential game if there exists a function* $\phi(\mathbf{w}, \mathbf{a})$ *such that*

$$\forall\, j,\ \forall\, \mathbf{a}_{-j},\ \arg\max_{a_j} \phi(\mathbf{w}, a_j, \mathbf{a}_{-j}) = \arg\max_{a_j} \tilde{R}_j(\mathbf{w}, a_j, \mathbf{a}_{-j})$$

Definition 4.4.3.5 (State-dependent pseudo-potential games). *the game* $\mathcal{G}(\mathbf{w})$ *is a pseudo-potential game if there exists a function* $\phi(\mathbf{w}, \mathbf{a})$ *such that*

$$\forall\, j,\ \forall\, \mathbf{a}_{-j},\ \arg\max_{a_j} \phi(\mathbf{w}, a_j, \mathbf{a}_{-j}) \subseteq \arg\max_{a_j} \tilde{R}_j(\mathbf{w}, a_j, \mathbf{a}_{-j}).$$

A direct consequence is that a state-dependent potential game is a state-dependent best-response potential game which is a state-dependent pseudo-potential game.

We define a *robust pseudo-potential game* as follows.

Definition 4.4.3.6 (Robust pseudo-potential game). *The family of games indexed by* $\mathcal{W}$ *is an expected robust pseudo-potential game if there exists a function* ϕ *defined on* $\mathcal{W} \times \prod_j \mathcal{A}_j$ *such that*

$$\forall\, j,\ \forall\, \mathbf{a}_{-j},\ \arg\max_{a_j} \mathbb{E}_{\mathbf{w}} \phi(\mathbf{w}, \mathbf{a}) \subseteq \arg\max_{a_j} \mathbb{E}_{\mathbf{w}} \tilde{R}_j(\mathbf{w}, \mathbf{a})$$

Particular cases of robust pseudo-potential games are coordination games, pseudo-potential games, ordinal potential games (sign preserving in Definition 4.4.3.6) which are indexed by a singleton.

Proposition 4.4.3.7. *The power allocation game with the sum of the Shannon information rate per link as payoff is an aggregative robust pseudo-potential game.*

Proof. By rewriting the payoff as $\tilde{R}_j(\mathbf{w}, \mathbf{a}) = \phi(\mathbf{w}, \mathbf{a}) + \tilde{V}_j(\mathbf{w}, \mathbf{a}_{-j})$ where

$$\phi(\mathbf{w}, \mathbf{a}) = \sum_{l \in \mathcal{L}} \log_2 \left(N_{0,l} + \sum_{j \in \mathcal{N}} a_{j,l} w_{j,l} \right),$$

$$\tilde{V}_j(\mathbf{w}, \mathbf{a}_{-j}) = \sum_{l \in \mathcal{L}} \log_2 \left(N_{0,l} + \sum_{j' \in \mathcal{N} \setminus \{j\}} a_{j',l} w_{j',l} \right),$$

we deduce that the power allocation game is a *robust pseudo-potential* with potential function $\mathbf{a} \longmapsto \int_{\mathbf{w} \in \mathcal{W}} \phi(\mathbf{w}, \mathbf{a}) \, d\nu(\mathbf{w})$. $\qquad\qquad\square$

Moreover, the power allocation game is a robust potential game. Thus, the following proposition holds:

Proposition 4.4.3.8. *Assume that $\mathcal{W}$ is a compact set. Then,*

- *The robust power allocation has at least one equilibrium in pure strategies.*

- *CODIPAS-RLs (A), (BG), (IBG), (M-BG), (M-IBG), (HCODIPAS-RL) converge almost surely to a pair that is near an equilibrium-expected equilibrium payoff.*

Numerical Examples

We consider two transmitters and two receivers. The set of actions for each transmitter is $\{Rx1, Rx2\}$. Below we plot the empirical evaluations of the strategies under CODIPAS-RLs.

In Figures 4.2 and 4.3, we plot the empirical mixed strategy of P1 under the CODIPAS-RL. We observe that the convergence time is observed after 150 iterations.

In Figures 4.4, 4.5, and 4.6, we plot the empirical strategies, average payoffs, and estimated payoffs of both player under Boltzmann-Gibbs CODIPAS-RL.

In Figures 4.7 and 4.8, we represent the probability of choosing the action 1 under Imitative CODIPAS-RL.

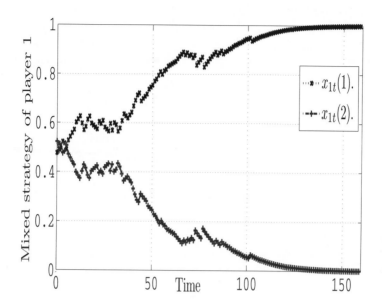

FIGURE 4.2
Mixed strategy of P1 under CODIPAS-RL BG.

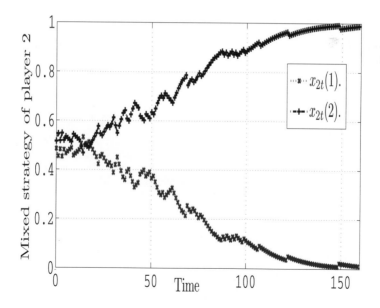

FIGURE 4.3
Mixed strategy of P2 under CODIPAS-RL BG.

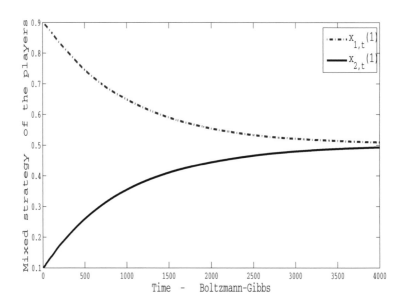

FIGURE 4.4
Probability to play the action 1 under Boltzmann-Gibbs CODIPAS-RL.

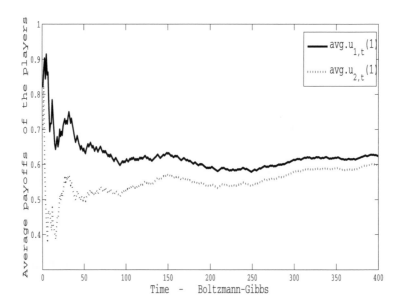

FIGURE 4.5
Average payoffs of action 1 under Boltzmann-Gibbs CODIPAS-RL.

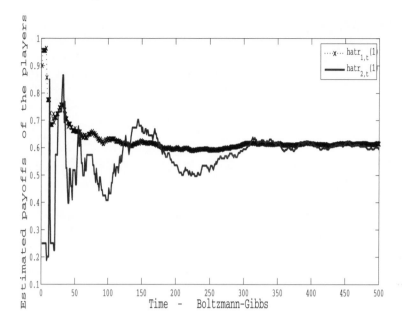

FIGURE 4.6
Estimated payoffs for action 1 under Boltzmann-Gibbs CODIPAS-RL.

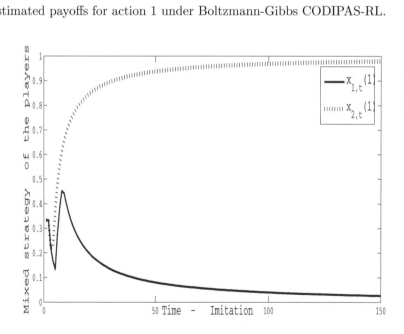

FIGURE 4.7
Probability to play the action 1 under Imitative CODIPAS-RL.

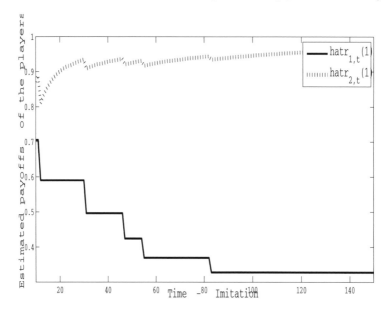

FIGURE 4.8
Estimated payoff for action 1 under Imitative CODIPAS-RL.

4.4.4   Wireless MIMO Systems

Multiple-Input Multiple-Output (MIMO) systems represent the key technology for the development of future generation wireless communication systems [152, 182, 186, 162]. Without considering the costs, the presence of multiple antennas brings multiple beneficial effects including improved reliability, larger capacity, and enhanced coverage. These benefits can be further exploited together with multiuser diversity gain in Multi-user MIMO systems. To further enhance the performance of wireless communication systems in terms of power efficiency and system throughput, various system architectures and advanced signal processing techniques are continuously being proposed and investigated for both uplink (multiple access channel) and downlink (broadcast channel) transmission. In particular, the following scenarios are receiving significant interest :

- cooperative networks, in which base stations from different cells cooperate to act as a large virtual base station;

- dense networks (femtocells), in which end users will be placing access points available to other users otherwise served by the base station; and

- massive antenna array networks, in which base stations are equipped with a large number of cheap antennas.

In all these intrinsically interference limited scenarios, multiuser MIMO plays a fundamental role in combatting, mitigating, or even annihilating the detrimental effect of interference by benefiting from the features of multi-antenna systems in various forms.

Deeper insights on multiuser MIMO potentials require the exploration of multifold directions spanning channel sounding and modeling to cooperation and cross-layer design via resource allocation, etc.

Here, we consider multiuser transmissions over multiple channels in multiple input multiple output (MIMO) wireless systems. Denote by $\tilde{x}_{j,t}$ the signal transmitted by the source j. The power constraint for each transmitter is

$$\mathbb{E}[\| \tilde{x}_{j,t} \|_2^2] = \text{tr}(Q_{j,t}) \leq c_j,$$

where $Q_{j,t}$ is the covariance matrix of the symbols transmitted by user j at time t, and c_j is the transmit power in units of energy per transmission. The maximum information rate [50] for a given collection of covariance matrix $(Q_{j,t})_{j \in \mathcal{N}}$ is given by

$$\log \det \left(I + \frac{1}{N_0} \sum_{j=1}^{n} H_{j,t} Q_{j,t} H_{j,t}^\dagger \right) - \log \det \left(I + \frac{1}{N_0} \sum_{l \neq j} H_{l,t} Q_{l,t} H_{l,t}^\dagger \right),$$

where $H_{j,t}$ is the channel matrix of j at time t, N_0 is noise, and $H_{l,t}^\dagger$ is the Hermittian of the matrix $H_{l,t}$.

4.4.4.1 Learning the Outage Probability

Consider a discrete power allocation problem under the unknown channel law. The success probability estimation is given by

$$\hat{r}_{j,t+1}(s_j) = \hat{r}_{j,t}(s_j) + \mu_{j,t} \mathbb{1}_{\{a_{j,t+1}=s_j\}} \left(r_{j,t+1} - \hat{r}_{j,t}(s_j) \right), \qquad (4.24)$$

for all $j \in \mathcal{N}$, $s_j \in \mathcal{A}_j$.

The action $s_j = Q_j$ is now a finite matrix. The matrix is fixed in a way such that $\text{tr}(Q_j) \leq c_j$, $\forall j$. The constant c_j is a fixed power constraint for user j. Let $r_{j,t} = \mathbb{1}_{\{R_j(H_t,Q_t) \geq R_j^*\}}$. When t goes to infinity, the above algorithm converges to $\mathbb{P}(R_j(H,Q) \geq R_j^*)$ without knowing the law of H_t, provided that all the matrices Q have been sufficiently explored. Thus, we deduce the outage probability by taking $1 - \mathbb{P}(R_j(H,Q) \geq R_j^*)$.

We now focus on the discrete power allocation game [24] for outage probability estimation. Using CODIPAS-RL to learn both payoffs and optimal strategies, one gets:

- Starting from the interior of the simplex (the vertices are fixed matrices Q_j), if $x_{j,t}$ converges, the outcome is an equilibrium of the game with payoff given by $\mathbb{P}(R_j(H,Q) \geq R_j^*)$.

- For a single user 1, the strategy $x_{1,t}$ converges almost surely to the optimal power.

- Using Boltzmann-Gibbs based CODIPAS-RL, any rest points are approximated equilibria and the estimated payoffs are close to the expected payoffs if all the actions have been sufficiently explored and exploited.

4.4.4.2   Learning the Ergodic Capacity

In this section, we focus on how to learn the optimal ergodic capacity

$$\mathbb{E}\log\left(\det(N_0 I + \sum_j H_{j,t} Q_{j,t} H_{j,t})\right)$$

without knowing the channel characteristics: To this end, we consider the information rate as

$$R_t = \log\left(\det(N_0 I + \sum_{j=1}^n H_{j,t} Q_{j,t} H_{j,t})\right)$$

and we assume that the ergodic theorem holds i.e the expectation is well defined: there exists a measure ν, such that $\mathbb{E}_H R = \langle f_1, \nu \rangle$ where

$$f_1(H, Q^*) = \log\det\left(N_0 I + \sum_j H_j Q_j^* H_j\right).$$

We use Riez's theorem to get the stationary distribution of the joint process $H = (H_j)_j$. Thus, CODIPAS-RL allows to learn the ergodic capacity without knowledge of the law of the time-varying channels.

4.5   Learning in Games with Continuous Action Spaces

Following the same idea as above, we can construct iterative methods for continuous action space. We assume that the set of action $\mathcal{A}_j$ is a nonempty and convex set. The payoff R_j is assumed to be concave in a_j (a_j is an element on a linear space, that can be a scalar, a vector, a matrix, an infinite dimensional path, etc.). An example is the stochastic gradient iterative method:

$$a_{j,t+1} = a_{j,t} + \lambda_{j,t} \alpha_j \left(D_{a_j} R_{j,t}(\mathbf{w}_t, a_t) + M_{j,t}\right) \tag{4.25}$$

where

- Each player chooses an action and observes an approximated gradient corresponding to its action,

- Each player j knows its learning rates $\lambda_{j,t}$, $\tilde{r}_j$.

4.5.1 Stable Robust Games

We say that the game is *stable robust game* if $\forall\, a, a'$,

$$\sum_{j=1}^{n} \mathbb{E}_{\mathbf{w}} \langle (D_{a_j} R_j(\mathbf{w}, a) - D_{a_j} R_j(\mathbf{w}, a')), (a_j - a'_j) \rangle \leq 0 \qquad (4.26)$$

We assume that for almost any $w_t \in \mathcal{W}$, the following holds: $\forall\, a_t, a'_t$,

$$\sum_{j=1}^{n} \langle D_{a_j} R_j(\mathbf{w}_t, a_t) - D_{a_j} R_j(\mathbf{w}_t, a'_t), a_{j,t} - a'_{j,t} \rangle \leq 0 \qquad (4.27)$$

Then, the game is stable robust game. This says that the vector of $-\mathbb{E}_{\mathbf{w}} DR_j(\mathbf{w}, .)$ is a monotone operator.

The strict monotonicity of $-DR_j(\mathbf{w}, .)$, $\forall j$ is sometimes referred to *Diagonally Strict Concavity* [138]. This subclass of games is called *strict stable games* [172, 167].

Proposition 4.5.1.1. *Assume that the payoffs are smooth and concave.* •
Under the strict monotonicity condition in Equation (4.26), the set of equilibria of the expected game is non-empty and convex. Moreover, uniqueness in pure strategies holds.

• *If $\mathcal{G}(\mathbf{w})$ is a strictly stable game for almost sure all $\mathbf{w}$, then the expected game is also a strict stable game. Moreover, the above iterative method (4.25) converges to the unique equilibrium of the expected game.*

• *If the continuous twice differentiable strict potential concave function, the unique equilibrium of a strictly stable game is globally asymptotically stable under the ordinary differential equation (ODE)*

$$\dot{a}_{j,t} = \alpha_j \mathbb{E}_{\mathbf{w}} D_{a_j} R_j(\mathbf{w}, a_t), \ j \in \mathcal{N}.$$

Proof. To prove the first two statements, we use [138] and we take the expectation. Since the expectation preserves the monotonicity inequality, one has a strict stable robust game. Now, we prove the last assertion. Denote by $l_j(\alpha_j, a) = \alpha_j \mathbb{E}_{\mathbf{w}} D_{a_j} R_j(\mathbf{w}, a)$, then,

$$\frac{d}{dt} l_j(.) = \langle D_a l_j, \ \dot{a} \rangle.$$

Thus, the quantity

$$\frac{1}{2} \frac{d}{dt} \parallel l(r, a) \parallel^2 = \langle l, \dot{l} \rangle = \frac{1}{2} \langle l, [D_a l_j + (D_a l_j)^t] \dot{a} \rangle$$

This implies that

$$\frac{1}{2} \frac{d}{dt} \parallel l(r, a) \parallel^2 = \frac{1}{2} \langle \dot{a}, [D_a l + (D_a l)^t] \dot{a} \rangle$$

Since the Hessian $[D_a l_j + (D_a l_j)^t]$ is definite negative, there exists $\epsilon' > 0$ such that

$$\langle \dot{a}, [D_a l + (D_a l)^t] \dot{a} \rangle \leq -\epsilon' \parallel \dot{a} \parallel^2$$

Thus, $\parallel l(\alpha, a_t) \parallel^2 \longrightarrow 0$ when t goes to infinity. This means that $a_t \longrightarrow a^*$, where a^* is the equilibrium of the expected game.

$\square$

Proposition 4.5.1.2. • *Assume that the actions spaces are compact and nonempty. Then, a robust pseudo-potential game with continuous function ϕ has least one Nash equilibrium in pure strategies.*

• *In addition, assume that the action spaces are convex. Then, every almost sure robust potential concave game with continuously differentiable potential is a stable robust game. If the potential function is strictly concave in the joint actions, then global convergence to the unique equilibrium holds.*

Proof. • Since $\mathbf{a} \longmapsto \phi(\mathbf{w}, \mathbf{a})$ is continuous on a compact and nonempty set, for any fixed value of $\mathbf{w}$, the function has a maximum. As in [116], a global maximizer of the function $\mathbf{a} \longmapsto \mathbb{E}_{\mathbf{w}} \phi(\mathbf{w}, \mathbf{a})$ is an equilibrium of the expected game.

• Now, assume that $\mathbf{a} \longmapsto \mathbb{E}_{\mathbf{w}} \phi(\mathbf{w}, \mathbf{a})$ is concave. Then, for any two action profiles $\mathbf{a}$ and $\mathbf{a}'$, one has

$$[D\{\mathbb{E}_{\mathbf{w}} \phi(\mathbf{w}, \mathbf{a})\} - D\{\mathbb{E}_{\mathbf{w}} \phi(\mathbf{w}, \mathbf{a}')\}] . [\mathbf{a} - \mathbf{a}'] \leq 0.$$

By reversing the order of $D\{\mathbb{E}_{\mathbf{w}}\}$ and using the fact that ϕ is a continuously differentiable function, one has that

$$[\mathbb{E}_{\mathbf{w}} D_a \phi(\mathbf{w}, \mathbf{a}) - \mathbb{E}_{\mathbf{w}} D_{a'} \phi(\mathbf{w}, \mathbf{a}')] . [\mathbf{a} - \mathbf{a}'] \leq 0.$$

Hence, it is a stable robust game. If the potential function is strictly concave in the joint actions, then one gets a strict stable robust game. Hence, global convergence to the unique equilibrium holds.

$\square$

Note that robust pseudo-potential games are more general pseudo-potential games. As a corollary, one has the following,

• Assume that the actions spaces are compact and nonempty. Then, a pseudo-potential game with continuous function ϕ has at least one Nash equilibrium in pure strategies.

• In addition, if the action spaces are convex, then the potential concave game is a stable game.

4.5.2 Stochastic-Gradient-Like CODIPAS

The general learning structure that we consider has the form:

$$
\begin{cases}
a_{j,t+1} &= a_{j,t} + f_j(d_{j,t}, a_{j,t}, \lambda_{j,t}) \\
\hat{\mathbf{r}}_{j,t+1} &= \hat{\mathbf{r}}_{j,t} + g_j(\hat{\mathbf{r}}_{j,t}, r_{j,t}, a_{j,t}, \mu_{j,t}) \\
& \qquad \forall j \in \mathcal{N},
\end{cases}
$$

For example, if an estimated gradient of the payoff is observed, measured or computed at each time slot, one gets:

$$
\begin{cases}
a_{j,t+1} &= a_{j,t} + \lambda_{j,t}(\nabla_j r_{j,t} + M_{j,t+1}) \\
\hat{\mathbf{r}}_{j,t+1} &= \hat{\mathbf{r}}_{j,t} + \mu_{j,t}(r_{j,t+1} - \hat{\mathbf{r}}_{j,t}) \\
& \qquad \forall j \in \mathcal{N},
\end{cases}
$$

Note that in this scheme the gradient is not known but an estimated value of the gradient, namely $\nabla_j r_{j,t} + M_{j,t+1}$, is observed by player j. The noise $M_{j,t+1}$ is assumed to satisfy $\mathbb{E}(M_{j,t+1} \mid \mathcal{F}_t) \longrightarrow 0$, where $\mathcal{F}_t$ is the filtration generated by the past event up to t.

Example 4.5.3 (Continuous power allocation). *We apply CODIPAS with continuous action space to MIMO systems with time-varying channels, without knowing the law of the channel $w_t = H_t$. (Here, we have assumed that each player has the numerical value of its estimated gradient or an approximated gradient with a small noise $M_{j,t}$.) We first use the stochastic best-response potential decomposition of the Shannon rate as payoff function. Then, we decompose R_j as*

$$
R_j(\mathbf{w}, a) = \mathcal{Y}(\mathbf{w}, a) + \mathcal{Z}_j(\mathbf{w}, a_{-j}).
$$

Now write

$$
r_{j,t} = \log \det \left(N_0 I + \sum_{j=1}^{n} H_{j,t} Q_{j,t} H_{j,t}^{\dagger} \right) - \log \det \left(N_0 I + \sum_{l \neq j} H_{l,t} Q_{l,t} H_{l,t}^{\dagger} \right),
$$

$\mathrm{tr}(Q_j) \leq c_j$. *The payoff $r_{j,t}$ has the form of $R_j(\mathbf{w}_t, \mathbf{a}_t)$ with $\mathbf{w}_t = (H_{1t}, \ldots, H_{n,t})$ and $\mathbf{a}_t = (Q_{1,t}, \ldots, Q_{n,t})$. In addition, $r_{j,t}$ is a best-response potential game with potential given by $\log \det \left(N_0 I + \sum_{j=1}^{n} H_{j,t} Q_{j,t} H_{j,t}^{\dagger} \right)$ which is concave. Then, we use the fact that every potential concave game is a stable game. Hence, the results in Subsection 4.5.1 apply.*

4.6 CODIPAS for Stable Games with Continuous Action Spaces

In this section, we develop a CODIPAS algorithm with imitation for stable games with continuous action spaces. We provide convergence bounds to approximated Nash equilibria and solutions of the variational inequalities. The convergence rate is in order of $O(\frac{c}{\sqrt{T}} + \frac{c'}{T}), c > 0, c' > 0$. The details of the proofs can be found in [154].

Let $\mathcal{X} = \prod_{j \in \mathcal{N}} \mathcal{X}_j$, with $\mathcal{X}_j$ be a $(k_j - 1)$-dimensional simplex of $\mathbb{R}^{k_j}$. Denote $d := \sum_{j \in \mathcal{N}} k_j$. Let $R_j(w, x)$ be the payoff function of player j. The expected payoff is $r_j(x) = \mathbb{E}_w R_j(w, x)$.

4.6.1 Algorithm to Solve Variational Inequality

We aim to approximate a solution to the variational inequality. Let

$$\epsilon_{NE}(x) = \sum_{j \in \mathcal{N}} \left[\max_{x'_j \in \mathcal{X}_j} r_j(x'_j, x_{-j}) - r_j(x) \right],$$

be the sum of improvements that can be obtained by unilateral deviation.

It is clear that $\epsilon_{NE}(x) \geq 0$ for any $x \in \mathcal{X}$.

Exercise 4.3. *Show that if $\epsilon_{NE}(x) = 0$ for a certain $x \in \mathcal{X}$, then one gets an equilibrium. Hints: sum of unsigned numbers.*

Monotonicity and regularity assumptions

Assumption C0: Assume that the function $x \longmapsto r_j(x)$ is Lipschitz continuous for any $j \in \mathcal{N}$, the function $x_j \longmapsto r_j(x_j, x_{-j})$ is concave, the function $x_{-j} \longmapsto r_j(x_j, x_{-j})$ is convex, and the sum function $x \longmapsto \sum_{j \in \mathcal{N}} r_j(x)$ is concave.

It follows that $r_j(., x_{-j})$ admits a subdifferential, and we choose $V_j(x) \in \partial_{x_j} r_j(x_j, x_{-j})$. The function $x \longmapsto -V(x)$ defines a monotone function.

Assumption C1: Assume that for any pair $(x, x') \in \mathcal{X}^2$, the function V satisfies

$$\| V(x) - V(x') \| \leq L \| x - x' \| + M$$

Let $\hat{V}_k(x)$ be the k-th component of the estimated value of $V(x)$. The CODIPAS algorithm for continuous action is described as follows.

Algorithm 4: CODIPAS for continuous action space.

---

foreach *Player j* **do**
 Initial x_0 in the relative interior of $\mathcal{X}$;
 Initialize to some estimations ζ_0^0, ζ_0^1;

for *t=1* **to** *T* **do**
 foreach *Player j* **do**
  Given ζ_{t-1}^1 set;
  $$\zeta_t^0[k] = \frac{\zeta_{t-1}^1[k]e^{\lambda_t \hat{V}_k(\zeta_{t-1}^1)}}{\sum_{k'} \zeta_{t-1}^1[k']e^{\lambda_t \hat{V}_{k'}(\zeta_{t-1}^1)}};$$
  $$\zeta_t^1[k] = \frac{\zeta_{t-1}^1[k]e^{\lambda_t \hat{V}_k(\zeta_t^0)}}{\sum_{k'} \zeta_{t-1}^1[k']e^{\lambda_t \hat{V}_{k'}(\zeta_t^0)}};$$
  $$x_t[k] = x_{t-1}[k] + \mu_t\left(\zeta_t^0[k] - x_{t-1}[k]\right);$$

The CODIPAS algorithm for continuous action space is then defined by the sequence of random variables

$$
\begin{cases}
\zeta_t^0[k] = \dfrac{\zeta_{t-1}^1[k]e^{\lambda_t \hat{V}_k(\zeta_{t-1}^1)}}{\sum_{k'} \zeta_{t-1}^1[k']e^{\lambda_t \hat{V}_{k'}(\zeta_{t-1}^1)}} \\[2ex]
\zeta_t^1[k] = \dfrac{\zeta_{t-1}^1[k]e^{\lambda_t \hat{V}_k(\zeta_t^0)}}{\sum_{k'} \zeta_{t-1}^1[k']e^{\lambda_t \hat{V}_{k'}(\zeta_t^0)}} \\[2ex]
x_t[k] = x_{t-1}[k] + \mu_t\left(\zeta_t^0[k] - x_{t-1}[k]\right) \\[2ex]
\mu_t = \dfrac{\lambda_t}{\sum_{t'=1}^t \lambda_{t'}}
\end{cases}
$$

4.6.2 Convergence to Variational Inequality Solution

Theorem 4.6.2.1 provides a convergence to a solution of the variational inequality with a very fast convergence rate. A proof can be obtained from [154].

Theorem 4.6.2.1. *Consider the variational inequality for state-independent Nash equilibria of the robust games. Let $T < +\infty$. We apply CODIPAS with a constant learning rate*

$$\lambda_t = \lambda = \min\left(\frac{1}{L\sqrt{3}}, \frac{\sqrt{2\log(d)}}{M}\sqrt{\frac{1}{10T}}\right), \ t \leq T.$$

Assume C0 and C1 hold.
Then,

$$\mathbb{E}\left[\epsilon_{NE}(x_T)\right] \leq \frac{7}{2T}L\log(d) + \frac{7M\log(d)}{\sqrt{T}}.$$

Moreover, for any $\chi > 0$, one has

$$\mathbb{P}\left(\epsilon_{NE}(x_T) > \rho_T + \frac{7M\chi\sqrt{\log(d)}}{\sqrt{2T}}\right) \leq e^{\frac{-\chi^2}{3}} + e^{-\chi T}$$

where $\rho_T = \frac{7}{2T}L\log(d)$.

4.7   CODIPAS-RL via Extremum-Seeking

Extremum-seeking algorithms have received significant attention recently in non-model-based online optimization problems involving dynamical systems. The basic algorithm, based on sinusoidal perturbations, has been treated in, e.g.,[9]. Recently, a time-varying version of the algorithm has been introduced, whose convergence, with probability one, has been proved in the presence of i.i.d measurement noise.

We start with the case where the action variables of the players are scalars. Assume that the payoff functions are third continuously differentiable and there exists at least one, possibly multiple, isolated stable Nash equilibria a^* such that $\frac{\partial}{\partial a_i} r_i(a^*) = 0$, $\frac{\partial^2}{\partial a_i^2} r_i(a^*) < 0$, and the matrix $\frac{\partial^2}{\partial a_i \partial a_j} r_i(a^*)$ is diagonally dominant and hence, nonsingular.

Specifically, when employing this algorithm, player j implements the strategy in continuous time:

$$a_{j,t} = \hat{a}_{j,t} + \mu_{j,t} \tag{4.28}$$

$$\frac{d}{dt}\hat{a}_{j,t} = k_j \mu_{j,t} r_{j,t} \tag{4.29}$$

$$\mu_{j,t} = l_j \sin(\tilde{w}_j t + \phi_j) \tag{4.30}$$

$$l_j, \tilde{w}_j, \phi_j > 0 \tag{4.31}$$

Under the above assumption, Krstic et al. have established a *local* convergence result: if the starting point is sufficiently close to the isolated equilibrium, the extremum-seeking algorithm (which is a fully distributed learning algorithm) is asymptotically close to the equilibrium with exponential decay.

We discuss how to extend our algorithm when an approximated gradient is not available. In other words:

Can we extend the CODIPAS-RL into dynamic robust games with continuous action spaces, nonlinear payoffs, and with the only observation of the numerical value of own-payoffs?

To answer to this question, we observe that if instead of the numerical value of the payoffs, a value of the gradient of the payoff is observed then a descent-ascent and projection based method can be used. Under monotone gradient-payoffs, the stochastic gradient-like algorithms are known to be convergent (almost surely or weakly depending on the learning rates). However, if the gradient is not available, these techniques cannot be used directly. Sometimes one need to estimate the gradient from the past numerical values as in Robbins-Monro, and the Kiefer-Wolfowitz stochastic approximation procedure.

If each player uses this CODIPAS with sinusoidal perturbation signals and if the players use distinct frequencies, then a local convergence result can be obtained [104] for non-stochastic payoff functions, i.e., single state case. The extension to stochastic payoff function is a challenging open issue.

Based on the sinusoidal perturbation algorithm, the following CODIPAS-RL scheme can be used for the unconstrained problems with small ranges of stochastic parameters:

$$\hat{Q}_{j,kl,t+1} = \hat{Q}_{j,kl,t} + \lambda_{j,t}\epsilon_j k_j \wp_{j,kl,t} r_{j,t} + \epsilon_j \sqrt{\lambda_{j,t}}\sigma_j Z_{j,t} \qquad (4.32)$$

$$\wp_{j,kl,t} = a_{j,kl}\sin(w_{j,kl}t + \phi_{j,kl}), \quad Q_{j,kl,t} = \hat{Q}_{j,kl,t} + \wp_{j,kl,t} \qquad (4.33)$$

$$\hat{r}_{j,t+1} = \hat{r}_{j,t} + \nu_t(r_{j,t} - \hat{r}_t) \qquad (4.34)$$

where $Q_{j,kl,t}$ denotes the entry (k,l) of the matrix $\mathcal{Q}_{j,t}$, $Z_{j,t}$ is an independent and identically distributed Gaussian process, and a_j, w_j, ϕ_j, are positive real-valued matrices.

> In CODIPAS-RL equilibrium-seeking algorithm, can we replace the sine function by another function?

The sinus in the expression of the CODIPAS can be replaced by another function under some regularities. Below, we explain how this can be done using averaging methods. The technique has been studied in detail in [104] for stochastic Nash seeking.

Let start with the single or empty state space case. Suppose that each player independently employs a CODIPAS with stochastic perturbation as follows.

Each player j implements the strategy in discrete time:

$$a_{j,t} = \hat{a}_{j,t} + \mu_{j,t}, \qquad (4.35)$$

$$\hat{a}_{j,t+1} = \hat{a}_{j,t} + \lambda_{j,t}k_j\mu_{j,t}r_{j,t}, \qquad (4.36)$$

$$\mu_{j,t} = l_j g_j(\eta_{j,t}), \qquad (4.37)$$

$$\eta_{j,t+1} = \eta_{j,t}(1 - \frac{\lambda_{j,t}}{\epsilon_j}) + \sigma_j\sqrt{\frac{\lambda_{j,t}}{\epsilon_j}}(\mathcal{B}_{t+1} - \mathcal{B}_t) \qquad (4.38)$$

$$l_j, \tilde{w}_j, \phi_j > 0, \epsilon_j > 0. \qquad (4.39)$$

---

where l_j is the perturbation amplitude, $k_j > 0$ is the adaptive gain, $\lambda_{j,t}$ is a learning rate, $\mathcal{B}_t$ is an independent one-dimensional standard Brownian motion on a complete probability space $(\Omega, \mathcal{F}, \mathcal{P})$ with the sample space Ω, filtration $\mathcal{F}$, and probability measure $\mathbb{P}$, $r_{j,t}$ is a perceived payoff value for player j at time t and g_j is a bounded smooth function that player j chooses, and η_j independent time homogeneous Markov ergodic processes chosen by player j. CODIPAS 4.35 can be seen as a discrete time version with step size $\lambda_{j,t}$ of the continuous time CODIPAS given by

Each player j implements the strategy in discrete time:

$$a_{j,t} = \hat{a}_{j,t} + \mu_{j,t}, \tag{4.40}$$

$$\frac{d}{dt}\hat{a}_{j,t} = k_j \mu_{j,t} r_{j,t}, \tag{4.41}$$

$$\mu_{j,t} = l_j g_j(\eta_{j,t}), \tag{4.42}$$

$$\epsilon_j \frac{d}{dt}\eta_{j,t} = -\eta_{j,t}dt + \sqrt{\epsilon_j}\sigma_j d\mathcal{B}_{j,t} \tag{4.43}$$

$$l_j, \tilde{w}_j, \phi_j > 0, \epsilon_j, \sigma_j > 0. \tag{4.44}$$

In order to establish a local convergence result for the continuous time CODIPAS 4.40, we use the work in [104] which considered a multi-input stochastic extremum-seeking algorithm to solve the problem of seeking Nash equilibria for a noncooperative game whose players seek to maximize their individual payoff functions. The payoff functions $r_j(.)$ are general, and their forms are not known to the players. The algorithm is a non-model-based approach or model-free for asymptotic attainment of the Nash equilibria. Different from classical optimization and classical game theory algorithms, where each player employs the knowledge of the functional form of his payoff and the knowledge of the other players' actions, a player employing the CODIPAS algorithm measures only his own payoff values, i.e., fully distributed, without knowing the functional form of his or other players' payoff functions. Using the stochastic differential equation perturbation, the authors prove local exponential (in probability) convergence of the algorithm. A similar local convergence in probability for discrete CODIPAS. An interesting open issue is to extend the methodology to robust games.

4.8   Designer and Users in an Hierarchical System

In this subsection we discuss how CODIPAS-RL can be used in hierarchical systems. This CODIPAS is a first step for learning in cross-layer systems and multi-level games in which there are games within games. It is a fact that in many wireless scenarios, due to uncertainty, the mobile devices cannot compute and does not play the equilibrium strategies. As a consequence, the designer who can be a local leader of the subsystem, must take this into account. We can modify the above stochastic description and adapt to this situation. At each play, there are two steps:

- The designer chooses a control strategy (for example, this can be a pricing mechanism).

- The mobile devices select their actions in response to this control.

Payoffs are received by all the designers and all the mobile devices. These payoffs depend on all the actions and the control of the designer. We look at the long-term payoff.

The designer observes her payoff from her control and from the actions of the other players. For example, in the power allocation case, an intelligent base station can play the role of designer. The intelligent base station receives all the signals and therefore computes the payoffs.

We assume that the strategies of the mobile devices are close to the best reaction of the mobile devices for t sufficiently large. Then, it is natural to ask:

How does the designer to optimize her payoff?

This type of problem is widely studied in Stackelberg markets. CODIPAS-RL solves this problem for some class of games, those with unique equilibrium in the secondary game, which is asymptotically stable for any fixed designer control. Since the choice of the designer determines which game to be played among the mobile devices, we fix the set of actions (control) of the designer to $\mathcal{A}_d$. With each control c_d, we associate a game $\mathcal{G}_{c_d}$. The designer receives a payoff $r_d(\mathcal{G}_{c_d}, \mathbf{a})$, where $\mathbf{a}$ is the strategy profile of the mobile devices.

Using the CODIPAS-RL, one has the following result:

Proposition 4.8.0.2. *Assume that for any fixed strategy c_d of the designer, the mobile devices learn in the game $\mathcal{G}_{c_d}$ would converges to a unique joint strategy of the mobile devices $a^*(c_d)$. Then, any optimization scheme that samples all the secondary games $\mathcal{G}_{c_d}$ infinity often, the designer's strategy converges to the c_d that optimizes $r_d(\mathcal{G}_{c_d}, a^*(c_d))$.*

The main difficulty in applying this result is the uniqueness and stability condition for the secondary game. In general, the secondary game may exhibit multiple equilibria but also unstable equilibrium.

Proof. The designer adjusts its strategy along a gradient with learning rate λ_t and the mobile devices adapt with learning rate μ_t such that $\frac{\lambda_t}{\mu_t} \longrightarrow 0$. The algorithm is then given by

$$\begin{cases} c_{d,t+1} &= c_{d,t} + \lambda_t(f_d(c_{d,t}, a_t) + M_{t+1}^{(1)}) \\ a_{t+1} &= a_t + \mu_t(g(c_{d,t}, a_t) + M_{t+1}^{(2)}) \end{cases}$$

with

$$\mathbb{E}\left(M_{t+1}^{(k)} \mid \mathcal{F}_t\right) \longrightarrow 0, \ \forall \, k \in \{1, 2\}.$$

The mapping is chosen such that a zero of f optimizes $c_d \longmapsto u_d(\mathcal{G}_{c_d}, a^*(c_d))$. By rewriting the first equation as $c_{d,t+1} = c_{d,t} + \mu_t \frac{\lambda_t}{\mu_t}(f_d(c_{d,t}, a_t) + M_{t+1}^{(1)})$, one gets that the sequences

$$(c_{d,t}, a_t)_t \longrightarrow \{(c_d, a^*(c_d)), \ c_d \in \mathcal{A}_d\}.$$

Now, consider the first equation $c_{d,t+1} = c_{d,t} + \lambda_t(f_d(c_{d,t}, a_t) + M_{t+1}^{(1)})$.

This equation can be approximated asymptotically by $c_{d,t+1} = c_{d,t} + \lambda_t(f_d(c_{d,t}, a^*(c_{d,t})) + M_{t+1}^{(3)})$ where $M_{t+1}^{(3)}$ is a noise. This last learning scheme has the same asymptotic pseudo-trajectory as the ODE

$$\dot{c}_d = f_d(c_d, a^*(c_d))$$

which zeros optimize the designer's payoff. $\qquad\square$

Designing Optimal Pricing in Parallel MAC

We consider a modified parallel MAC with the payoff function of the users as

$$\tilde{R}_j = \sum_{l \in \mathcal{L}} [b_l \log_2(\alpha_{j,l} SINR_{j,l}(\mathbf{h}, \mathbf{p})) - c_{j,l}p_{j,l}],$$

where $0 < p_{j,l,\min} \leq p_{j,l}$, $\forall j, l$, and $\forall j$ $\sum_l p_{j,l} \leq p_{j,\max}$, $\alpha_{j,l} > 0$, $b_l > 0$. There are two types of participants of the game:

- a set $\mathcal{N} = \{1, 2, \ldots, n\}$ of users.

- A designer chooses a pricing function (namely, the coefficients of the matrix $C = (c_{j,l})_{j \in \mathcal{N}, l \in \mathcal{L}}$)

We observe that

- The game is a concave game (each payoff function $\tilde{R}_j$ is concave in the j−th argument $\mathbf{p}_j = (p_{j,l})_{l \in \mathcal{L}}$).

- The payoff function can be written as

$$\begin{aligned}
\tilde{R}_j(\mathbf{h}, \mathbf{p}) &= \sum_l b_l \log_2 \alpha_{j,l} + \sum_l b_l \log_2 |h_{j,l}|^2 \\
&+ \sum_l [b_l \log_2(p_{j,l}) - c_{j,l}p_{j,l}] \\
&+ \sum_l b_l \log_2(N_{0,l} + \sum_{j' \neq j} |h_{j',l}|^2 p_{j',l}).
\end{aligned}$$

- The above form is exactly a potential plus a dummy part plus an (action-independent) random process.

- Thus, the potential function is

$$V(\mathbf{p}) = \sum_j \sum_l [b_l \log_2(p_{j,l}) - c_{j,l}p_{j,l}].$$

- Importantly, $V(\mathbf{p})$ is independent of the random channel gains $|h_{j,l}|^2$.

- The function $\mathbf{p} \longmapsto V(\mathbf{p})$ is strictly concave.

- Thus, the auxiliary game between users with payoff function V has a unique equilibrium that we denote by $a^*(c)$.

- The designer objective is to maximize some function $\tilde{R}^d(c, a^*(c))$ by finding the optimal pricing. This leads to an hierarchical system with two levels. For example, if $\tilde{R}^d$ is the Shannon sum-rate, i.e.,

$$\tilde{R}^d = \sum_j \sum_l b_l \log_2 \left(1 + SINR_{j,l}\right).$$

The equilibrium of the auxiliary game coincides with the global optimal of Shannon sum-rate if the pricing coefficients are fixed to be inversely proportional to optimal power. More details can be found in Theorem 3 of [44].

- Using the above combined learning, we attempt to find the joint optimal pricing-equilibrium strategies.

4.9 From Fictitious Play with Inertia to CODIPAS-RL

In robust games with a large number of actions, players are inherently faced with limitations in both their observational and computational capabilities. Accordingly, players in such games need to make their decisions using algorithms that accommodate limitations in information gathering and processing. This disqualifies some of the well-known decision-making models such as Fictitious Play, in which each player must monitor the individual actions of every other player and must optimize over a high dimensional probability space.

In this section we relax some of the observation assumptions in the fictitious play with inertia in the context of dynamic robust games. The algorithm is described as follows.

Algorithm 5: Generic representation of the fictitious-based CODIPAS-RL

---

foreach *Player j* **do**
  | Initial action $a_{j,0}$;
  | Initialize to some estimations $\hat{\mathbf{r}}_{j,0}$,;

for *t=1* **to** *max* **do**
  | **foreach** *Player j* **do**
  | | Choose an action $a_{j,t}$ with inertia $\epsilon \in (0,1)$:;
  | | $a_{j,t} \in \arg\max_{s'_j} \hat{\mathbf{r}}_{j,t}(s'_j)$ with probability $(1-\epsilon)$;
  | | $a_{j,t} = a_{j,t-1}$ with probability ϵ;
  | | Observe a numerical value of its noisy payoff $r_{j,t}$;
  | | Update its payoff estimation via
  | | $\hat{\mathbf{r}}_{j,t+1}(s_j) = \hat{\mathbf{r}}_{j,t}(s_j) + \nu_t \mathbb{1}_{\{a_{j,t+1}=s_j\}}(r_{j,t+1} - \hat{\mathbf{r}}_{j,t}(s_j))$;

The factor $\nu_t \geq 0$ can be chosen to be a small constant or a time-varying parameter. The following proposition gives the convergence of the algorithm in robust potential games.

Proposition 4.9.0.3. *The fictitious-based CODIPAS-RL converges almost surely to equilibria in robust potential games.*

Proof. It is a direct extension of the proof in [193, 107, 108] to robust games.
□

Example 4.9.1. *Our motivating example for high complexity and imperfect observation is a large-scale congestion game. We consider a distributed traffic routing, in which a large number of vehicles make daily routing decisions to optimize their own objectives in response to their own observations. In this setting, observing and responding to the individual actions of all vehicles on a daily basis would be a formidable task for any individual driver. A more realistic measurement on the information tracked and processed by an individual driver is the daily aggregate congestion on the roads that are of interest to that driver. Therefore, the standard fictitious play is not applicable. Using the fictitious-based CODIPAS-RL, one attempts to predict the optimal routing as well as the associated minimal cost.*

4.10   CODIPAS-RL with Random Number of Active Players

In this section, we discuss how CODIPAS-RL can be extended to games with an unknown and random number of interacting players. The number of players in interaction may vary in time, and the distribution is unknown. We consider a system with a maximum n of potential players. Each player j can be active or not, depending on some independent random process. Each player j has a finite number of actions, denoted by $\mathcal{A}_j$. The payoff of player j is a random variable $R_{j,t}$ with realized value denoted by $r_{j,t}$ at time t. The random variable is indexed by $\mathbf{w}$ in some set $\mathcal{W}$ and a random set of active players $\mathcal{B}(t)$. We assume that the players

- do not observe the past strategies played by the others (imperfect monitoring),
- do not know the state of the game, i.e., the pair $(\mathbf{w}_t, \mathcal{B}(t))$ at time t,
- do not know the expression of their payoff functions,
- we assume that the transition probabilities between the states $\mathbf{w}$ are independent and unknown to the players.

> ⚠ You don't have to know you are playing game to be in a game!

At each time t, each player chooses an action and receives a numerical noisy value of its payoff (*perceived payoff*) at that time. The perceived payoff is not necessarily obtained by feedback, ACK/NACK, etc. The perceived payoff can be interpreted as a measured metric in the random environment. The stochastic game [71, 198] with random number of interacting players is described by

$$\mathcal{G} = (\mathcal{N}, (\mathcal{A}_j)_{j \in \mathcal{N}}, \mathcal{W}, (R_j(\mathbf{w}, .))_{j \in \mathcal{N}, \mathbf{w} \in \mathcal{W}}, \mathcal{B})$$

where

- $\mathcal{N} = \{1, 2, \ldots, n\}$ denotes the set of potential players. This set is unknown to the players. $\mathcal{B}(t) \subseteq \mathcal{N}$ is the set of interacting players (active players) at time t.

- $j \in \mathcal{N}$, $\mathcal{A}_j$, is the set of actions available to player j. In each state, the same action spaces are available. Player j does not know the action spaces of the other players.

- *No information on the current state:* $\mathcal{W}$ is the set of possible states, a subset of $\mathbb{R}^k$ for some integer k. We assume that the state space $\mathcal{W}$ and the probability transition on the states are both *unknown* to the players. The state is represented by independent random variable (the transitions between the states are independent of the chosen actions). We assume that the current state w_t is unknown to the players.

- *Payoff functions are unknown:*

$$R_j^{\mathcal{B}} : \mathcal{W} \times \prod_{j=1}^{n} \mathcal{X}_j \longrightarrow \mathbb{R}$$

Information	Known	Unknown
Set of active players	Variable	Yes
Existence of being in a game	No	
Variable number of active players	No	
Transition probabilities	No	
Existence of opponents	No	
Current game data	No	Yes
Current state	Not observed	
Nature of the noise	No	Yes
Action spaces of the others	No	Yes
Last action of the others	No	Yes
Payoff functions of the others	No	Yes
Own-payoff function	No	Yes
Number of potential players	No	Yes
Measurement own-payoff	Yes	No
Last own-action	Yes	No
Capability	Arithmetic operations	

TABLE 4.2
CODIPAS: information and computation assumptions

where $\mathcal{X}_j := \Delta(\mathcal{A}_j)$ is the set of probability distributions over $\mathcal{A}_j$, i.e.,

$$\mathcal{X}_j := \left\{ \mathbf{x}_j : \ x_j(s_j) \in [0,1], \ \sum_{s_j \in \mathcal{A}_j} x_j(s_j) = 1 \right\}; \qquad (4.45)$$

- We denote the stage game by

$$\mathcal{G}(\mathbf{w}_t) = \left(\mathcal{N}, (\mathcal{A}_j)_{j \in \mathcal{N}}, (R_j^{\mathcal{B}}(\mathbf{w}_t, .))_{j \in \mathcal{N}} \right).$$

Table 4.2 summarizes the information available to the active players.

Since the players do not observe the past actions of the other players, we consider strategies used by both players to be only dependent on their current perceived own payoffs. Denote by $x_{j,t}(s_j)$ the probabilities of player j choosing s_j at time t, and let $\mathbf{x}_{j,t} = [x_{j,t}(s_j)]_{s_j \in \mathcal{A}_j} \in \mathcal{X}_j$ the mixed state-independent strategy of player j. The payoffs $R_{j,t}$ are random variables; the payoff functions are unknown to the players. We assume that the distribution or the law of the possible payoffs is also unknown. The long-run game is described as follows.

- At time slot $t = 0$, $\mathcal{B}(0)$ is the set of active players. Each player $j \in \mathcal{B}(0)$ chooses an action $a_{j,0} \in \mathcal{A}_j$ and perceives a numerical value of its payoff that corresponds to a realization of the random variables depending on the actions of the other players and the state of the nature, etc. He initializes its estimation to $\hat{\mathbf{r}}_{j,0}$. The non-active players get zero.

Data of the game	Learnable
Law of the payoffs	Possible under specific payoffs
Law of the opponents choices	Riez's theorem links the expected payoff and law of the opponents choices
Equilibrium payoff (own)	Learnable in many convergent behavior

TABLE 4.3
CODIPAS: learnable data.

- At time slot t, each player $j \in \mathcal{B}(t)$ has an estimation of his or her payoffs, chooses an action based his or her own-experiences and experiments with a new strategy. Each player j receives an output $r_{j,t}$. Based on this target $r_{j,t}$ the player j updates his or her estimation vector $\hat{\mathbf{r}}_{j,t}$ and builds a strategy $\mathbf{x}_{j,t+1}$ for next time slot. The strategy $\mathbf{x}_{j,t+1}$ is a function only of $\mathbf{x}_{j,t}, \hat{\mathbf{r}}_{j,t}$ and the target value. Note that the exact value of the state of the nature $\mathbf{w}_t$ at time t and the past strategies $\mathbf{x}_{-j,t-1} := (\mathbf{x}_{k,t-1})_{k\neq j}$ of the other players and their past payoffs $\mathbf{r}_{-j,t-1} := (r_{k,t-1})_{k\neq j}$ are unknown to player j at time t.

- The game moves to $t+1$.

Table 4.3 summarizes some learnable data of the game.

Remark 4.10.1. *If the set of active players is constant, i.e., $\mathcal{B}(t) = \mathcal{N}$, one gets the model of [161, 162] developed for n-player dynamic games. This case covers, in particular, matrix games with i.i.d random entries.*

We focus on the limiting of the average payoff, i.e., $F_{j,T} = \frac{1}{T}\sum_{t=1}^{T} r_{j,t}$. A player's information consists of his past own-actions and perceived own-payoffs. A private history of length t for player j is a collection

$$h_{j,t} = (a_{j,0}, r_{j,0}, a_{j,1}, r_{j,1}, \ldots, a_{j,t-1}, r_{j,t-1}) \in H_{j,t} := (\mathcal{A}_j \times \mathbb{R})^t.$$

A behavioral strategy for player j at time t is a mapping

$$\tilde{\tau}_{j,t} : H_{j,t} \longrightarrow \mathcal{X}_j.$$

The set of complete histories of the dynamic game after t stages is $H_t = (\mathcal{B} \times \mathcal{W} \times (\prod_j \mathcal{A}_j) \times \mathbb{R}^n)^t$, it describes the set of active players, the states, the chosen actions and the received payoffs for all the players at all past steps before t. A strategy profile $\tilde{\tau} = (\tilde{\tau}_j)_{j\in\mathcal{N}}$ and a initial state $\mathbf{w}$ induce a probability distribution $P_{\mathbf{w},\tilde{\tau}}$ on the set of plays $H_\infty = (\mathcal{W} \times (\prod_j \mathcal{A}_j) \times \mathbb{R}^n)^\infty$. Given an initial state $\mathbf{w}$ and a strategy profile $\tilde{\tau}$, the payoff of player j is the superior limiting of the Cesaro-mean payoff $\mathbb{E}_{\mathbf{w},\tilde{\tau},\mathcal{B}} F_{j,T}$. We assume that $\mathbb{E}_{\mathbf{w},\tilde{\tau},\mathcal{B}} F_{j,T}$ has a limit.

We will develop fully distributed learning procedures to learn the expected payoffs as well as the optimal strategies associated with the learned payoffs. Note that, since the state is unknown, the players do not know which game they are playing. They even do not know if they are playing a game or not! We have a sequence of games generated by the value of random variable $\mathcal{W}$ which determines the states in which the number interacting players varies over time. The distribution of the number of players is unknown. We define the expected game as

$$\left(\mathcal{N}, (\mathcal{X}_j)_{j \in \mathcal{N}}, \mathbb{E}_{\mathbf{w}, \mathcal{B}} R_j^{\mathcal{B}}(\mathbf{w}, .)\right).$$

A state-independent strategy profile $(\mathbf{x}_j)_{j \in \mathcal{N}} \in \prod_{j=1}^n \mathcal{X}_j$ is a (mixed) stationary *equilibrium* for the expected game if and only if

$$\forall j \in \mathcal{N}, \quad \mathbb{E}_{\mathbf{w}, \mathcal{B}} R_j^{\mathcal{B}}(\mathbf{w}, \mathbf{y}_j, \mathbf{x}_{-j}) \leq \mathbb{E}_{\mathbf{w}, \mathcal{B}} R_j^{\mathcal{B}}(\mathbf{w}, \mathbf{x}_j, \mathbf{x}_{-j}), \ \forall \, \mathbf{y}_j \in \mathcal{X}_j. \quad (4.46)$$

The existence of a solution of Equation (4.46) is equivalent to the existence of a solution of the following *variational inequality problem*: find $\mathbf{x}$ such that

$$\langle \mathbf{x} - \mathbf{y}, V(\mathbf{x}) \rangle \geq 0, \ \forall \mathbf{y} \in \prod_j \mathcal{X}_j,$$

where $\langle ., . \rangle$ is the inner product, $V(\mathbf{x}) = [V_1(\mathbf{x}), \ldots, V_n(\mathbf{x})]$,

$$V_j(\mathbf{x}) = [\mathbb{E}_{\mathbf{w}, \mathcal{B}} R_j^{\mathcal{B}}(\mathbf{w}, \mathbf{e}_{s_j}, \mathbf{x}_{-j})]_{s_j \in \mathcal{A}_j}.$$

Lemma 4.10.1.1. *The stochastic game with unknown state and variable number of interacting players has at least one stationary state-independent equilibrium.*

The existence of such equilibrium points are guaranteed since the mappings $r_j : (\mathbf{x}_j, \mathbf{x}_{-j}) \longmapsto \mathbb{E}_{\mathbf{w}, \mathcal{B}} R_j^{\mathcal{B}}(\mathbf{w}, \mathbf{x}_j, \mathbf{x}_{-j})$ are jointly continuous, quasi-concave in $\mathbf{x}_j$, the spaces $\mathcal{X}_j$, are non-empty, convex and compact. The result follows by using the Kakutani fixed point theorem.

Exercise 4.4 (Two jammers and one regular user). *Consider the following scenario between three users (players). Each user has two choices which can be the set {transmit, stay quiet} or { transmit to receiver 1, transmit to receiver 2}. User 1 chooses a row, mobile 2 chooses a column and user 3 chooses one of the two matrices. User 1's objective is to jam the communication of user 2. User 2's objective is to jam the communication of user 3. User 3 is a regular node, his/her objective is to avoid a collision with user 1. The cycling nature of the game creates a lot of difficulties in learning.*

- *Compute the Nash equilibria of the finite game in Figure 4.9.*

- *Apply fictitious play in this game. What are the frequencies generated by the fictitious play algorithm? Specify the limiting behavior.*

- *The CODIPAS-RL-based fictitious play described in Section 4.9.*

	1	2		1	2
1	$(\mu,\mu,0)$	$(0,0,0)$	1	$(\mu,0,\mu)$	$(0,\mu,\mu)$
2	$(0,\mu,\mu)$	$(\mu,0,\mu)$	2	$(0,0,0)$	$(\mu,\mu,0)$

FIGURE 4.9
Two jammers and one regular node.

- *Apply the Boltzmann-Gibbs-based CODIPAS-RL*
    - *with same learning rates,*
    - *with fixed ratio learning rates*
    - *with two slow learners and one fast learner.*
    - *with the payoff-RL learning rates faster than the learning rates of the strategy-learning*
- *Order the learning rates*

$$0 < \lambda_{1,t} < \lambda_{2,t} < \lambda_{3,t},$$

such that $\lambda_j \in l^2 \backslash l^1$ *and*

$$\frac{\lambda_{1,t}}{\lambda_{2,t}} \longrightarrow 0, \quad \frac{\lambda_{2,t}}{\lambda_{3,t}} \longrightarrow 0.$$

   - *Prove that the asymptotic trajectory of user 3 depends only on user 1's behavior. As long as user 1's strategy is fixed, then user 3's strategy will converge to*

$$\mathbf{x}_3 = \beta_{3,\epsilon}(\mathbf{x}_1, \mathbf{x}_2) = \beta_{3,\epsilon}(\mathbf{x}_1).$$

   - *Show that user 2's strategy converges to the unique smooth best response* $\beta_{2,\epsilon}(\mathbf{x}_3)$
   - *Study the ordinary differential equation*

$$\dot{\mathbf{x}}_1 = \beta_{1,\epsilon}\left(\beta_{2,\epsilon}(\mathbf{x}_1, \beta_{3,\epsilon}(\mathbf{x}_1)), \beta_{3,\epsilon}(\mathbf{x}_1)\right) - \mathbf{x}_1.$$

   - *Show that the Nash equilibrium is a global attractor under the above dynamics.*
- *Conclude.*

4.11 CODIPAS for Multi-Armed Bandit Problems

The multi-armed bandit (MAB) is a celebrated problem in the field of sequential prediction. In the MAB problem, the forecaster has to choose, over some time sequence, among a finite set of available actions and the only information he gets at each step is the instantaneous loss he suffers for the selected action, i.e., he receives a numerical value of its cost. A standard criterion to assess the performance of a given strategy for action selection is the difference between the average over time of the instantaneous losses called *time-average cost* and the minimal (over the set of actions) average loss. This gap is also known as the *regret term*. The statistical theory is then devoted to estimating how this risk function can be controlled in terms of the number K of possible actions and the length T of the time sequence.

In this section, we briefly mention the basic idea of the MAB in a stochastic setup first introduced by Robbins (1952, [133]).

> Description of the problem:
> Let $\mathcal{A} = \{s_1, s_2, \ldots, s_K\}$ be the set of K available actions, $2 \leq K < +\infty$. There is a single player. At each time $t \in \mathbb{Z}_+$, the player chooses sequentially an action $a_t \in \mathcal{A}$, and observable a loss c_t which is a stochastic process. Then the average loss up to time horizon T is $\bar{C}_T = \frac{1}{T} \sum_{t=1}^{T} c_t$. The player aims to minimize $\bar{C}_T$.

Next, we define history and strategy for the player. A history for the player corresponds to sequence action-loss up to the current time. A strategy is a sequence of mapping for the choice of an action $a_{t'}$ depending on the past events $(a_1, c_1, \ldots, a_{t'}, c_{t'})$. Any strategy generates a flow of $\sigma-$algebras $\mathcal{F}_{t'} = \sigma(a_1, c_1, \ldots, a_{t'}, c_{t'})$, $t' \geq 1$.

We provide a convergence in mean and get the rate of convergence of $\mathbb{E}(\bar{C}_T)$ to the minimum loss as fast as possible. In particular, we provide non-asymptotic upper bounds for the expected excess risk that are close, up to logarithmic factors.

Recall that the length of the horizon T is unknown to the player. Therefore all the parameters are designed without the value of T. For any vector $c = (c_1, \ldots, c_K)$, we denote by $\beta_\epsilon(c)$ the Boltzmann-Gibbs distribution defined by the probability vector

$$\tilde{\beta}_{\epsilon,j}(c) = \frac{e^{-\frac{c_j}{\epsilon}}}{\sum_{k=1}^{K} e^{-\frac{c_k}{\epsilon}}},$$

where ϵ is a strictly positive parameter (not necessarily small).

We propose a CODIPAS with randomized action in which, at each step $t + 1$, the action a_{t+1} is drawn according to a distribution $x_t = (x_t(s_1), \ldots, x_t(s_K))$ over $\mathcal{A}$ where

$$x_t(s_j) = \mathbb{P}\left(a_{t+1} = s_j \mid \mathcal{F}_t\right), \ s_j \in \mathcal{A}, \ t \geq 1.$$

This is a stochastic algorithm. For deterministic MAB problems, see [12]. Here, each player updates the distribution x_t over time in an iterative manner. The result of this section are mainly based on [154].

The algorithm is described as follows.

Algorithm 6: Boltzmann-Gibbs-based CODIPAS for Multi-Armed Bandit

---

Fix $\mathbf{x}_0$;

Initialize $\hat{c}_0 = 0 \in \mathbb{R}^K$;

for *t=1* **to** T **do** do;

Draw an action $a_t = s_{j_t}$ with random j_t distributed according to $x_t.$;

Observe its current payoff $c_t = c(a_t)$;

Update its estimation via

$$\hat{c}_{t+1}(s_{j_t}) = \hat{c}_t(s_{j_t}) + \nu_t \left(\frac{1}{x_t(s_{j_t})}[c_t + \delta_t] + \mathbb{1}_{\{a_t = s_{j_t}\}} \right);$$

Update via Boltzmann-Gibbs: $\mathbf{x}_{t+1}(s_{j_t}) = \tilde{\beta}_{\epsilon_t, s_{j_t}}(\hat{c}_{t+1})$;

We are going to tune the parameters $\nu_t, \delta_t, \epsilon_t$ involved in the algorithm. The following is for the case where the loss function is positive, and $\delta_t = 0$ which corresponds to the exponentiated stochastic gradient algorithm.

Theorem 4.11.0.2. *Let* $\nu_t = 1$, $\delta_t = 0$, $\epsilon_t = \epsilon_0 \sqrt{t+1}$, $\epsilon_0 = \sigma \sqrt{\frac{K}{\log K}}$, *where* σ^2 *is a bound for conditional second moment of the cost/loss function:*

- *We assume that the second conditional moment of the cost/loss function is almost surely bounded by a constant:* $\mathbb{E}\left(c_t^2 \mid \mathcal{F}_{t-1}, a_t\right) \leq \sigma^2 < +\infty.$

- *We assume that with probability one, the conditional expectations satisfy*

$$\mathbb{E}\left(c_t \mid \mathcal{F}_{t-1}, a_t = s_j\right) = \bar{c}(s_j)$$

where $\bar{c}(s_j)$ *are unknown deterministic time-independent values.*

- *The costs are nonnegative* $c_t \geq 0$ *almost surely.*

Then, for any horizon $T \geq 2$, *one has*

$$\mathbb{E}\left(\bar{C}_T\right) - \min_{s_j \in \mathcal{A}} \bar{c}(s_j) \leq 2\sigma \frac{\sqrt{T+1}}{T} \sqrt{K \log K}$$

Application to single user cognitive radio environment

We apply the CODIPAS arm-bandit scheme in cognitive radio environment where a single user wants to learn the quality of experience/service under uncertain environment as well as the action that maximizes the expected quality. The quality of service and the quality of experience depend on the environment state (channel state) and the action picked by user which can be the access point or cognitive station.

To do so, we define the indicator of the quality by an Acknowledgment.

The user adapts her choice based on the received ACK/NACK. The above CODIPAS scheme provides a bound on the error over horizon T. Theorem 4.11.0.2 says that the gap between the expected quality obtained with this algorithm and the maximum quality is less than ϵ if $T \geq \frac{2K \log K}{\epsilon^2}$.

Next we provide a useful bound for unsigned loss functions.

Theorem 4.11.0.3. *Let* $\nu_t = 1$, $\delta_t = \frac{\sigma}{2\kappa} \log t$, $\epsilon_t = \epsilon_0 \sqrt{t+1} \log(t+e)$, $\epsilon_0 = \tilde{\epsilon}_0 \sigma \sqrt{\frac{K}{\log K}}$, *where* σ^2 *is a bound for the conditional second moment of the loss function: we assume that*

- *(i)* $\mathbb{E}\left(c_t^2 \mid \mathcal{F}_{t-1}, a_t\right) \leq \sigma^2 < +\infty$.

- *(ii) With probability one, the conditional expectations satisfy*

$$\mathbb{E}\left(c_t \mid \mathcal{F}_{t-1}, a_t = s_j\right) = \bar{c}(s_j)$$

where $\bar{c}(s_j)$ *are unknown deterministic time-independent values.*

- *(iii) The cost functions are unsigned but we assume that there exists* $\kappa, \tilde{\kappa}_c > 0$ *such that* $\forall t \geq 1$;

$$\mathbb{E}\left(|c_t| e^{-\frac{\kappa}{\sigma} c_t} \mid \mathcal{F}_{t-1}, a_t\right) \leq \tilde{\kappa}_c \sigma.$$

Then, for any horizon $T \geq 2$, *one has*

$$\mathbb{E}\left(\bar{C}_T\right) - \min_{s_j \in \mathcal{A}} \bar{c}(s_j) \leq \sigma \frac{\sqrt{T+1}}{T} \sqrt{K \log K} \times$$

$$\left((\tilde{\epsilon}_0 + \frac{1}{2\kappa^2 \tilde{\epsilon}_0}) \log(T+e) + \frac{2}{\tilde{\epsilon}_0} + \frac{2\tilde{\kappa}_c}{\sqrt{K \log K}} \right) \qquad (4.47)$$

By minimizing the right hand side (rhs) of the equation (4.47) in $\tilde{\epsilon} > 0$, one can find both the optimal parameter $\tilde{\epsilon}$ and the corresponding optimal upper bound (then $\tilde{\epsilon}$ will depend on the time horizon T, however). In case T is unknown, we can fix $\tilde{\epsilon} = \frac{1}{\kappa}$ or $\tilde{\epsilon} = 1$ if κ is unknown.

4.12   CODIPAS and Evolutionary Game Dynamics

The notion of revision protocol is fundamental in population games and evolutionary game dynamics. In this section we define a revision protocol (of actions) for finite normal form games as well as robust games and population games.

4.12.1 Discrete-Time Evolutionary Game Dynamics

Given a state profile or subpopulation profile $\mathbf{x} \in \prod_{j \in \mathcal{N}} \Delta(\mathcal{A}_j)$ and a payoff function $V(\mathbf{x}) \in \mathbb{R}^{\sum_{j \mathcal{N}} |\mathcal{A}_j|}$, we define a discrete time revision protocol as

$$\mathcal{L}_{j,s_j s_j'} : \left(\prod_{j \in \mathcal{N}} \Delta(\mathcal{A}_j) \right) \times \mathbb{R}^{\sum_{j \in \mathcal{N}} |\mathcal{A}_j|} \longrightarrow [0,1],$$

the probability to switch from action s_j to the action s_j' by player j.

The strategy-learning scheme is given by

$$x_{j,t+1}(s_j) = x_{j,t}(s_j) + \sum_{s_j'} \mathcal{L}_{j,s_j' s_j}(\mathbf{x}_t, V(\mathbf{x}_t)) x_{j,t}(s_j') - x_{j,t}(s_j) \sum_{s_j'} \mathcal{L}_{j,s_j s_j'}(\mathbf{x}_t, V(\mathbf{x}_t))$$

where $j \in \mathcal{N}$, $(s_j, s_j') \in \mathcal{A}_j^2$, and the transitions $\mathcal{L}_{j,s_j' s_j}(.)$ are normalized such that

$$\sum_{s_j'} \mathcal{L}_{j,s_j' s_j}(.) \leq 1,$$

and the transition that player j decides not to switch is given by

$$\mathcal{L}_{j,s_j s_j}(.) = 1 - \sum_{s_j'} \mathcal{L}_{j,s_j' s_j}(.) \geq 0$$

Example 4.12.2 (Imitative dynamics). *A discrete time revision protocol is said to imitative if for every player j, the transition can be written as*

$$\mathcal{L}_{j,s_j s_j'}(.) = x_j(s_j') \hat{\mathcal{L}}_{j,s_j s_j'}(.).$$

Example 4.12.3. *For replicator dynamics, the function $\hat{\mathcal{L}}$ can be chosen in the form*

$$\tilde{\mathcal{L}}_{j,s_j s_j'}(\mathbf{x}, V(\mathbf{x})) = \frac{[V_{j,s_j'}(\mathbf{x}) - V_{j,s_j}(\mathbf{x})]_+}{c + \langle \mathbf{x}_j, V_j(\mathbf{x}) \rangle}, \quad c > 0$$

where $[y]_+ = \max(0,y) \geq 0$. This means that the replicator dynamics is an imitative dynamics. The dynamics is also uncoupled in terms of payoffs in the sense that $\hat{\mathcal{L}}_{j,s_j s_j'}(\mathbf{x}, V(\mathbf{x}))$ does not depend on the function V_{-j}.

4.12.4 CODIPAS-Based Evolutionary Game Dynamics

In the context of CODIPAS, the players do not have access to the expected payoff function $V(x)$. Each player j needs to learn her or his expected payoff V_j via a payoff-learning given by $\hat{r}_{j,t+1}$. Therefore, we need to incorporate the estimated payoff into the discrete time revision protocol. The CODIPAS-based revision protocol is therefore given by

$$\hat{\mathcal{L}}_{j,s_j s_j'} : (\mathbf{x}_j, \hat{\mathbf{r}}_j) \in \Delta(\mathcal{A}_j) \times \mathbb{R}^{|\mathcal{A}_j|} \longrightarrow \hat{\mathcal{L}}_{j,s_j s_j'}(\mathbf{x}_j, \hat{\mathbf{r}}_j) \in [0,1],$$

which represents the probability to switch from action s_j to the action s'_j by player j.

Note that player j does not need the multivector $\mathbf{x}_{-j}$ to compute the transition $\hat{\mathcal{L}}_{j,s_j s'_j}$. Similarly, player j do not need $\hat{\mathbf{r}}_{-j}$ in order to compute the transition $\hat{\mathcal{L}}_{j,s_j s'_j}$.

The CODIPAS-based evolutionary game dynamics is given by

$$
\begin{cases}
x_{j,t+1}(s_j) = x_{j,t}(s_j) + \sum_{s'_j} \mathcal{L}_{j,s'_j s_j}(\mathbf{x}_{j,t}, \hat{\mathbf{r}}_{j,t}) x_{j,t}(s'_j) - x_{j,t}(s_j) \sum_{s'_j} \mathcal{L}_{j,s_j s'_j}(\mathbf{x}_{j,t}, \hat{\mathbf{r}}_{j,t}) \\
\hat{r}_{j,t+1}(s_j) = \hat{r}_{j,t}(s_j) + \nu_t \mathbb{1}_{\{a_{j,t+1}=s_j\}}(r_{j,t+1} - \hat{r}_{j,t}(s_j)) \\
\mathbf{x}_{j,0} \in \Delta(\mathcal{A}_j) \\
j \in \mathcal{N}, \; s_j \in \mathcal{A}_j,
\end{cases}
$$

Note that the interaction between the players is indirectly in the measured payoff $r_{j,t+1}$.

4.13   Fastest Learning Algorithms

In this section, we address speed of convergence and running time of simple classes of learning algorithms. The running time analysis is a familiar problem in learning in games as well as in machine learning.

In order to introduce the problem of convergence time, we first start with a classical problem in statistics: Given a target population, how can we obtain a representative sample?

In the context of learning in games, this question can be seen as follows: Given a list measurement (such as perceived payoffs), can we obtain a useful information such as best-response strategy or expected payoff distribution?

We consider the class of CODIPAS schemes that generate irreducible aperiodic Markov chains. Let $\mathbf{x}_t$ be an irreducible aperiodic Markov chain with invariant probability distribution π, having support $\Omega \subseteq \prod_{j \in \mathcal{N}} \mathcal{A}_j$ and let $\mathbb{L}^t$ denote the distribution of $\mathbf{x}_t | \mathbf{x}_0$ for $t \geq 1$, that is

$$
\mathbb{L}^t(x, \Gamma) = \mathbb{P}(x_t \in \Gamma \mid x_0 = x).
$$

Then, given any $\eta > 0$, can we find an integer t^* such that

$$
\| \mathbb{L}^t(x, .) - \pi \|_{tv} \leq \eta, \; \forall t \geq t^*
$$

where tv denotes the total variation norm.

Note that, under the above assumptions, $\| \mathbb{L}^t(x, .) - \pi \|_{tv}$, is nonincreasing in t. This means that every draw past t will also be within a range η of π, thus providing a representative sample if we keep only the draws after t^*. For the Gibbs distributions/Glauber dynamics, there is an enormous amount of research on this problem for a wide variety of Markov chains leading a class of learning schemes in games. Unfortunately, there is apparently little that can be said generally about this problem so that we are forced to analyze each learning scheme chain individually or at most within a limited class of models or situations such as potential, geometric, etc.

To simplify the analysis, we focus on the reversible Markov chain case, this is, for example, satisfied by $\mathcal{L}_2$. If $\mathcal{L}_{a,a'}(x)$ denotes the transition matrix and $m = \prod_{j \in \mathcal{N}} |\mathcal{A}_j|$ the number of action profiles, it is wellknown that the convergence time to reach the stationary distribution is governed by the second highest eigenvalue [52] of the matrix $(\mathcal{L}_{a,a'})$ after 1, Let

$$1 = eig_1(\mathbb{L}) \geq eig_2(\mathbb{L}) \geq \ldots \geq eig_m(\mathbb{L}) \geq -1.$$

The speed of convergence is given by the $\frac{1}{1 - eig_2(\mathbb{L})}$. The smaller $eig_2(\mathbb{L})$ is, the faster the Markov chain $\mathbf{x}_t$ approaches π. We refer the reader to [35, 52] for more details on the fastest mixing Markov chains.

Based on this observation, we define the fastest learning algorithm along the class satisfying the above assumptions as follows:

$$\inf_{\mathbb{L} \geq 0} eig_2(\mathbb{L}) \tag{4.48}$$

$$\pi_a \mathbb{L}_{a,a'} = \pi_{a'} \mathbb{L}_{a',a} \tag{4.49}$$

$$\sum_{a' \in \mathcal{A}} \mathbb{L}_{a,a'} = 1, \ \forall a \in \Omega. \tag{4.50}$$

This is an optimization problem among the learning schemes. Since $eig_2(.)$ is continuous and the set of possible transition matrices constraint is compact, there is at least one optimal transition matrix; the inf can be by min,i.e., an optimal (for the convergence time to π) learning scheme among the class of CODIPAS satisfying the above assumptions exists.

Since we have the existence result, we need to explain how to find this optimal CODIPAS algorithm. This leads to the question of solvability of (4.48). Since the eigenvalue $eig_1(.) = 1$ with eigenvector $(1, 1, \ldots, 1)$, we can write the eigenvalue $eig_2(\mathbb{L})$ as an optimization of a quadratic term over vectors:

$$eig_2(\mathbb{L}) = \sup\{\langle v, \mathbb{L}v \rangle \mid \sum_{a \in \Omega} v_a = 0, \ \| v \| \leq 1\}$$

From [52], we know that Glauber like dynamics satisfies

$$\| \mathbb{L}^t(x, .) - \pi \|_{tv} \leq \sqrt{m} eig_2(\mathbb{L})^t.$$

As a consequence of [35, 52], the convergence time for Boltzmann-Gibbs-like CODIPAS falls within a range η to π is less than $\tilde{c}(m \log m + m \log(\frac{1}{\eta}))$, $\tilde{c} > 0$.

Exercise 4.5 (CODIPAS). *This exercise proposes to review the basic properties and results of CODIPAS.*

(L0) What is CODPAS?

(L1) Provide the main properties of CODIPAS: information requirement, computational capabilities assumptions, structure of the learning patterns, possible outcome.

(L2) Propose a CODIAS for two-by-two games.

(L3) Propose a CODIAS for two-by-three games.

(L4) How can CODIPAS be extended to robust games?

(L5) How can CODIPAS-RL be extended to robust games with asynchronous updates, noisy measurements?

(L6) Propose two CODIPASs for games with continuous action spaces,

(L7) Propose a CODIPAS algorithm that leads to state-independent equilibria in robust games with monotone gradient payoffs.

(L8) Propose a CODIPAS algorithm for population games

(L4) Propose a convergent CODIPAS algorithm for stable population games. Specify the convergence rate and the convergence time of the CODIPAS.

(L5) Propose a learning scheme to approximate the gradient of a (unknown) continuous differentiable function in $\mathbb{R}$.

- *Examine the bilevel CODIPAS for the design of wireless network,*

- *Apply CODIPAS-RL with multi-armed-bandit problems,*

- *Find the fastest learning schemes among the following CODIPASs: Boltzmann-Gibbs-based, imitation-based, multiplicative weighted-based, multiplicative weighted-and-imitation-based.*

Chapter review

This chapter develops combined fully distributed payoff and strategy reinforcement learning. The results of this chapter can be summarized as follows:

> - The consequences influence behavior.
> - The behaviors influence the outcomes.

This fixed-point of behaviors leads to novel classes of game dynamics including composed dynamics and heterogeneous dynamics that help to understand the behavior of cycling games from a dynamical system viewpoint.

The statement of the above system is that *the consequences influence behavior.* It means that players choose actions because they know other outcomes will follow. Thus, depending upon the type of consequence that follows, players will produce some behaviors and avoid others (eventually in a probabilistic way), which in consequence influences their decisions as in the second statement. Hence, the idea of CODIPAS-RL is realistic and pretty simple.

Using appropriated two-time-scaling technique one can reduces the analysis to the following system:

- The average of consequences $\Longrightarrow$ expected outcome.

- The behaviors influence the outcomes.

Exercise 4.4 shows that a variant of CODIPAS-RLs could be used to better understand the behavior of cycling games. Later, we extend the CODIPAS-RL based on extremum-seeking technique to continuous action space with local stability conditions. The methodology is adapted to leader-follower systems where the leader learning pattern influences the followers' learning process and vice-versa. The CODIPAS is flexible enough to incorporate a variable number of interacting players and random updates.

Interestingly, the imitative CODIPAS provides a good convergence rate in stable robust games. Under mild assumptions, the convergence rate to a solution of the variational inequality is given by

$$\mathbb{E}\left[\epsilon_{NE}(x_T)\right] \leq \frac{7}{2T} L \log(d) + \frac{7M \log(d)}{\sqrt{T}}.$$

The chapter also shows that CODIPAS can be applied to Multi-Armed Bandit Problems and convergence rate bounds are explicitly given by

$$\frac{\sqrt{T+1}}{T} \sqrt{n \log n} \left((\tilde{\epsilon}_0 + \frac{1}{2\kappa^2 \tilde{\epsilon}_0}) \log(T+e) + \frac{2}{\tilde{\epsilon}_0} + \frac{2\tilde{\kappa}_c}{\sqrt{n \log n}} \right)$$

Finally, a mathematical program is formulated to find the fastest learning algorithms. The program suggests looking at the learning patterns that have second highest eigenvalues, which are as small as possible.

5

Learning under Delayed Measurement

Education is what survives when what has been learned has been forgotten, B. F. Skinner

5.1 Introduction

In this chapter, we present robust strategies under the imperfectness of the state information and outdated measurement from the players' side. We develop fully distributed reinforcement learning schemes under uncertain state and delayed feedbacks. We provide the asymptotic pseudo-trajectories of the delayed schemes. Considering imperfectness in term of payoff measurement from the players' side, we propose a delayed COmbined fully DIstributed Payoff and Strategy Reinforcement Learning (delayed CODIPAS-RL, [162, 175, 89]) in which each player learns her expected payoff function as well as the associated optimal reaction strategies. Under the delayed CODIPAS-RL, the player can use different learning patterns (heterogeneous learning) and different learning speeds (different time-scaling). We give sufficient conditions for the convergence of the heterogeneous learning to ordinary differential equations (ODEs). Finally, we examine convergence properties of the resulting ODEs as in the previous chapters. The methodology is illustrated in power allocation in a Multiple-Input-Multiple-Output systems. Interestingly, this provides a fully distributed learning scheme, a CODIPAS that achieve an equilibrium in Multiple-Input-Multiple-Output interference dynamic environment without own-channel state information (CSI).

5.2 Learning under Delayed Imperfect Payoffs

We consider the dynamic robust game indexed by $\mathcal{G}(\mathbf{w}_t)$, $t \geq 0$. Since the players do not observe the past actions of the other players, we consider strategies used by both players to be only dependent on their delayed perceived own-payoffs and own-past histories. Denote by $x_{j,t}(s_j)$ the probabilities of player j choosing the power allocation s_j at time t, and let $\mathbf{x}_{j,t} = [x_{j,t}(s_j)]_{s_j \in \mathcal{A}_j} \in \mathcal{X}_j$ be the mixed state-independent strategy of player j. The payoffs $R_{j,t}$ are random variables, and the payoff functions are unknown to the players. We assume that the distribution or the law of the possible payoffs is also unknown. We do not use any Bayesian assumption on the initial beliefs formed over the possible states.

Our idea is to learn the expected payoffs simultaneously with the optimal strategies during a long-run interaction: a dynamic game. The dynamic game is described as follows.

- From time slot $t = 0$ to τ_j each player j chooses an action $a_{j,0} \in \mathcal{A}_j$. At time τ_j, layer j perceives a numerical noisy value of its payoff of $t = 0$ which corresponds to a realization of the random variables depending on the actions of the other players and the state. He initializes its estimation to $\hat{\mathbf{r}}_{j,\tau}$.

- At time slot t, each player j has an estimation of her own-payoffs, chooses an action $a_{j,t}$ based its own-experiences and experiments a new strategy. Each player j receives a delayed output (target) $r_{j,t-\tau_j}$ from her old experiment. Based on this target $r_{j,t-\tau_j}$, the player j updates her estimation vector $\hat{\mathbf{r}}_{j,t}$ and builds a strategy $\mathbf{x}_{j,t+1}$ for next time slot. The strategy $\mathbf{x}_{j,t+1}$ is a function only of $\mathbf{x}_{j,t}, \hat{\mathbf{r}}_{j,t}$ and the target value. Note that the exact value of the state $\mathbf{w}_t$ at time t is unknown to the player j, and the exact value of the delayed own-payoffs is unknown; the past strategies $\mathbf{x}_{-j,t-1} := (\mathbf{x}_{k,t-1})_{k \neq j}$ of the other players and their past payoffs $\mathbf{r}_{-j,t-1} := (\mathbf{r}_{k,t-1})_{k \neq j}$ are also unknown to player j.

- The game moves to $t + 1$.

Histories under delayed measurement

A player's information consists of his past own-actions and perceived delayed own-payoffs. A private history of length t for player j is a collection

$$h_{j,t} = (a_{j,0}, \ldots, a_{j,\tau_j}, \ r_{j,1}, \ldots, a_{j,t}, r_{j,t-\tau_j}) \in \mathcal{M}_{j,t} = \mathcal{A}_j^{\tau_j} \times (\mathcal{A}_j \times \mathbb{R})^{t-\tau_j+1}.$$

Behavioral strategies under delayed measurement

A strategy for player j at time t is a mapping

$$\tilde{\tau}_{j,t} : \mathcal{M}_{j,t} \longrightarrow \mathcal{X}_j.$$

In order to avoid interpretation problems with the meaning of the notation $\bigcup_{t\geq 0} \mathcal{M}_{j,t}$ which needs the completion of the spaces via canonical injections, we work with the collection of mappings $\tilde{\tau}_j = (\tilde{\tau}_{j,t})_t$. The set of complete histories of the dynamic game after t stages is $\mathcal{M}_t = \mathcal{W}^t \times \prod_j \mathcal{M}_{j,t}$; it describes the states, the chosen actions and the received delayed payoffs for all the players at all past stages before t.

A strategy profile $\tilde{\tau} = (\tilde{\tau}_j)_{j\in\mathcal{N}}$ and an initial state $\mathbf{w}$ induce a probability distribution $P_{\mathbf{w},\tilde{\tau}}$ on the set of plays

$$\mathcal{M}_\infty = \mathcal{W}^\infty \times \prod_j \mathcal{M}_{j,\infty}.$$

Cesaro-mean payoff

Given an initial state $\mathbf{w}$ and a strategy profile $\tilde{\tau}$, the payoff of player j is the superior limiting of the Cesaro-mean payoff $\mathbb{E}_{\mathbf{w},\tilde{\tau}} F_{j,T}$.

5.2.1 CODIPAS-RL under Delayed Measurement

We develop heterogeneous, delayed, and combined fully distributed payoff and strategy reinforcement learning (CODIPAS-RL) framework for the discrete action dynamic games under *uncertainty* and *delayed feedback*. The general learning pattern has the following form: for each player $j \in \mathcal{N}$,

$$\begin{cases} \mathbf{x}_{j,t+1}(s_j) = f_{j,s_j}(\lambda_{j,t}, a_{j,t}, r_{j,t-\tau_j}, \hat{\mathbf{r}}_{j,t}, \mathbf{x}_{j,t}), \\ \hat{\mathbf{r}}_{j,t+1}(s_j) = g_{j,s_j}(\mu_{j,t}, r_{j,t-\tau_j}, \mathbf{x}_{j,t}, \hat{\mathbf{r}}_{j,t}) \\ \\ \forall\, j,t,\ a_{j,t} \in \mathcal{A}_j, s_j \in \mathcal{A}_j \end{cases}$$

where

- The functions f and λ are based on estimated payoff and perceived measured payoff (with delay) such that the invariance of the simplex is preserved. The function f_j defines the strategy learning pattern of player j, and λ_j is its learning rate. If at least two the functions f_j are different then we refer to *heterogeneous learning* in the sense that the learning schemes of the players are different. We will assume that

$$\lambda_{j,t} \geq 0, \sum_t \lambda_{j,t} = \infty, \sum_t \lambda_{j,t}^2 < \infty.$$

That is, $\lambda_j \in l^2 \backslash l^1$. If all the functions f_j are identical but the learning rates λ_j are different, we refer to *learning with different speed*: slow learners, fast learners, etc.

- The functions g, and μ are well-chosen in order to have a good estimation of the payoffs. We assume that $\mu \in l^2 \backslash l^1$.

- $\tau_j \geq 0$ is a feedback delay associated with the payoff of player j.

The generic algorithm can be summarized as follows:

Algorithm 7: Generic representation of the delayed CODIPAS-RL

---

foreach *Player j* **do**
  Initial strategy $\mathbf{x}_{j,0}$;
  Initialize to some estimations $\hat{\mathbf{r}}_{j,0}$,;

for *t=1* **to** *max* **do**
  foreach *Player j* **do**
    Choose an action s_j with probability $\mathbf{x}_{j,t}(s_j)$;
    Observe its a numerical value of its delayed noisy payoff $r_{j,t-\tau_j}$;
    Update its payoff estimation via
    $\hat{\mathbf{r}}_{j,t+1} = g_j(\mu_{j,t}, r_{j,t-\tau_j}, \mathbf{x}_{j,t}, \hat{\mathbf{r}}_{j,t})$;
    Update its learning pattern strategy f_j;
    Compute the distribution over $a_{j,t+1}$ from
    $\mathbf{x}_{j,t+1} = f_j(\lambda_{j,t}, a_{j,t}, r_{j,t-\tau_j}, \hat{\mathbf{r}}_{j,t}, \mathbf{x}_{j,t})$;

Examples of CODIPAS-RL under Delayed Measurement

Let us give some examples of delayed fully distributed learning algorithms under delayed measurement/

$$
\text{(Delayed-BG)} \begin{cases} \mathbf{x}_{j,t+1} = (1 - \lambda_{j,t})\mathbf{x}_{j,t} + \lambda_{j,t}\tilde{\beta}_{j,\epsilon}(\hat{\mathbf{r}}_{j,t}) \\ \hat{\mathbf{r}}_{j,t+1}(s_j) = \hat{\mathbf{r}}_{j,t}(s_j) + \mu_{j,t}\mathbb{1}_{\{a_{j,t}=s_j\}}\left(r_{j,t-\tau_j} - \hat{\mathbf{r}}_{j,t}(s_j)\right) \\ j \in \mathcal{N}, \; s_j \in \mathcal{A}_j \end{cases}
$$

$$(5.1)$$

$$
\text{(Delayed-A)} \begin{cases} x_{j,t+1}(s_j) = x_{j,t}(s_j) + \lambda_{j,t}r_{j,t-\tau_j}\left(\mathbb{1}_{\{a_{j,t}=s_j\}} - x_{j,t}(s_j)\right) \\ \hat{\mathbf{r}}_{j,t+1}(s_j) = \hat{\mathbf{r}}_{j,t}(s_j) + \mu_{j,t}\mathbb{1}_{\{a_{j,t}=s_j\}}\left(r_{j,t-\tau_j} - \hat{\mathbf{r}}_{j,t}(s_j)\right) \\ j \in \mathcal{N}, \; s_j \in \mathcal{A}_j \end{cases}
$$

$$(5.2)$$

where

$$
\tilde{\beta}_{j,\epsilon}(\hat{\mathbf{r}}_{j,t})(s_j) = \frac{e^{\frac{1}{\epsilon_j}\hat{r}_{j,t}(s_j)}}{\sum_{s_j'} e^{\frac{1}{\epsilon_j}\hat{r}_{j,t}(s_j')}},
$$

is the Boltzmann-Gibbs strategy.

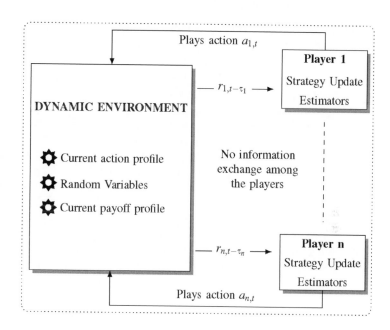

FIGURE 5.1
A delayed CODIPAS-RL algorithm.

Heterogeneous Delayed CODIPAS-RL

An example of heterogeneous learning with two players is then obtained by combining the above two patterns: (Delayed -HCODIPAS-RL)

$$\begin{cases} \mathbf{x}_{1,t+1} = (1 - \lambda_{1,t})\mathbf{x}_{1,t} + \lambda_{1,t}\tilde{\beta}_{1,\epsilon}(\hat{\mathbf{r}}_{1,t}) \\ \hat{\mathbf{r}}_{1,t+1}(s_1) = \hat{\mathbf{r}}_{1,t}(s_1) + \mu_{1,t}\mathbb{1}_{\{a_{1,t}=s_1\}}\left(r_{1,t-\tau_1} - \hat{\mathbf{r}}_{1,t}(s_1)\right) \\ \mathbf{x}_{2,t+1} = \mathbf{x}_{2,t} + \lambda_{2,t}r_{2,t-\tau_2}\left(\mathbb{1}_{\{a_{2,t}=s_2\}} - \mathbf{x}_{2,t}\right) \\ \hat{\mathbf{r}}_{2,t+1}(s_2) = \hat{\mathbf{r}}_{2,t}(s_2) + \mu_{2,t}\mathbb{1}_{\{a_{2,t}=s_2\}}\left(r_{2,t-\tau_2} - \hat{\mathbf{r}}_{2,t}(s_2)\right) \end{cases} \quad (5.3)$$

5.3 Reacting to the Interference

The problem of competitive Shannon rate maximization is an important signal-processing problem for power-constrained multiuser systems. It involves solving the power allocation problem for mutually interfering transmitters operating across multiple frequencies. The classical approach to Shannon rate maximization has been finding globally optimal solutions based on waterfilling [50]. However, the major drawback of this approach is that these solutions require centralized control. These solutions are inherently unstable in a competitive multi-user scenario, since a gain in performance for one transmitter may result in a loss of performance for others. Instead, a distributed game-theoretic approach is desirable and is being increasingly considered only over the past decade. The seminal works on competitive Shannon rate maximization use a game-theoretic approach to design a decentralized algorithm for two-user dynamic power control. These works proposed a sequential iterative waterfilling algorithm for reaching the Nash equilibrium in a distributed manner. A Nash equilibrium of the rate-maximization game is a power allocation configuration such that given the power allocations of other transmitters, no transmitter can further increase the achieved information rate unilaterally. However, most of the existing works on power allocation games assume perfect channel state information (CSI). This is a very strong requirement and generally cannot be met by practical wireless systems. The traditional game-theoretic solution for systems with imperfect information is the Bayesian game model, which uses a probabilistic approach to model the uncertainty in the system. However, a Bayesian approach is often intractable and the results strongly depend on the nature of the probability distribution functions. Thus, a relaxation of the use of the initial probability distribution is needed. we propose a robust game approach to solve this problem. The motivations to consider imperfect channel state are the following:

- The channel state information (CSI), is usually estimated at the receiver by using a training sequence, or semi-blind estimation methods. Obtaining channel state information at the transmitter requires either a feedback channel from the receiver to the transmitter, or exploiting the channel reciprocity such as in time division duplexing systems.

- While it is a reasonable approximation to assume perfect channel state information at the receiver, usually channel state information at the transmitter cannot be assumed perfect, due to many factors such as inaccurate channel estimation, erroneous or outdated feedback, and time delays or frequency offsets between the reciprocal channels.

- Therefore, the imperfectness of channel state from the transmitter-side has to be taken into consideration in many practical communication systems.

How to learn without own-CSI?

There are two classes of models frequently used to characterize imperfect channel state information: the stochastic and the deterministic models. One of the deterministic approaches is the pessimistic or Maximin robust approach modeled by an extended game in which *Nature* chooses the channel states. The pessimistic model consists of seeing *Nature* as a player who minimizes over all the possible states (the worst case for the transmitters). The pessimistic approach has been studied in [20, 184]. A similar approach of incomplete information finite games has been modeled as a distribution-free robust game where the transmitters use a robust approach to bounded payoff uncertainty. This robust game model also introduced a distribution-free equilibrium concept called the robust-optimization equilibrium. However, the results in [1] for the robust game model are limited to finite games, which need to be adapted to continuous power allocation (when the action space is a compact set). The difficulty in characterizing equilibria is that best response correspondence may not be conytractive and multiple equilibria are obtained [4, 158]. In that case, the best response dynamics need to be modified to be, for example, a Mann iteration based or an Ishikawa iteration based best response.

In contrast to the classical approaches in which the payoff is assumed to be perfectly measured and without time delays, we examine imperfectness and time delays in payoff measurement. Delayed evolutionary game dynamics have been studied in [158, 170] but in continuous time. The authors in [158, 165, 171, 170] showed that an evolutionarily stable strategy (which is robust to invasions by small fraction of users) can be unstable for time delays, and they provided sufficient conditions of stability of delayed Aloha-like systems.

The objective is heterogeneous learning under uncertain channel state and with delayed feedback. Our motivation for heterogeneous learning is the following: The rapid growth in the density of wireless access devices, accompanied by increasing heterogeneity in wireless technologies, requires adaptive allocation of resources. The traditional learning schemes use homogeneous behaviors and do not allow for a different behavior associated with the type of users and technologies.

These shortcomings call for new ways to implement learning schemes, for example, the one by *different speed of learning* and the one by *different learning patterns*. The heterogeneity leads to a new class of evolutionary game dynamics with different behaviors. We propose heterogeneous fully distributed learning framework for robust power allocation games. Our algorithm considers imperfectness in the measurement of payoffs (the payoff functions are unknown to the transmitters) and time delays. The instantaneous noisy payoffs are not available [174]. Each transmitter has only a numerical value of its delayed noisy payoff. Although transmitters observe only the delayed noisy payoffs of the specific action chosen on a past time slot, the observed values depend on some specific dependency parameters determined by the other players choices revealing implicit information about the system. The natural

Information	Known	Unknown
Existence of being in a game	no	
Existence of opponents	no	
Channel state information	no	yes
Own-CSI	not observed	
Law of own-CSI	no	
Power allocation of the others	no	yes
Shannon information rate of the others	no	yes
Number of users in the system	no	yes
Measurement delayed own-payoff	yes	no
Last own-action	yes	no
Capability	arithmetic operations	

TABLE 5.1
Assumptions for games under channel uncertainty.

question is whether such a learning algorithm based on a minimal piece of information may be sufficient to induce coordination and stabilize the system to an equilibrium. The answer to this question is positive for power allocation games under channel uncertainty, and access point selection games under suitable assumptions.

Our interest in studying these robust power allocation scenarios stems from the fact that, in decentralized parallel multiple access channel, the robust equilibrium does not use the perfect CSI assumption and it can increase the network spectral efficiency compared to Bayesian equilibrium.

Table 5.1 summarizes the information available to the players.

5.3.1 Robust PMAC Games

We consider networks that can be modeled by a parallel multiple access channel (PMAC), which consists of several orthogonal multiple access channels. The sets of transmitters and channels are respectively denoted by $\mathcal{N}$ and $\mathcal{L}$. The cardinality of $\mathcal{N}$ is n, and the cardinality of $\mathcal{L}$ is L. The $L-$dimensional vector of received signals is denoted by

$$y_t = (y_{1,t}, \ldots, y_{L,t})$$

and defined by

$$y_t = \sum_{j=1}^{n} H_{j,t} \tilde{s}_{j,t} + z_t,$$

where t is the time index (the associated rate depends on whether fast fading or slow fading channels are considered), $\forall j \in \mathcal{N}$, $H_{j,t}$ is the channel transfer matrix from the transmitter j to the receiver, $\tilde{s}_{j,t}$ is the vector of symbols transmitted by transmitter j at time t, and the vector z_t represents the noise observed at the receiver, which is assumed to be centered, Gaussian, and distributed as

$$\mathbb{E}\left(z_t z_t^\dagger\right) = \text{diag}\left(N_{0,1}, \ldots, N_{0,L}\right)$$

. We will exclusively deal with the scenario where $\forall j \in \mathcal{N}$, matrix H_j is an m-dimensional diagonal matrix, i.e.,

$$H_{j,t} = \text{diag}\left(h_{j,t}(1), \ldots, h_{j,t}(L)\right),$$

whose entries $h_{j,t}(l)$, $\forall l \in \mathcal{L}$, are i.i.d. complex Gaussian random variables with independent real and imaginary parts, each with zero-mean and variance $N_{0,l}$. This scenario models, for instance, the case of intersymbol interference channels that have been diagonalized in the frequency domain by using a precyclic prefix and the Fourier transform, or the case of the uplink of a network with several access receivers operating over non-overlapping flat-fading channels. We denote by $\mathbf{h}_t = (h_{1,t}, \ldots, h_{n,t})$ where $h_{j,t} = (h_{j,t}(1), \ldots, h_{j,t}(L))$. The vector of transmitted symbols $\tilde{s}_{jt}$, $\forall j \in \mathcal{N}$ is characterized in terms of power by the covariance matrix

$$P_{j,t} = \mathbb{E}\left(\tilde{s}_{j,t} \tilde{s}_{j,t}^\dagger\right) = \text{diag}\left(p_{j,t}(1), \ldots, p_{j,t}(L)\right).$$

As a matter of fact, $p_{j,t}(l)$, $\forall (j,l) \in \mathcal{N} \times \mathcal{L}$, represents the transmit power allocated by the transmitter j over the channel l. Now, since transmitters are power-limited, we have that

$$\forall j \in \mathcal{N}, \ \forall t \geq 0, \ \sum_{l=1}^{L} p_{j,t}(l) \leq p_{j,\text{max}}. \tag{5.4}$$

Since $p_{j,\text{max}} > 0$, we can normalize the last equation $p_{j,\text{max}}$ and one gets

$$\forall j \in \mathcal{N}, \ \forall t \geq 0, \ \sum_{l=1}^{L} \theta_{j,t}(l) \leq 1. \tag{5.5}$$

where $\theta_{j,t}(l) = \frac{p_{j,t}(l)}{p_{j,\text{max}}}$.

We define a power allocation vector for transmitter $j \in \mathcal{N}$ as a vector $p_{j,t} = (p_{j,t}(1), \ldots, p_{j,t}(L))$ with non-negative entries satisfying (5.4). We denote by $p_t = (p_{1,t}, \ldots, p_{n,t})$. At last, the band of channel $l \in \mathcal{L}$ will be denoted by b_l and the band of the system by $b = \sum_{l=1}^{L} b_l$ [Hz], then, the individual spectral efficiency of transmitter j, denoted by $R_j(\mathbf{h}_t, \mathbf{p}_t)$ for user $j \in \mathcal{N}$ is

$$R_j(\mathbf{h}_t, \mathbf{p}_t) = \sum_{l \in \mathcal{L}} \frac{b_l}{b} \log_2\left(1 + \gamma_{j,t}(l)\right) \text{ [bps/Hz]},$$

where $\gamma_{j,t}(l)$ is the signal to interference plus noise ratio (SINR) seen by transmitter j over its link l at time t, i.e.,

$$\gamma_{j,t}(l) = \frac{p_{j,t}(l) g_{j,t}(l)}{N_{0,l} + \sum_{j' \in \mathcal{N} \setminus \{j\}} p_{j',t}(l) g_{j',t}(l)}, \text{ with } g_{j,t}(l) = |h_{j,t}(l)|^2. \tag{5.6}$$

We develop a fully learning framework (Figure 5.1) for the dynamic robust power allocation game.

5.3.2 Numerical Examples

In this subsection, we provide some numerical results illustrating our theoretical findings. We start with the two receivers case and illustrate the convergence to global optimum under the heterogeneous CODIPAS-RL (HCRL). Next, we study the impact of delayed feedback of the system in the three receivers case.

5.3.2.1 Two Receivers

In order to illustrate the algorithm, a simple example with two transmitters and two channels is considered. The discrete set of actions for each transmitter is described as follows. Each transmitter chooses among two possible actions $s_1 = [p_{\max}, 0]$, $s_2 = [0, p_{\max}]$. Each transmitter follows the CODIPAS-RL algorithm as described in Section 5.2. The only one-step delay feedback received by the transmitter is the noisy payoff. A mixed strategy $\mathbf{x}_{j,t}$ in this case corresponds to the probabilities of selecting elements in $\mathcal{A}_1 = \mathcal{A}_2 = \{s_1, s_2\}$ while the payoff perceived by transmitter j, $\hat{r}_{j,t}$, is the achievable capacity. We normalize the parameters b_l such that the payoffs are $[0, 1]$. The transmitters will learn the expected payoff and strategy via the scheme described with $\tilde{\beta}_{j,\epsilon}$, the Boltzmann-Gibbs distribution with $\epsilon_j = 0.1$. The learning rates λ_t and μ_t are given by $\lambda_t = \frac{1}{1+t}$ and $\mu_t = \frac{1}{(1+t)^{0.6}}$. It is clear that the game has many equilibria: (s_1, s_2), (s_2, s_1), and $((\frac{1}{2}, \frac{1}{2}), (\frac{1}{2}, \frac{1}{2}))$. The action profiles (s_1, s_2) and (s_2, s_1) are global optima of the normalized expected game. Below, we show the convergence to one of the global optima using heterogeneous learning.

Heterogeneous Learning CODIPAS-RL: HCRL

The ODE convergence of the strategy and payoff is shown in Figures 5.2 (and its zoom), 5.5 and 5.4, respectively, where the game is played several times. For one run, we observe that when the two transmitters use different learning patterns as in (HCRL), the convergence times are different as well as the outcome of the game.

Note that in this example the CODIPAS-RL converges to the global optimum of the robust game, which is also a strong equilibrium (resilient to any coalition of transmitters of any size).

5.3.2.2 Three Receivers

In this subsection, we illustrate the learning algorithm with two transmitters and three channels. The discrete set of actions for each transmitter is described as follows. Transmitter j's action space is $\mathcal{A}_j = \{s_1^*, s_2^*, s_3^*\}$, where

$s_1^* = [p_{j,\max}, 0, 0]$, $s_2^* = [0, p_{j,\max}, 0]$, $s_3^* = [0, 0, p_{j,\max}]$. Each transmitter chooses among three possible actions s_1^*, s_2^* or s_3^*. These actions correspond to the case where each transmitter uses its total power to one of the channels from a set of three available channels. Each transmitter follows the CODIPAS-RL algorithm as described in Section 5.2. The only one-step delay feedback received by the transmitter is the noisy payoff, which was obtained after allocating power to the pair of receivers. A mixed strategy $\mathbf{x}_{j,t}$ in this case corresponds to the probability of selecting an element in $\mathcal{A}_j = \{s_1^*, s_2^*, s_3^*\}$ while the payoff perceived by the transmitter j, $\hat{r}_{j,t}$, is the imperfect achievable capacity.

We fix the following parameters

$$n = 2, L = 3, T = 300, \tag{5.7}$$

$$b_l/b = \frac{1}{3}, \lambda_t = 2/T, \ \tau_j = 1. \tag{5.8}$$

In Figure 5.5 we represent the strategy evolution of transmitters Tx1 and Tx2. We observe that CODIPAS-RL converges to a global optimum of the expected long-run interaction. The total number of iterations needed to guarantee a small error tolerance is relatively small. In the long-term, transmitter Tx1 will transmit its maximum power with frequency 1, which corresponds to action s_1^*, and Tx2 will be using frequency 3. At a small fraction of time, frequency 2 is used. Thus, the transmitters will not interfere and the equilibrium is learned.

Impact of Time Delayed Noisy Payoffs

We keep the same parameters as above, but we change the time delays to be $\tau_j = 2$. In figure 5.5, we represent the strategy evolution of transmitters Tx1 and Tx2 under delayed CODIPAS-RL. As we can see the convergence time as well as the stability of the system changed. The transmitters use more the action s_2^* compared to the scenario of Figure 5.5. This is because the estimated payoff under two-step time delays is uncertain, and the prediction is not good enough compared to the actual payoffs. The length of the horizon for a good prediction is much bigger than the first scenario (2000 versus 300). This scenario tells us how much the feedback delay is important at the transmitter: the time delay τ can change the outcome of the interaction.

How to eliminate delay effect in discrete time process?

This question is important because the above analysis show that delayed and outdated measurements can destabilize the system. Suppose now that all the time delays are natural numbers and known. Then one can construct a learning with synchronized data i.e., the delayed learning scheme given above will be employed not at time t but at time $t + \tau$. Thus, each player keeps his or her recent measurements in memory and use the synchronized updates. The resulting scheme is thus similar to the scheme of the previous chapter and the delay effect is avoided. However, this technique seems not appropriated for large delays because it needs larger memory which could be expensive.

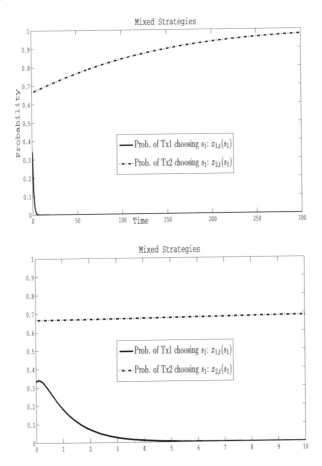

FIGURE 5.2
Heterogeneous CODIPAS-RL: Convergence of the ODEs of strategies.

5.3.3   MIMO Interference Channel

Multiple-input-multiple-output (MIMO) links use antenna arrays at both ends of a link to transmit multiple streams at the same time and frequency chan-

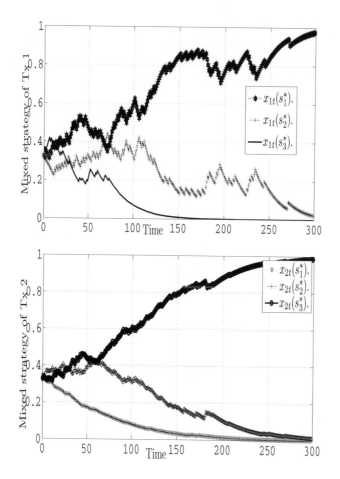

FIGURE 5.3
CODIPAS-RL: Convergence to global optimum equilibria.

nel. Signals transmitted and received by array elements at different physical locations are spatially separated by array processing algorithms. Depending on the channel conditions, MIMO links can yield large gains in capacity of wireless systems. Multiple links, each with different transmitter-receiver pairs, are allowed to transmit in a given range possibly through multiple streams per link. Such a multi-access network with MIMO links is referred to as a MIMO interference system.

We expect that *robust game theory* and *robust optimization* are more appropriate to analyze the achievable equilibrium rates under imperfectness and time-varying channel states. We consider $n-$link communication network that can be modeled by a MIMO Gaussian interference channel. Each link is as-

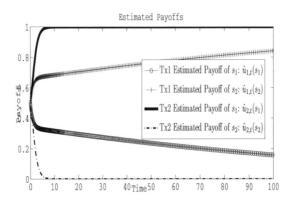

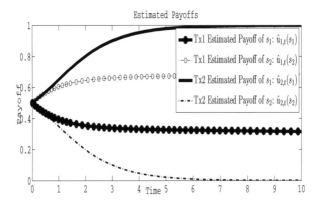

FIGURE 5.4
CODIPAS-RL: Convergence of payoff estimations.

sociated with a transmitter-receiver pair. Each transmitter and receiver are equipped with n_t and n_r antennas, respectively.

The sets of transmitters is denoted by $\mathcal{N}$. The cardinality of $\mathcal{N}$ is n. The transmitter j transmits a complex signal vector $\tilde{s}_{j,t} \in \mathbb{C}^{n_t}$ of dimension n_t. Consequently, a complex baseband signal vector of dimension n_r denoted by $\tilde{y}_{j,t}$ is received. The vector of received signals from j is defined by

$$\tilde{y}_{j,t} = H_{jj,t}\tilde{s}_{j,t} + \sum_{j' \neq j} H_{jj',t}\tilde{s}_{j',t} + z_{j,t},$$

where t is the time index, $\forall j \in \mathcal{N}$, $H_{jj',t}$ is the complex channel matrix of dimension $n_r \times n_t$ from the transmitter j to the receiver j', and the vector $z_{j,t}$ represents the noise observed at the receivers; it is a zero-mean circularly sym-

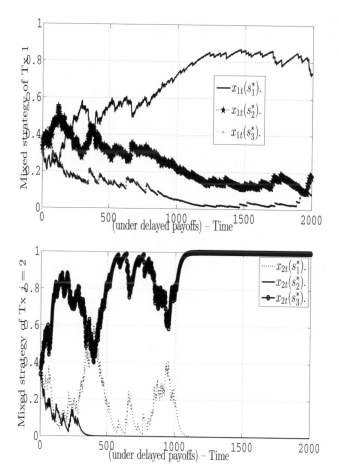

FIGURE 5.5
CODIPAS-RL under two-step delayed payoffs.

metric complex Gaussian noise vector with arbitrary non-singular covariance matrix R_j

For all $j \in \mathcal{N}$, the matrix $H_{jj,t}$ is assumed to be non-zero. We denote by $\mathbf{H}_t = (H_{jj',t})_{j,j'}$. The vector of transmitted symbols $\tilde{s}_{jt}$, $\forall \ j \in \mathcal{N}$ is characterized in terms of power by the covariance matrix $Q_{j,t} = \mathbb{E}\left(\tilde{s}_{j,t}\tilde{s}_{j,t}^\dagger\right)$ which is an Hermitian positive semi-definite matrix. Now, since transmitters are power-limited, we have that

$$\forall \ j \in \mathcal{N}, \ \forall t \geq 0, \quad \mathrm{tr}(Q_{j,t}) \leq p_{j,\max}. \tag{5.9}$$

We define a transmit power covariance matrix vector for transmitter $j \in \mathcal{N}$ as a matrix $Q_j \in \mathcal{M}_+$ satisfying (5.9), where $\mathcal{M}_+$ denotes the set of Hermitian positive matrices. The payoff function of j is its mutual information $I(\tilde{s}_j; \tilde{y}_j)(\mathbf{H}, Q_1, \ldots, Q_n)$. Under the above assumption, the maximum information rate is

$$\log \det \left(I + H_{jj}^{\dagger} \Gamma_j^{-1}(Q_{-j}) H_{jj} Q_j \right),$$

where

$$\Gamma_j(Q_{-j}) = R_j + \sum_{j' \neq j} H_{jj'} Q_{j'} H_{jj'}^{\dagger},$$

is the multi-user interference plus noise observed at j and $Q_{-j} = (Q_{j'})_{j' \neq j}$ is the collection of users' covariance matrices, except the j-th one. The robust individual optimization problem of player j is then

$$j \in \mathcal{N}, \quad \sup_{\mathbf{Q}_j \in \mathcal{Q}_j} \inf_{\mathbf{H}} I(\tilde{s}_j; \tilde{y}_j)(\mathbf{H}, Q_1, \ldots, Q_n)$$

where

$$\mathcal{Q}_j := \{Q_j \in \mathbb{C}^{n_t \times n_t} \mid Q_j \in \mathcal{M}_+, \ \operatorname{tr}(Q_j) \leq p_{j,\max}\}.$$

5.3.3.1 One-Shot MIMO Game

We examine the strategic MIMO game. The action of a player is typically a covariance matrix. The payoff function is the maximum transmission rate. The one-shot strategic form (or normal form) is represented by a collection:

$$\mathcal{G} = (\mathcal{N}, \{\mathcal{Q}_j\}_{j \in \mathcal{N}}, \{R_j\}_{j \in \mathcal{N}}).$$

In the case of a discrete action space (a set of fixed covariance matrices from the set $\mathcal{Q}_j$), when player $j \in \mathcal{N}$ chooses an action $s_j \in \mathcal{Q}_j$ according to a probability distribution $\mathbf{x}_j = (x_j(s_j))_{s_j \in \mathcal{Q}_j}$ over $\mathcal{Q}_j$ the choice of $\mathbf{x}_j$ is called a mixed strategy of the one-shot game. When $\mathbf{x}_j$ is on a vertex of the simplex $\mathcal{X}_j = \Delta(\mathcal{Q}_j)$, the mixed strategy boils down to a pure strategy, i.e., the deterministic choice of an action. Since there are random variables $\mathbf{H}_{j,j'}$ which determine the game, we add a state space $\mathcal{H}$ and the payoff function will be defined on product space: $\mathcal{H} \times \prod_j \mathcal{X}_j$. We denote by $\mathcal{G}(\mathbf{H})$ the normal-form game

$$(\{\mathbf{H}\}, \mathcal{N}, \{\mathcal{Q}_j\}_{j \in \mathcal{N}}, \{R_j(\mathbf{H}, .)\}_{j \in \mathcal{N}}).$$

For the use of notation, the payoff function will be defined on $\prod_j \mathcal{X}_j$ in the discrete action case and, on $\prod_j \mathcal{Q}_j$ for the continuous (non-empty compact convex) action space. We will write

$$R_j(\mathbf{H}, \mathbf{x}_1, \ldots, \mathbf{x}_n) = \mathbb{E}_{\mathbf{x}_1, \ldots, \mathbf{x}_n} R_j(\mathbf{H}, Q_1, \ldots, Q_n).$$

Then, an action profile $(Q_1^*, \ldots, Q_n^*) \in \prod_{j \in \mathcal{N}} \mathcal{Q}_j$ is a (pure) Nash equilibrium of the one-shot game $\mathcal{G}(\mathbf{H})$ if

$$\forall j \in \mathcal{N}, \quad R_j(\mathbf{H}, Q_j, \mathbf{Q}_{-j}^*) \leq R_j(\mathbf{H}, Q_j^*, \mathbf{Q}_{-j}^*), \ \forall Q_j \in \mathcal{Q}_j. \tag{5.10}$$

A strategy profile $(\mathbf{x}_1, \ldots, \mathbf{x}_n) \in \prod_{j \in \mathcal{N}} \mathcal{X}_j$ is a Nash equilibrium of one-shot game $\mathcal{G}(\mathbf{H})$ if

$$\forall j \in \mathcal{N}, \quad R_j(\mathbf{H}, \mathbf{y}_j, \mathbf{x}_{-j}) \le R_j(\mathbf{H}, \mathbf{x}_j, \mathbf{x}_{-j}), \quad \forall \mathbf{y}_j \in \mathcal{X}_j. \tag{5.11}$$

> Karusk-Kuhn-Tucker (KKT) conditions help to compute the best-responses of the one-shot game for a fixed state.

The KKT conditions were originally named after Harold W. Kuhn, and Albert W. Tucker [95]. The necessary conditions for this problem had been stated by William Karush [86].

Consider the following maximization problem:

$$\begin{cases} \text{Max}_{x \in \mathbb{R}^d} & \tilde{f}(x) \\ & \tilde{g}_k(x) \ge 0, \ k \in \mathcal{I} = \{1, \ldots, c_1\}, c_1 \in \mathbb{Z}_+ \\ & \tilde{h}_{k'}(x) = 0, \ k' \in \mathcal{E} = \{1, \ldots, c_2\}, c_2 \in \mathbb{Z}_+ \end{cases}$$

where $\mathcal{I}$ is the set of inequality constraints, $\mathcal{E}$ is the set of equality constraints, $\tilde{f} : \mathbb{R}^d \longrightarrow \mathbb{R}$, $\tilde{g} : \mathbb{R}^l \longrightarrow \mathbb{R}$ and $\tilde{h} : \mathbb{R}^m \longrightarrow \mathbb{R}$ being continuously differentiable functions, $d, l, m \ge 1$.

⚠ The symbol *Max* is a generic term that indicates that the objective is to maximize. It does not mean that the problem has a maximizer. In that case the supremum will be taken.

Define the *Lagrange* function of the problem as

$$L(x, \tilde{\lambda}, \tilde{\mu}) = \tilde{f}(x) + \langle \tilde{g}(x), \tilde{\lambda} \rangle + \langle \tilde{h}(x), \tilde{\mu} \rangle,$$

Define inf sup as

$$\bar{v} = \inf_{\tilde{\lambda}, \tilde{\mu} \succeq 0} \ \sup_{x \in \mathbb{R}^d} \ L(x, \tilde{\lambda}, \tilde{\mu})$$

Then

$$\inf_{\tilde{\lambda}, \tilde{\mu} \succeq 0} \ L(x, \tilde{\lambda}, \tilde{\mu}) \le \bar{v} \le \sup_{x \in \mathbb{R}^d} \ L(x, \tilde{\lambda}, \tilde{\mu})$$

Remark 5.3.4. *$\bar{v}$ can be seen as the upper value in a zero-sum game.*

Below we provide the basic Karusk-Kuhn-Tucker conditions.

- First order condition: $\partial_x \tilde{f}(x) + \sum_{k \in C_1} \tilde{\lambda}_k \partial_x \tilde{g}_k + \sum_{k' \in C_2} \tilde{\mu}_{k'} \partial_x \tilde{h}_{k'} = 0$
- Feasibility of the primal problem:

$$\tilde{g}_k(x) \ge 0, \ k \in \mathcal{I}$$

$$\tilde{h}_{k'}(x) = 0, \ k' \in \mathcal{E} = \{1, \ldots, c_2\},$$

- Positivity (Feasibility of the dual problem) $\tilde{\lambda}_k \geq 0$

- Complementary slackness: $\tilde{\lambda}_k g_k(x) = 0$, $\forall k \in C_1$

⚠ To be a maximizer, additional constraint qualifications may be required even for smooth functions.

In some cases, the necessary conditions are also sufficient for optimality (for example, for smooth concave function, if the derivative is zero it is a local extremum, which implies a global extremum). However, in general, the necessary conditions are not sufficient for optimality and additional information is necessary, such as the second-order sufficient conditions. For smooth functions, it involve the second derivatives.

We are ready to use the KKT conditions for the one-shot game. Note that the sum of the payoffs is not concave in the joint argument. Since each of the payoff functions $\tilde{R}_j$ is concave with the respect to Q_j, given $\mathbf{H}$, the Nash equilibrium of the $\mathcal{G}(\mathbf{H})$ is given by the MIMO waterfilling solution described as follows:

- The term $H_{jj}^{\dagger}\Gamma_j^{-1}(Q_{-j})H_{jj}$ is written as $E_j^{\dagger}(\mathbf{Q}_{-j})D_j(\mathbf{Q}_{-j})E_j(\mathbf{Q}_{-j})$ by eigen-decomposition, where $E_j(\mathbf{Q}_{-j}) \in \mathbb{C}^{n_t \times n_t}$ is a unitary matrix containing the eigenvectors and $D_j(\mathbf{Q}_{-j})$ is a diagonal matrix with n_t positive eigenvalues.

- The Nash equilibrium is given by $(Q_1^*, \ldots, Q_n^*)$ solution of the MIMO waterfilling operator

$$\mathrm{WF}_j(\mathbf{H}, \mathbf{Q}_{-j}) = E_j(Q_{-j})[\mu_j I - D_j^{-1}(\mathbf{Q}_{-j})]^+ E_j(\mathbf{Q}_{-j})$$

where μ is chosen in order to satisfy

$$\mathrm{tr}\left([\mu_j I - D_j^{-1}(\mathbf{Q}_{-j})]^+\right) = p_{j,\max},$$

where $x^+ = \max(0, x)$.

- Note that $\mathrm{WF}_j(\mathbf{H}, \mathbf{Q}_{-j})$ is exactly the best response to $\mathbf{Q}_{-j}$ i.e $\mathrm{BR}_j(\mathbf{H}, \mathbf{Q}_{-j}) = \mathrm{WF}_j(\mathbf{H}, \mathbf{Q}_{-j})$. This operator is continuous in the sense of norm 2 or norm-sup (in a finite dimensional vectorial space, they are topologically equivalent).

- The existence of solution $\mathbf{Q}^*$ of the above fixed point equation is guaranteed by Brouwer fixed point theorem[1] which states that a continuous mapping from a non-empty, compact convex set into itself has at least one fixed point i.e $\exists\, Q^* \in \mathbb{C}^{n_t \times n_t}$, $Q_j^* \in \mathcal{Q}_j$, and $Q_j^* = \mathrm{WF}_j(\mathbf{H}, \mathbf{Q}_{-j}^*)$.

---

[1] see also Kakutani, Glicksberg, Fan, Debreu fixed point theorem for set-valued fixed points

- In special cases where the mapping WF is strict contracting, one can use the Banach-Picard iterative procedure to show the convergence. However, In general, the best response (namely the iterative Waterfilling-based methods: simultaneous, sequential or asynchronous versions) may not converge. A simple example of non-convergence is obtained when considering a cycling behavior between the receivers. Let

$$J = 3, n_r = 2, p_{1,\max} = p_{2,\max} = p_{\max}$$

and the transfer matrices are

$$H_1 = H_2 = \begin{pmatrix} 1 & 0 & 2 \\ 2 & 1 & 0 \\ 0 & 2 & 1 \end{pmatrix}, \ \bar{R}_1 = (\sigma^2), \bar{R}_2 = (\sigma^2 + p_{\max}).$$

Then, we observe that starting from the first channel, the three players cycle between the two channels indefinitely. We show how to eliminate this cycling phenomena using CODIPAS.

- In order to a convergent behavior one can Isikhawa-like iteration.

5.3.4.1 MIMO Robust Game

We develop two approaches in the one-shot robust power allocation game:

- The first one is based on the expectation over the channel states, called *expected game*, with a payoff function defined by

$$r_j^1(\mathbf{x}_1, \ldots, \mathbf{x}_n) := \mathbb{E}_\mathbf{H} \mathbb{E}_{\mathbf{x}_1, \ldots, \mathbf{x}_n} R_j(\mathbf{H}, Q_1, \ldots, Q_n)$$

$$:= \int_{\mathbf{H} \in \mathcal{H}} \left[\sum_{Q_1, \ldots, Q_n} \left(\prod_{j' \in \mathcal{N}} \mathbf{x}_{j'}(Q_{j'}) \right) R_j(\mathbf{H}, Q_1, \ldots, Q_n) \right] \nu(d\mathbf{H})$$

in the discrete power allocation, and

$$v_j^1(Q_1, \ldots, Q_n) = \mathbb{E}_\mathbf{H} R_j(\mathbf{H}, Q_1, \ldots, Q_n),$$

for continuous power allocation. We will denote the associated static games by $\mathcal{G}^{1,d}$ for the discrete power allocation and $\mathcal{G}^{1,c}$ for the continuous power allocation. A strategy profile $(\mathbf{x}_1, \ldots, \mathbf{x}_n) \in \prod_{j \in \mathcal{N}} \mathcal{X}_j$ is a Nash equilibrium of the expected game $\mathcal{G}^{1,d}$ if

$$\forall j \in \mathcal{N}, \ \mathbb{E}_\mathbf{H} R_j(\mathbf{H}, \mathbf{y}_j, \mathbf{x}_{-j}) \le \mathbb{E}_\mathbf{H} R_j(\mathbf{H}, \mathbf{x}_j, \mathbf{x}_{-j}), \ \forall \mathbf{y}_j \in \mathcal{X}_j. \quad (5.12)$$

The existence of solution of Equation (5.12) is equivalent to the existence of solution of the following *variational inequality problem*: find $\mathbf{x}$ such that

$$\langle \mathbf{x} - \mathbf{y}, V(\mathbf{x}) \rangle \ge 0, \ \forall \mathbf{y} \in \prod_j \mathcal{X}_j,$$

where $\langle .,. \rangle$ is the inner product, $V(\mathbf{x}) = [V_1(\mathbf{x}), \ldots, V_n(\mathbf{x})]$, $V_j(\mathbf{x}) = [\mathbb{E}_{\mathbf{H}} R_j(\mathbf{H}, e_{s_j}, \mathbf{x}_{-j})]_{s_j \in \mathcal{Q}_j}$. An equilibrium of the expected $\mathcal{G}^{1,c}$ is similarly defined.

- The second approach is a pessimistic approach also called *worst-case* approach or Maximin robust approach. It consists of considering the payoff function as

$$r_j^2(\mathbf{x}) = \inf_{\mathbf{H}} \mathbb{E}_{\mathbf{x}_1, \ldots, \mathbf{x}_n} R_j(\mathbf{H}, Q_1, \ldots, Q_n),$$

$$v_j^2(\mathbf{Q}) = \inf_{\mathbf{H}} R_j(\mathbf{H}, \mathbf{Q}).$$

We denote the associated static maximin robust games by $\mathcal{G}^{2,d}$ for the discrete power allocation and $\mathcal{G}^{2,c}$ for the continuous power allocation. A profile $(\mathbf{x}_j, \mathbf{x}_{-j}) \in \mathcal{X}_j$ is a *maximin robust equilibrium* of $\mathcal{G}^{2,d}$ if

$$\forall j \in \mathcal{N}, \ \inf_{\mathbf{H}} R_j(\mathbf{H}, \mathbf{y}_j, \mathbf{x}_{-j}) \leq \inf_{\mathbf{H}} R_j(\mathbf{H}, \mathbf{x}_j, \mathbf{x}_{-j}), \ \forall \mathbf{y}_j \in \mathcal{X}_j.$$

We now study the static robust power allocation games. By rewriting the payoff as

$$R_j(\mathbf{H}, \mathbf{p}) = \phi(\mathbf{H}, \mathbf{Q}) + \tilde{R}_j(\mathbf{H}, \mathbf{Q}_{-j}) \text{ where,} \tag{5.13}$$

$$\phi(\mathbf{H}, \mathbf{Q}) = \log \det \left(\underline{R} + \sum_{j \in \mathcal{N}} H_{j,t} Q_{j,t} H_{j,t}^\dagger \right),$$

$$\tilde{R}_j(\mathbf{H}, \mathbf{Q}_{-j}) = -\log \det \left(\underline{R} + \sum_{l \neq j} H_{l,t} Q_{l,t} H_{l,t}^\dagger \right),$$

we deduce that the power allocation game is a *robust pseudo-potential* game with potential function ϕ.

⚠ The expression of ϕ is not necessarily known to the players. Hence, the players may not know that the game is a best-response robust potential game.

Corollary 5.3.4.2. *The robust power allocation is a robust pseudo-potential game with potential function given by*

$$\xi: \ \mathbf{Q} \longmapsto \mathbb{E}_{\mathbf{H}} \phi(\mathbf{H}, \mathbf{Q}) = \int_{\mathbf{H}} \phi(\mathbf{H}, \mathbf{Q}) \, \nu(d\mathbf{H}).$$

Proof. By taking the expectation of Equation (5.13), one has that

$$\forall j \in \mathcal{N}, \ \forall \mathbf{Q}_{-j}, \ \arg\max_{Q_j} \mathbb{E}_{\mathbf{H}} I(\tilde{s}_j; \tilde{y}_j)(\mathbf{H}, \mathbf{Q}) = \arg\max_{Q_j} \ \mathbb{E}_{\mathbf{H}} \phi(\mathbf{H}, \mathbf{Q})$$

□

Using the fact that the function $\log \det$ is concave on positive matrices, the function ϕ continuously differentiable and concave in $\mathbf{Q} = (Q_1, \ldots, Q_n)$, the following corollary holds:

Corollary 5.3.4.3 (Existence in $\mathcal{G}^{1,d}, \mathcal{G}^{1,c}$). *Assume that the action spaces are reduced to a compact set and all the random variables $\mathbf{H}_{j,j'}$ have a compact support in $\mathbb{C}^{n_r \times n_t}$. Then, the robust power allocation has at least one pure expected robust equilibrium.*

Proof. By compactness assumptions and by the continuity of the function $\phi(.,.)$ the mapping ξ defined by $\xi : \mathbf{Q} \longmapsto \mathbb{E}_{\mathbf{H}} \phi(\mathbf{H}, \mathbf{Q})$, is continuous over $\prod_j \mathcal{Q}_j$. Thus, ξ has a maximizer Q^* which is a pure expected robust equilibrium. □

We now focus on the maximin robust solutions. A maximin robust solution is a solution of

$$\sup_{Q_j \in \mathcal{Q}_j} \inf_{\mathbf{H}} I(\tilde{s}_j; \tilde{y}_j)(\mathbf{H}, \mathbf{Q}), j \in \mathcal{N}. \tag{5.14}$$

Proposition 5.3.4.4 (Existence in $\mathcal{G}^{2,d}, \mathcal{G}^{2,c}$). *Assume that all the random variables $\mathbf{H}_{j,j'}$ have a compact support in $\mathbb{C}^{n_r \times n_t}$, then the power allocation has at least one pure maximin robust equilibrium.*

Proof. By compactness assumptions,

$$\inf_{\mathbf{H}} I(\tilde{s}_j; \tilde{y}_j)(\mathbf{H}, \mathbf{Q}) = \min_{\mathbf{H}} I(\tilde{s}_j; \tilde{y}_j)(\mathbf{H}, \mathbf{Q}) \tag{5.15}$$

$$= I(\tilde{s}_j; \tilde{y}_j)(\mathbf{H}^*, \mathbf{Q}) \tag{5.16}$$

$$= v_j^2(\mathbf{Q}) \tag{5.17}$$

and by continuity of the function $\tilde{\xi} : \mathbf{Q} \longmapsto \phi(\mathbf{H}^*, \mathbf{Q})$ over $\prod_j \mathcal{Q}_j$. Thus, there is a maximizer $\tilde{Q}^*$ which is also a pure maximin robust equilibrium. □

Limitations of the maximin robust approach

A worst-state case approach is sometimes useful when the game is a one-shot game and the uncertainty set is known but not the distribution of states. However the uncertainty set needs to be well-chosen. In Equation (5.14), if the uncertainty set $\mathcal{H}$ contains the null matrix, then the inf is achieved for $\mathbf{H} = 0$ and the payoff function is always 0. Thus, every strategy profile is an equilibrium and a global optimum which does not reflect the maximization of the mutual information problem. To avoid this type of trivial solution, many authors has suggested to restrict the uncertainty set to be unbounded away from 0. The questions of why one set instead of other and how to choose that set remains open.

5.3.4.5 Without Perfect CSI

Much of the work in the interference control literature has been done under the following assumptions:

(i) assumption of perfect channel state information at both transmitter and receiver sides of each link assumed;

(ii) each receiver is also assumed to measure with no errors the covariance matrix of the noise plus multiple-user-interference generated by the other transmitters.

However, these assumptions (i)-(ii) may not be satisfied in many wireless scenarios. It is natural to ask if some of these assumptions can be relaxed without compromising performance. We address the following question:

> *Given a MIMO interference dynamic environment, is there a way to achieve equilibria without coordination and with minimal information?*

To answer to this question, we adopt learning and dynamical system approaches. Recall that a learning scheme is said to be fully distributed (model-free) if the updating rule of player needs only own-action and own-perceived-payoff (a numerical value). In particular, player j does not know his payoff function and he/she does not know the exact state of the environment. The actions and payoffs of the other players are unknown too. Under this framework the question becomes:

> *Is there a fully distributed learning scheme to achieve an equilibrium or global optima in MIMO interference dynamic environment?*

We would like to mention that most often game theoretical models assume rational players and that rationality is common knowledge. Here, each player is assumed to follow a learning scheme but does not need to know whether the other players are rational or not.

We develop dynamic robust power allocation games; that is, players play several times and all under channel state uncertainty. The general case where each player j chooses the probability distribution $\mathbf{x}_{j,t}$ at each time slot t based on its history up to t is considered. In the dynamic game with uncertainty, the joint channel states change randomly from time slot to another. In the robust power allocation scenarios, the state is the joint channel state e.g., the matrix of channel state matrices $\mathbf{H}_t = (H_{j,j',t})_{j,j'} \in \mathcal{H}$. In that case we will denote the instantaneous payoff function by $R_j(\mathbf{H}_t, \mathbf{x}_t)$. In our setting, the payoff function of player j is $I(\tilde{s}_j; \tilde{y}_j)(\mathbf{H}, Q_1, \ldots, Q_n)$ the mutual information.

The CODIPAS-RL scheme for MIMO is then given by (delayed-CODIPAS-RL)

$$\begin{cases} \mathbf{x}_{j,t+1} = f_j(\lambda_{j,t}, Q_{j,t}, r_{j,t-\tau_j}, \hat{\mathbf{r}}_{j,t}, \mathbf{x}_{j,t}), \\ \hat{\mathbf{r}}_{j,t+1} = g_j(\mu_{j,t}, r_{j,t-\tau_j}, \mathbf{x}_{j,t}, \hat{\mathbf{r}}_{j,t}) \\ \qquad\qquad \forall\, j, t,\ Q_{j,t} \in \mathcal{Q}_j. \end{cases}$$

The functions f, g are chosen such that the ergodic payoff

$$\mathbb{E}_{\mathbf{H}} I(\tilde{s}_j; \tilde{y}_j)(\mathbf{H}, Q_1, \ldots, Q_n),$$

is learned with the optimal strategies $\mathbf{x}^*$ over covariance matrices. By corollary 5.3.4.2, the answer to the above question is positive.

Notes

In this section, we have proposed novel robust game theoretical formulations to solve one of the challenging and unsolved power allocation problems in wireless systems:

How to allow in a decentralized way communications over parallel multiple access channels among multiple transmitters, under uncertain channel state and delayed noisy feedback Shannon rates (delayed imperfect payoffs)?

We provided heterogeneous, delayed, combined, fully distributed payoff and strategy reinforcement learning algorithm (CODIPAS-RL) for the corresponding robust games. We have provided an ODE approach and illustrated the CODIPAS-RL numerically. A number of further issues are under consideration. It would be of great interest to develop theoretical bounds for the rate of convergence of CORDIPAS-RLs. Also, it would be natural to extend the analysis of our CODIPAS-RL algorithms to more classes of wireless games including non-potential games, outage probability games under uncertain channel states and the dynamic robust games with the energy efficiency function as payoff function.

Extension of CODIPAS

It would be interesting to generalize the CODIPAS in the context of Itô's stochastic difference equation. Typically, the case where the strategy learning has the following form:

$$\mathbf{x}_{t+1} = \mathbf{x}_t + \lambda_t(f(\mathbf{x}_t, \hat{\mathbf{r}}_t) + M_{t+1}) + \sqrt{\lambda_t}\sigma(\mathbf{x}_t, \hat{\mathbf{r}}_t)\xi_t,$$

where ξ_t is random variable with zero-mean and finite second moment, can be seen as a Euler scheme of Itô's stochastic differential equation (SDE):

$$d\mathbf{x}_{j,t} = f_j(\mathbf{x}_t, \hat{\mathbf{r}}_t)dt + \sigma_j(\mathbf{x}_t, \hat{\mathbf{r}}_t)d\mathbb{B}_{j,t},$$

where $\mathbb{B}_{j,t}$ is a standard Brownian motion in $\mathbb{R}^{|Q_j|}$. Note that the distribution of the above SDE can be expressed as a solution of a Fokker-Planck-Kolmogorov forward equation. This question will be discussed in Chapter 9.

6

Learning in Constrained Robust Games

6.1 Introduction

This chapter is concerned with the concept of constrained equilibrium and satisfactory solution in constrained-like games. In addition to the Nash equilibrium, where no player can improve its payoff by unilateral deviation, which can be seen as players are interested in selfishly maximizing its payoff function, we present the concept of satisfactory solution, which models the case where players only aim to guarantee a minimum satisfaction level. At a satisfactory solution, whenever it exists, each player is able to guarantee its minimum satisfaction level which corresponds to a certain quality of experience (application layer) or a quality of service (network layer) needed by the player. Then, we investigate an efficient satisfactory solution and establish a connection with the constrained equilibrium. Under this setting, we develop a fully distributed algorithm to be close to the set of satisfactory solutions in long-run interactions.

6.2 Constrained One-Shot Games

The focus of this section is on how to model and solve constrained games. We distinguish two type of constraints: *Orthogonal Constraints* and *Coupled Constraints*.

6.2.1 Orthogonal Constraints

An orthogonal constraint is a constraint that is independent of the others actions. This is also referred as *individual constraint* since it is specific to that player. Constrained games with individual constraints and independent constraints can be transformed to standard normal form games without constraints (the constraints are integrated in the actions' set leading to the feasible actions set).

6.2.2   Coupled Constraints

A constraint of player j is said to be coupled (with the others) if it depends not only on the choice of that player but also on the actions of the others. This type of constraints is frequently met in engineering problems. Examples include

- Price evolution in energy market

$$p_{t+1} = p_t + \left[\bar{p}_t - \frac{d}{n} \sum_{j=1}^{n} a_{j,t} \right] + \sigma \left(\mathcal{B}_{t+1} - \mathcal{B}_t \right).$$

where $d > 0$ is a parameter that influences the demand function, $\bar{p}_t$ is a referent price parameter such that no consumer will be interested in joining the system at that price, σ is a real number, n is the number of operators/producers, $a_{j,t}$ is the action of operator j at time t, $\mathcal{B}_t$ is a standard Brownian motion (to capture to stochastic nature of the environment) and p_t is the market price (a real number). This solution needs to be projected to $\mathbb{R}_+$. Here the price constraint is a shared constraint for all the players.

- Consider a multi-user access channel with capacity region given by

$$\sum_{j \in \mathcal{J} \in 2^{\mathcal{N}}} \alpha_j \leq \log \left(1 + \sum_{j \in \mathcal{J}} \frac{P_j |h_j|^2}{N_0} \right)$$

where $\alpha_j \geq$ is the rate of user j, P_j is a fixed positive power of j, N_0 is a (positive) noise parameter, $|h_j|^2 > 0$ is a channel gain and $\mathcal{N}$ is the set of users. Suppose now that α_j a choice variable. Then, the collection of actions should be the capacity constraint which is a shared constraint.

The theorems proving the existence of general equilibrium in a competitive economy, which necessarily involved specifying the conditions under which such an equilibrium would exist, are an extraordinary achievement of twentieth-century economics. The discovery is commonly attributed to the paper by Kenneth Arrow and Gerard Debreu, entitled, Existence of an Equilibrium for a Competitive Economy, which was published in the July 1954 issue of Econometrica [51, 10]. See also [110].

> A challenging open issue is to know how a coupled-constrained game will be played. Since the actions are coupled and should be chosen simultaneously, how a player will choose his or her action?

The next theorem provides a sufficient condition for existence of constrained equilibria in concave games.

Theorem 6.2.2.1 (Rosen 1965 Theorem 1, concave games). *Let* $\mathcal{C} \subseteq \mathbb{R}^m$, $m = m_1 + \ldots + m_n$ *be nonempty, convex, closed and bounded.* $\tilde{R}_j :$

$\mathcal{C} \longrightarrow \mathbb{R}$ be continuous on $\mathcal{C}$ and concave in a_j for every fixed a_{-j} such that $a = (a_j, a_{-j}) \in \mathcal{C}$. Then, the game admits a Nash equilibrium in pure strategies in $\mathcal{C}$.

Note that under continuity and the compactness assumptions, the unconstrained game has a Nash equilibrium in mixed strategies.

Proof. Define

$$\phi(a,b) = \sum_{j=1}^{n} \tilde{R}_j(b_j, a_{-j}).$$

ϕ is continuous in $\mathcal{C} \times \mathcal{C}$. Let

$$R(a) = \{b \in \mathcal{C} \mid \phi(a,b) \geq \phi(a,c), \ \forall c \in \mathcal{C}\}$$

be the reaction set. The correspondence $R : \mathcal{C} \longrightarrow 2^{\mathcal{C}}$ is upper semi-continuous in $\mathcal{C}$, which is closed, non-empty, convex and bounded in finite dimension. By Kakutani Fixed Point Theorem (FPT), there exists a point $a \in \mathcal{C}$ such that $a \in R(a)$. Hence, an equilibrium exists in pure strategies in $\mathcal{C}$. □

Consider the Nikaido-Isoda [123, 130] function ϕ defined by

$$\phi(a,b) = \sum_{j=1}^{n} \tilde{R}_j(b_j, a_{-j}).$$

It is not difficult to see that $a \in \mathcal{C}$ is a Nash equilibrium if and only if $\phi(a,a) \geq \phi(a,c), \ \forall c \in \mathcal{C}$.

In the next section, we propose, learning approach in quality of service (QoS) and quality of experience (QoE) problems. QoE and QoS satisfaction are not exactly constrained games. However, the satisfaction set can be linked with the constraints. Then, the learning question is how to reach a satisfactory solution. The answer to this question is not obvious because there are only few works in strategic learning for constrained games [131].

6.3   Quality of Experience

Service providers are looking at new applications to compensate for declining revenues from traditional voice and flat-rate high-speed data services. These are often multimedia applications involving text, voice, pictures, mixed media and can be provided by either the service provider or an Internet-based application provider. Next-generation multimedia networks need to deliver applications with a high quality of experience (QoE) for users. QoE is the measure of how well a system or an application meets the user's expectations.

Many network elements provide the building blocks for service delivery, and element managers provide performance data for specific network elements. However, this discrete measurement data is not sufficient to assess the overall end user experience with particular applications.

In today's highly competitive environment, users have the option of choosing from a plethora of service providers such as wireline, wireless, and cable operators. Therefore, it is not enough to simply make the services available to users, service providers must deliver those services in such a way that users fully enjoy a rich experience at a reasonable price.

It is not enough to measure performance. One has to take a step beyond to determine QoE and manage it effectively. This requires a paradigm shift from a network centric to a customer centric view. Managing QoE is essential. It is one thing to monitor and measure it, but another thing to maximize it (to increase profit). Maximizing QoE requires early detection of potential QoE-affecting problems and solving them as soon as possible, before they reach the customer complaint stage.

The concept of QoE is different from quality of service (QoS), which focuses on measuring performance from a network perspective. For instance, QoE focuses on user-perceived effects, such as degradation in voice or video quality, whereas QoS focuses on network effects such as end-to-end delays or jitter. Of course, QoE is directly related to QoS, but the challenge for a service provider is to have the right set of tools and processes to map the QoS at the network level to the QoE at the user and session levels and have the ability to control it. Another important point to note is that measurements in individual nodes may indicate acceptable QoS, but end users may still be experiencing unacceptable QoE.

6.4   Relevance in QoE and QoS satisfaction

In the last decade, game theory has played a central role in the analysis and design of autonomous and self-configuring wireless networks. These kinds of networks can be modeled as noncooperative games as long as players (radio devices) autonomously set up their transmission configuration (actions) to selfishly maximize their own quality of service (QoS) level (utility function). As a consequence, the concept of Nash equilibrium has been widely used. In the context of autonomous and self-configuring wireless networks, a Nash equilibrium is the network state where none of the radio devices improves its QoS by unilaterally changing its transmission scheme. A (Cournot/Nash) equilibrium is then a strategy profile where each radio device attains the highest achievable QoS level given the transmission schemes of all the other counterparts. However, from a practical point of view, a radio device might be more interested in guaranteeing a minimum satisfaction level (QoS) rather than attaining the highest achievable one. First, because a reliable communication becomes possible only when certain parameters meet some specific conditions (minimum QoS requirement), e.g., minimum signal to interference plus noise ratio (SINR), minimum delay, etc. Second, because higher QoS levels often imply greater efforts from the players, e.g., higher transmit power levels, more complex signal processing, etc. Third, because increasing the QoS for one communication often decreases the QoS of other communications. This reasoning implies that in practical terms, the Nash equilibrium concept might fail to model the network performance.

In the presence of minimum QoS requirements, a more suitable solution is the constrained equilibrium concept (see Debreu (1951, 1952)). In the context of autonomous and self-configuring networks, a constrained equilibrium is also a steady state where players satisfy their QoS constraints at the same time that their performance cannot be improved by feasible unilateral deviations. Nonetheless, the constrained equilibrium might be too demanding and depending on the QoS metrics and network configurations, it might not necessarily exist. Moreover, when it does, a player always ends up achieving the highest achievable payoff, which is often costly, as described above.

In the most general case, one can consider that players aim to be above its reference (satisfaction level) instead of considering that players aim to maximize their own payoff subject to a set of constraints. This reasoning leads to the concept of a satisfactory solution: any strategy profile of a given game where all players are above their satisfactory level. To each action, we assume an effort quantification.

In this order of ideas, we introduce the concept of efficient satisfactory solution. An efficient satisfactory solution is a steady state solution where all players satisfy their constraints by using the feasible action that requires the lowest effort. Assuming that the satisfaction profile (the demand) is feasible, i.e., the minimum satisfaction requirements can be simultaneously supported by the network, one can show that at least one efficient satisfactory solution exists if the set of actions is discrete and finite. Similarly, assuming the feasibility of the satisfaction levels, the existence of at least one efficient satisfactory solution is also ensured when the set of actions is compact and the payoff function is continuous over a linear space with finite dimension containing the set of satisfactory solution profiles. Finally, we present a fully distributed algorithm that allows a set of players to achieve a satisfactory solution using only perceived own-payoffs and own-actions.

6.5   Satisfaction Levels as Benchmarks

When one imposes constraints over the payoffs that each player obtains or over the action that a player can choose in the game $\mathcal{G}$, the Nash equilibrium is replaced by a constrained (Nash) equilibrium. In the presence of constraints, the set of actions each player can take reduces to the set of actions which verifies the constraints given the actions adopted by the other players. Let us determine such a set of available actions by the correspondence $\tilde{\mathcal{C}}_j : \mathcal{A}_{-j} \to 2^{\mathcal{A}_j}$ [1] for each player $j \in \mathcal{N}$ and denote the constrained game by $\mathcal{G}_c = (\mathcal{N}, \mathcal{A}_j, \tilde{R}_j, \tilde{\mathcal{C}}_j)$. A widely used solution to the game $\mathcal{G}_c$ is known as the generalized Nash equilibrium or constrained equilibrium which we define as follows:

Definition 6.5.0.2 (Constrained equilibrium). An action profile $\boldsymbol{a}^* \in \mathcal{A}$ is a constrained Nash equilibrium (CNE) of the game $\mathcal{G}_c$ if and only if

$$\forall j \in \mathcal{N}, \quad a_j^* \in \tilde{\mathcal{C}}_j \left(\mathbf{a}_{-j}^* \right) \text{ and}$$

$$\forall j \in \mathcal{N} \text{ and } \forall a_j \in \tilde{\mathcal{C}}_j \left(\mathbf{a}_{-j}^* \right), \ \tilde{R}_j(a_j^*, \mathbf{a}_{-j}^*) \geq \tilde{R}_j(a_j, \mathbf{a}_{-j}^*).$$

Note that the classical definition of Nash equilibrium is obtained from Def. 6.5.0.2, when the constrained action set of player $j \in \mathcal{N}$ does not depend on the actions of the other players, i.e., $\forall j \in \mathcal{N}$ and $\forall \mathbf{a}_{-j} \in \mathcal{A}_{-j}, \tilde{\mathcal{C}}_j \left(\mathbf{a}_{-j} \right) = \mathcal{A}_j$, which means that no constraint is imposed.

6.6 Satisfactory Solution

Consider now that the players in game $\mathcal{G}_c$ are interested in satisfying their own level. Here, the idea of satisfaction becomes intuitive: a player is said to be satisfied if his or her perceived payoff is above his or her satisfaction level. Once a player satisfied it has no interest on changing its action, and thus, an steady state is observed if all players are simultaneously satisfied. We refer to this solution as, satisfactory solution, and we define it as follows,

Definition 6.6.0.3 (Satisfactory solution). An action profile $\mathbf{a}^+$ is a satisfactory solution if

$$\forall j \in \mathcal{N}, \quad a_j^+ \in \tilde{\mathcal{C}}_j \left(\mathbf{a}_{-j}^+ \right). \tag{6.1}$$

Note that if one defines the function $\tilde{\mathcal{C}}_j$, for all $j \in \mathcal{N}$, as

$$\tilde{\mathcal{C}}_j(\mathbf{a}_{-j}) = \left\{ a_j \in \mathcal{A}_j : \tilde{R}_j \left(a_j, \mathbf{a}_{-j} \right) \geq \Gamma_j \right\},$$

where Γ_j is the satisfaction level required by player j, then, the notion can be easily interpreted as a minimum quality requirement or an admission control condition.

---

[1]The notation $2^{\mathcal{A}}$ represents the set of all possible subsets of the set $\mathcal{A}$, the empty set and $\mathcal{A}$ itself.

6.7 Efficient Satisfactory Solution

Assume now that players care about the cost or effort of using a given action. For instance, using a higher transmit power level or using a more complex modulation scheme (in the sense of the size of the constellation) might require a higher energy consumption and thus, reduce the battery life time of the transmitters. Hence, high transmit power levels and complex modulations can be considered as costly actions. If players are able to measure their effort incurred when using a specific action, then it becomes natural to think that players would aim to be satisfied incurring by the minimum effort. Following this reasoning, we define the efficient satisfactory solution as in [126].

Definition 6.7.0.4 (Efficient Satisfactory Solution). *Define a mapping* $c_j :$ $\mathcal{A}_j \to [0,1]$ *for all* $j \in \mathcal{N}$ *and consider the game* $\mathcal{G}$. *For all* $\left(j, a_j^*, a_j'\right) \in$ $\mathcal{N} \times \mathcal{A}_j^2$, *the action* a_j' *is said to be more costly than the action* a_j^* *if and only if* $c_j\left(a_j'\right) > c_j\left(a_j\right)$. *An action profile* $\mathbf{a}^* \in \mathcal{A}$ *is an efficient satisfactory solution if and only if*

$$\forall j \in \mathcal{N}, \quad \mathbf{a}_j^* \in \arg\min_{a_j \in \mathcal{C}_j\left(\mathbf{a}_{-j}^*\right)} c_j\left(a_j\right). \tag{6.2}$$

Then, $\mathbf{a}^*$ *is one of the efficient satisfactory solution of the game* $\mathcal{G}$.

From Definition 6.7.0.4, the implication is that the set of efficient satisfactory solution of the game $\mathcal{G}$ is identical to the set of constrained equilibrium of a non-cooperative game in normal-form denoted by

$$\mathcal{G}' = \left\{\mathcal{N}, \{\mathcal{A}_j\}_{j\in\mathcal{N}}, \{c_j\}_{j\in\mathcal{N}}, \{\mathcal{C}_j\}_{j\in\mathcal{N}}\right\},$$

where players aim to minimize their respective cost functions c_j subject to the set of constraints imposed over their payoff functions $\tilde{R}_j$ and represented by the function $\mathcal{C}_j$.

6.8 Learning a Satisfactory Solution

Now, we focus on the design of fully decentralized algorithms for allowing autonomous self-configuring wireless networks to achieve any satisfactory solution (not necessarily an efficient one) when the QoS constraints can be written as

$$\mathcal{C}_j(\mathbf{a}_{-j}) = \left\{a_j \in \mathcal{A}_j : \tilde{R}_j\left(a_j, \mathbf{a}_{-j}\right) \geq \Gamma_j\right\},$$

where Γ_j is the satisfactory level of player j. In addition, we assume that a player knows her own set of actions and is able to observe her own payoff at each stage. Denote by s_j an action of player j. Assume that player $j \in \mathcal{N}$ chooses her action at instant $t > 0$ following the probability distribution $\mathbf{x}_{j,t} = (\mathbf{x}_{j,t}(s_j))_{s_j \in \mathcal{A}_j}$, where $\mathbf{x}_{j,t}(s_j)$ is the probability with which the player j chooses its action s_j at instant t. Using this notation, we present the a fully algorithm, a slightly modified version of the algorithm presented in [194] (for the case of NE), where the benchmark is replaced by the satisfaction level:

1. At time $t = 0$, all players $j \in \mathcal{N}$ set up their initial action $a_{j,0}$, following an arbitrary chosen probability distribution $\mathbf{x}_{j,0}$.

2. At each time $t > 0$, each player $j \in \mathcal{N}$ computes $b_{j,t+1} = \frac{M_j + r_{j,t} - \Gamma_j}{2M_j}$, where $r_{j,t}$ is the observed payoff corresponds to chosen action and M_j is the highest payoff player j can achieve. Then, she updates her strategy for the next stage $t + 1$ as follows

$$a_{j,t+1} = \begin{cases} a_{j,t} & \text{if} \qquad r_{j,t} - \Gamma_j \geq 0 \\ a_{j,t} \sim \mathbf{x}_{j,t} & \text{otherwise.} \end{cases}$$

and the probability distribution is as follows, $\forall s_j \in \mathcal{A}_j$,

$$\mathbf{x}_{j,t+1}(s_j) = \begin{cases} \mathbf{x}_{j,t}(s_j), & \text{if } r_{j,t} - \Gamma_j \geqslant 0 \\ \mathbf{x}_{j,t}(s_j) + \lambda_{j,t} b_{j,t} \left(\mathbb{1}_{\{a_{j,t}=s_j\}} - \mathbf{x}_{j,t}(s_j) \right) & \text{otherwise,} \end{cases}$$

Here, where $\forall j \in \mathcal{N}$, $\lambda_{j,t}$ is a learning rate of player j.

3. If convergence is not achieved, then return to step (2).

It is important to remark that players do not change their action dumbly. Conversely, at each action change, players update their probability distribution so that higher probabilities are allocated to the actions that bring higher payoffs and thus, reduces the time of convergence.

Definition 6.8.0.5 (Securing Action). *In the game, a player $j \in \mathcal{N}$ is said to have a securing action a_j if and only if*

$$\forall \, \mathbf{a}_{-j} \in \mathcal{A}_{-j}, \quad a_j \in \mathcal{C}_j(\mathbf{a}_{-j}). \tag{6.3}$$

Once a player plays his or her securing action, it remains indifferent to the actions of all the other players, since it is always satisfied. The existence of securing actions in the game $\mathcal{G}$ might inhibit the convergence to satisfactory solution of the other players.

Remark 6.8.1 (Non-convergence of satisfactory solution). Assume the existence of at least one player with a securing action in the game and denote it by $a_j \in \mathcal{A}_j$ for player j. Then, if there exists a player $j' \in \mathcal{N} \setminus \{j\}$, for which $\mathcal{C}_{j'}(a_j, a_{-\{j',j\}}) = \emptyset$, $\forall \mathbf{a}_{-\{j',j\}} \in \mathcal{A}_{-\{j',j\}}$, the algorithm may not converge to a satisfactory solution with strictly positive probability.

The above remark follows from the fact that at time t before convergence, if the probability of the securing action a_j is strictly positive, and thus player j might play it. If so, by definition, there exists a player $j' \neq j$ who is never satisfied when player j plays his or her securing action. Then, the procedure does not converge (it is absorbed by the securing action). On the contrary, if none of the players possesses a securing strategy, the procedure converges to a satisfactory solution with high probability. This result comes from the fact that in the absence of securing actions, there always exists a nonzero probability of visiting all possible action profiles. Once a satisfactory action profile is visited, none of the players changes his or her actions, and convergence is observed.

Example 6.8.2. *Consider a weakly dominated game with two players. Each player has two choices.*

- *When player 1 chooses action 1, she always gets a payoff 1 whatever the other player does.*

- *When player 1 chooses the second action, she can get 0 or 1 depending on the choice of the second player.*

In the table below, player 1 chooses a row, and player 2 chooses a column.

Player 1\Player 2	s_1'	s_2'
s_1	$(1,0)$	$(1,0)$
s_2	$(0,0)$	$(1,1)$

The main challenging task here is to design a fully distributed learning algorithm that converges (or approaches in some sense) to the global optima of the game, i.e., the outcome $(1,1)$. As we can see, the action s_1 is a securing action of player 1.

6.8.3   Minkowski-Sum of Feasible Sets

The question of expectation of feasible sets can be identified with the Minkowski-sum of feasible sets, i.e., the mixture of the feasible sets when mixed strategies are allowed in the game.

6.9   From Nash Equilibrium to Satisfactory Solution

Lemma 6.9.0.1. *Consider the payoff function*

$$\tilde{v}_j(x^*) = \mathbb{1}_{\{x_j^* \in F_j(x_{-j}^*)\}}, j \in \mathcal{N}$$

where F_j is a generic satisfaction set of player j and $|\mathcal{N}| \geq 2$. Then, x^* is a Nash equilibrium with payoff 1 for each of the players if and only if it is a satisfactory solution of the problem with satisfaction set F_j.

Proof. Consider the one-shot game

$$\mathcal{G} = (\mathcal{N}, (\mathcal{X}_j, \tilde{v}_j(.))_{j \in \mathcal{N}}).$$

Suppose that the game $\mathcal{G}$ has at least one Nash equilibrium with payoff 1 for each player (this means, in particular, that the sets $F_j(.)$ are non-empty).

$$\tilde{v}_j(x^*) = 1, \quad \forall j \in \mathcal{N}. \tag{6.4}$$

$$\mathbb{1}_{\{x_j^* \in F_j(x_{-j}^*)\}} = 1, \forall\, j \in \mathcal{N}. \tag{6.5}$$

$$\Longleftrightarrow x_j^* \in F_j(x_{-j}^*), \forall\, j \in \mathcal{N}. \tag{6.6}$$

$$\Longleftrightarrow \text{ All the players are satisfied.} \tag{6.7}$$

$$\Longleftrightarrow x^* \text{ is a satisfactory solution.} \tag{6.8}$$

$\square$

Lemma 6.9.0.2. *Consider non-empty, convex, and compact set $\mathcal{X}_j$. Assume that a player is satisfied when he/she plays a best response, i.e., the payoff function $\tilde{v}_j(x^*) = \mathbb{1}_{\{x_j^* \in BR_j(x_{-j}^*)\}}$ where BR_j is the best response correspondence of some game. Then, x^* is a Nash equilibrium means that the payoff is equal to 1 for each of the players if and only if it is a satisfactory solution of the problem with satisfaction set BR_j.*

Proof. The proof is decomposed in two-steps: (1) pure equilibria, (2) partially and fully mixed equilibria

- For pure equilibria, x^* is a Nash equilibrium of the one-shot game $\mathcal{G} = (\mathcal{N}, \mathcal{A}_j, \tilde{R}_j)$ if and only if $a_j^* \in \widetilde{BR_j}(a_{-j}^*)$ i.e $\tilde{v}_j(a^*) = 1, \forall\, j$. This states precisely that each player meets his satisfaction set.

- For partially mixed equilibria, we need the Minkowski sum of the satisfaction sets. The strategy profile x^* is a Nash equilibrium if and only $\forall\, j$, $x_j^* \in BR_j(x_{-j}^*)$, i.e., $v_j(x^*) = 1, \forall\, j \in \mathcal{N}$, x^* is a satisfactory solution with $F_j = BR_j$.

$\square$

6.10   Mixed and Near-Satisfactory Solution

In this section, we introduce the notion of mixed and near-satisfactory solutions.

Definition 6.10.0.3. *We say that a mixed strategy profile x^* is a mixed satisfactory solution if and only for any player j, $\mathbb{P}_{x^*}(a_j \in F_j(a_{-j})) = 1$ where $\mathbb{P}_{x^*}$ denotes the probability under the probability under the measure $\mu(a_1, \ldots, a_n) = \prod_{j \in \mathcal{N}} x_j^*(a_j)$.*

This means that we have, an almost surely satisfactory solution. As we can see, this is a hard constraint. We can relax the constraint by considering soft inequalities. We say that a strategy profile x^* is an $\epsilon-$satisfactory solution if the strategy profile such that each of the players is satisfied with probability at least $1 - \epsilon$.

$$\forall j \in \mathcal{N}, \mathbb{P}_{x^*}\left(a_j \in F_j(a_{-j})\right) \geq 1 - \epsilon.$$

This is equivalent to

$$\inf_{j \in \mathcal{N}} \mathbb{P}_{x^*}\left(a_j \in F_j(a_{-j})\right) \geq 1 - \epsilon.$$

QoS Satisfaction in Medium Access Control

Consider a wireless network with L access points and a finite number of users. Each user j's demand is $\rho_j \geq 0$. The payoff of user j is

$$r_j(x) = \sum_{l \in \mathcal{L}} x_{j,l} \prod_{j' \neq j} (1 - x_{j'l}).$$

The satisfaction problem is then formulated as

$$\forall j, \ \sum_{l \in \mathcal{L}} x_{jl} \prod_{j' \neq j} (1 - x_{j'l}) \geq \rho_j.$$

Parallel Multiple Access Channel: Hard Constraint

In this example, we propose a power minimization under QoS constraints. The problem can be formulated as a satisfactory solution [128]. To do so, we define the satisfaction level of each player j at each link l by r_{jl}, i.e., the player j is satisfied if his or her rate weakly dominates the profile $(r_{jl})_{l \in \mathcal{L}}$. The problem is then written as

$$\begin{cases} \inf_{\mathbf{p}_j} \sum_{l \in \mathcal{L}} p_{j,l} \text{ s.t} \\ \log_2\left(1 + \text{SINR}_{jl}(\mathbf{h}_l, \mathbf{p}_l)\right) \geq r_{j,l} \\ j \in \mathcal{N}, \ l \in \mathcal{L}. \end{cases}$$

Power MIMO games: Hard Constraint

$$\begin{cases} \inf_{Q_j} \text{ tr}(Q_j) \text{ s.t} \\ I(\tilde{s}_j; \tilde{y}_j)(\mathbf{H}, Q_1, \ldots, Q_n) \geq r_j \end{cases}$$

Power MIMO games: Soft Constraint

- Indicator as payoff: At each time, each player gets 0 or 1. For the player j, the number 0 means that its mutual information is greater than r_j. If the frequency of success is greater than $1 - \epsilon_j$, then

$$\mathbb{P}\left(I(\tilde{s}_j; \tilde{y}_j)(\mathbf{H}, Q_1, \ldots, Q_n) \geq r_j\right) \geq 1 - \epsilon_j.$$

The objective of player j is to find an efficient satisfactory solution, i.e.,

$$\begin{cases} \inf_{Q_j} \ \mathrm{tr}(Q_j) \ \text{s.t} \\ \mathbb{P}\left(I(\tilde{s}_j; \tilde{y}_j)(\mathbf{H}, Q_1, \ldots, Q_n) \geq r_j\right) \geq 1 - \epsilon_j, \end{cases}$$

where $\gamma_j, \epsilon_j \geq 0$. Propose a CODIPAS-RL based fully learning for Pareto optimal solutions, constrained equilibria and global optima.

- Mutual information as payoff:

$$\begin{cases} \inf_{Q_j} \ \mathrm{tr}(Q_j) \ \text{s.t} \\ \mathbb{E}_{\mathbf{H}, \mathbf{x}} I(\tilde{s}_j; \tilde{y}_j)(\mathbf{H}, Q_1, \ldots, Q_n) \geq r_j \end{cases}$$

- Minimum delay under QoS-requirement: bi-objective

$$\begin{cases} \inf_{Q_j} \ \mathrm{Delay}_j(Q_j, \mathbf{Q}_{-j}) \ \text{s.t} \\ \mathbb{E}_{\mathbf{H}, \mathbf{x}} I(\tilde{s}_j; \tilde{y}_j)(\mathbf{H}, Q_1, \ldots, Q_n) \geq r_j \end{cases}$$

6.11 CODIPAS with Dynamic Satisfaction Level

According to the theory of satisfaction, a decision maker wants satisfaction rather than optimization. Instead of solving an optimization problem, a decision maker has a satisfactory level, and searches for a better alternative until she finds the one of whose payoff exceeds the satisfaction level. Once she finds such an alternative, she quits his search and sticks to it.

We examine the behavior of players (without full rationality assumptions) who play a dynamic game according to a simple CODIPAS: she continues to play the same action if the action generates one instant payoff that exceeds the satisfaction level, which summarizes a history according to the average of the past payoffs, and switches to the other action with a positive probability otherwise.

Example 6.11.1 (Emergence of cooperation in the long-run). *Consider two players who play a two-by-two game with action space $\{C, D\}$, i.e., the players can cooperate or defect. If a_t denotes the action profile chosen at time t, each player j observes a numerical value $r_{j,t}$ of this payoff and computes the average payoff using the payoff learning algorithm:*

$$\bar{r}_{j,t+1} = \bar{r}_{j,t} + \frac{1}{t+1}(r_{j,t+1} - \bar{r}_{j,t})$$

Using this scheme, player j summarizes a history of the average payoff in the past, suppresses much information that could have been used for making a choice, such as the expected outcomes.

The dynamic satisfactory level is the level of payoff that player j expects from the action $a_{j,t}$ in period t, based on his own past experience. The dynamic satisfactory level at time t is $\bar{r}_{j,t}$.

If the current payoff $r_{j,t}$ from player $j's$ action $a_{j,t} \in \{C, D\}$ exceeds the current satisfaction level $\bar{r}_{j,t}$, then player j takes $a_{j,t+1} = a_{j,t}$ in the $(t+1)$st stage of the dynamic interaction. If $r_{j,t} < \bar{r}_{j,t}$ holds, then player j takes a different action in the next period according to some strategy.

The algorithm is given by

$$a_{j,t+1} = \begin{cases} a_{j,t} & \text{if } r_{j,t} \geq \bar{r}_{j,t} \\ b_{j,t} \in \{C, D\} & \sim x_{j,t} \in \Delta(\{C, D\}) \end{cases}$$

where $x_{j,t}$ is of full support.

The same methodology can be extended to dynamic robust games and the asymptotic pseudo-trajectories can be studied with stochastic approximation schemes to locate stable and unstable sets.

6.12   Random Matrix Games

In this section we study finite matrix games with random entries which we call *random matrix games* (RMGs). Given random payoff matrices, the questions arise as what is meant by playing the random matrix game (RMG) in an optimal way. Because now the actual payoffs of the game depend not only on the action profile picked by the players but also on the sample point of realized state of the nature. Therefore the players cannot guarantee themselves a certain payoff level or satisfaction level. The players are forced to gamble depending a certain random variable. The question of how one gambles in an optimal way needs to be defined. One of the approaches to this type of games is to replace the random matrices by its expectation leading to *expected game* and then solve deterministic matrix game. Second important approach is considering a satisfactory criterion consisting of the probability that the payoff is above a certain threshold. Third approach is to incorporate the variance or higher moments of the payoffs in the response of the players, a simple criterion could be to maximize the mean payoff and to minimize the *variance* of the random payoff simultaneously. The latter leads to a multiobjective criterion whose solutions are related to Pareto boundaries depending on the behavior of the other players.

6.12.1 Random Matrix Games Overview

Random matrix game theory is not well-investigated in terms of analytical results. To a given game matrix, a variety of formulae and algorithms may be applied to yield limit cycles, equilibrium strategies and equilibrium payoffs, etc. But if the game matrix is random, these will be random variables. Their distribution is not easy to determine analytically for any fixed matrix size, still less their asymptotics as the matrix size tends to infinity as it is usually done in random matrix theory. The results of random matrix theory does not apply directly to RMG due to interactive behavior of the players.

We briefly review some works on random matrix games. Most of the existing results focus on the probability of the size of the support of equilibria and probability for the value of zero-sum random matrix games to be in a given interval under fixed distribution such as *Cauchy* distribution, *Gaussian* distribution or *Uniform* distribution. In [79] Kaplansky examined of size of the support of equilibria for a generic antisymmetric RMG. In [148] Stanford have studied the number of pure Nash equilibria in random coordination games. In [65], Goldman examined the probability of a saddle point in RMG. The work of Roberts [135] analyzes kernel Sizes of zero-sum random matrix games when the entries follow a given distribution. The author in [136] studies Nash equilibria of random zero-sum and coordination RMG under Cauchy distribution. The above works provided formula for the chance that a random l_1-by-l_2 coordination game has exactly k pure Nash equilibria.

In terms of applications, RMG are widely observed in evolutionary biology, economics and management sciences, engineering, etc. Matrix games are well-used in evolutionary biology. These types of games can model for example, pairwise interaction between animals, food competition, Hawk-Dove, sex ratio. In many cases, the resulting matrices are influenced by the environment, the locations, the season, the types of the individuals etc which can be seen as a random variable. Example of works in that direction include [31, 54, 77].

RMG models are used in economics and management sciences. By use of satisficing methods,[45] discussed the solution of games with random payoffs.

RMGs are present in many networking and communication problems. RMG is particularly useful tool for modeling and studying interactions between cognitive radios envisioned to operate in future communications systems. Such terminals will have the capability to adapt to the context and uncertain environment they operate in, through possibly power and rate control as well as channel selection.

Users performance metrics such as successful reception condition, through-put/connectivity are generally random quantities because they are influenced by random channel states and environment states. Most of the game-theoretic analysis in communication networks are done only in the context static games (by fixing some parameters) which may not capture the stochastic nature of wireless communication systems. RMG turns out to be a more appropriated tool for modelling interaction in communication networks. It allows to model interactive decision-making problems of the nodes and evaluate network performances under random channel states, fading, shadowing, measurement noise, background noise in situations where finite number of choices are available to the players (users, operators, mobile devices, base stations, protocol etc).

6.12.2 Zero-Sum Random Matrix Games

We consider the most basic class of games. There are two players, the row player and the column player. A game between them is determined by an $l_1 \times l_2$ matrix $M = (m_{ij})_{i,j}$ of real numbers. The row player has l_1 pure strategies, corresponding to the l_1 rows of M. The column player has l_2 pure strategies, corresponding to the l_2 columns of M. If the row player plays his i-th strategy and the column player plays his j−th strategy then the column player pays the row player the real number m_{ij}. The player are allowed to randomize their actions i.e. they can choose mixed strategies as well. This means the row player's move is to choose a row vector $x_1 = (x_{11}, \ldots, x_{1l_1})$ with the x_{1k} nonnegative and summing to one. The column player's move is to choose a column vector $x_2 = (x_{21}, \ldots, x_{2l_2})$ with the $x_{2k'}$ likewise nonnegative and summing to one. The players make their moves independently and the game is concluded by the column player paying the expected number $x_1' M x_2$ to the row player.

von Neumann minmax theorem: We now focus on solution concepts of such games. A triple (x_1, x_2, v) consisting of a row probability l_1−vector x_1, a column probability l_2−vector x_2, and a number v is called a solution of the zero-sum[2] if it satisfies the following conditions:

$$\forall y_1, \ y_1' M x_2 \leq x_1' M x_2 = v \leq x_1' M y_2, \ \forall y_2,$$

In that case the pair (x_1, x_2) is called *saddle point*. Thus (x_1, x_2, v) is a classical solution (in the sense of von Neumann) and (x_1, x_2) is a saddle point if and only if unilateral deviation from (x_1, x_2) by either player will never result in an improved outcome for that player. It is known that from von Neumann minmax theorem that every zero-sum game of this type with matrix M has at least one saddle point. Moreover for all these saddle points (if many), the number v is the same, this number being called the value of the game. In

---

[2]the results extend to two-player constant-sum games

addition the saddle points are interchangeable i.e., if (x_1, x_2) and (y_1, y_2) are two saddle points then (x_1, y_2) and (y_1, x_2) are also saddle points.

We now introduce randomness to the coefficients of the matrix M. To consider random games, we fix an entry measure for each entry of the matrix $m_{ij}(w) \in \mathbb{R}$, where the state w is driven by a probability measure ν on a finite dimensional real space $\mathcal{W}$. The collection

$$(\{1, 2\}, \mathcal{W}, \nu, \{1, \dots, l_1\}, \{1, \dots, l_2\}, M(w), -M(w))$$

is a zero-sum *random matrix game* (RMG) where $w \sim \nu$. The triplet $(x_1(w), x_2(w), v(w))_{w \in \mathcal{W}}$ is a state-dependent saddle point of RMG if for each $w \in \mathcal{W}, \forall y_1(w) \in \mathcal{X}_1$,

$$y_1'(w)M(w)x_2(w) \le x_1'(w)M(w)x_2(w) = v(w) \le x_1'(w)M(w)y_2(w), \quad (6.9)$$

$\forall y_2(w) \in \mathcal{X}_2$, where $\mathcal{X}_j$ is the $(l_j - 1)$−dimensional simplex of $\mathbb{R}^{l_j}$.

Relevance in Networks

There are many interesting applications of zero-sum random matrix games in both wired and wireless networks. The randomness allows to study the robustness of the network with the respect to malicious attack, failure, noisy measurements and environment perturbation. Below we provide a basic example of application in network security.

Example 6.12.3 (Network Security & Reliability). *Many wireless networks are inherently vulnerable to physical attacks, such as jamming and destruction of nodes and links. From a topological point of view, a common characteristic of these networks is about communication and information dissemination between a source node and destination nodes i.e the goal of the nodes is to communicate with a designated node. The network topology can be seen as a graph $(\mathcal{V}, \mathcal{E})$. The goal of the network operator is to keep the nodes of the network connected to the base-station, while the goal of the adversary is to separate as many nodes as possible from the base-station. The interaction between the network operator and the jammer can be modeled as a two player, zero-sum game. The network operator chooses a spanning tree to be used for communications. The mixed strategy of the network operator is a distribution on the set of spanning trees. The adversary chooses a subset of edge to be attacked. The mixed strategy of the jammer is a distribution on the subset of edge of the graph. The payoff for the adversary is $\tilde{w}$ times the number of nodes from which there is no path to the destination d in the tree $a_1 \backslash a_2$, where a_2 is the subset of edges chosen by the jammer, and $\tilde{w}$ is a random that influences the failure state of the tree. We model the support of random variable as $\{0, 1\}$. If $a_2 \bigcap a_1 = \emptyset$, or if $\tilde{\alpha} = 0$ then the payoff of the jammer is zero. Let $M_2(a_1, a_2)$ be the number of nodes that are disconnected from d if state variable is $\tilde{w}$ the operator uses the tree a_1 and the jammer attacks the list of* *links in a_2. The network operator maximizes $-\tilde{w}M_2(a_1, a_2)$ and the attacker minimizes $\tilde{w}M_2(a_1, a_2)$ i.e. the maximum damage to the operator. We assume that the probability of failure i;e $\mathbb{P}(\tilde{w} = 1) = \alpha > 0$.*

This game is clearly a RMG because the entries of the payoffs depend on the realized value of the random variable $\tilde{w}$ which is generated by the environment.

6.12.4 NonZero Sum Random Matrix Games

We consider a two-player non-zero sum game. As above there is a row player and a column player. A game between them is determined by a random state variable w and two matrices of size $l_1 \times l_2$, $M_1(w) = (m_{1,ij}(w))$ and $M_2(w) = (m_{2,ij}(w))$ of real numbers. The row player has l_1 pure strategies, corresponding to the l_1 rows of $M_1(w)$. The column player has l_2 pure strategies, corresponding to the l_2 columns of $M_2(w)$. If the row player plays his i-th strategy and the column player plays his j-th strategy then the row player receives the real number $m_{1,ij}(w)$ and the column player receives the real number $m_{2,ij}(w)$.

Since the coefficients have to be finite our key assumption is following:

Assumption A0: Finite expectation for the matrix entries: $\forall i', i, j$, one has $-\infty < \mathbb{E}_{\mathbf{w}} m_{i',ij}(w) < +\infty$.

Definition 6.12.4.1 (State-robust equilibrium). *A strategy profile (x_1, x_2) is a state-robust equilibrium if it is an equilibrium for any state w.*

These strategies are state-independent and distribution independent.

Lemma 6.12.4.2. *A state-robust equilibrium may not exists in RMG.*

To prove this Lemma, consider two games G_1 and G_2 such that in the first game, the first action is a dominant action and in the second game the second action is a dominant action. Now consider a random variable w which takes value in $\{O_1, O_2\}$. Suppose that the support of w is $\{O_1, O_2\}$. For $w = O_i$ the corresponding game is G_i. It is clear that the resulting RMG has no state-robust equilibrium.

Definition 6.12.4.3 (State-dependent equilibrium). *A state-dependent equilibrium $(x_1(w), x_2(w))_w$ is a collection of equilibrium profile per state.*

Lemma 6.12.4.4. *Under assumption A0, there exists at least one state-dependent (mixed) equilibrium of the RMG.*

The proof of this Lemma is obtained by concatenation of the equilibrium of each realized game which has at least one equilibrium in mixed strategies (Nash theorem).

Definition 6.12.4.5 (State-independent equilibrium). *A strategy profile (x_1, x_2) is state-independent (mixed) equilibrium if x_1 and x_2 are independent of the realized value of the state and they form an equilibrium of the expected game.*

Lemma 6.12.4.6. *Under assumption A0, there exists at least one state-independent (mixed) equilibrium in the expected RMG.*

To prove this Lemma, we apply Nash theorem to the expected game which is a bi-matrix game.

Example 6.12.5 (Two-by-two expected RMG). *Consider two players. The action spaces of each player is independent on the state. Player I has two actions $\{Up, Down\}$ in any state and Player II's action set is $\{Left, Right\}$ in any state. We transform the random matrix game into expected game by taking the expectation of the payoff. where $\bar{a}_1 = \mathbb{E}_{\mathbf{w}} m_{1,11}(w) = \mathbb{E}_w \tilde{M}_1(w, Up, Left)$.*

Decisions of player I \ Decisions of player II	$Left$	$Right$
Up	$\bar{a}_1, \bar{a}_2$	b_1, b_2
$Down$	$\bar{c}_1, \bar{c}_2$	d_1, d_2

TABLE 6.1
2×2 expected robust game.

The other coefficients are defined in a similar way.

Computation of state-independent equilibria. The state-independent pure equilibria are obtained by simply checking the above inequalities. It is easy to show that the optimal completely state-independent mixed strategy (if it exists) is given by

$$x^* = \frac{\bar{d}_1 - \bar{b}_1}{\bar{d}_1 - \bar{b}_1 + \bar{a}_1 - \bar{c}_1},$$

$$y^* = \frac{\bar{d}_2 - \bar{b}_2}{\bar{d}_2 - \bar{b}_2 + \bar{a}_2 - \bar{c}_2}.$$

Now, how to compute equilibrium when the coefficients are random and their laws are unknown to the players? Under which conditions, a global optimum can be approximated?

In the next section we describe how to learn these through iterative process and discuss the relationship between learning outcomes, equilibria and global optima.

6.12.5.1 Relevance in Networking and Communication

Example 6.12.6 (Ergodic Rate and Energy Constraint). *Consider two mobile stations (MSs) and two base stations (BSs). Each mobile station can transmit to one of the base stations. If mobile station $MS1$ chooses a different base station than mobile station $MS2$ then, mobile station $MS1$ gets the payoff $\log_2\left(1 + \frac{p_1|h_1|^2}{N_0}\right)$, and mobile station 2 gets $\log_2\left(1 + \frac{p_2|h_2|^2}{N_0}\right)$ where p_1 and p_2 are positive transmit powers, h_1, h_2 are channel states (random), and N_0 is a background noise parameter. If both MSs transmit at the same base station, there is an interference; mobile station $MS1$ gets $\log_2\left(1 + \frac{p_1|h_1|^2}{N_0 + p_2|h_2|^2}\right)$,* and mobile station $MS2$ gets $\log_2\left(1+\frac{p_2|h_2|^2}{N_0+p_1|h_1|^2}\right)$. *The following table summarizes the different configurations. Mobile station $MS1$ chooses a row (row player) and $MS2$ chooses a column (column player). The first component of the payoff is for $MS1$ and the second component is for $MS2$.*

We distinguish four configurations:

- $(h_1, h_2) \neq (0,0)$

- $(0, h_2), \ h_2 \neq 0.$

- $(h_1, 0), \ h_1 \neq 0.$

- $(0,0)$

The tables below represent the four configurations.

$MS1\backslash MS2$	$BS1$	$BS2$
$BS1$	$\log_2\left(1+\frac{p_1\mid h_1\mid^2}{N_0+p_2\mid h_2\mid^2}\right),$ $\log_2\left(1+\frac{p_2\mid h_2\mid^2}{N_0+p_1\mid h_1\mid^2}\right)$	$\log_2\left(1+\frac{p_1\mid h_1\mid^2}{N_0}\right),$ $\log_2\left(1+\frac{p_2\mid h_2\mid^2}{N_0}\right)$
$BS2$	$\log_2\left(1+\frac{p_1\mid h_1\mid^2}{N_0}\right),$ $\log_2\left(1+\frac{p_2\mid h_2\mid^2}{N_0}\right)$	$\log_2\left(1+\frac{p_1\mid h_1\mid^2}{N_0+p_2\mid h_2\mid^2}\right),$ $\log_2\left(1+\frac{p_2\mid h_2\mid^2}{N_0+p_1\mid h_1\mid^2}\right)$

$MS1\backslash MS2$	$BS1$	$BS2$
$BS1$	$0,$ $\log_2\left(1+\frac{p_2\mid h_2\mid^2}{N_0}\right)$	$0,$ $\log_2\left(1+\frac{p_2\mid h_2\mid^2}{N_0}\right)$
$BS2$	$0,$ $\log_2\left(1+\frac{p_2\mid h_2\mid^2}{N_0}\right)$	0 $\log_2\left(1+\frac{p_2\mid h_2\mid^2}{N_0}\right)$

$MS1\backslash MS2$	$BS1$	$BS2$
$BS1$	$\log_2\left(1+\frac{p_1\mid h_1\mid^2}{N_0}\right),$ 0	$\log_2\left(1+\frac{p_1\mid h_1\mid^2}{N_0}\right),$ 0
$BS2$	$\log_2\left(1+\frac{p_1\mid h_1\mid^2}{N_0}\right),$ 0	$\log_2\left(1+\frac{p_1\mid h_1\mid^2}{N_0}\right)$ 0

$MS1\backslash MS2$	$BS1$	$BS2$
$BS1$	$(0,0)$	$(0,0)$
$BS2$	$(0,0)$	$(0,0)$

Next we give a detailed analysis of the game in each configuration in $\{1,2,3,4\}$.

- *Configuration one: In the first configuration, all the parameters are strictly positive. This is an anticoordination game. The game has two pure equilibria which consists of $(BS1, BS2)$ and $(BS2, BS1)$. There is also one fully mixed equilibrium. Thus, each game in this configuration has in total three equilibria.*

- *Configuration two: The games in this configuration have a state in the form of $(0, h_2)$, $h_2 \neq 0$. These games have a continuum of equilibria. Any strategy of MS 1 is an equilibrium strategy.*

-

- *Configuration three: A game within this category has its state in the form $(h_1, 0)$, $h_1 \neq 0$. These games have continuum of equilibria.*

- *Configuration four: In this configuration, any strategy profile is an equilibrium and a global optimum because the entries are trivially zero.*

6.12.7 Evolutionary Random Matrix Games

We consider evolutionary games described by many local pairwise interactions. In each pairwise interaction, players are involved in a RMG. The resulting evolutionary game dynamics are stochastic dynamics. Using stochastic approximations, the asymptotic pseudotrajectories of these dynamics can be studied with ordinary differential equations with the expected payoffs. An example is the replicator dynamics given by

$$\frac{d}{dt}x_{1,t}(r') = x_{1,t}(r') \left[(\mathbb{E}M_1(w)x_{2,t})_{r'} - \sum_{r''} x_{1,t}(r'')(\mathbb{E}M_1(w)x_{2,t})_{r''} \right] \quad (6.10)$$

$$\frac{d}{dt}x_{2,t}(c) = x_{2,t}(c) \left[(\mathbb{E}x'_{1,t}M_2(w))_c - \sum_{c'} x_{2,t}(c')(\mathbb{E}x_{1,t}M_2(w))_{c'} \right] \quad (6.11)$$

The set of stationary points (also called rest points) of this dynamics contains the set of state-independent Nash equilibria of the RMG. It contains also the set of pure global optima of the expected game.

6.12.8 Learning in Random Matrix Games

Let $x_{1,t+1}(s_1)$ be the probability of the row player to choose the row $r' = s_1$ at time iteration $t + 1$ and $x_{2,t+1}(s_2)$ be the probability of the column player to choose the column $c = s_2$ at time iteration $t + 1$. The parameter $\lambda_{j,t}$ is a positive real number and represents the learning rate for strategy-dynamics of player j. $\mu_{j,t}$ is a positive real number which represents the learning rate for payoff-dynamics. Below we provide the imitative CODIPAS which is well-adapted to RMGs. $\hat{m}_{j,t}$ is the estimated payoff vector of player j at time t. This is a l_j- dimensional vector.

$$
\begin{array}{l}
\textbf{Initialization} \\[4pt]
\hat{r}_{1,0} = (\hat{r}_{1,0}(1), \ldots, \hat{r}_{1,0}(l_1)) \\
x_{1,0} = (x_{1,0}(1), \ldots, x_{1,0}(l_1)) \\
\hat{r}_{2,0} = (\hat{r}_{2,0}(1), \ldots, \hat{r}_{2,0}(l_2)) \\
x_{2,0} = (x_{2,0}(1), \ldots, x_{2,0}(l_2))
\end{array}
$$

$$
\textbf{Learning pattern of the row player}
$$
$$
s_1 \in \{1, 2 \ldots, l_1\}
$$
$$
x_{1,t+1}(s_1) = \frac{x_{1,t}(s_1)(1+\lambda_{1,t})^{\hat{r}_{1,t}(s_1)}}{\sum_{s_1'} x_{1,t}(s_1')(1+\lambda_{1,t})^{\hat{r}_{1,t}(s_1')}}
$$
$$
\hat{r}_{1,t+1}(s_1) = \hat{r}_{1,t}(s_1) + \mu_{1,t}\mathbb{1}_{\{a_{1,t}=s_1\}}(r_{1,t} - \hat{r}_{1,t}(s_1))
$$

$$
\textbf{Learning pattern of the column player}
$$
$$
s_2 \in \{1, 2 \ldots, l_2\}
$$
$$
x_{2,t+1}(s_2) = \frac{x_{2,t}(s_2)(1+\lambda_{2,t})^{\hat{r}_{2,t}(s_2)}}{\sum_{s_2'} x_{2,t}(s_2')(1+\lambda_{2,t})^{\hat{r}_{2,t}(s_2')}},
$$
$$
\hat{r}_{2,t+1}(s_2) = \hat{r}_{2,t}(s_2) + \mu_{2,t}\mathbb{1}_{\{a_{2,t}=s_2\}}(r_{2,t} - \hat{m}_{2,t}(s_2))
$$

6.12.9 Mean-Variance Response

Consider a RMG with 2 players. The row player has a state-dependent payoff function given by $x_1' M(w) x_2$. We introduce two key performance metrics: the *mean of the payoff* and the variance of the payoff. The objective here is not to consider the expected payoff as performance but to include also the variance which captures the notion of risk.

We define the mean-variance response (MVR) of the player 1 to a strategy x_2 as follows: $x_1 \in MVR_1(x_2)$ if there is no strategy y_1 for which the following inequalities are simultaneously true:

$$
\mathbb{E}_{w\sim\nu} y_1' M(w) x_2 \geq \mathbb{E}_{w\sim\nu} x_1' M(w) x_2, \quad (6.12)
$$
$$
\mathbb{E}_w \left[y_1' M(w) x_2 - y_1' \mathbb{E}_w M(w) x_2 \right]^2 \leq \mathbb{E}_w \left[x_1' M(w) x_2 - x_1' \mathbb{E}_w M(w) x_2 \right]^2
$$

where at least one inequality is strict.

Similarly, we define the MVR of the column player as $x_2 \in MVR_2(x_1)$ if there is no strategy y_2 for which the following inequalities are simultaneously true:

$$
\mathbb{E}_{w\sim\nu} x_1' M(w) y_2 \leq \mathbb{E}_{w\sim\nu} x_1' M(w) x_2, \quad (6.13)
$$
$$
\mathbb{E}_w \left[x_1' M(w) y_2 - x_1' \mathbb{E}_w M(w) y_2 \right]^2 \geq \mathbb{E}_w \left[x_1' M(w) x_2 - x_1' \mathbb{E}_w M(w) x_2 \right]^2
$$

where at least one inequality is strict.

This formulation can be seen as a Pareto optimality response to x_2 in the sense that one cannot improve one of the component without degrading the other component. This is a vector optimization criterion where the first objective is to be maximized and the second objective (variance) is to be minimized. In the context of games, one has a fixed point of MVR.

The mean variance solution is a saddle point of the MVR correspondence.

We now formulate the nonzero-sum case. **Mean-variance solution** We define a mean-variance solution as a fixed point of the MVR correspondence i.e., a strategy profile x such that $x_1 \in MVR_1(x_2)$ and $x_2 \in MVR_1(x_1)$.

Example 6.12.10 (Energy market). *We consider an energy market with large population of consumers and finitely many producers (players), each having a certain production and an individual production cost per unit of energy. This cost is determined by producers' type. The price of the energy market is determined by supply and demand, with supply (the total quantity produced) driven by producer's costs and market structure, and demand driven by customers' tastes and state of nature (aka weather). We classify the interactions at different levels:*

- *Producers/generators (of different types of renewable energy). Each producer knows its own production cost, does not know the costs of the other producers and does not know (exact) weather condition in advance*

- *Distributors: Each of them has a contract with producers and and contract with at least a subset of users*

- *Users/Consumers, A user sees its realized prices per period/slot, Users interact also through a network, application in the Cloud Expectation over type distribution, over environment distribution (weather).*

There are clearly RMGs at all the layers of this energy market interaction. To simplify the presentation we consider two producers. Each producer decides how much unit to produce (the action set is a finite set of positive numbers) and a price $p(\xi, q_1, q_2)$ to be a realization of the stochastic process

$$\max \left\{ 0, \bar{p} - D \left(\sum_{j=1}^{n} q_j \right) + \sigma \xi \right\},$$

where $n = 2$, $\bar{p}, \sigma$ are positive parameters, D is a demand function and ξ is a stochastic process and $\sum_{j=1}^{n} q_j$ is the total supply which is supposed to estimate the demand of the consumers in the large population. The payoff of producer 1 is $q_1 p(\xi, q_1, q_2) - \theta_1 q_1$ and the payoff of producer 2 is $q_2 p(\xi, q_1, q_2) - \theta_2 q_1$ where θ_1 and θ_2 are random variables that determines the type of producers. θ_j can be seen as the unit production cost of producer j. This defines clearly RMG between producers.

For this RMG we list two important issues:

- *How to flatten the peaks and met the demand?*

- *How the incorporates the risk with the producers preferences?*

All these issues can be addressed in the mean-variance approach proposed above.

6.12.11 Satisfactory Solution

Consider a two-player zero-sum RMG. In the satisfactory criterion of optimality, the row player maximizes her probability of winning a certain amount no matter what strategy the other player used. Formally, this is captured by the following minmax problem

$$\sup_{x_1} \inf_{x_2} \mathbb{P}\left(x_1' M(w) x_2 \geq \beta\right),$$

where $\beta \in \mathbb{R}$.

> ### RMG and related games

- RMG can be seen as particular case of stochastic games with independent and identically distributed states.

- RMG is the class of games under state uncertainty also called *robust games*.

The framework developed here can be extended to

- Random poly-matrix games,

- Games with finitely many interacting players and random payoffs.

- Random matrix population games with more than three players in each local interaction.

- Dynamic robust games with discounted payoffs.

6.13 Mean-Variance Response and Demand Satisfaction

In this section we generalize the mean-variance response and demand satisfaction to finitely many players. Consider a robust game with n players. Each player j has a state-dependent payoff function given by $r_j(w, x_j, x_{-j})$. We introduce two key performance metrics: the *mean of the payoff* and the *variance of the payoff*. The objective here is not to consider the expected payoff as performance but to include also the variance which captures the notion of risk as we will see in last Chapter.

> ### Mean-variance response viewed as coupled constraint

We define the mean-variance response (MVR) of player j to a strategy x_{-j} as follows: $x_j \in MVR(x_{-j})$ if there is no strategy x'_j for which the following inequalities are simultaneously true:

$$\bar{r}_j(x'_j, x_{-j}) = \mathbb{E}_w r_j(w, x'_j, x_{-j}) \geq \mathbb{E}_w r_j(w, x_j, x_{-j}) \text{ and} \tag{6.14}$$

$$\mathbb{E}_w \left[r_j(w, x'_j, x_{-j}) - \bar{r}_j(x'_j, x_{-j}) \right]^2 \leq \mathbb{E}_w \left[r_j(w, x_j, x_{-j}) - \bar{r}_j(x_j, x_{-j}) \right]^2, \tag{6.15}$$

where at least one inequality is strict.

This formulation can be seen as a Pareto optimality response to x_{-j} in the sense that one cannot improve one of the component without degrading the other component. This is a vector optimization criterion where the first objective is to be maximized and the second objective (variance) is to be minimized. In the context of games, one has a fixed point of MVR.

Mean-variance solution

We define a mean-variance solution as a fixed point of the MVR correspondence i.e a strategy profile x such that for every player j, x_j is a mean-variance response to x_{-j}.

Next we address the question of how to learn the variance payoff from the measurements. We know how to learn the expected payoff with the risk-neutral CODIPAS which is $\hat{m}_{j,t}$. We combine this approach with a learning scheme for the variance of the payoffs. The variance learning scheme is constructed as follows. The estimator of the variance is

$$\hat{v}_{j,t} = \frac{1}{t} \sum_{t'=1}^{t} \left(r_{j,t'} - \hat{r}_{j,t'} \right)^2,$$

which satisfies the recursive stochastic difference equation:

$$\hat{v}_{j,t+1}(s_j) = \hat{v}_{j,t}(s_j) + \tilde{\mu}_{j,t} \left((r_{j,t} - \hat{r}_{j,t})^2 - \hat{v}_{j,t}(s_j) \right).$$

where $\tilde{\mu}_{j,t} = \frac{1}{t+1}$.

7

Learning under Random Updates

7.1 Introduction

In a situation such as a dynamic environment under uncertainty, one would like to have a learning and adaptive procedure that does not require any information about the other players' actions or payoffs, and less memory (a small number of parameters in terms of past own-experiences and observed data) as possible. In the previous chapters, we have called such a rule *fully distributed*. In a dynamic unknown environment, such fully distributed learning schemes are essential for applications of game theory and distributed optimization. In dynamic scenarios with

- a random set of active players, and where

- the traffic, the topology, and the states of the environment may vary over time and the communications between players are difficult and may be noisy and delayed,

it is important to investigate games under uncertainty (also called robust games [1, 159]) and fully distributed learning algorithms that demand a minimal amount of information from the other players.

In this chapter, we study dynamic games under uncertainty and a variable number of interacting players. The set of potential players is finite. At each time, each player can be in an active mode or in sleep mode. The set of active players is generated by a random variable independent of the state process. We develop a combined fully distributed payoff and strategy reinforcement learning for such games and prove their almost sure convergence to non-autonomous dynamics under suitable scaling and regularity conditions. Nonconvergence and convergence of heterogeneous learning to equilibria are examined in dynamic robust potential games.

As we have seen in Chapter 2, if the player knows the payoff function and the past actions played by the others one can use the fictitious play-like algorithms [41], the best response dynamics, the gradient-based algorithms, etc. Even under these observation of past actions and payoffs, these algorithms may not converge [159]. Additional learning difficulties arise when the players do not know the payoff function nor the last actions played by the other players. Now, each player needs to interact with the dynamic environment to find out its optimal payoff and its optimal strategy.

The players build their strategies and update them by using only their own actions and estimated own payoffs.

The objective of this chapter is to extend the COmbined fully DIstributed PAyoff-reinforcement and Strategy-Reinforcement Learning algorithms (CODIPAS-RL) developed in Chapter 4 to general-sum dynamic games with a variable number of interacting players and *random updates*.

Model-Free Learning.

Under the CODIPAS-RLs, players can learn their expected payoffs and their optimal strategies by using some simple iterative techniques based own-experiences. The strategies that give good performances are reinforced and new strategies are explored. Such approach is called model of *reinforcement learning* [42]. When the player can also directly learn about its optimal policy without knowing the payoff function, the approach is called *model-free reinforcement learning*. The proposed CODIPAS-RL scheme is in the class of model-free reinforcement learning for both strategies and payoffs learning (also called Q-value learning [188, 189, 19]).

Compared to the standard reinforcement learning algorithms [42, 179, 11, 32, 139] in which the expected payoff function is not learned, CODIPAS-RL has the advantage of estimating the optimal expected payoffs in parallel with strategy-learning. The estimated payoffs give the expected payoff function if all the strategies have been sufficiently explored. CODIPAS-RL is a well-adapted learning scheme for games under uncertainty or robust games. We use stochastic approximation techniques to show that, under suitable conditions, combined learning schemes with different rate of learning can be studied by their deterministic ordinary differential equation (ODE) counterparts. The resulting ordinary differential equations differ from the standard game dynamics (the replicator dynamics, the (myopic) best response (BR) dynamics, the logit dynamics and fictitious play dynamics, etc.). We then develop unified heterogeneous learning schemes in which each player adopts a distinct learning scheme in the dynamic interactions under incomplete information and stochastic processes: state process, random set of active players, local clocks.

We briefly overview recent literature on fully distributed learning related to the framework developed in this chapter. In robust games with a large number of actions, players are inherently faced with limitations in both their observational and computational capabilities. Accordingly, players in such games need to make their decisions using algorithms that accommodate limitations in information gathering and processing. This disqualifies some of the well known decision making models such as *fictitious play*, in which each player must monitor the actions of every other player and must optimize over a high dimensional probability space (Cartesian product of the action spaces). The authors in [108] proposed a modified version of the fictitious play called joint fictitious play with inertia and proved its convergence in congestion games using the finite path improvement property.

Interactive trial-and-error learning is a recent version of standard trial-and-error learning studied in [194] that takes into account the interactive and dynamic nature of the learning environment. In ordinary trial and error learning, users occasionally try out new actions and accept them if and only if they lead to higher payoffs. In an interactive situation, however, "errors" can arise in two different ways:

- the active errors, those due to trying some new action that turns out to be not better (in terms of satisfaction) than what was chosen, or

- the passive errors, those due to continuing to keep the old strategy which turns out to be worse than it was before.

In [194], it is shown that the interactive trial and error learning implements Nash equilibrium behavior in any game with generic payoffs and which has at least one pure Nash equilibrium. The interactive trial and error learning is completely uncoupled learning rule such that, when used by all players in a game, period-by-period play comes close to pure Nash equilibrium play a high proportion of the time, provided that the game has such an equilibrium and the payoffs satisfy an interdependency condition. Most of the convergence proofs of reinforcement learning algorithms rely on the theory of stochastic approximation. Those use the ordinary differential method, which relates the inhomogeneous, random, discrete time process to deterministic, continuous time dynamical systems. Using the development of the ordinary differential equation (ODE) approach to stochastic approximation studied in [105, 96, 34, 26, 28], most of the strategy learning algorithms can be written in the form of Robbin-Monro-like iterative schemes [134, 93, 29]

$$x_{t+1} = x_t + \lambda_t(F(x_t) + M_{t+1})$$

approximated by the ODE

$$\dot{x} = F(x)$$

under standard assumptions, where F is a vectorial correspondence, M is a noise and $\lambda \geq 0$, $\lambda \in l^2 \backslash l^1$ is a learning rate. Stochastic fully distributed reinforcement learning schemes have been studied in [179, 56, 139]. These works used stochastic approximation techniques to derive ordinary differential equations (ODEs) equivalent to the adjusted replicator dynamics [151]. By studying the orbits of the replicator dynamics, one can get some convergence, divergence and stability properties of the system. However the replicator dynamics may not lead to *approximate equilibria* even in simple games [185, 159]. Convergence properties in special classes of games such as weakly acyclic games and best-response potential games can be found in [109, 115]. Recently, distributed learning algorithms and feedback-based update rules have been extensively developed in networking and communication systems. However, payoff-reinforcement learning (Q-value learning) is not examined in most of these models. Learning in zero-sum stochastic games and their application to network security problems have been studied in [197]. Evolutionary games with random number of players have been studied in [170] in the case where the distribution of the number of interacting players is *known and fixed*. The framework has been applied in the context of resource allocation and access control in wireless networks. However, the evolutionary game model developed in [170] does not take into account the nonautonomous consideration due to random activity and asynchronous updates. In this work we take into account the random update of strategies.

Inspired by stochastic approximation techniques developed by Ljung (1977, [105]), Kushner and Clark (1978,[96]), Benaim (1999, [26]) and the multiple time-scales stochastic techniques developed by Borkar (1997, [34]), Collins and Leslie (2003, [103, 101, 102]), we construct a class of combined learning frameworks for both payoff and strategies (joint multiple objective learning). We prove the almost sure convergence of the CODIPAS-RL to differential equations under standard assumptions on the learning rates. In contrast to standard game dynamics, the limiting behavior of heterogeneous learning under variable number of players is a *non-autonomous (time-dependent)* differential equation. Global asymptotic convergence to the set of equilibria is shown in dynamic robust potential games with variable number of interacting players and asynchronous updates. This is our first attempt to formulate the heterogeneous learning convergence in the context of incomplete information, imperfect observations robust games with random number of interacting players.

The rest of the chapter is structured as follows. In the next section we present the model and provide some results on existence of solutions. Section 7.3 presents the main results on almost sure convergence for

- strategy-learning,

- payoff learning,

- combined fully distributed payoff and strategy reinforcement learning.

A connection between the limiting behavior of nonautonomous dynamics and Nash equilibria are established in robust potential games and applied to dynamic routing with random traffic.

7.2 Description of the Random Update Model

In this section, we describe the random update game model and the assumptions required for our learning scheme to be applied to a given dynamic interactive scenario.

We consider a system with a maximum of n potential players. Each player j can be in active or in sleep mode depending on some random process. Each player j has a finite number of actions (pure strategies), denoted by $\mathcal{A}_j$. The payoff of player j is a random variable $R_{j,t} \in \mathbb{R}$ with a realized value denoted by $r_{j,t} \in \mathbb{R}$ at time t. The random variable is indexed by $\mathbf{w}$ in some set $\mathcal{W}$ and a random set of active players $\mathcal{B}(t)$, independent of $\mathbf{w}_t \in \mathcal{W}$. We assume that the players do not observe the past strategies employed by the others (imperfect monitoring). We assume that the players do not know the state of the game, i.e., the pair $(\mathbf{w}_t, \mathcal{B}(t))$ at time t. We assume that the players do not know the expression of their payoff functions. At each time t, each player chooses an action and receives a numerical noisy value of his payoff (*perceived payoff*) at that time. The perceived payoff can be interpreted as a measured metric in the random environment. We assume that the transition probabilities between the states $\mathbf{w}$ are independent and unknown [1] to the players. The strategic form representation of this stochastic game under uncertainty or robust game is given by

$$\mathcal{G} = \left(\mathcal{N}, (\mathcal{A}_j)_{j \in \mathcal{N}}, \mathcal{W}, (R_j(\mathbf{w}, .))_{j \in \mathcal{N}, \mathbf{w} \in \mathcal{W}}, \mathcal{B}\right),$$

where

- $\mathcal{N} = \{1, 2, \ldots, n\}$ denotes the set of potential players. This set is unknown to the players. $\mathcal{B}(t) \subseteq \mathcal{N}$ is the set of interacting players (active players) at time t.

- $j \in \mathcal{N}$, $\mathcal{A}_j$, is the set of all the possible actions of player j. In each state, the same action spaces are available. Player j does not know the action spaces of the other players.

- *No information on the current state:* $\mathcal{W}$ is the set of possible states, a subset of $\mathbb{R}^k$ for some integer k. We assume that the state space $\mathcal{W}$ and the probability transition on the states are both *unknown* to the players. The state is represented by an independent random variable (the transitions between the states are independent of the chosen actions). We assume that the current state w_t is unknown by the players.

- *Payoff functions are unknown:*

$$R_j^{\mathcal{B}} : \mathcal{W} \times \prod_{j \in \mathcal{N}} \mathcal{X}_j \longrightarrow \mathbb{R},$$

$\mathcal{X}_j := \Delta(\mathcal{A}_j)$ is the set of probability distributions over $\mathcal{A}_j$ i.e

$$\mathcal{X}_j := \left\{ \mathbf{x}_j : \ x_j(s_j) \in [0,1], \ \sum_{s_j \in \mathcal{A}_j} x_j(s_j) = 1 \right\}; \qquad (7.1)$$

In most of the analysis, we will assume that the payoff of inactive players (players in sleep mode) is zero.

---

[1] We do not use Shapley operator because the transition probabilities are unknown

- We denote the stage game by

$$\mathcal{G}(\mathbf{w}_t) = \left(\mathcal{N}, (\mathcal{A}_j)_{j \in \mathcal{N}}, (R_j^{\mathcal{B}}(\mathbf{w}_t, .))_{j \in \mathcal{N}} \right).$$

Since the players do not observe the past actions of the other players, we consider strategies used by both players to be only dependent on their current perceived own-payoffs.

The players do not need to know their own action space in advance. Each player should learn his action space (using, for example, exploration techniques). We need to add an exploration phase or a progressive exploration during the game.

> The result of single player in independent and identically distributed environment is that if all the actions have been explored and sufficiently exploited and if the learning rate is well chosen then the prediction can be "good" enough.

For this the player may not need to know his action space in advance. We provide an example where the players can learn the action space progressively.

> At each time t, if player j chooses a_{j,k_j} then he automatically learn the existence of the actions $a_{j,k_j-1}, a_{j,k_j}, a_{j,k_j+1}$, i.e., up-down exploration method (this exploration method can be generalized to a neighborhood of an action in a random graph). We use an exploration method such that each player learns its action space.

Denote by $x_{j,t}(s_j)$ the probabilities of player j choosing s_j at time t, and let the vector $\mathbf{x}_{j,t} = [x_{j,t}(s_j)]_{s_j \in \mathcal{A}_j} \in \mathcal{X}_j$ be the mixed state-independent strategy of player j at time t.

7.2.1 Description of the Dynamic Robust Game

The idea is to learn the expected payoffs simultaneously with the optimal strategies during a long-run interaction. The dynamic robust game is described as follows:

- At time slot $t = 0$, $\mathcal{B}(0)$ is the set of active players. Each player $j \in \mathcal{B}(0)$ chooses an action $a_{j,0} \in \mathcal{A}_j$ and perceives a numerical value of his payoff that corresponds to a realization of the random variables depending on the actions of the other players and the state of the nature, etc. He initializes its estimation to $\hat{\mathbf{r}}_{j,0}$. The non-active players get zero.

- At time slot t, each player $j \in \mathcal{B}(t)$ has an estimation of his payoffs, chooses an action based his own-experiences and experiments a new strategy. Each player j receives an output $r_{j,t}$. Based on this target $r_{j,t}$ the player j updates its estimation vector $\hat{\mathbf{r}}_{j,t}$ and builds a strategy $\mathbf{x}_{j,t+1}$ for next time slot. The strategy $\mathbf{x}_{j,t+1}$ is a function only of $\mathbf{x}_{j,t}, \hat{\mathbf{r}}_{j,t}$ and the target value. Note that the exact value of the state of the nature $\mathbf{w}_t$ at time t and the past strategies $\mathbf{x}_{-j,t-1} := (\mathbf{x}_{k,t-1})_{k \neq j}$ of the other players and their past payoffs $\mathbf{r}_{-j,t-1} := (r_{k,t-1})_{k \neq j}$ are unknown to player j at time t. The game moves to $t+1$.

Remark 7.2.2. *This model is a stochastic game [144] with incomplete information, and variable and unknown number of interacting players. It covers some interesting well-studied problems:*

- *The class of matrix games with i.i.d random entries in the form $\tilde{R}(\mathbf{a}) = \tilde{D}(\mathbf{a}) + \tilde{S}$ (a deterministic part and a stochastic part with $\mathbb{E}\tilde{S} = 0$.)*

- *If the set of active players is constant, i.e., $\mathcal{B}(t) = \mathcal{N}$, one gets the model of [159] developed for n-player dynamic robust games. Detailed analysis with observation of a public signal can be found in [155, 156].*

- *If $\mathcal{B}(t) = \{j_{1,t}, j_{2,t}\}, j_{k,t} \in \mathcal{N}$, one gets the pairwise interaction learning developed in [61].*

We focus on the limiting of the average payoff, i.e., $v_{j,T} = \frac{1}{T} \sum_{t=1}^{T} r_{j,t}$. The long-term payoff reduces to

$$\frac{1}{\sum_{t=1}^{T} \mathbb{1}_{\{j \in \mathcal{B}(t)\}}} \sum_{t=1}^{T} r_{j,t} \mathbb{1}_{\{j \in \mathcal{B}(t)\}},$$

when considering only the activity of player j.

Well-Posedness

The above sum is ill-posed because the denominator $\sum_{t=1}^{T} \mathbb{1}_{\{j \in \mathcal{B}(t)\}}$ could be zero if player j has not been active yet up to T. In that case, we just need to to specify the indetermination "$\frac{0}{0}$" in place of the payoff.

We assume that we do not have short-term players or equivalently the probability for a player j to be active is strictly positive at any time.

Histories

A player's information consists of his past own activities, own actions, and perceived own payoffs. A private history of length t for player j is a collection

$$h_{j,t} = (b_{j,0}, a_{j,0}, r_{j,0}, b_{j,t}, a_{j,1}, r_{j,1}, \ldots, b_{j,t-1}, a_{j,t-1}, r_{j,t-1}, b_{j,t}).$$

where $b_{j,t}$ is the output of $\mathbb{1}_{j \in \mathcal{B}(t)}$. The set of private history of player j at time t is denoted by

$$H_{j,t} := (\{0,1\} \times \mathcal{A}_j \times \mathbb{R})^t \times \{0,1\}.$$

Behavioral Strategy

A behavioral strategy for player j at time t is a mapping $\tilde{\tau}_{j,t} : H_{j,t} \longrightarrow \mathcal{X}_j$. In order to avoid the meaning of the notation $\bigcup_{t \geq 0} H_{j,t}$ which needs the completion of the spaces via canonical injections, we work with the collection of mappings $\tilde{\tau}_j = (\tilde{\tau}_{j,t})_t$. We denote by Σ_j the set of behavioral strategies of player j.

The set of complete histories of the dynamic game after t stages is

$$H_t = (\mathcal{B} \times \mathcal{W} \times \prod_j \mathcal{A}_j \times \mathbb{R}^n)^t \times \mathcal{B},$$

it describes the set of active players, the states, the chosen actions and the received payoffs for all the players at all past stages before t. A strategy profile $\tilde{\tau} = (\tilde{\tau}_j)_{j \in \mathcal{N}} \in \prod_j \Sigma_j$ and a initial state $\mathbf{w} \in \mathcal{W}$ induce a probability distribution $P_{\mathbf{w}, \tilde{\tau}}$ on the set of plays $H_\infty = (\mathcal{W} \times \prod_j \mathcal{A}_j \times \mathbb{R}^n)^\infty$. Given an initial state $\mathbf{w}$ and a strategy profile $\tilde{\tau}$, the payoff of player j is the superior limiting of the Cesaro-mean payoff $\mathbb{E}_{\mathbf{w}, \tilde{\tau}, \mathcal{B}} v_{j,T}$. We assume that $\mathbb{E}_{\mathbf{w}, \tilde{\tau}, \mathcal{B}} v_{j,T}$ has a limit.

We will develop a fully distributed learning procedure to learn the expected payoffs as well as the optimal strategies associated with the learned payoffs. Note that, since the state is unknown, the players do not know which game they are playing. We have a sequence of games generated by the value of random variable $\mathcal{W}$, which determines the states in which the number of interacting players vary over time. In the case where only a single state is available (the Dirac distribution case), the exact payoff function is learned in the long term.

Definition 7.2.2.1 (Expected robust game). *We define the expected robust game as*

$$\left(\mathcal{N}, (\mathcal{X}_j)_{j \in \mathcal{N}}, \mathbb{E}_{\mathbf{w}, \mathcal{B}} R_j^{\mathcal{B}}(\mathbf{w}, \cdot)\right).$$

Definition 7.2.2.2 (Equilibrium of the expected robust game). *A strategy profile* $(\mathbf{x}_j)_{j \in \mathcal{N}} \in \prod_{j=1}^n \mathcal{X}_j$ *is a (mixed) state-independent equilibrium for the expected robust game if and only if*

$$\forall j \in \mathcal{N}, \quad \mathbb{E}_{\mathbf{w}, \mathcal{B}} U_j^{\mathcal{B}}(\mathbf{w}, \mathbf{y}_j, \mathbf{x}_{-j}) \leq \mathbb{E}_{\mathbf{w}, \mathcal{B}} R_j^{\mathcal{B}}(\mathbf{w}, \mathbf{x}_j, \mathbf{x}_{-j}), \ \forall \mathbf{y}_j \in \mathcal{X}_j. \quad (7.2)$$

The existence of a solution to Equation (7.2) is equivalent to the existence of solution of the following *variational inequality problem*: find $\mathbf{x}$ such that

$$\langle \mathbf{x} - \mathbf{y}, V(\mathbf{x}) \rangle \geq 0, \ \forall \mathbf{y} \in \prod_j \mathcal{X}_j,$$

where $\langle .,. \rangle$ is the inner product,

$$V(\mathbf{x}) = [V_1(\mathbf{x}), \ldots, V_n(\mathbf{x})],$$

$$V_j(\mathbf{x}) = [\mathbb{E}_{\mathbf{w},\mathcal{B}} R_j^{\mathcal{B}}(\mathbf{w}, e_{s_j}, \mathbf{x}_{-j})]_{s_j \in \mathcal{A}_j}.$$

Note that an equilibrium of the expected robust game may not be an equilibrium at each time slot. This is because $\mathbf{x}$ being an equilibrium for expected robust game (with random activity) does not implies that $\mathbf{x}$ is an equilibrium of the game $\mathcal{G}(\mathbf{w}, \mathcal{B})$ for some state $(\mathbf{w}, \mathcal{B})$ and the set of active players $\mathcal{B}$ may vary.

Lemma 7.2.2.3. *Assume that $\mathcal{W}$ is compact. Then, the expected robust game with unknown state and variable number of interacting players has at least one equilibrium.*

The existence of such equilibrium points is guaranteed since the mappings

$$r_j : (\mathbf{x}_j, \mathbf{x}_{-j}) \longmapsto \mathbb{E}_{\mathbf{w},\mathcal{B}} R_j^{\mathcal{B}}(\mathbf{w}, \mathbf{x}_j, \mathbf{x}_{-j}),$$

are jointly continuous, quasi-concave in $\mathbf{x}_j$, the spaces $\mathcal{X}_j$, are non-empty, convex and compact. The result follows by using the Kakutani fixed point theorem [85].

> How to find an approximate solution of the variational inequality?

In Section 4.6, we provided an algorithm to compute a solution to variational inequalities (VI). The algorithm extends to robust games with random updates. In the next section, we present fully distributed learning algorithms with random updates.

7.3 Fully Distributed Learning

In this section, we develop fully distributed schemes for dynamic robust game with random set of interacting players. We start with distributed strategy-learning algorithms.

7.3.1 Distributed Strategy-Reinforcement Learning

Let

$$\tilde{F} : \mathbb{R}^{\sum_{j \in \mathcal{N}} |\mathcal{A}_j|} \longrightarrow \mathbb{R}^{\sum_{j \in \mathcal{N}} |\mathcal{A}_j|}$$

be a continuous map. Consider a discrete time process $\{\mathbf{x}_t\}_{t \in \mathbb{Z}_+}$ living in $\mathbb{R}^{\sum_{j \in \mathcal{N}} |\mathcal{A}_j|}$ whose general form can be written as

$$\mathbf{x}_{t+1} = \mathbf{x}_t + \lambda_t [\tilde{F}(\mathbf{x}_t) + M_{t+1}] \tag{7.3}$$

where

- $\{\lambda_t\}_t$ is a sequence of nonnegative numbers such that $\lambda \notin l^1$ and $\lim_t \lambda_t = 0$.

- $M_t \in \mathbb{R}^{\sum_{j \in \mathcal{N}} |\mathcal{A}_j|}$ is a perturbation.

The form (7.3) is well studied in standard fully distributed strategy-reinforcement learning, which has the following form:

$$(\textbf{Strategy-RL}) \begin{cases} x_{j,t+1}(s_j) = x_{j,t}(s_j) + \lambda_t \left[\tilde{F}_{j,s_j}(\mathbf{x}_t) + M_{j,t+1}(s_j) \right] \\ t \geq 0 \\ j \in \mathcal{N}, \ s_j \in \mathcal{A}_j, \\ \mathbf{x}_0 = \mathbf{x}(0) \in \prod_j \mathcal{X}_j \end{cases}$$

The objects $\tilde{F}, M$ are chosen in a way such that almost surely $x_{j,t} \in \mathcal{X}_j$, $\forall t \geq 0$. The function $\tilde{F}$ is assumed to be a deterministic function. It could be, for example $\tilde{F}_{j,s_j}(x) = \mathbb{E}_w \tilde{G}_{j,s_j}(w, x)$ for some integrable function $\tilde{G}$ with the respect to the law of w. The following Lemma follows from the definition of behavioral strategies and provides the well-posedness of (**Strategy-RL**) and its use in games.

Lemma 7.3.1.1. *The above strategies generated by the strategy-learning process are in the set Σ_j.*

Proof. Let us start with an initial state w_0 with some probability $\mathbb{P}_{w_0}$. The initial state x_0 is drawn with some measure $\mathbb{P}_{x_0}$. Then, x_1 depends on (x_0, w_0, M_1), and hence it is a mapping from H_1 to the set $\prod_{j \in \mathcal{N}} \mathcal{X}_j$. Using the iterative stochastic equation, one has that $x_{j,t}$ is adapted to the filtration generated by the previous variables up to $t-1$ and the martingale $M_{t'}$, $t' \leq t$. This means that

$$x_{j,t} \in \sigma((x_{t'})_{t' \leq t-1}, \ (w_{t'})_{t' \leq t}, (M_{t'})_{t' \leq t}).$$

The stated result follows. $\qquad \square$

The iterative equation (7.3) can be considered as a perturbation version of a variable step-size Euler-scheme for numerical implementation of the ordinary differential equation

$$\dot{\mathbf{x}}_t = \tilde{F}(\mathbf{x}_t), \tag{7.4}$$

i.e., the sequence

$$\mathbf{y}_{t+1} = \mathbf{y}_t + \lambda_t \tilde{F}(\mathbf{y}_t). \tag{7.5}$$

We compare the behavior of a sample path $\{\mathbf{x}_t\}_{t\geq 0}$ with the trajectories of the flow induced by the vector field $\tilde{F}$. To do this, we denote

$$T_0 = 0, \ T_k = \sum_{t=1}^{k} \lambda_t, \ k \geq 1.$$

We define the continuous time affine and piecewise interpolated process $X :$ $\mathbb{R}_+ \longrightarrow \mathbb{R}^{\sum_{j\in\mathcal{N}} |\mathcal{A}_j|}$

$$X(T_k + s') = \mathbf{x}_k + s' \frac{\mathbf{x}_{k+1} - \mathbf{x}_k}{T_{k+1} - T_k}, \ \forall \ k \in \mathbb{Z}_+, \ 0 \leq s' \leq \lambda_{k+1}.$$

Theorem 7.3.1.2. *Let $\tilde{F}$ be a locally Lipschitz vector field on the product of the simplex. Assume that*

$$(H1) \ \forall \ T, \ \lim_{t \longrightarrow \infty} \ \sup_{0 \leq \Delta t \leq T} \ \| \int_t^{t+\Delta t} \tilde{M}(s) \ ds \| = 0 \qquad (7.6)$$

Then, for t sufficiently large enough, we have that

$$\sup_{0 \leq \Delta t \leq T} \ \| \ X(t + \Delta t) - \Phi_{\Delta t}(X(t)) \ \|$$

$$\leq C_T \left[\sup_{s \in [t, t+T]} \tilde{\lambda}(s) + \sup_{0 \leq \Delta t \leq T+1} \| \int_{t-1}^{t-1+\Delta t} \tilde{M}(s) \ ds \| \right]$$

where C_T depends only on T and $\tilde{F}$, and $\Phi_{\Delta t}$ satisfies

$$\dot{\Phi}_{\Delta t}(.) = \tilde{F}(\Phi_{\Delta t}(.)), \ t \geq \Delta t.$$

In particular, the interpolated process X is an asymptotic pseudo-trajectory of the flow Φ induced by $\tilde{F}$, i.e.,

$$\forall \ T > 0, \ \lim_{t \longrightarrow +\infty} \ \left[\sup_{0 \leq \Delta t \leq T} \ \| \ X(t + \Delta t) - \Phi_{\Delta t}(X(t)) \ \| \right] = 0$$

Theorem 7.3.1.2 says that even under random and asynchronous updates, the interpolated process of strategy-learning has an asymptotic pseudo-trajectory that can be identified with a solution of the ordinary differential equation defined by the expected changes (in one unit) of the stochastic difference equation.

Next we provide some preliminaries for an outline of the proof of Theorem 7.3.1.2.

Let $\tilde{F} : \mathbb{R}^{\sum_j |\mathcal{A}_j|} \longrightarrow \mathbb{R}^{\sum_j |\mathcal{A}_j|}$ be a continuous map. Consider a discrete time process $\{\mathbf{x}_t\}_{t \in \mathbb{Z}_+}$ living in $\mathbb{R}^{\sum_j |\mathcal{A}_j|}$ whose general form can be written as

$$\mathbf{x}_{t+1} = \mathbf{x}_t + \lambda_t [\tilde{F}(\mathbf{x}_t) + M_{t+1}] \tag{7.7}$$

where

- $\{\lambda_t\}_t$ is a sequence of nonnegative numbers such that $\lambda \notin l^1$ and $\lim_t \lambda_t = 0$.

- $M_t \in \mathbb{R}^{\sum_j |\mathcal{A}_j|}$ is a perturbation.

The iterative equation (7.7) can be considered as a perturbation version of a variable step-size Euler-scheme for numerical implementation of the ordinary differential equation

$$\dot{\mathbf{x}}_t = \tilde{F}(\mathbf{x}_t), \tag{7.8}$$

i.e., the sequence

$$\mathbf{y}_{t+1} = \mathbf{y}_t + \lambda_t \tilde{F}(\mathbf{y}_t). \tag{7.9}$$

We compare the behavior of a sample path $\{\mathbf{x}_t\}_{t \geq 0}$ with the trajectories of the flow induced by the vector field $\tilde{F}$. Denote $T_0 = 0$, $T_k = \sum_{t=1}^k \lambda_t$, $k \geq 1$. We define the continuous time affine and piecewise interpolated process $X :$ $\mathbb{R}_+ \longrightarrow \mathbb{R}^{\sum_j |\mathcal{A}_j|}$,

$$X(T_k + s') = \mathbf{x}_k + s' \frac{\mathbf{x}_{k+1} - \mathbf{x}_k}{T_{k+1} - T_k}, \ \forall \, k \in \mathbb{Z}_+, \ 0 \leq s' \leq \lambda_{k+1}.$$

Define $\tilde{T} : t \longmapsto \sup\{k, T_k \leq t\}$ the left-hand-inverse of the map $k \longmapsto T_k$. By the assumption $\forall t, \lambda_t \geq 0$, $\lambda \notin l^1$, one has that $T_k, \tilde{T}$ are unbounded. Denote by $\tilde{M}, \tilde{\lambda}$ the continuous time process defined by

$$\begin{cases} \tilde{M}(T_k + s') = M_k, \ \forall \, k \in \mathbb{Z}_+, \ 0 \leq s' \leq \lambda_{k+1} \\ \tilde{\lambda}(T_k + s') = \lambda_k, \ \forall \, k \in \mathbb{Z}_+, \ 0 \leq s' \leq \lambda_{k+1} \end{cases}$$

Equation (7.7) can be written as

$$X(t) - X(0) = \int_0^t \left[\tilde{F}(\tilde{X}(s)) + \tilde{M}(s) \right] \, ds. \tag{7.10}$$

Recall that if $\tilde{F}$ is bounded and locally Lipschitz then $\tilde{F}$ has a unique integral curv. Next we give a sufficient condition for convergence of the Robbin-Monro's-like algorithms. We are now ready to prove the Theorem 7.3.1.2. In this proposition, d denotes a metric on the space of the continuous maps from $\mathbb{R}$ to $\prod_j \mathcal{X}_j$. The process X is extended to $\mathbb{R}_-$ by setting $X(t) = X(0)$ if $t \leq 0$.

$$d(f, g) = \sum_{k \in \mathbb{Z}_+} \frac{1}{2^k} \min\{1, d_k(f, g)\} \tag{7.11}$$

$$d_k(f, g) = \sup_{t \in [-k,k]} \tilde{d}(f(t), g(t)) \tag{7.12}$$

Denote by

$$\tilde{\epsilon}_{t,T} = \sup_{0 \le \Delta t \le T} \| \int_t^{t+\Delta t} \tilde{M}(s) \, ds \| .$$

Assumption $H1$ says that $\lim_{t \to \infty} \tilde{\epsilon}_{t,T} = 0$. The discrete time version of Assumption $H1$ is exactly $\forall \, T > 0$,

$$\lim_{k \to \infty} \sup \left\{ \| \sum_{t=k}^{L'-1} \lambda_{t+1} M_{t+1} \| : \; L' \in \{k+1, k+2, \dots, \tilde{T}(T_k + T)\} \| \right\} = 0 \tag{7.13}$$

Note that this assumption is automatically satisfied if M_{t+1} is a martingale difference with the respect to the filtration $\mathcal{F}_t = \sigma(\mathbf{x}_s, \; s \le t)$ and $\lambda \succeq 0$, $\lambda \in l^2$. A formal proof of these statements can be found in Kushner (1978), Benaim (1999), Borkar (2008).

We are now ready for a sketch of the proof of Theorem 7.3.1.2.

Proof. By continuity of $\tilde{F}$ and the fact that $\{\mathbf{x}_t\}_t \subset \prod_j \mathcal{X}_j$, which is a convex, non-empty and compact set, there exists a constant $c > 0$ such that

$$\| \tilde{F}(\mathbf{x}_t) \| \le c, \, \forall \, t \ge 0.$$

From Equation (7.10), one has that,

$$\sup_{\Delta t \in [0,T]} \| X(t + \Delta t) - X(t) \| \le \int_{[0,T]} \sup_{s \in [0,T]} \| \tilde{F}(X(s)) \| + \tilde{\epsilon}_{t,T}$$

This implies by H1 that

$$\lim_{t \to +\infty} \sup_{\Delta t \in [0,T]} \| X(t + \Delta t) - X(t) \| \le cT.$$

Hence, the interpolated process is uniformly continuous. We now rewrite (7.10) as

$$X(t + \Delta t) \;=\; X(t) + \int_0^{\Delta t} \tilde{F}(X(t + s')) \, ds' \tag{7.14}$$

$$+ \int_t^{t+\Delta t} [\tilde{F}(\tilde{X}(s')) - \tilde{F}(X(s'))] \, ds' \tag{7.15}$$

$$+ \int_t^{t+\Delta t} \tilde{M}(s') \, ds' \tag{7.16}$$

By Assumption H1, $\lim_{t \to \infty} \int_t^{t+\Delta t} \tilde{M}(s') \, ds' \equiv 0$, where the equality is in the sense of the space of continuous functions from $\mathbb{R}$ to $\mathbb{R}^{\sum_j |\mathcal{A}_j|}$. Again, by (7.10), one gets that, for any $T > 0$ and any $s \in [t, t + T]$,

$$\| X(s) - \tilde{X}(s) \| = \| \int_{T_{\tilde{T}(s)}}^{s} \tilde{F}(\tilde{X}(s')) + \tilde{M}(s') \ ds' \|$$

$$\leq c\tilde{\lambda}(s) + \| \int_{T_{\tilde{T}(s)}}^{s} \tilde{M}(s') \ ds' \| .$$

Since $\tilde{\lambda}(.) \longrightarrow 0$, one has that $\tilde{\lambda}(s) < 1$ for s large enough.

$$\| \int_{T_{\tilde{T}(s)}}^{s} \tilde{M}(s') \ ds' \| \leq \| \int_{t-1}^{T_{\tilde{T}(s)}} \tilde{M}(s') \ ds' \| + \| \int_{t-1}^{s} \tilde{M}(s') \ ds' \| \leq 2\tilde{\epsilon}_{t-1,T+1}.$$

Thus, $\sup_{s \in [t, t+T]} \| X(s) - \tilde{X}(s) \| \leq 2\tilde{\epsilon}_{t-1,T+1} + \sup_{s \in [t, t+T]} c\tilde{\lambda}(s)$. Since $\tilde{F}$ is uniformly continuous on $\prod_j \mathcal{X}_j$, the second term (7.15) satisfies

$$\lim_{t \longrightarrow \infty} \int_{t}^{t+\Delta t} \left[\tilde{F}(\tilde{X}(s')) - \tilde{F}(X(s')) \right] \ ds' \equiv 0$$

where the equality is in the sense of the space of the continuous functions from $\mathbb{R}$ to $\mathbb{R}^{\sum_j |\mathcal{A}_j|}$. Let X_* denote an omega-limit of $\{X(t + .)\}$. Then,

$$X_*(s) = X_*(0) + \int_{0}^{s} \tilde{F}(X_*(s')) \ ds'.$$

By a unique integral curve, this implies that $X_* = \Phi(X_*)$ and X is an asymptotic pseudo-trajectory of Φ.

It remains to establish the bound of convergence. First, we estimate the norm of $\int_{t}^{t+\Delta t} [\tilde{F}(\tilde{X}(s')) - \tilde{F}(X(s'))] \ ds'$. Let $\Delta t \in [0, T]$ and L is a Lipschitz constant of the function $\tilde{F}$. Then,

$$a_0(\Delta t) := \| \int_{t}^{t+\Delta t} [\tilde{F}(\tilde{X}(s')) - \tilde{F}(X(s'))] \ ds' \| \leq LT \sup_{s' \in [0,T]} \| \tilde{X}(s')) - X(s') \|$$

$$\leq LT \left[c \sup_{s \in [t, t+T]} \tilde{\lambda}(s) + 2\tilde{\epsilon}_{t-1,T+1} \right]$$

Thus,

$$\| X(t+\Delta t) - \Phi_{\Delta t}(t) \| \;\leq\; L \int_0^{\Delta t} \| X(t+s') - \Phi_{s'}(X(t)) \| \; ds' + a_0(\Delta t)$$

$$+ \; \| \int_t^{t+\Delta t} \tilde{M}(s') \; ds' \|$$

$$\leq \; L \int_0^{\Delta t} \| X(t+s') - \Phi_{s'}(X(t)) \| \; ds'$$

$$+ LT \left[c \sup_{s \in [t, t+T]} \tilde{\lambda}(s) + 2\tilde{\epsilon}_{t-1,T+1} \right] + \tilde{\epsilon}_{t-1,T+1}$$

We now use Gronwall's inequality to get the stated bound. Let $\tilde{f}, \tilde{g}, \tilde{h}$ be continuous functions on the compact interval $[a, \Delta t]$ It states that if $\tilde{g}$ is non-negative and if $\tilde{h}$ satisfies the inequality

$$\tilde{f}(t) \leq \tilde{h}(t) + \int_0^t \tilde{g}(s')\tilde{f}(s') \; ds', \; \forall t \in [0, \Delta t]$$

Then,

$$\tilde{f}(t) \leq \tilde{h}(t) + \int_a^t \tilde{h}(s')\tilde{g}(s')e^{\int_{s'}^t \tilde{g}(s'') \; ds''} \; ds'$$

In addition, if $\tilde{h}(.)$ is constant, then

$$\tilde{f}(t) \leq \tilde{h}e^{\int_a^t \tilde{g}(s') \; ds'}, \; t \in [0, \Delta t]$$

This completes the proof. □

In presence of the i.i.d random variables $\mathbf{w}_t, \mathcal{B}(t)$, the discrete time dynamics writes

$$
\begin{cases}
x_{j,t+1}(s_j) & = \quad x_{j,t}(s_j) + \lambda_t \mathbb{1}_{\{j \in \mathcal{B}(t)\}} \times \\
& \left[\mathbb{E}_{\mathbf{w},\mathcal{B}} F_{j,s_j}(\mathbf{x}_t, R_j^{\mathcal{B}}(\mathbf{w}, ., \mathbf{x}_{-j,t})) + M_{j,t+1}^{(0)}(s_j) + M_{j,t+1}(s_j) \right] \\
& \quad j \in \mathcal{N}, \; s_j \in \mathcal{A}_j, \; t \geq 0, \; \mathbf{x}_0 = \mathbf{x}(0) \in \prod_j \mathcal{X}_j
\end{cases}
$$

where $M_{j,t+1}^{(0)}(s_j) = F_{j,s_j}(\mathbf{x}_t, R_j^{\mathcal{B}(t)}(\mathbf{w}, ., \mathbf{x}_{-j,t})) - \mathbb{E}_{\mathbf{w},\mathcal{B}} F_{j,s_j}(\mathbf{x}_t, R_j^{\mathcal{B}}(\mathbf{w}, ., \mathbf{x}_{-j,t}))$ which is a martingale difference. Note that the noise is not additive here.

7.3.2 Random Number of Interacting Players

The above stochastic approximation approach we have used for strategy-reinforcement learning with synchronous update does not apply automatically anymore and some work is needed for the asynchronous update (the players update only when they are active). To each player, we associate his own clock

$$\theta_j(t) = \sum_{t'=1}^t \mathbb{1}_{\{j \in \mathcal{B}(t')\}}.$$

The random number θ_j is the number of times the player j has been involved in an interaction until t. Player j knows his own-timer $\theta_j(t)$, so he need not know the global timer of the system $\{t\}$.

The discrete time dynamics is (**Random-RL**):

$$
\begin{cases}
x_{j,t+1}(s_j) &= \quad x_{j,t}(s_j) + \lambda_{[\theta_j(t)]} \mathbb{1}_{\{j \in \mathcal{B}(t)\}} \times \\
& \qquad \left[F_{j,s_j}(\mathbf{x}_t, R_j^{\mathcal{B}(t)}(\mathbf{w}_t, ., \mathbf{x}_{-j,t})) + M_{j,t+1}(s_j) \right] \\
& \quad j \in \mathcal{N}, \ s_j \in \mathcal{A}_j, \ t \geq 0, \ \mathbf{x}_0 = \mathbf{x}(0) \in \prod_j \ \mathcal{X}_j
\end{cases}
$$

Note that the nonactive players $\mathcal{B}(t)$ at time t do not update. We assume that

$$[H2], \ \forall j \in \mathcal{N}, \ \liminf_{t \longrightarrow \infty} \ \frac{\theta_j(t)}{t} > 0 \qquad (7.17)$$

This assumption is satisfied if $\mathcal{B}(t)$ is an irreducible Markov chain over $2^{\mathcal{N}} := \{J, \ J \subseteq \mathcal{N}\}$. In particular, [H2] ensures that $\forall j$, $\theta_j(t) \longrightarrow +\infty$ when $t \longrightarrow \infty$, i.e., each player j is active infinitely often

$$\sum_{t \geq 0} \mathbb{1}_{\{j \in \mathcal{B}(t)\}} = +\infty.$$

(Recall that no player knows when the other players will be active). Assume that $R_j^{\mathcal{B}(t)}(\mathbf{w}, ., \mathbf{x}_{-j,t}))$ depends only on the strategies of the set of active players i.e., $(\mathbf{x}_{j,t})_{j \in \mathcal{B}(t)}$. Note that if the total number of activities of a player j^* is finite: $\sup_t \theta_{j^*}(t) \leq b_{j^*} < \infty$, then this player can be omitted in the analysis of the dynamic game. Such player is called a *short-term player*. It suffices to consider the game starting from $t_0 = b_{j^*} + 1$ and player j^* does not influence the game (since $\forall t \geq t_0$, $j^* \notin \mathcal{B}(t)$ and j^* does not influence the long-term payoffs). Define

$$\mathcal{F}_t = \sigma(\mathbf{x}_{t'}, M_{t'}, \mathcal{B}(t'), \ t' \leq t), \ t \geq 0.$$

Assume that

$$[H3] \ \lambda_t \geq 0, \ \lambda \in l^2 \backslash l^1, \ \mathbb{E}(M_{j,t+1} \mid \mathcal{F}_t) = 0,$$

$$\forall j, \ \mathbb{E}\left(\| M_{j,t+1} \|^2\right) \leq c_1 \left[1 + \sup_{t' \leq t} \| \mathbf{x}_{t'} \|^2 \right],$$

where $c_1 > 0$ is a constant. Let

$$\lambda_t^* = \max_{j \in \mathcal{B}(t)} \ \lambda_{[\theta_j(t)]}, \ t \geq 0.$$

Next we show that the random learning rates $\{\lambda_t^*\}_{t \in \mathbb{Z}_+}$ satisfy the standard sufficiency conditions in Q-learning.

Lemma 7.3.2.1. *The random learning rates $\{\lambda_t^*\}_{t \in \mathbb{Z}_+}$ satisfy almost surely*

$$\lambda_t^* \geq 0, \ \sum_{t \in \mathbb{Z}_+} \lambda_t^* = +\infty, \ and \ \sum_{t \in \mathbb{Z}_+} (\lambda_t^*)^2 < +\infty.$$

Proof. For any fixed $j \in \mathcal{N}$,

$$+\infty = \sum_{t \in \mathbb{Z}_+} \lambda_t = \sum_{t \in \mathbb{Z}_+} \lambda_{[\theta_j(t)]} \mathbb{1}_{\{j \in \mathcal{B}(t)\}} \leq \sum_{t \in \mathbb{Z}_+} \max_{j \in \mathcal{B}(t)} \lambda_{[\theta_j(t)]} \leq \sum_{t \in \mathbb{Z}_+} \lambda_t^*.$$

Thus, $\sum_{t \in \mathbb{Z}_+} \lambda_t^* = +\infty$, and

$$\sum_{t \in \mathbb{Z}_+} (\lambda_t^*)^2 \leq \sum_{t \in \mathbb{Z}_+} \sum_{j \in \mathcal{N}} \lambda_{[\theta_j(t)]}^2 \mathbb{1}_{\{j \in \mathcal{B}(t)\}} \leq |\mathcal{N}| \sum_{t \in \mathbb{Z}_+} \lambda_t^2 < +\infty.$$

$\square$

We consider the player who has the maximum learning rate among all the active players as a reference clock. This is given by the random variable λ_t^*. The discrete-time dynamics can be written as a function of λ_t^* :

$$\begin{cases} x_{j,t+1}(s_j) & = & x_{j,t}(s_j) + \lambda_t^* \dfrac{\lambda_{[\theta_j(t)]}}{\lambda_t^*} \mathbb{1}_{\{j \in \mathcal{B}(t)\}} \\ & & \left[F_{j,s_j}(\mathbf{x}_t, R_j^{\mathcal{B}(t)}(\mathbf{w}, ., \mathbf{x}_{-j,t})) + M_{j,t+1}(s_j) \right] \\ & & j \in \mathcal{N}, \ s_j \in \mathcal{A}_j, \ t \geq 0, \ \mathbf{x}_0 = \mathbf{x}(0) \in \prod_j \mathcal{X}_j \end{cases}$$

Denote by $q_{j,t} = \frac{\lambda_{[\theta_j(t)]}}{\lambda_t^*} \mathbb{1}_{\{j \in \mathcal{B}(t)\}} \in [0,1]$, $\forall t$. Under the above assumptions, $T_k^* = \sum_{t'=0}^{k} \lambda_{t'}^*$ goes to infinity with k. We define the continuous time affine and piecewise interpolated process $X : \mathbb{R}_+ \longrightarrow \mathbb{R}^{\sum_{j \in \mathcal{N}} |\mathcal{A}_j|}$

$$X^*(T_k^* + s') = \mathbf{x}_k + s' \frac{\mathbf{x}_{k+1} - \mathbf{x}_k}{T_{k+1}^* - T_k^*}, \ \forall \ k \in \mathbb{Z}_+, \ 0 \leq s' \leq \lambda_{k+1}^*$$

For $j \in \mathcal{N}$, $g_j(T_k^* + s') := q_{j,T_k^*+s'}$, $\forall \ k \in \mathbb{Z}_+$, $0 \leq s' \leq \lambda_{k+1}^*$. Let $G(t) :=$ $\text{diag}(g_j(t))_{j \in \mathcal{N}} \in [0,1]^{n \times n}$. Let $\Phi_{\Delta t}^*(t)$, $t \geq \Delta t$ be the unique solution of the nonautonomous differential equation

$$\dot{\Phi}_{\Delta t}^*(t) = G(t) F^*(\Phi_{\Delta t}^*(t)), \ F^* = \mathbb{E}F$$

Lemma 7.3.2.2. *Assume that [H2] and [H3] hold. Then, for any $T > 0$,*

$$\lim_{\Delta t \longrightarrow +\infty} \left[\sup_{t \in [\Delta t, \Delta t + T]} \| X^*(t) - \Phi_{\Delta t}^*(t) \| \right] = 0 \ almost \ surely$$

Proof. The proof follows the same lines as in Theorem 7.3.1.2 by replacing the deterministic learning rate λ_t by the random learning rate λ_t^*. $\square$

We are now ready for the main convergence result to nonautonomous differential equations:

Theorem 7.3.2.3. *Almost surely, any limit of a subsequence of $s' \longmapsto$ $X(\Delta t + s')$ in the space $C^0([0, +\infty), \mathbb{R}^{\sum_j |\mathcal{A}_j|})$ as $\Delta t \longrightarrow +\infty$ is a solution of a nonautonomous differential equation*

$$\dot{\mathbf{y}}(t) = G(t) F^*(\mathbf{y}(t))$$

Proof of Theorem 7.3.2.3. We first need some preliminary notions. Denote by $L_2([0,t], \mathbb{R}^n)$ the space of measurable functions $q : [0,t] \longrightarrow \mathbb{R}^n$ such that

$$\int_0^t \|q(s)\|_2^2 \, ds = \int_0^t \langle q(s), q(s) \rangle \, ds < +\infty.$$

It is well-known that $L_2([0,t], \mathbb{R}^n)$ is a Hilbert space (thus a Banach space). The space $C^0([0,t], \mathbb{R}^{\sum_j |\mathcal{A}_j|})$ is a Banach space under the sup-norm or infinity-norm

$$\|q\|_\infty = \sup_{s \in [0,t]} \|q(s)\|.$$

The space $C^0([0,+\infty), \mathbb{R}^{\sum_j |\mathcal{A}_j|})$ is not a Banach space. This space will be equipped with weaker convergence. We say that $q_k \longrightarrow q$ in $C^0([0,+\infty), \mathbb{R}^{\sum_j |\mathcal{A}_j|})$ if for any $t > 0$ the restriction of $(q_k - q)$ to $C^0([0,t], \mathbb{R}^{\sum_j |\mathcal{A}_j|})$ goes to zero.

We now prove the Theorem 7.3.2.3 which states: Almost surely, any limit of a subsequence of $s' \longmapsto X(\Delta t + s')$ in the space $C^0([0,+\infty), \mathbb{R}^{\sum_j |\mathcal{A}_j|})$ as $\Delta t \longrightarrow +\infty$ is a solution of a nonautonomous differential equation

$$\dot{\mathbf{y}}(t) = G(t) F^*(\mathbf{y}(t)).$$

Since the map $G(.)$ is a diagonal matrix of functions from $[0,+\infty)$ to $[0,1]$, it is equivalent to a mapping from $[0,+\infty)$ to $[0,1]^n$ with a respect a topology that makes continuous the map:

$$\gamma(.) \longmapsto \int_0^t \langle q(s), \gamma(s) \rangle \, ds, \ \forall t > 0, q(.) \in L^2([0,t], \mathbb{R}^n)$$

We denote by E the space of such functions. This is a compact and metrizable space with the metric

$$d^*(\gamma_1(.), \gamma_2(.)) = \sum_{t \in \mathbb{Z}_+} \sum_{n' \geq 1} \frac{1}{2^{t+n'}} \ \min \left\{ 1, | \int_0^t \langle \gamma_1(s) - \gamma_2(s), e_{n',t}^*(s) \rangle| \, ds \right\}$$

where $e_{n',t}^*(.)$ is a complete orthonormal basis for $L^2([0,t], \mathbb{R}^{\sum_j |\mathcal{A}_j|})$. Using the above result and Ascoli's theorem, the process $X^*(\Delta t+.)$, $\Delta t \geq 0$ is relatively compact in $C^0([0,+\infty), \mathbb{R}^{\sum_j |\mathcal{A}_j|})$. Consider a subsequence $\alpha_1(\Delta t) \longrightarrow +\infty$ such that $X^*(\alpha_1(\Delta t) + .)$ goes to $X^*(.)$ in $C^0([0,+\infty), \mathbb{R}^{\sum_j |\mathcal{A}_j|}}$. Consider a bijection α_2 and a subsequence such that $g_j(\alpha_1 \circ \alpha_2(\Delta t) + .) \longrightarrow g_j^*(.)$ (convergence in the sense of E) when $\Delta t \longrightarrow \infty$. Let $G^*(.) = \text{diag}(g_j^*(.))_j$. By Theorem 7.3.2.2,

$$X^*(\alpha_1 \circ \alpha_2(\Delta t) + t) - X^*(\alpha_1 \circ \alpha_2(\Delta t) + \epsilon)$$

$$= \int_\epsilon^t G(\alpha_1 \circ \alpha_2(\Delta t) + s') F^*(X^*(\alpha_1 \circ \alpha_2(\Delta t) + s')) \, ds' + o(1),$$

for $t > \epsilon > 0$. When $\Delta t \longrightarrow \infty$, one has

$$X^*(t) - X^*(0) = \int_0^t G^*(s')F^*(X^*(s')) \, ds', \ t > 0.$$

i.e., $X^* = \Phi^*(X^*)$. This completes the proof. $\qquad\square$

7.3.3 CODIPAS-RL for Random Updates

We consider CODIPAS-RL with asynchronous updates in the following form:
$(CODIPAS - RL)$

$$\begin{cases} \mathbf{x}_{j,t+1} - \mathbf{x}_{j,t} &= \mathbb{1}_{j \in \mathcal{B}(t)} K_j^1(\lambda_{j,\theta_j(t)}, a_{j,t}, r_{j,t+1}, \hat{\mathbf{r}}_{j,t}, \mathbf{x}_{j,t}) \\ \hat{\mathbf{r}}_{j,t+1} - \hat{\mathbf{r}}_{j,t} &= \mathbb{1}_{j \in \mathcal{B}(t)} K_j^2(\nu_{j,\theta_j(t)}, a_{j,t}, r_{j,t+1}, \mathbf{x}_{j,t}, \hat{\mathbf{r}}_{j,t}) \\ & \qquad\qquad j \in \mathcal{N}, t \geq 0, a_{j,t} \in \mathcal{A}_j. \end{cases}$$

where

- $\hat{\mathbf{r}}_{j,t} = (\hat{r}_{j,t}(s_j))_{s_j \in \mathcal{A}_j} \in \mathbb{R}^{|\mathcal{A}_j|}$ is a vector payoff estimation of player j at time t.

- The functions K^1 and λ are based on estimated payoff and perceived measured payoff such that the invariance of simplex is preserved. The function K_j^1 defines the strategy learning pattern of player j, and $\lambda_{j,\theta_j(t)}$ is his strategy learning rate. If at least two of the functions f_j are different then we refer it as *heterogeneous learning* in the sense that the learning schemes of the players are different. If all the K_j^1 are identical but the learning rates λ_j are different, we refer to *learning with different speed*: slow learners, medium learners and fast learners.

- The functions K^2, and ν are well-chosen in order to have a good estimation of the payoffs.

We introduce the local clock in payoff and strategy-learning and examine the following form of CODIPAS-RL:

$$\begin{cases} x_{j,t+1}(s_j) &= x_{j,t}(s_j) + \lambda_t^* \frac{\lambda_{[\theta_j(t)]}}{\lambda_t^*} \mathbb{1}_{\{j \in \mathcal{B}(t+1)\}} \left[F_{j,s_j}(\mathbf{x}_{j,t}, \hat{\mathbf{r}}_{j,t})) + M_{j,t+1}(s_j) \right] \\ \hat{r}_{j,t+1}(s_j) &= \hat{r}_{j,t}(s_j) + \nu_t^* \frac{\nu_{[\theta_j(t)]}}{\nu_t^*} \mathbb{1}_{\{j \in \mathcal{B}(t+1), a_{j,t+1} = s_j\}} (r_{j,t+1} - \hat{r}_{j,t}(s_j)) \\ & \qquad j \in \mathcal{N}, \ s_j \in \mathcal{A}_j, \ t \geq 0, \ \mathbf{x}_0 = \mathbf{x}(0) \in \prod_j \mathcal{X}_j \end{cases}$$

$$[H4] \ \lambda_t \geq 0, \nu_t \geq 0, \ (\lambda, \nu) \in (l^2 \backslash l^1)^2, \ \frac{\lambda_t}{\nu_t} \longrightarrow 0 \qquad (7.18)$$

As in Theorem 7.3.2.3, one has an almost sure convergence to combined non-autonomous dynamics under [H4].

Theorem 7.3.3.1. *Assume [H1-H4]. Then, CODIPAS-RL scheme with variable number of players has the asymptotic pseudo trajectory of the following non-autonomous system:*

$$\begin{cases} \dot{\mathbf{x}}_{j,t}(s_j) = g_j(t) F_{j,s_j}(\mathbf{x}_{j,t}, \mathbb{E}_{\mathbf{w},\mathcal{B}} R_j^{\mathcal{B}}(\mathbf{w}, ., \mathbf{x}_{-j,t})) \\ x_j(s_j) > 0 \Longrightarrow \hat{\mathbf{r}}_{j,t}(s_j) \longrightarrow \mathbb{E}_{\mathbf{w},\mathcal{B}} R_j^{\mathcal{B}}(\mathbf{w}, \mathbf{e}_{s_j}, \mathbf{x}_{-j}) \end{cases}$$

Proof. The proof is obtained by combining the argument of Theorem 7.3.2.3 and the argument of multiple time-scales given in appendix. □

7.3.4 Learning Schemes Leading to Multi-Type Replicator Dynamics

The Multiplicative Weight Imitative Boltzmann-Gibbs (M-IBG) CODIPAS-RL is given by

$$(\textbf{M-IBG}) \begin{cases} x_{j,t+1}(s_j) = \dfrac{x_{j,t}(s_j)(1+\lambda_{[\theta_j(t)]})^{\hat{r}_{j,t}(s_j)}}{\sum_{s'_j \in \mathcal{A}_j} x_{j,t}(s'_j)(1+\lambda_{[\theta_j(t)]})^{\hat{r}_{j,t}(s'_j)}} \\ \hat{r}_{j,t+1}(s_j) = \hat{r}_{j,t}(s_j) + \nu_{j,t} \mathbb{1}_{\{a_{j,t+1}=s_j\}}(r_{j,t+1} - \hat{r}_{j,t}(s_j)) \\ \qquad\qquad j \in \mathcal{B}(t), \; s_j \in \mathcal{A}_j, \; t \geq 0 \end{cases}$$

For $j \notin \mathcal{B}(t)$, $(x_{j,t+1}, \hat{r}_{j,t+1}) = (x_{j,t}, \hat{r}_{j,t})$.

Proposition 7.3.4.1. *If the learning rates are well-chosen, the learning scheme (M-IBG) converges almost surely to a nonautonomous multiple-type replicator dynamics [151] given by*

$$\dot{x}_{j,t}(s_j) = g_j(t) x_{j,t}(s_j) \left[\mathbb{E}_{\mathbf{w},\mathcal{B}} R_j^{\mathcal{B}}(\mathbf{w}, \mathbf{e}_{s_j}, \mathbf{x}_{-j,t}) \right.$$
$$\left. - \sum_{s'_j \in \mathcal{A}_j} x_{j,t}(s'_j) \mathbb{E}_{\mathbf{w},\mathcal{B}} R_j^{\mathcal{B}}(\mathbf{w}, \mathbf{e}_{s'_j}, \mathbf{x}_{-j,t}) \right]$$

Proof. We compute the drift (the expected changes in one-time slot) for (M-IBG):

$$\begin{aligned} \text{drift} &= \frac{x_{j,t+1}(s_j) - x_{j,t}(s_j)}{\lambda_t^*} \\ &= \mathbb{1}_{\{j \in \mathcal{B}(t)\}} \frac{1}{\lambda_t^*} \left[\frac{x_{j,t}(s_j)(1+\lambda_{[\theta_j(t)]})^{\hat{r}_{j,t}(s_j)}}{\sum_{s'_j \in \mathcal{A}_j} x_{j,t}(s'_j)(1+\lambda_{[\theta_j(t)]})^{\hat{r}_{j,t}(s'_j)}} - x_{j,t}(s_j) \right] \\ &= \frac{x_{j,t}(s_j) \mathbb{1}_{\{j \in \mathcal{B}(t)\}}}{\sum_{s'_j \in \mathcal{A}_j} x_{j,t}(s'_j)(1+q_{j,t}\lambda_t^*)^{\hat{r}_{j,t}(s'_j)}} \times \\ &\qquad \left[\frac{(1+q_{j,t}\lambda_t^*)^{\hat{r}_{j,t}(s_j)}-1}{\lambda_t^*} - \sum_{s'_j \in \mathcal{A}_j} x_{j,t}(s'_j)\frac{(1+q_{j,t}\lambda_t^*)^{\hat{r}_{j,t}(s'_j)}-1}{\lambda_t^*} \right] \end{aligned}$$

Using the fact that $\lambda_t^* \longrightarrow 0$, the term $\frac{(1+q_{j,t}\lambda_t^*)^{\hat{r}_{j,t}(s_j)}-1}{\lambda_t^*} \longrightarrow q_{j,t}\hat{r}_{j,t}(s_j)$ and taking the conditional expectation, we conclude that the drift goes to the multi-type replicator equation:

$$\dot{\mathbf{x}}_{j,t}(s_j) = g_j(t)\mathbf{x}_{j,t}(s_j)\left[\mathbb{E}\hat{r}_{j,t}(s_j) - \sum_{s_j' \in \mathcal{A}_j}\mathbb{E}\hat{r}_{j,t}(s_j')\mathbf{x}_{j,t}(s_j')\right].$$

Hence, the ODE of imitative Boltzmann-Gibbs CODIPAS-RL with multiplicative weights is given by **(ODE: M-IBG)** :

$$\begin{cases} \dot{\mathbf{x}}_{j,t}(s_j) &= g_j(t)\mathbf{x}_{j,t}(s_j)\left[\hat{r}_{j,t}(s_j) - \sum_{s_j' \in \mathcal{A}_j}\hat{r}_{j,t}(s_j')\mathbf{x}_{j,t}(s_j')\right] \\ \frac{d}{dt}\hat{\mathbf{r}}_{j,t}(s_j) &= \tilde{g}_j(t)\mathbf{x}_{j,t}(s_j)\left[\mathbb{E}_{\mathbf{w},\mathcal{B}}R_j^{\mathcal{B}}(\mathbf{w},\mathbf{e}_{s_j},\mathbf{x}_{-j,t}) - \hat{\mathbf{r}}_{j,t}(s_j)\right] \end{cases}$$

Combining this with the assumption [H4] on ν and λ, one gets the stated result. $\qquad\square$

Corollary 7.3.4.2. *If the dynamic game is a robust potential game, i.e., if there exists a continuously differentiable function $\tilde{V}$ defined on $\mathbb{R}^{\sum_j |\mathcal{A}_j|}$ such that $\nabla\tilde{V}(\mathbf{x}) = [\mathbb{E}_{\mathbf{w},\mathcal{B}}R_j^{\mathcal{B}}(\mathbf{w},.,\mathbf{x}_{-j})]_j$. Then, global convergence to equilibria of the expected robust game holds.*

Note that the class of robust potential games is more general than the class of finite exact potential games introduced in [116]. The model of finite potential game is obtained when there is a single state and $\mathcal{B}(t) = \mathcal{N}$.

Proof. From [140], we know that the replicator is positively correlated (PC). Since

$$\frac{d}{dt}\tilde{V}(\mathbf{x}_t) = \sum_{j\in\mathcal{N}}\sum_{s_j}[\frac{d}{dt}x_{j,t}][\frac{\partial}{x_{j,t}(s_j)}\tilde{V}(\mathbf{x}_t)] \tag{7.19}$$

$$= \sum_{j\in\mathcal{N}}g_j(t)\sum_{s_j}x_{j,t}(s_j)[\frac{\partial}{x_{j,t}(s_j)}\tilde{V}(\mathbf{x}_t)]^2$$

$$-[\sum_{j\in\mathcal{N}}g_j(t)\sum_{s_j}x_{j,t}(s_j)\frac{\partial}{x_{j,t}(s_j)}\tilde{V}(\mathbf{x}_t)]^2 \geq 0 \tag{7.20}$$

with strict inequality if $\nabla\tilde{V} \neq 0$. In the last inequality (Jensen), we have used the convexity of y^2 and positivity of g. This means that $-\tilde{V}$ serves as a Lyapunov function to the non-autonomous dynamics. $\qquad\square$

Note that the replicator dynamics may have some rest points that are non-Nash equilibria (in the faces of the simplex). However, the interior rest points of the replicator dynamics (7.19) are Nash equilibria of the expected robust game with a variable number of interacting players. Nonconvergence to linearly unstable equilibria can be found in [76, 124, 27]. See also [179, 178].

7.3.5 Heterogeneous Learning with Random Updates

We consider five standard learning mechanisms well-used in the literature and classify the players into 5 groups of learners. Denote by I_l the set of players with learning patterns $l \in \{1, 2, 3, 4, 5\}$. The set of players can be written as

$$\mathcal{N} = I_1 \cup I_2 \cup I_3 \cup I_4 \cup I_5.$$

This leads to a heterogeneous CODIPAS-RL with random updates.

$$
\begin{cases}
\mathbf{x}_{j_1,t+1} - \mathbf{x}_{j_1,t} = \lambda_t^* q_{j_1,t} \left[f_{j_1}^{I_1,\mathcal{B}(t)}(a_{j_1,t}, \mathbf{x}_{j_1,t}, r_{j_1,t}) + M_{t+1}^{(1)} \right], \quad j_1 \in I_1 \\
\qquad\qquad f^{I_1} : \text{Bush-Mosteller} \\
\mathbf{x}_{j_2,t+1} - \mathbf{x}_{j_2,t} = \lambda_t^* q_{j_2,t} \left[f_{j_2}^{I_2,\mathcal{B}(t)}(a_{j_2,t}, \mathbf{x}_{j_2,t}, r_{j_2,t}) + M_{t+1}^{(2)} \right], \quad j_2 \in I_2 \\
\qquad\qquad f^{I_2} : \text{Borger-Sarin} \\
\mathbf{x}_{j_3,t+1} - \mathbf{x}_{j_3,t} = \lambda_t^* q_{j_3,t} \left[f_{j_3}^{I_3,\mathcal{B}(t)}(a_{j_3,t}, \mathbf{x}_{3,t}, \hat{\mathbf{r}}_{j_3,t}) - \mathbf{x}_{j_3,t} + M_{t+1}^{(3)} \right], \quad j_3 \in I_3 \\
\qquad\qquad f^{I_3} : \text{Boltzmann-Gibbs} \\
\mathbf{x}_{j_4,t+1} - \mathbf{x}_{j_4,t} = \lambda_t^* q_{j_4,t} \left[f_{j_4}^{I_4,\mathcal{B}(t)}(\mathbf{x}_{j_4,t}, \hat{\mathbf{r}}_{j_4,t}) - x_{j_4,t} + M_{t+1}^{(4)} \right], \quad j_4 \in I_4 \\
\qquad\qquad f^{I_4} : \text{Imitative-BG} \\
\mathbf{x}_{j_5,t+1} - \mathbf{x}_{j_5,t} = \lambda_t^* q_{j_5,t} \left[f_{j_5}^{I_5,\mathcal{B}(t)}(\mathbf{x}_{j_5,t}, \hat{\mathbf{r}}_{j_5,t}) - x_{j_5,t} + M_{t+1}^{(5)} \right], \quad j_5 \in I_5 \\
\qquad\qquad f^{I_5} : \arg\max_{s'_{j_5}} \hat{\mathbf{r}}_{j_5,t}(s'_{j_5}) \\
\hat{r}_{j,t+1}(s_j) = \hat{r}_{j,t}(s_j) + \nu_t^* \frac{\nu_{\theta_j(t)}}{\nu_t^*} \mathbb{1}_{\{j \in \mathcal{B}(t), a_{j,t}=s_j\}} (r_{j,t} - \hat{r}_{j,t}(s_j)) \\
\qquad\qquad \text{Payoff-learning}
\end{cases}
$$

where $\nu_t^* = \sup_{j \in \mathcal{B}(t)} \nu_{\theta_j(t)}$.

- I_1 is the group of players with Bush-Mosteller [42],

$$f_{j_1,s_{j_1}}^{I_1,\mathcal{B}(t)} = \frac{r_{j_1,t} - \Gamma_{j_1}}{\sup_{\mathbf{a},\mathbf{w}} |\tilde{R}_{j_1}(\mathbf{w},\mathbf{a}) - \Gamma_{j_1}|} \left(\mathbb{1}_{\{a_{j_1,t}=s_{j_1}\}} - \mathbf{x}_{j_1,t}(s_{j_1}) \right)$$

where Γ_{j_1} is reference level of j_1.

- I_2 is the group of players with Borger-Sarin [32] or Arthur [11]:

$$f_{j_2,s_{j_2}}^{I_2,\mathcal{B}(t)} = r_{j_2,t} \left(\mathbb{1}_{\{a_{j_2,t}=s_{j_2}\}} - \mathbf{x}_{j_2,t}(s_{j_2}) \right)$$

- I_3 is the group of players with Boltzmann-Gibbs [62]

$$f_{j_3,s_{j_3}}^{I_3,\mathcal{B}(t)} = \tilde{\beta}_{j_3,\epsilon_{j_3}}(\hat{\mathbf{r}}_{j_3,t})(s_{j_3}) = \frac{e^{\frac{1}{\epsilon_{j_3}}\hat{r}_{j_3,t}(s_{j_3})}}{\sum_{s'_{j_3} \in A_{j_3}} e^{\frac{1}{\epsilon_{j_3}}\hat{r}_{j_3,t}(s'_{j_3})}}$$

- I_4 is the group of players with Imitative Boltzmann-Gibbs (IBG)

$$f^{I_4,\mathcal{B}(t)}_{j_4,s_{j_4}} = \tilde{\sigma}_{j_4,\epsilon_{j_4}}(\hat{\mathbf{r}}_{j_4,t})(s_{j_4}) = \frac{x_{j_4,t}(s_{j_4})e^{\frac{1}{\epsilon_{j_4}}\hat{\mathbf{r}}_{j_4,t}(s_{j_4})}}{\sum_{s'_{j_4}\in\mathcal{A}_{j_4}} x_{j_4,t}(s'_{j_4})e^{\frac{1}{\epsilon_{j_4}}\hat{\mathbf{r}}_{j_4,t}(s'_{j_4})}}$$

- I_5 is the group of players with weakened-fictitious-play based on payoff estimations response

$$f^{I_5,\mathcal{B}(t)}_{j_5,s_{j_5}} \in \epsilon_{[\theta_{j_5}(t)]}\delta_{\max_{s'_{j_5}}\hat{\mathbf{r}}_{j_5,t}(s'_{j_5})} + (1 - \epsilon_{[\theta_{j_5}(t)]})\frac{\mathbb{1}}{|\mathcal{A}_{j_5}|}$$

- Simultaneously with strategy learning, the players employ payoff learning when they are active.

⚠ Some of the CODIPAS dynamics are not differentiable.

The players update only when they are active. Only the active players can interact with each other, i.e., the players in sleep mode at time t do not interfere in the payoff/outcome of the active players. Under this setting, the asymptotic pseudo-trajectory of the payoff-learning is a nonautonomous differential equation. Given an interior point $\mathbf{x}$, the payoff learning with random updates converges to a time-dependent ODE:

$$\frac{d}{dt}\hat{r}_{j,t}(s_j) = \tilde{g}_j(t)\mathbf{x}_j(s_j)\left[\mathbb{E}R_j - \hat{r}_{j,t}(s_j)\right] \tag{7.21}$$

We now provide the convergence time of the dynamics (7.21). The next lemma provides an explicit solution of each equation of the system.

Lemma 7.3.5.1. *The function*

$$t \longrightarrow \hat{r}_{j,t} = \mathbb{E}R_j + [\hat{r}_{j,0} - \mathbb{E}R_j]\,e^{-\int_0^t g_j(s)\,ds}$$

is the solution of the nonautonomous differential equation (7.21).

Proof. The equation is in the form

$$\dot{y}_j(t) = \hat{g}_j(t)(a_j - y_j(t))$$

for some constant a_j and positive function $\hat{g}_j$. Denote by $z_j(t) = y_j(t) - a_j$. Then, $\dot{z}_j = \dot{y}_j$ and the time-dependent ODE becomes

$$\dot{z}_j(t) = -\hat{g}_j(t)z_j(t).$$

The solution of this last ODE is given by $z_j(t) = z_j(0)e^{-\int_0^t \hat{g}_j(s)\,ds}$. This implies that

$$y_j(t) = a_j + z_j(t) = a_j + z_j(0)e^{-\int_0^t \hat{g}_j(s)\,ds} = a_j + (y_j(0) - a_j)e^{-\int_0^t \hat{g}_j(s)\,ds}.$$

By taking $a_j = \mathbb{E}R_j$, $y_j(t) = \hat{r}_{j,t}$, $\hat{g}_j(t) = \tilde{g}_j(t)x_j(s_j)$, one gets the stated result. □

Proposition 7.3.5.2. *The convergence time to close to the expected payoff with error tolerance η is at most*

$$\max_j (G^j)^{-1} \left(\log[\frac{error(0)}{\eta}] \right)$$

where $(G^j)^{-1}$ is the inverse of the mapping $t \longmapsto G^j(t) = \int_0^t g_j(s) \, ds$ and $error_j(0) := \| \hat{r}_{j,0} - \mathbb{E}R_j \|$, $error = \max_j error_j$. In particular for $g_j = 1$ (almost active case), the convergence time is of order

$$\log \left[\frac{error(0)}{\eta} \right].$$

Notice that the convergence is in order of $\log \left[\frac{1}{\eta} \right]$ is equivalent to say that we have exponential convergence, i.e., the convergence rate is exponential decay in time.

Proof. From Lemma 7.3.5.1, one gets

$$\| \hat{r}_{j,t} - \mathbb{E}R_j \| = error_j e^{- \int_0^t g_j(s) \, ds}.$$

From the assumptions, the primitive function G^j is a bijection and $\| \hat{r}_{j,t} - \mathbb{E}R_j \| \leq \eta$ if $t \geq \max_j (G^j)^{-1} \left(\log[\frac{error(0)}{\eta}] \right)$. The last assertion is obtained for $g_j = 1$. This completes the proof. $\square$

As a corollary of Theorem 7.3.3.1 (which uses [H4]), one has that the following heterogeneous combined dynamics

$$\begin{cases} \dot{x}_{j_1,t}(s_{j_1}) = g_{j_1}(t) \left[\mathbb{E}F_{j_1}^{I_1,s_{j_1}}(\mathbf{x}_{j_1,t}, \mathbb{E}_{\mathbf{w},\mathcal{B}} R_{j_1}^{\mathcal{B}}(\mathbf{w}, \mathbf{e}_{s_{j_1}}, \mathbf{x}_{-j_1,t})) \right], \ j_1 \in I_1 \\ \dot{\mathbf{x}}_{j_2,t} = g_{j_2}(t) x_{j_2,t}(s_{j_2}) \left[\mathbb{E}_{\mathbf{w},\mathcal{B}} R_{j_2}^{\mathcal{B}}(\mathbf{w}, \mathbf{e}_{s_{j_2}}, \mathbf{x}_{-j_2,t}) \right. \\ \qquad \left. - \sum_{s'_{j_2} \in \mathcal{A}_{j_2}} x_{j_2,t} \mathbb{E}_{\mathbf{w},\mathcal{B}} R_{j_2}^{\mathcal{B}}(\mathbf{w}, \mathbf{e}_{s_{j_2}}, \mathbf{x}_{-j_2,t}) \right], \ j_2 \in I_2 \\ \dot{\mathbf{x}}_{j_3,t} = g_{j_3}(t) \left[\tilde{\beta}_{j_3,\epsilon_{j_3}} \left(\mathbb{E}_{\mathbf{w}} R_{j_3}(\mathbf{w}, \mathbf{e}_{s_{j_3}}, \mathbf{x}_{-j_3,t}) - \mathbf{x}_{j_3,t} \right) \right], \ j_3 \in I_3 \\ \dot{\mathbf{x}}_{j_4,t} = g_{j_4}(t) \left[\tilde{\sigma}_{j_4,\epsilon_{j_4}} \left(\mathbb{E}_{\mathbf{w},\mathcal{B}} R_{j_4}^{\mathcal{B}}(\mathbf{w}, \mathbf{e}_{s_{j_4}}, \mathbf{x}_{-j_4,t}) - \mathbf{x}_{j_4,t} \right) \right], \ j_4 \in I_4 \\ \dot{\mathbf{x}}_{j_5,t} \in g_{j_5}(t) \left[\widetilde{\mathrm{BR}}_{j_5}(\mathbf{x}_{-j_5,t}) - x_{j_5,t} \right], \ j_5 \in I_5 \\ \hat{r}_{j,t}(s_j) \longrightarrow \mathbb{E}_{\mathbf{w},\mathcal{B}} R_j^{\mathcal{B}}(\mathbf{w}, \mathbf{e}_{s_j}, \mathbf{x}_{-j,t}) \end{cases}$$

is an asymptotic pseudo-trajectory of the heterogeneous learning, where $\widetilde{\mathrm{BR}}$ denotes the best response correspondence of the expected robust game i.e.,

$$\widetilde{\mathrm{BR}}_j(\mathbf{x}_{-j}) = \arg \max_{\mathbf{x}_j} \mathbb{E}_{\mathbf{w},\mathcal{B}} R_j^{\mathcal{B}}(\mathbf{w}, \mathbf{x}_j, \mathbf{x}_{-j}).$$

7.3.6 Constant Step-Size Random Updates

Our proof extends to the constant learning rate cases but the results are weaker. The almost sure convergence becomes a convergence in law or in distribution.

We compare the behavior of a sample path $\{\mathbf{x}_t\}_{t\geq 0}$ with the trajectories of the flow induced by the vector field $\tilde{F}$. To do this, we denote by

$$T_0 = 0, \ T_k = \sum_{t=1}^{k} \lambda = k\lambda, \ k \geq 1.$$

We define the continuous time affine and piecewise interpolated process $X :$ $\mathbb{R}_+ \longrightarrow \mathbb{R}^{\sum_j |\mathcal{A}_j|}$

$$X(T_k + s') = \mathbf{x}_k + s' \frac{\mathbf{x}_{k+1} - \mathbf{x}_k}{\lambda}, \ \forall \, k \in \mathbb{Z}_+, \ 0 \leq s' \leq \lambda$$

By the assumption $\forall t, \lambda_t = \lambda > 0,$ one gets $\lambda \notin l^1$. This means that T_k and $\tilde{T}$ are unbounded. Then, the weak convergence bound to the asymptotic pseudo-trajectory is in order of λ assuming that the the process $\| \mathbf{x}_k \|^2$ is uniformly integrable. For example the asymptotic pseudo-trajectory result of Theorem 7.3.2.2 becomes

$$\forall T > 0 \ \lim_{\Delta t \longrightarrow +\infty} \mathbb{E} \left[\sup_{s \in [\Delta t, \Delta t + T]} \| X(\Delta t + s) - \Phi_{\Delta t}(s) \| \right] = O(\lambda)$$

where the learning rate is constant $\lambda_t = \lambda$. For $\lambda = b > 0$, one gets Theorem 3.1 in [179], which is a weak convergence result (convergence in law).

7.3.7 Revision Protocols with Random Updates

Consider the following CODIPAS-RL with imitation, local timer, and random updates in a $K-$subpopulation system.

$$\begin{cases} \mathbf{x}_{j,t+1}^p(s_j) - \mathbf{x}_{j,t}^p(s_j) = \\ \mathbb{1}_{\{j \in \mathcal{B}(t)\}} \left[\dfrac{\mathbf{x}_{j,t}^p(s_j)(1+\lambda_{j,[\theta_j(t)]}^p)^{\hat{\mathbf{r}}_{j,t}^p(s_j)}}{\sum_{s_j' \in \mathcal{A}_j^p} \mathbf{x}_{j,t}^p(s_j')(1+\lambda_{j,[\theta_j(t)]}^p)^{\hat{\mathbf{r}}_{j,t}^p(s_j')}} - \mathbf{x}_{j,t}^p(s_j) \right] + \sqrt{\lambda_{j,[\theta(t)]}}\,\tilde{f}W_t, \\ \hat{\mathbf{r}}_{j,t+1}^p(s_j) = \hat{\mathbf{r}}_{j,t}^p(s_j) + \mu_{j,[\theta_j(t)]}^p \mathbb{1}_{\{j \in \mathcal{B}(t), a_{j,t}=s_j\}} \left(r_{j,t}^p - \hat{\mathbf{r}}_{j,t}^p(s_j) \right), \\ \theta_j(t+1) = \theta_j(t) + \mathbb{1}_{\{j \in \mathcal{B}(t)\}}, \ s_j \in \mathcal{A}_j^p, \ \forall \, t \geq 0, \ \mathcal{B}(t) \subseteq \mathcal{N}, \\ \mathbf{x}_{j,0}^p \in \Delta(\mathcal{A}_j^p), \hat{\mathbf{r}}_{j,0}^p \in \mathbb{R}^{|\mathcal{A}_j^p|}, \\ p \in \mathcal{P} = \{1, 2, \ldots, K\}. \end{cases}$$

where $\mathcal{A}^p := \cup_{j \in \mathcal{N}} \mathcal{A}_j^p$ is finite and W_t is a standard Brownian motion. The asymptotic pseudo-trajectory of the above scheme leads to a stochastic dynamics.

7.4   Dynamic Routing Games with Random Traffic

Our illustrating example in this section for high complexity and imperfect observation is a large-scale congestion game with random traffic. We consider a distributed traffic routing in a network with parallel links, in which a large number of vehicles make routing and data allocation decisions in random frames to optimize their own objectives in response to their own observations and local clocks. In this setting, observing and responding to the individual choice of all vehicles on a frame basis would be a formidable task for any individual vehicle. A more realistic measurement for the information tracked and processed by an individual vehicle is the measurement aggregate congestion (received signals) on the roads that are of interest to that vehicle in a given frame. Therefore, the standard fictitious play and the iterative best reply are not directly applicable. We apply the CODIPAS-RL to this scenario. The state of the environment is denoted by $w_{j,t}(r)$ represents the characteristics at time t (weather conditions, channel gain, etc.). The payoff function of a player $j \in \mathcal{B}(t)$ is a function $\mathbf{w}_t = (w_{j,t}(r))_{j,r}$ but also the weighted distribution of vehicles on the road. More the road r is congested, more the payoff decreases.

We assume that players can have:

- a noisy measurement of the payoff at the chosen road and

- an estimate of the alternative routes by using $\hat{\mathbf{r}}$.

Players make a decision based on this rough information by using a randomized decision rule to revise the strategies. The payoffs of the chosen alternative are then perceived and are used to update the strategy for that particular route. When active, each player experiments only the payoffs of the selected route on that time frame, and uses this information to adjust his strategy for that particular route. This scheme is repeated every slot, generating a discrete time stochastic learning process. Basically we have four fundamental ingredients:

- the state $\mathbf{w}_t$;

- the set of active players $\mathcal{B}(t)$;

- the rule for revision of strategies from the state to randomized actions $\mathbf{x}_t$; and,

- a rule to update the new estimations, $\hat{\mathbf{r}}_t$.

Although players observe only the payoffs of the specific route chosen on the given time slot, the observed values depend on the congestion levels determined by the other active players choices revealing implicit information on the entire system. The natural question is whether such a learning algorithm based on a minimal piece of information may be sufficient to induce coordination (note that we do not assume coordination between players, there is not central authority, and there is no signaling scheme to the players), and make the system stabilize to an equilibrium. The answer to this question is positive for dynamic routing games with parallel links and random number of interacting players.

We denote the set of routes by $\mathcal{R}$ and set $\mathcal{A}_j = \mathcal{R}, \ \forall j$. An estimation of player j at time is a vector $\hat{\mathbf{r}}_{j,t} = (\hat{r}_{j,t}(\underline{r}))_{\underline{r} \in \mathcal{R}}$ where $\hat{r}_{j,t}(\underline{r})$ represents player j's estimate of the average payoff of route r. When he is active i.e., $j \in \mathcal{B}(t)$, player j updates his strategy using the imitative Boltzmann-Gibbs with multiplicative weight $\tilde{\sigma}_{j,\epsilon}(\hat{\mathbf{r}}_j)$ to use route $\underline{r}$ with probability $x_{j,t}(\underline{r})$, We consider the payoff in the form $\tilde{R}_j^{\mathcal{B}}(\mathbf{w}, \underline{r}, \mathbf{a}_{-j}) = \phi_{\underline{r}}(\mathbf{w}, I_{\underline{r}}(\mathbf{w}, \mathbf{a})) \mathbb{1}_{\{j \in \mathcal{B}\}}$ where

$$I_r(\mathbf{w}, \mathbf{a}) = \sum_{j \in \mathcal{N}} w_{j,t}(\underline{r}) \mathbb{1}_{\{j \in \mathcal{B}(t), a_{j,t} = \underline{r}\}}$$

is the cumulative interference at the route $\underline{r}$. The congestion is captured by the inequality

$$\forall t, \ \phi_r(\mathbf{w}, I_{\underline{r}}'(\mathbf{w}, \mathbf{a})) \leq \phi_{\underline{r}}(\mathbf{w}, I_{\underline{r}}(\mathbf{w}, \mathbf{a})), \ I_{\underline{r}}(\mathbf{w}, \mathbf{a}) \leq I_{\underline{r}}'(\mathbf{w}, \mathbf{a}),$$

for some map $\phi_{\underline{r}}(.)$. Since this structure of the congestion game is a potential game [116], using the CODIPAS-RLs developed in the previous section, one gets a global convergence to the set of equilibria. Using Corollary 7.3.4.2, the result is extended to robust potential game.

Notes

We have presented a combined and fully distributed payoff and strategy reinforcement learning algorithm (CODIPAS-RL) for a particular class of stochastic games with incomplete information, and a variable number of players. Using multiple-scale stochastic approximations, we proved almost sure convergence of CODIPAS-RLs to non-autonomous differential equations as well as weak convergence with explicit bounds. However some interesting questions remain open:

- Extension to different speed of learning

- How to design heterogeneous, efficient, and fully distributed learning algorithm leading to *global optima* in simple classes of games with a variable number of interacting players?

- What is the effect/impact of delayed and noisy payoff measurements?

- What happens if the players do not observe the numerical value of their experiences?

7.5   Extensions

In this section, we briefly present some interesting extensions of the learning framework developed above.

7.5.1   Learning in Stochastic Games

What is a stochastic game?

There are different models of stochastic games. Stochastic games in the sense of Shapley, stochastic differential games, stochastic difference games etc. Here we present the first type. A stochastic game, introduced by Lloyd Shapley in the 1950s [144], is a dynamic game with probabilistic transitions played by one or several players. The game is played in a sequence of runs. At the beginning of each step (time slot), the game is in some state (which can be unknown to the players). The players select actions, and each player receives a payoff that depends on the current state and the chosen actions. The game then moves to a new random state whose distribution depends on the previous state and the actions chosen by the players (Markovian model). However, the strategies do not need to be Markovian. The procedure is repeated at the new state and play continues for a finite or infinite number of stages. The total payoff, discounted payoff or the limit inferior of the averages of the stage payoffs to a player are often considered.

The game framework developed in the previous chapters is already in the class of stochastic games since there is a generic state which probabilistically changes in time. Howeve,r we have only considered the case where the transitions probabilities are not controlled i.e., do not depend on the actions. A more general model of stochastic games is the case where the state dynamics depends on the actions.

Stochastic games have become an appealing framework to model the multi-player system that involves dynamical interactions among players in a stochastic environment. The practical computational algorithms and tools findings for stochastic games are essential in their applications. Since the seminal paper by Shapley [144], many numerical algorithms have been proposed for finding solutions to Bellman's dynamic programming equation [22], Shapley's operator and fixed-point equations. Based on the concept of dynamic programming, the value iteration method, and policy iteration have been proposed to compute equilibria and optimal strategies.

Is it possible to extend CODIPAS to stochastic games?

The difficulty when extending the learning framework developed in this book to stochastic games those transitions are action-dependent is coming from the fact that playing a Nash equilibrium in each round may not to be Nash equilibrium of the stochastic game. Similarly, playing a global optimum in each state does not necessarily lead to a global optimum in the long term.

Example 7.5.2. *To illustrate the above statement, consider the following example: There are two actions a_1 and a_2 and three states, w_1, w_2, and w_3; If the action a_1 is chosen in state w_1, then the system moves to w_2, If a_2 is chosen, then the system moves to w_3. Both w_2 and w_3 are followed by w_1 regardless of the action played. In state w_1, the action a_1 gives payoff $\wp > 0$ and the action a_2 gives the payoff 0. In state w_2, payoffs are 0 regardless of the action played, while in state w_3 payoffs are identically equal to $5\wp$. State-by-state optimization leads the player to choose a_1 in state w_1, but this is not optimal if $T > T_0$. Thus, the equilibrium play for the game $\mathcal{G}(w_1)$ need not be consistent with the overall equilibrium, as players consider not only their current period payoff but also the payoff in the next periods.*

a_1	$\tilde{R} = \wp$, $q_{1a_1} \sim (0,1,0)$
a_2	$\tilde{R} = 0$, $q_{1a_2} \sim (0,0,1)$

$\mathcal{G}(w_2):$
$\tilde{R} = 0$, $q_2 \sim (1,0,0)$

$\mathcal{G}(w_3):$
$\tilde{R} = 5\wp$, $q_3 \sim (1,0,0)$

The key message is that being optimum at each time is not optimal in the long-run Markov decision process (MDP).

7.5.2.1   Nonconvergence of Fictitious Play

Here we give a nonconvergence result in two-player, two-actions per player and two-states stochastic game. The game is described in the Table below.

	a_1	a_2		a_1	a_2
a_1	$(2,1)$ $q \sim \left(\frac{9}{10}, \frac{1}{10}\right)$	$(4,0)$ $q \sim \left(\frac{9}{10}, \frac{1}{10}\right)$	a_1	$(0,1)$ $q \sim \left(\frac{9}{10}, \frac{1}{10}\right)$	$(2,0)$ $q \sim \left(\frac{9}{10}, \frac{1}{10}\right)$
a_2	$(0,0)$ $q \sim \left(\frac{1}{10}, \frac{9}{10}\right)$	$(7,1)$ $q \sim \left(\frac{1}{10}, \frac{9}{10}\right)$	a_2	$(2,0)$ $(q \sim \left(\frac{1}{10}, \frac{9}{10}\right)$	$(4,1)$ $q \sim \left(\frac{1}{10}, \frac{9}{10}\right)$

Notice that the transition probabilities in this game depend only on the action of player 1 and they are independent of the state. Furthermore, the game is irreducible, which means that irrespective of the players' strategies both states will be visited infinitely often with probability 1 and the limiting average payoff of the game does not depend on the starting state s_0. The following result has been proved by Schoenmakers, Flesch, Thuijsman (2007, [57]).

Theorem 7.5.2.2. *In the above stochastic game with two players, two actions per player in each state and two states, the fictitious play does not need to converge.*

7.5.2.3   Q-learning in Zero-Sum Stochastic Games

As pointed out by Bellman in Markov decision processes, many of these methods suffer from the so-called *curse of dimensionality*, that is, as the dimension of the state space increases, the complexity of dynamic programming equation grows exponentially. The second drawback of these methods is that most of them require knowledge of transition law and its distributions, which are not available in many interesting scenarios. Therefore, Q-learning [83, 188, 189] haS been proposed. Q-learning is a specific reinforcement learning technique that allows online updates of the game value and strategies in a stochastic environment. In the single player case, Q-learning has been widely investigated and many theoretical results on the convergence and error bounds of the algorithm have been investigated in the finite state space case. However, fewer results have been known for its counterpart in stochastic games with more than two players. A specific class of stochastic games is the class of *two-player zero-sum stochastic games*. In that case a minmax Q-table can be used to approximate the value.

In the single player case, the approximation technique used in temporal difference learning (TD-learning) reduces the curse of dimensionality by finding a parameterized approximated solution within a prescribed finite-dimensional function class. Using stochastic approximation techniques, one can show that the TD-learning that approximates the value functions converges almost surely under suitable conditions. We describe a discounted stochastic game [144, 168] by $\Gamma_\delta = (\mathcal{N}, \mathcal{W}, \mathcal{A}_j, \tilde{R}_j(\mathbf{w}, a), q, \delta)$ where

- $WA = \{(\mathbf{w}, a) \mid (\mathbf{w}, a) \in \mathcal{W}, \ a_{j'} \in \mathcal{A}_{j'}(\mathbf{w})\}$

- $\tilde{R}_j : \ (\mathbf{w}, a) \in WA \longrightarrow \mathbb{R}$,

- $\delta \in (0, 1)$

- $q : WA \longrightarrow \Delta(\mathcal{W})$.

The stochastic game is a zero-sum $\sum_j R_j(\mathbf{w}, a) = c$, where c is a constant. We focus on two-player zero-sum stochastic i.e., $\tilde{R}_1(.,.) = -\tilde{R}_2(.,.) = \tilde{R}(.,.)$.

Let $(v_{\delta,\mathbf{w}})_{\mathbf{w} \in \mathcal{W}}$ be the discounted-value of the stochastic game with discount factor $\delta \in (0, 1)$. Then, v_δ is a fixed point of the Shapley operator

$$
\begin{aligned}
v_{\delta,\mathbf{w}} &= \sup_{\mathbf{x}_1(\mathbf{w})} \inf_{\mathbf{x}_2(\mathbf{w})} \left[(1 - \delta) R(\mathbf{w}, \mathbf{x}_1(\mathbf{w}), \mathbf{x}_2(\mathbf{w})) + \delta \sum_{\mathbf{w}'} q_{\mathbf{w},\mathbf{x},\mathbf{w}'} v_{\delta,\mathbf{w}'} \right] \\
&= \inf_{\mathbf{x}_2(\mathbf{w})} \sup_{\mathbf{x}_1(\mathbf{w})} \left[(1 - \delta) R(\mathbf{w}, \mathbf{x}_1(\mathbf{w}), \mathbf{x}_2(\mathbf{w})) + \delta \sum_{\mathbf{w}'} q_{\mathbf{w},\mathbf{x},\mathbf{w}'} v_{\delta,\mathbf{w}'} \right]
\end{aligned}
$$

Recall that, given a finite polymatrix $Q(\mathbf{w}, s_1, s_2)$, the zero-sum game with payoffs defined by $Q(\mathbf{w}, ., .)$ has max min = min max called value of the game in $\mathbf{w}$. The corresponding strategies leading to the value are called *saddle points*.

- The saddle points are interchangeable i.e., $(\mathbf{x}_1, \mathbf{x}_2)$, $(\mathbf{x}_1', \mathbf{x}_2')$ are saddle points then $(\mathbf{x}_1', \mathbf{x}_2)$, $(\mathbf{x}_1, \mathbf{x}_2')$ are both saddle points. This property does not holds for non-zero-sum since we know that the Nash equilibria can have different payoffs.

- We denote by *val* the value operator. Denote by $\mathcal{H}_\mathbf{w}$ the vector space of bounded Q-functions at the state $\mathbf{w}$ equipped with the supremum norm. $\mathcal{H}_\mathbf{w}$ is a Banach space. Then, it is known that the value operator has the non-expensive property, i.e.,

$$|val(Q_1(\mathbf{w}, ., .)) - val(Q_2(\mathbf{w}, ., .))| \leq \sup_{s_1, s_2} |Q_1(\mathbf{w}, s_1, s_2) - Q_2(\mathbf{w}, s_1, s_2)|$$

Let the Q-function be

$$Q(\mathbf{w}, s_1, s_2) = \left[(1 - \delta) R(\mathbf{w}, s_1, s_2) + \delta \sum_{\mathbf{w}'} q_{\mathbf{w}, s_1, s_2, \mathbf{w}'} v_{\delta, \mathbf{w}'} \right]$$

Then, the optimal Q-function Q^* satisfies $v_\delta = val(Q^*(\mathbf{w}, ., .)$ Then,

$$Q^*(\mathbf{w}, s_1, s_2) = \left[(1 - \delta) R(\mathbf{w}, s_1, s_2) + \delta \sum_{\mathbf{w}'} q_{\mathbf{w}, s_1, s_2, \mathbf{w}'} v_{\delta, \mathbf{w}'}^* \right]$$

and the Shapley operator becomes

$$Q^*(\mathbf{w}, s_1, s_2) = \left[(1 - \delta) R(\mathbf{w}, s_1, s_2) + \delta \sum_{\mathbf{w}'} q_{\mathbf{w}, s_1, s_2, \mathbf{w}'} val[Q^*(\mathbf{w}', ., .)] \right]$$

The iterative Q-learning algorithm writes

$$Q_{t+1}(\mathbf{w}, s_1, s_2) = Q_t(\mathbf{w}, s_1, s_2)$$

$$+ \lambda_t(\mathbf{w}, s_1, s_2) \left[(1 - \delta) R_t + \delta \sum_{\mathbf{w}'} val[Q_t(\mathbf{w}', ., .)] - Q_t(\mathbf{w}, s_1, s_2) \right] \quad (7.22)$$

Note that this recursion does not require the knowledge of transition probability q.

Lemma 7.5.2.4. *Let $\lambda_t \geq 0$ be a step size satisfying*

$$\sum_t \lambda_t = \infty, \quad \sum_t \lambda_t^2 < \infty.$$

The Q-learning algorithm in Equation (7.22) converges almost surely to the value.

7.5.3 Connection to Differential Dynamic Programming

Q-learning is a technique used to compute an optimal policy for a controlled Markov chain based on observations of the system controlled using a non-optimal policy. Many interesting results have been obtained for models with finite state and action space. Recently, [111] establishes connections between Q-learning and nonlinear control of continuous-time models with general pay-off functions. The authors show that the Hamiltonian appearing in nonlinear control theory is essentially the same as the *Q-function* that is the object of interest in Q-learning. They established a close connection between Q-learning and differential dynamic programming [83] state space and general action space.

7.5.4 Learning in Robust Population Games

7.5.4.1 Connection with Mean Field Game Dynamics

In this section, we establish a relation between CODIPAS with a random number of interacting players and the mean field game dynamics in large population [153, 164, 177]. Usually, in population games, the player plays in pure actions[2] The consequence of this assumption is that, mass behavior is in finite dimension if the set of actions is finite. The standard evolutionary game dynamics do not cover the scenario that we have described above in which a player may use a mixed action x_t.

We consider a large population of players containing a finite number of subpopulations. A typical scenario is depicted in Figure 7.1.

We assume that players from different subpopulations adopt different learning patterns. We denote the set of subpopulations by $\mathcal{P} = \{1, \ldots, K\}$ where K is the total number of learning schemes. The set of learning schemes is denoted by $\mathcal{L} = \{\mathcal{L}_1, \ldots, \mathcal{L}_K\}$. A player from subpopulation p adopts a learning scheme $\mathcal{L}_p$ and his set of pure actions is a finite set $\mathcal{A}^p$. The drift of the learning scheme $\mathcal{L}_p$ is denoted by $\tilde{F}^p$. i.e

$$\mathbf{x}_{t+1}^p = \mathbf{x}_t^p + \lambda_t \tilde{F}_t^p(\mathbf{x}_t^p, \hat{\mathbf{r}}_t^p), \ p \in \mathcal{P}.$$

Denote by F_t^p the drift limit of $\tilde{F}_t^p$. Therefore, the mixed actions evolution are given by

$$\dot{\mathbf{x}}_t^p = F_t^p(\mathbf{x}_t), \ p \in \mathcal{P}, \ \mathbf{x}_0^p \in \Delta(\mathcal{A}^p)$$

Here we do not restrict ourselves to pure actions. Our objective is to derive the mean field game dynamics, i.e., the evolution of the density of the sub-population playing mixed actions. The mean field dynamics for *mixed actions* are derived from the different learning dynamics. The density is defined by

$$m^p : \Delta(\mathcal{A}^p) \times \mathbb{R}_+ \longrightarrow [0, 1],$$

---

[2]Revision protocols in pure actions.

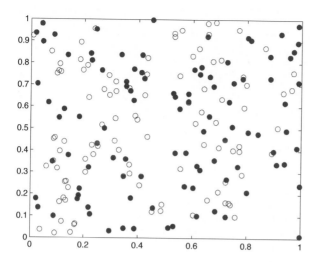

FIGURE 7.1
Large population of users.

$$\int_{\Delta(\mathcal{A}^p)} m_t^p(\mathbf{x}^p)) \; d\mu(\mathbf{x}^p) = 1, \; m_t^p(\mathbf{x}^p)) \geq 0, \; \forall(t,p),$$

where μ is a measure of the set of mixed actions. In the single subpopulation case (homogeneous population), the population of players is represented by a density function $m_t(\mathbf{x}) \; d\mu(\mathbf{x})$. The total mass of players using mixed strategies in an open set $O \subset int\Delta(A)$ is $\int_O m_t(\mathbf{x}) \; d\mu(\mathbf{x})$. The rate of changes of mass in O is

$$\frac{d}{dt} \int_O m_t(\mathbf{x}) \; d\mu(\mathbf{x}) = \int_O \frac{\partial}{\partial t} m_t(\mathbf{x}) \; d\mu(\mathbf{x}).$$

Let us examine the incoming flux and the outgoing flux of O across the boundary ∂O. The population learning process is represented by

$$\dot{\mathbf{x}}_t = F_t(\mathbf{x}_t).$$

The flux of mass out of O is given by the vector field $\langle F, \vec{n} \rangle m_t$ on ∂O ($\vec{n}$ normal of ∂O). The aggregate flux of mass into O across its boundary is

$$-\int_{\partial O} \langle F_t, \vec{n} \rangle m_t(\mathbf{x}) \; d\text{Area}$$

where *Area* is an elementary measure at ∂O. Recall that the *Divergence theorem* states:

$$\int_{\partial O} \langle F_t, \vec{n} \rangle m_t(\mathbf{x}) \; d\text{Area} = \int_O \text{div}[F_t(\mathbf{x}) m_t(\mathbf{x})] \; d\mu(\mathbf{x}).$$

Thus,

$$\int_O \frac{\partial}{\partial t} m_t(\mathbf{x}) \, d\mu(\mathbf{x}) = -\int_O \text{div}(F_t m_t) \, d\mu(\mathbf{x}),$$

i.e.,

$$\int_O \left[\frac{\partial}{\partial t} m_t(\mathbf{x}) + \text{div}(F_t m_t) \right] d\mu(\mathbf{x}) = 0.$$

Mean field dynamics for single population is

$$\frac{\partial}{\partial t} m_t(\mathbf{x}) + \text{div}(F_t(\mathbf{x}) m_t(\mathbf{x})) = 0, \qquad (7.23)$$

$$\dot{\mathbf{x}}_t = F(\mathbf{x}_t), \qquad (7.24)$$

$$\mathbf{x}_0 \in \Delta(A), m_0(.) \in \Delta(\Delta(A)). \qquad (7.25)$$

We now focus on the multiple subpopulations case. Let $m^p : \Delta(A^p) \times \mathbb{R}_+ \longrightarrow [0,1]$ be the time-dependent probability density function over the space of mixed strategies for subpopulation p. If B_p is a measurable set, then

$$m_t^p(B_p) = \int_{B_p} m_t^p(\mathbf{x}^p) \, d\mu(\mathbf{x}^p),$$

is the mass of players using their mixed action in B_p. Following the same reasoning, the mean strategy of subpopulation p at time t is then

$$\bar{\mathbf{x}}_t^p := \int_{\mathbf{y}^p \in \Delta(A^p)} \mathbf{y}^p m_t^p(\mathbf{y}^p) \, d\mu(\mathbf{y}^p).$$

Proposition 7.5.4.2. *The mean field game dynamics of CODIPAS-RL is given by*

$$\begin{cases} \frac{\partial}{\partial t} m_t^p(\mathbf{x}^p) + div[\Gamma_t^p(\mathbf{x}^p, m_t^{-p}) m_t^p(\mathbf{x}^p)] = 0 \\ \dot{\mathbf{x}}_t^p = F_t^p(\mathbf{x}_t^1, \ldots, \mathbf{x}_t^K) \end{cases} \quad \mathbf{x}_0^p \in \Delta(A^p), \ m_0^p \in \Delta(\Delta(A^p)),$$

where

$$\Gamma_t^p(\mathbf{x}^p, m_t^{-p}) = \int_{\mathbf{x}^{-p}} \sum_{(s^1, \ldots, s^K)} \left(\prod_{p=1}^K \mathbf{x}^p(s^p) \right) F_{t,s}^p(\mathbf{x}_t) \prod_{p' \neq p} m_t^{p'}(\mathbf{x}^{p'}) d\mu(\mathbf{x}^{p'})$$

and $m_t^{-p} = (m_t^{p'})_{p' \neq p}$.

The above mean field game dynamics is related to the transport equations [146, 145] also called continuity equations.

Proof. We provide a brief proof of Proposition 7.5.4.2. We show that the forward (respectively backward) equations for the mass of the $K-$subpopulation learning patterns generate a system of mean field game dynamics. To prove the result, we first recall the divergence theorem: For any test function ϕ,

$$\int_{X^p} div(\phi^p F^p) \, d\mu = \int_{\partial X^p} \langle \vec{n}_p, Z^p \rangle \phi^p \, d\mu,$$

where $\vec{n}_p$ the normal to $X^p = \Delta(A^p)$, Z^p is a vector field and ∂X^p is the boundary of X^p. We now establish the relation between the divergence and the gradient. By using the identity $div(\phi^p Z^p) = \phi^p div(Z^p) + \langle \nabla \phi^p, Z^p \rangle$ and taking the integral, one gets,

$$\int_{X^p} \langle \nabla \phi^p, Z^p \rangle \, d\mu = \int_{X^p} div(\phi^p Z^p) \, d\mu - \int_{X^p} \phi^p div(Z^p) \, d\mu$$

$$= \int_{\partial X^p} \langle \vec{n}_p, Z^p \rangle \phi^p \, d\mu - \int_{X^p} \phi^p div(Z^p) \, d\mu.$$

Since $\phi^p = 0$ over the boundary, one has,

$$\int_{X^p} \langle \nabla \phi^p, Z^p \rangle \, d\mu = - \int_{X^p} \phi^p div(Z^p) \, d\mu \tag{7.26}$$

$$\bar{\phi}^p_{t+\lambda_t}$$

$$= \sum_s \int_{(\mathbf{x}^p, \mathbf{x}^{-p})} \phi^p(\mathbf{x}^p) \left[\left(\prod_{p=1}^{K} \mathbf{x}^p(s^p) \right) m_t^p(.) d\mu(.) \right] (\mathbf{x}_{t+1}) \times$$

$$\prod_{p' \neq p} m^{p'}(\mathbf{x}^{p'}) d\mu(\mathbf{x}^{p'})$$

$$= \sum_s \int_{(\mathbf{x}^p, \mathbf{x}^{-p})} \phi^p(\mathbf{x}_t^p + \lambda_t F_{s,t}^p) \left(\prod_{p=1}^{K} \mathbf{x}^p(s^p) \right) m_t^p(\mathbf{x}^p) d\mu(\mathbf{x}^p) \times$$

$$\prod_{p' \neq p} m^{p'}(\mathbf{x}^{p'}) d\mu(\mathbf{x}^{p'})$$

We use a Taylor development of ϕ^p :

$$\phi^p(\mathbf{x}_t^p + \lambda_t F_{s,t}^p) = \phi^p(\mathbf{x}_t^p) + \lambda_t \langle \nabla \phi^p, F_{s,t}^p(\mathbf{x}_t) \rangle + o(\lambda_t).$$

Thus,

$$\bar{\phi}^p_{t+\lambda_t} :=$$

$$\sum_s \int_{(\mathbf{x}^p, \mathbf{x}^{-p})} \phi^p(\mathbf{x}_t^p) \left(\prod_{p=1}^{K} \mathbf{x}^p(s^p) \right) m_t^p(\mathbf{x}^p) d\mu(\mathbf{x}^p) \prod_{p' \neq p} m^{p'}(\mathbf{x}^{p'}) d\mu(\mathbf{x}^{p'})$$

$$+ \lambda_t \sum_s \int_{(\mathbf{x}^p, \mathbf{x}^{-p})} F_{s,t}^p \left(\prod_{p=1}^{K} \mathbf{x}^p(s^p) \right) m_t^p(\mathbf{x}^p) d\mu(\mathbf{x}^p) \prod_{p' \neq p} m^{p'}(\mathbf{x}^{p'}) d\mu(\mathbf{x}^{p'})$$

Hence, the following holds:

$$\int_{X^p} \phi^p(\mathbf{x}^p) \frac{1}{\lambda_t} \left\{ m_{t+\lambda_t}^p(\mathbf{x}^p) - m_t^p(\mathbf{x}^p) \right\} d\mu(\mathbf{x}^p)$$

$$= \sum_s \int_{\mathbf{x}^p} \int_{\mathbf{x}^{-p}} \langle \nabla \phi^p(\mathbf{x}^p), F_{s,t}^p(\mathbf{x}) \rangle \left(\prod_{p=1}^K \mathbf{x}^p(s^p) \right) \times$$

$$m_t^p(\mathbf{x}^p) d\mu(\mathbf{x}^p) \prod_{p' \neq p} m^{p'}(\mathbf{x}^{p'}) d\mu(\mathbf{x}^{p'})$$

$$= \int_{\mathbf{x}^p} \int_{\mathbf{x}^{-p}} \langle \nabla \phi^p(\mathbf{x}^p), \sum_s F_{s,t}^p(\mathbf{x}) \left(\prod_{p=1}^K \mathbf{x}^p(s^p) \right) \rangle m_t^p(\mathbf{x}^p) d\mu(\mathbf{x}^p) \times$$

$$\prod_{p' \neq p} m^{p'}(\mathbf{x}^{p'}) d\mu(\mathbf{x}^{p'})$$

$$= \int_{\mathbf{x}^p} \int_{\mathbf{x}^{-p}} \langle \nabla \phi^p(\mathbf{x}^p), \underline{F}_t^p(\mathbf{x}) \rangle m_t^p(\mathbf{x}^p) d\mu(\mathbf{x}^p) \times$$

$$\prod_{p' \neq p} m^{p'}(\mathbf{x}^{p'}) d\mu(\mathbf{x}^{p'})$$

where $\underline{F}_t^p(\mathbf{x}) = \sum_s F_{s,t}^p(\mathbf{x}) \left(\prod_{p=1}^K \mathbf{x}^p(s^p) \right)$. By taking the limit when $\lambda_t \longrightarrow 0$, one gets,

$$\int_{X^p} \phi^p(\mathbf{x}^p) \frac{\partial}{\partial t} m_t^p(\mathbf{x}^p) d\mu(\mathbf{x}^p)$$

$$= \int_{\mathbf{x}^p} \langle \nabla \phi^p(\mathbf{x}^p), \int_{\mathbf{x}^{-p}} \underline{F}_t^p(\mathbf{x}) m_t^p(\mathbf{x}^p) \prod_{p' \neq p} m^{p'}(\mathbf{x}^{p'}) d\mu(\mathbf{x}^{p'}) \rangle d\mu(\mathbf{x}^p).$$

We now use Equation (7.26), and gets,

$$\int_{X^p} \phi^p(\mathbf{x}^p) \frac{\partial}{\partial t} m_t^p(\mathbf{x}^p) d\mu(\mathbf{x}^p) = \int_{\mathbf{x}^p} \langle \nabla \phi^p(\mathbf{x}^p), \Gamma^p(\mathbf{x}^p, m_t^{-p}) m_t^p(\mathbf{x}^p) \rangle d\mu(\mathbf{x}^p)$$

$$= - \int_{\mathbf{x}^p} \phi^p(\mathbf{x}^p) div \left[\Gamma^p(\mathbf{x}^p, m_t^{-p}) m_t^p(\mathbf{x}^p) \right] d\mu(\mathbf{x}^p)$$

where $m_t^{-p} = (m_t^{p'})_{p' \neq p}$. We conclude that

$$\int_{X^p} \phi^p(\mathbf{x}^p) \left[\frac{\partial}{\partial t} m_t^p(\mathbf{x}^p) + div \left(\Gamma_t^p(\mathbf{x}^p, m_t^{-p}) m_t^p(\mathbf{x}^p) \right) \right] d\mu(\mathbf{x}^p) = 0$$

for any test function vanishing that the boundary ∂X^p. Hence,

$$\begin{cases} \frac{\partial}{\partial t} m_t^p(\mathbf{x}^p) + div[\Gamma_t^p(\mathbf{x}^p, m_t^{-p}) m_t^p(\mathbf{x}^p)] = 0 \\ \dot{\mathbf{x}}_t^p = F_t^p(\mathbf{x}_t^1, \ldots, \mathbf{x}_t^K), \end{cases} \quad \mathbf{x}_0^p \in \Delta(\mathcal{A}^p), \; m_0^p \in \Delta(\Delta(\mathcal{A}^p)),$$

This completes the proof. $\qquad\square$

The computation of $\Gamma^p_t(\mathbf{x}^p, m_t^{-p})$ for a revision protocol of mixed leading to the replicator dynamics such as (M-IBG) or the learning schemes developed in Borger-Sarin [32] or in Arthur [11] is easily obtained by incorporating the value of

$$F^p_{t,s}(\mathbf{x}_t) = \mathbf{x}^p_t(s^p) \left[\hat{\mathbf{r}}^p_t(s^p) - \sum_{s'^p \in \mathcal{A}^p} \hat{\mathbf{r}}^p_t(s'^p) \mathbf{x}^p_t(s'^p) \right]$$

For the two subpopulations case,

$$\Gamma^1_t(\mathbf{x}^1, m_t^2) = \int_{\mathbf{x}^2} \sum_{s^1, s^2} x^1(s^1) x^2(s^2) F^{(1)}_{s^1 s^2, t}(\mathbf{x}_t) m_t^2(\mathbf{x}^2) \, d\mu(\mathbf{x}^2)$$

$$= \sum_{s^1} x^1(s^1) \left[e_{s^1} R^1 \int \mathbf{x}^2 m_t^2(\mathbf{x}^2) d\mu(\mathbf{x}^2) - \langle \mathbf{x}^1, R^1 \int \mathbf{x}^2 m_t^2(\mathbf{x}^2) d\mu(\mathbf{x}^2) \rangle \right]$$

where R^1 is the payoff matrix of subpopulation 1.

7.5.5 Simulation of Population Games

In this subsection, we examine a simulation of mean field dynamics in large population game [153, 165].

Apart from the highly unrealistic assumptions regarding players capacity to compute a perfect best reply to a given population state there is also the drawback that the best-response dynamics may define a differential inclusion, i.e. a set-valued function. The best responses may not be unique and multiple trajectories can emerge from the same initial conditions. Here we consider finite population games and we want to understand the evolution of the population state (stochastic) by simulation. We consider the rock-scissor-paper (RSP) game in a population with 6000 players. The individual state of a player can be in the set $\{r, p, s\}$ i.e., the actions. The generic payoffs are given by the following matrix:

$$\begin{pmatrix} 0 & \delta & -\epsilon \\ -\epsilon & 0 & \delta \\ \delta & -\epsilon & 0 \end{pmatrix}$$

The transition probabilities between the states are given by the system state and the payoff functions which depend on the other players. Thus, the transition of an individual depends on the state of the other players. The rock beats scissor which beats paper which beats rock. The payoff of the winner is $+\delta$ and the payoff of the loser is $-\epsilon$. The transitions are payoff-dependent:

$$\mathcal{L}_{xx'}(m) = m_{x'} \max(0, r_{x'}(m) - r_x(m)).$$

We distinguish between the good RSP and the bad RSP (opposite sign in the payoff difference). The following figures are simulations of the mean field process. We fix the parameters $\delta = \epsilon = 1$. The evolutionary RSP game has a unique equilibrium $m^* = \frac{1}{3}(1, 1, 1)$, which is unstable as illustrated in the figures 7.2, 7.3. As we can see in both figures there is a limit cycle.

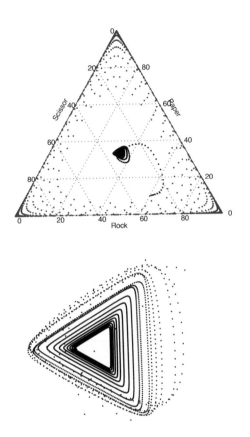

FIGURE 7.2
Bad RSP: Zoom around the stationary point.

7.6 Mobility-Based Learning in Cognitive Radio Networks

Wireless technology is proliferating rapidly, and the vision of pervasive wireless computing and communications offers the promise of many industrial and individual benefits. While consumer devices such as cell phones, personal data assistants (PDAs), and laptops receive a lot of attention, the impact of wireless technology is much broader, e.g., through sensor networks for safety applications, vehicular networks, smart grid control, home automation, medical wearable and embedded wireless devices, and entertainment systems. This explosion of wireless applications creates an ever-increasing demand for more radio spectrum. However, most easily usable spectrum bands have been allo-

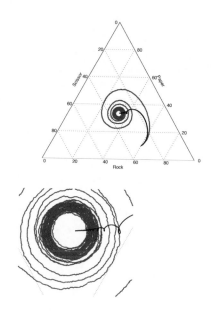

FIGURE 7.3
Mean field simulation of good RSP ternary plot and zoom.

cated, although many measurements, nevertheless, show that at any specific time and location, these spectrum bands are significantly *underutilized* and a lot of energy is wasted.

This manifests itself through voids in either energy, time, space or spectrum.

Cognitive radios offer the promise of being a disruptive technology innovation that will enable the future wireless world. Cognitive radios are fully programmable wireless devices that can *sense, learn* their environment and dynamically adapt their transmission waveform, energy management policy, channel access method, spectrum use, and networking protocols as needed for good network and application performance.

Cognitive radio aims at exploiting these voids in order to accommodate extra radio devices referred to as *secondary users* (SUs) in a network whose first function is to serve the primary users (PUs), e.g. licensed to use the spectrum. The key requirement is that the primary users have to be as little affected as possible by the presence of secondary users.

The protocols are a carrier-sensing-based mechanism for secondary users to exploit the spectrum left over by the primary user. Each primary transmitter has a dynamic protection zone. If a user located in this zone transmits at the same time, he interferes with the primary user. Any user located in this protection zone is hence called a contender of this transmitter. In the general setting, the union of all primary transmitter protection zones does not cover the whole space. Thus, one can accommodate secondary users in the remaining space to better utilize the spatial resource under energy constraints.

In this section, we study learning, mobility effect, and power saving in cognitive radio networks (see also Example 2.3.3.3). We consider two types of users: primary users and secondary users. When active, each secondary transmitter-receiver uses carrier sensing and is subject to a long-term energy constraint. We formulate the interaction between primary and large number of secondary users as a hierarchical mean field difference game. In contrast to the classical large-scale approaches based on stochastic geometry, percolation theory, and large random matrices, the proposed learning-based mean field framework allows one to describe the evolution of the density distribution and the associated performance metrics using coupled partial differential equations. We provide explicit formulas and learning scheme based power management for both primary and secondary users. A complete characterization of the optimal distribution of energy and probability of success is given.

The present section concentrates on exploiting mobility and energy consumption [88] by considering cognitive radio in a wireless network with users distributed in the Euclidean plane. In order to save energy it is important to control the transmission power and the sensing threshold of secondary users in order to further limit the interference they cause to primary users. The users are energy limited and each secondary user decides on their transmission power depending on the primary signal strength and its remaining battery state.

Different mobility models (Random Walk, Levy Flight, Random Waypoint) of wireless nodes and human mobility have been studied. However, most of the studies in the wireless literature are static and stationary in the sense they assume

- a constant power scheme,

- a fixed intensity of interaction,

- independence of the virtual received powers, etc.

As a consequence, the dynamic nature of the network is not captured by the "snapshot" network models.

Below, we relax some of these stationarity assumptions. We introduce battery-state power management in the large-scale network. The transmitted power is to be adapted in response to the localized interference, state, and powers of the primary users, the channel gains and the remaining energy for the future. Due to these additional assumptions, the wireless network is *dynamic and interactive*. For example, the evolution of the optimal energy management (we put the power $p = 0$ for user in sleep mode or silent mode) will depend on the optimal trajectory of the density distribution of battery state as well as the channel gain and spatial distribution of the user at the current time.

7.6.1 Proposed Cognitive Network Model

The mean field approach provides a natural way of defining and computing macroscopic properties of dynamic wireless networks, by averaging over all spatial-temporal patterns for the nodes. Modeling wireless networks in terms of *mean field interaction* seems particularly relevant for large scale interacting networks. The idea consists in treating such a network as a collection of interacting nodes characterized by dynamic random processes in the whole Euclidean plane or space and analyzing it in a stochastic optimizations, strategic decision-making and probabilistic way. In particular, the channel gains and the locations of the network elements are seen as the realizations of some random processes. A key feature of this approach is the asymptotic indistinguishability per class property, that is,

> *Under the power control strategies, asymptotically, the law of any collection of the states become invariant by permutation of the index of the users.*

When the underlying random model satisfies the asymptotic indistinguishability per class property, the stochastic control and large deviation analysis provide a way of estimating the spatial mean field, which often captures the key dependencies of the network performance characteristics (access probability, connectivity, probability of success, capacity, energy-efficiency etc.) as functions of a relatively small number of parameters. Typically, these are the densities of the underlying controlled random processes.

One of the main advantage of the mean field game approach in wireless network (compared to classical approaches in large-scale networks such as *stochastic geometry, percolation theory, large random matrices, etc*) is that it describes not only the optimal power management but also the evolution of density of users and the evolution of SINR in time as solution of controlled partially differential equation (PDE).

However for CRN, the classical mean field game model is not sufficient. We need to extend the mean field game framework to *hierarchical interacting mean field system* in order to accommodate CRN.

On the Validity of the Proposed Mean Field Model

In wireless networks, communication between users transmission of data required a certain amount of time (mini-slot). In order to take this to consideration, we propose a *difference game model* (discrete time game) instead of a differential game model. By doing this, the power allocation decisions are made per time unit (or frame). When the mini-slot length is sufficiently small, one gets a differential game between primary and secondary users.

We analyze different performance metrics: access probability, coverage probability, energy of remaining energy per area, outage probability, information rate, etc.

> • We compute the *access probability* of users. The density distribution of the underlying processes is obtained by the solutions of Fokker-Planck-Kolmogorov equation.
> • The *coverage probability* of user for a given receiver position is defined as the probability of the signal-to-interference-plus-noise ratio to be greater than a certain threshold. This is a key performance metric since it is involved in the throughput, outage probability, and queue dynamics. The evolution of the outage probability is completely characterized by a PDE.
> • The evolution of the remaining energy distribution is provided.

7.6.2 Cognitive Radio Network Model

We start with a very basic model that we refer to as *model 1*. We consider two types of transmitter-receiver pairs (users): *primary* and *secondary users* distributed over a domain $\mathcal{D}$ in Euclidean space. The physical layer setting considers a signal model with one primary user (a transmitter-receiver pair) and n secondary users. Interference is treated as noise, and the transmission is successful if the signal-to-interference-plus-noise ratio (SINR) is larger than a certain threshold (constant) and the access conditions are met.

For the MAC layer setting, we use a carrier sense SINR-based model. When a user is active, he senses the medium to check whether there is any primary contender. The tagged secondary user transmits only if he sees that the network free of primary contenders. To do so, a user has to choose a transmit power depending on the state of the battery (if no energy is available, the user cannot transmit). Whenever he is active, a primary user can access the channel. Then, a primary user has to design his transmit power. For secondary users, because of interference, not every transmission attempt is successful. Figure 7.4 represents a typical CRN scenario. Black nodes are active SUs, white nodes are inactive SUs, and the plus sign is a PU.

7.6.2.1 Mobility of Users

Let $\delta_n > 0$ be the duration of the time slot. The mobility of the PU is described by

$$\tilde{x}_{PU,k+1} = \tilde{x}_{PU,k} + \tilde{v}_{PU}(\tilde{x}_{PU,k+1})\delta_n + \sigma_{PU}\left(\tilde{\mathcal{B}}_{k+1} - \tilde{\mathcal{B}}_k\right), \qquad (7.27)$$

where $\tilde{\mathcal{B}}_k$ is a standard Brownian motion, where $\tilde{x}_{PU,k} \in \mathcal{D}$ for all $k \in \mathbb{Z}_+$ and σ_{PU} is positive constant, $v_{PU}(.)$ being a velocity function which is assumed to

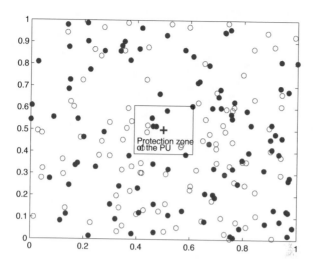

FIGURE 7.4
Typical cognitive radio scenario under consideration.

be a smooth and uniformly Lipschitz function. When $\delta_n \longrightarrow 0$, the mobility of the primary user is described by

$$dx_{PU,t} = v_{PU}(x_{PU,t})dt + \sigma_{PU}d\mathcal{B}_t \tag{7.28}$$

where $x_{PU,t} \in \mathcal{D}$ for all $t \in \mathcal{T}$ and σ_{PU} is positive constant, and the mobility of jth SU by

$$dx_{j,t} = v_{SU}(x_{j,t}, x_{PU,t})dt + \sigma_{SU}d\mathcal{B}_{j,t}, \tag{7.29}$$

where $x_{j,t} \in \mathcal{D}$ for $j \in \mathcal{N}_{SU}$ and $t \in \mathcal{T}$ and σ_{SU} is a positive constant, $v_{SU}(.)$ is a velocity function which we assumed to be a smooth and uniformly Lipschitz function.

The processes $\mathcal{B}_{j,t}$ are standard Brownian motion (also called Wiener process) in $\mathbb{R}$, independent across the users and across time and defined over a measure space $(\Omega, \mathcal{F}, \mathbb{P})$. We assume that for any user j, $\mathcal{B}_{j,0}$ is independent of the initial distribution of users. The mobility process in (7.29) is understood as a classical Ito's stochastic differential equation (SDE). These conditions on the coefficients ensures that the trajectory of the SDE is well-posed for a given initial point. An example of generic Brownian mobility is illustrated in Figure 7.5.

When $\delta_n \longrightarrow 0$, the density of the above mobility model can be characterized by a partial differential equation (PDE) known as Fokker-Planck-Kolmogorov equation (FPK):

$$\partial_t \nu_t(x) + \partial_x(v_{SU}(x)\nu_t(x)) - \frac{\sigma_{SU}^2}{2}\partial_{xx}^2\nu_t(x) = 0. \tag{7.30}$$

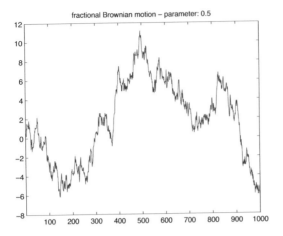

FIGURE 7.5
A generic Brownian mobility.

Recall that a Gaussian with mean $\bar{m}$ and variance σ_{SU}^2 has a density given by $\nu_0(x) = \frac{1}{\sqrt{2\pi\sigma_{SU}^2}} e^{-\frac{(x-\bar{m})^2}{\sigma_{SU}^2}}$. For $\nu_{SU} = 0$, the fundamental solution of (7.30) is explicitly given by a Gaussian density with (time-varying) variance $\sigma_{SU}^2 t$

$$\nu_t(x) = \frac{1}{\sqrt{2\pi\sigma_{SU}^2 t}} \exp\left(-\frac{x^2}{2\sigma_{SU}^2 t}\right).$$

We are also interested in optimizing the energy (battery) usage of all the users, from a initial value of $E_0 > 0$. We consider a model with renewable energy (e.g., solar powered devices). In that case, a stochastic model is well justified.

7.6.3 Power Consumption

The battery state of each user is modeled as a stochastic process that depends on the power consumption. A generic player starts initially with energy $E_0 > 0$ (full battery level).

$$\tilde{E}_{j,k+1} = \tilde{E}_{j,k} - p_{j,k}\delta_n + \sigma_e\left(\tilde{\mathcal{B}}_{k+1} - \tilde{\mathcal{B}}_k\right), \ j \in \mathcal{N}. \tag{7.31}$$

The energy $\tilde{E}_{j,k}$ is subject to nonnegativity constraint. Similarly, when $\delta_n \longrightarrow 0$, the energy usage dynamics is described by Ito's SDE

$$dE_{j,t} = -p_{j,t}dt + \sigma_e d\mathcal{B}_{j,t}, \ j \in \mathcal{N}. \tag{7.32}$$

with constraint $E_{j,t} \geq 0$, for all $t \in \mathcal{T}$, where $p_{j,t} \geq 0$ is the power consumption of user j at time t and $\sigma_e > 0$ is constant. The presence of noise term is to capture the stochastic nature of the anticipated prediction and the fact that under renewable energy, the battery may be recharged if the weather conditions allow. Moreover, $p_{j,t} = 0$ if $E_{j,t} \leq \epsilon_E$, where ϵ_E is a fixed minimum energy threshold for the device to be usable. We will provide the distribution of energy in Subsection 7.6.6 using Fokker-Planck-Kolmogorov equation. A typical trajectory of energy consumption with small noise is depicted in Figure 7.6.

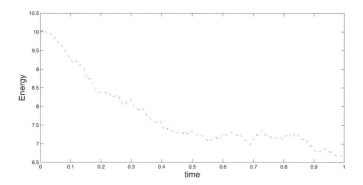

FIGURE 7.6
Evolution of remaining energy.

The variance of (7.32) becomes very big when $t \longrightarrow +\infty$ while the game stopped at T. To control this, we modify the dynamics (7.32) to be

$$dE_{j,t} = -p_{j,t}dt + \sigma_e E_{j,t}d\mathcal{B}_{j,t}, \quad j \in \mathcal{N}. \tag{7.33}$$

Note that we can restrict the instantaneous $p_{j,t}$ in a certain interval $[0, p_{\max}]$ or a remaining-energy dependent constraint $p_{j,t} \leq \delta E_{j,t}$. Then, all the power solutions provided below need to be projected in these sets.

For the case where a user can transmit to multiple receivers Rx, the remaining energy state will be written as

$$\tilde{E}_{j,k+1} = \tilde{E}_{j,k} - \delta_n \left[\sum_{r \in Rx(k)} p_{jr,k} \right] + \sigma_e \tilde{E}_{j,k} \left(\tilde{\mathcal{B}}_{k+1} - \tilde{\mathcal{B}}_k \right),$$

where is the $p_{jr,k}$ is the power allocated to receiver r, and the limiting energy dynamics is

$$dE_{j,t} = - \left[\sum_{r \in Rx(t)} p_{jr,t} \right] dt + \sigma_e E_{j,t} \, d\mathcal{B}_{j,t} \tag{7.34}$$

7.6.4   Virtual Received Power

The virtual received power from a user i located at $x_{i,t}$ at a receiver located at $y_{j,t}$ is defined as

$$r_{ij,t} := \frac{p_{j,t} g_{i,j,t}}{l(\|x_{i,t} - y_{j,t}\|)} \tag{7.35}$$

for $(i,j) \in \mathcal{N}^2$ where $p_{j,t}$ is the transmit power, $g_{i,j,t}$ is the channel gain (positive random process), and $l(\|x_{i,t} - y_{j,t}\|)$ is such that

$$l(d) = c(d^2 + \epsilon^2)^{\alpha/2}$$

where $\alpha \geq 2$ is the pathloss exponent, $c > 0$. The term ϵ is the component of a hidden constant dimension, that is, every receiver location has components $(y_{j,t}^1, y_{j,t}^2, \epsilon)$ and every sender has components in the form $(x_{j,t}^1, x_{j,t}^2, 0)$. Then, we avoid a vanishing denominator.

Another function l could be

$$l_2(d) = [\max(d, d_0)]^{\alpha/2}, \ d_0 > 0$$

We define the state of the user i at time t as $s_{i,t} := (E_{i,t}, x_{i,t}, g_{i,\cdot,t}) \in \mathcal{S} = \mathbb{R}_+ \times \mathcal{D} \times \mathbb{R}_+$.

7.6.5   Scaled SINR

We now define the *scaled signal to interference plus noise ratio (sSINR)* from SU i to receiver j at time t as

$$sSINR_{i,j,t} := \frac{r_{i,j,t}}{N_0(y_{j,t}) + \frac{1}{n-1}\sum_{i' \neq i} r_{i',j,t} + r_{PU,j,t}} \tag{7.36}$$

for $(i,j) \in \mathcal{N}_{SU}^2$ where $N_0 : \mathcal{D} \longrightarrow \mathbb{R}_+$, $N_0(y_{j,t}) > 0$ is the ambient noise at the receiver location and $r_{PU,j,t}$ is the interference generated by the primary user to the receiver located at $y_{j,t}$. The term $\frac{1}{n-1}$ in the denominator is to take into consideration the number of other SUs than i.

A transmission is successful from SU i to receiver j if

1. $sSINR_{i,j,t} \geq \beta_{SU}$, where β_{SU} is a fixed minimum quality of the signal, and

2. $r_{i,PU,t} \leq \gamma$, where γ is a fixed quantity, that is, the primary user should not get more than certain amount of noise from the transmission.

The sSINR from primary user to a receiver j is defined as

$$sSINR_{PU,j,t} := \frac{r_{PU,j,t}}{N_0(y_{j,t}) + \frac{1}{n}\sum_{i'=1}^{n} r_{i',j,t}} \tag{7.37}$$

where the factor $\frac{1}{n}$ averages with the total number of SUs. Note that the scaled interference form is widely used in Code Division Multiple Access (CDMA) networks.

We say that a secondary user i belongs to the primary user's protection zone if the virtual received power is greater than some threshold $\gamma > 0$. The access condition in presence of PU is then $A_{i,t} = 1$ where

$$A_{i,t} = \mathbb{1}_{\left\{\frac{p_{i,t}g_i,PU,t}{l(\|x_{i,t}-y_{PU,t}\|)}\leq\gamma\right\}}.$$

7.6.6 Asymptotics

First observe that the only quantity in $sSINR_{PU,j,t}$ (7.37) which takes importance in n is $\frac{1}{n}\sum_{i'=1}^{n}r_{i',j,t}$. Considering indistinguishable SU nodes, we can model the system with a mean field interaction.

The scaled interference from SUs to a receiver j located at $y_{j,t} = y$ at time t is

$$\frac{1}{n}\sum_{i=1}^{n}r_{i,j,t} = \frac{1}{n}\sum_{i=1}^{n}\frac{p_{j,t}g_{i,j,t}}{l(\|x_{i,t}-y\|)} \tag{7.38}$$

$$= \int_{\mathcal{S}}\frac{\bar{p}_{SU,t}(s,y,t)\bar{g}(x,y,t)}{l(\|x-y\|)}M_t^n(ds) \tag{7.39}$$

$$\xrightarrow{n\to+\infty} \int_{\mathcal{S}}K_t(s,y)m_t(ds) \tag{7.40}$$

where $K_t : \mathcal{S} \times \mathcal{D} \to \mathbb{R}_+$ is the kernel $\bar{p}.\bar{g}/l$, $M_t^n(\cdot,x,\cdot) = \frac{1}{n}\sum_{i=1}^{n}\delta_{x-x_{i,t}}$ is the occupancy measure of the n SUs. $\bar{p}_{SU}$ is the control for SUs at time t, $\bar{p}_{SU} : \mathcal{S} \times \mathcal{D} \times \mathcal{T} \to \mathbb{R}_+$. $\bar{g}_t$ is a positive random process, defined by $\bar{g} = |h|^2$.

Hence, the following hold:

Lemma 7.6.6.1. *Under the above assumptions, the scaled interference from the SUs converges weakly to*

$$\bar{I}_t(y) := \int_{\mathcal{S}}K_t(s,y)m_t(ds)$$

The dynamics of h_t are given by

$$dh_t = \sigma_h d\mathcal{B}_t \tag{7.41}$$

and using Itô's rule we get

$$d\bar{g}_t = \frac{2\sigma_h^2}{2}dt + \sigma_h\partial_h\bar{g}_t d\mathcal{B}_t = \sigma_h^2 dt + 2\sigma_h h_t d\mathcal{B}_t \tag{7.42}$$

Then, the dynamics of s_t are described by

$$ds_t = \begin{pmatrix} -p_t \\ v_t \\ \sigma_h^2 \end{pmatrix} dt + \begin{pmatrix} \sigma_{PU} & 0 & 0 \\ 0 & \sigma_E & 0 \\ 0 & 0 & 2\sigma_G h_t \end{pmatrix} d\mathcal{B}_t \tag{7.43}$$

$$= fdt + \sigma d\mathcal{B}_t \tag{7.44}$$

Remark 7.6.7. *Note that the variance of h_t goes to infinite with time, and this can be interpreted as the channel becoming better and better. In order to avoid that, we can use the following Ornstein-Uhlenbeck dynamics*

$$dh_t = \frac{1}{2}(\mu_h - h_t)dt + \sigma_h dB_t \qquad (7.45)$$

Then, the stationary distribution of h_t is Gaussian with mean μ_h and variance σ_h^2. In this case,

$$dg_t = (0 + 2h_t \frac{1}{2}(\mu_h - h_t) + \frac{\sigma_h^2}{2} 2)dt + \sigma_h 2 h_t dB_t \qquad (7.46)$$

$$= h_t(\mu_h - h_t)dt + 2\sigma_h h_t dB_t \qquad (7.47)$$

The probability density function $m_h(h,t)$ of the Ornstein-Uhlenbeck process satisfies the Fokker-Planck-Kolmogorov equation

$$\frac{\partial m_h}{\partial t} = \frac{1}{2}\frac{\partial}{\partial h}[(h - \mu_h)m_h] + \frac{\sigma^2}{2}\frac{\partial^2 m_h}{\partial h^2}$$

The solution of this equation, taking $\mu_h = 0$ for simplicity, and the initial condition ξ_0 is,

$$m_h(h,t) = \sqrt{\frac{1}{2\pi\sigma^2(1 - e^{-t})}} \exp\left\{\frac{-1}{2\sigma^2}\left[\frac{(h - \xi_0 e^{-\frac{1}{2}t})^2}{1 - e^{-t}}\right]\right\}$$

The stationary solution of this equation is the limit for time tending to infinity, which is a Gaussian distribution with mean μ_h and variance σ^2,

$$m_{h,st}(h) = \sqrt{\frac{1}{2\pi\sigma^2}}\, e^{-\frac{1}{2\sigma^2}(h - \mu_h)^2}.$$

Following similar lines the explicit distribution of a solution of an Ornstein-Uhlenbeck process

$$dx_t = \frac{\beta}{2}(\mu - x_t)dt + \sigma dB_t,$$

is given by

$$m_t(x|x_0) = \frac{1}{\sqrt{\frac{2\pi}{\beta}(1 - e^{-\beta t})}} e^{-\frac{\left(x - \mu + (x_0 - \mu)e^{-\frac{\beta}{2}t}\right)^2}{\frac{\sigma^2}{\beta}(1 - e^{-\beta t})}}$$

Another channel model that takes into consideration finite variance is the log-normal channel. The well-known log-normal channel can be modeled as $g_t = e^{\eta_t}$, where η_t is solution of a linear SDE.

Now, we define the continuous virtual received power from user j by a receiver located at y,

$$\bar{r} : \mathcal{S} \times \mathcal{D} \times \mathcal{T} \to \mathbb{R}_+$$

$$\bar{r}_.(s_j, y, t) := \frac{\bar{p}_.(s_j, y, t)\bar{g}(x_j, y)}{l(\|x_j - y\|)} \tag{7.48}$$

where $\cdot$ can be PU or SU, and x_j is the location of user j.

Note that if $\bar{p}_{.,t}$ is integrable w.r.t. the measure m_t, then $\bar{r}_{.,t}$ is integrable and the resulting integral is the scaled interference $\bar{I}_{SU} : \mathcal{D} \times \mathcal{T} \to \mathbb{R}$

$$\bar{I}_{SU}(y, t) = \int_{s \in \mathcal{S}} \bar{r}_{SU}(s, y, t)m_t(ds) \tag{7.49}$$

and for PU,

$$\bar{I}_{PU}(y, t) = \int_{s \in \mathcal{S}} \bar{r}_{PU,t}(s, y, t)\delta_{s_{PU}}(ds) \tag{7.50}$$

Now we define the *mean SINR, (mSINR)* for PU and for SUs, given a receiver at location y

$$mSINR_{PU}(y, t) := \frac{\bar{r}_{PU}(s_{PU}, y, t)}{N_0(y) + \bar{I}_{SU}(y, t)} \tag{7.51}$$

$$mSINR_x(y, t) := \frac{\bar{r}_{SU}(s_x, y, t)}{N_0(y) + \bar{I}_{SU}(y, t) + \bar{I}_{PU}(y, t)} \tag{7.52}$$

for $x \in \mathcal{D}$.

7.6.8 Performance of a Generic User

7.6.8.1 Access Probability

Given the position of the primary receiver at y_{PU}, and that of the i-th secondary user located at $x_t = x$ at time t, the *access probability* of the latter at time t is

$$\mathbb{P}(A_{i,t} = 1) = \mathbb{E}(A_{i,t}) \tag{7.53}$$

$$= \mathbb{P}\left(\frac{\bar{p}(s_{SU}, y_{PU}, t)\bar{g}(x, y_{PU}, t)}{l(\|x - y_{PU,t}\|)} \leq \gamma\right) \tag{7.54}$$

Lemma 7.6.8.2. *The access probability of PU at time t for a given power $\bar{p} > 0$ is*

$$\mathbb{P}\left(\frac{\bar{g}(x, y_{PU}, t)}{l(\|x - y_{PU,t}\|)} \leq \frac{\gamma}{\bar{p}}\right) = F_{Z_t}(\frac{\gamma}{\bar{p}})$$

where the random variable $Z_t = \frac{\bar{g}(x, y_{PU}, t)}{l(\|x - y_{PU,t}\|)}$, and F_{Z_t} is the cumulative distribution of Z_t which is the ratio distribution.

Proof. Given $\bar{p} > 0$, we write $A_{i,t} = 1$, which is $\frac{\bar{g}(x, y_{PU}, t)}{l(\|x - y_{PU,t}\|)} \leq \frac{\gamma}{\bar{p}}$, which gives the cumulative function of Z_t at the point $\frac{\gamma}{\bar{p}}$. $\square$

In order to compute F_{Z_t} we use the ratio distribution formula. Recall that the distribution of the ratio between two random variables X_t and Y_t with continuous density is given by

$$\bar{f}_{Z_t}(z) = \bar{f}_{X_t/Y_t}(z) = \int_{\mathbb{R}} |y| \, \bar{f}_{(X_t, Y_t)}(zy, y) \, dy$$

Exercise 7.1. *Find the Jacobian of the mappings*

$$g_1 : (a, b) \longmapsto (a/b, b)$$

$$g_2 : (a, b) \longmapsto (ab, b).$$

Find the determinant of the Jacobians.

The ratio distribution of two independent random Gaussian variables with parameters $\mathcal{N}(\mu'_x, \sigma'^2_x)$, $\mathcal{N}(\mu'_y, \sigma'^2_y)$, is a well-known result and is explicitly given by

$$p_Z(z) = \frac{b'(z) \cdot c'(z)}{a'^3(z)} \frac{1}{\sqrt{2\pi}\sigma'_x\sigma'_y} \left[2\Phi\left(\frac{b'(z)}{a'(z)}\right) - 1 \right]$$

$$+ \frac{1}{a'^2(z) \cdot \pi\sigma'_x\sigma'_y} e^{-\frac{1}{2}\left(\frac{\mu'^2_x}{\sigma'^2_x} + \frac{\mu'^2_y}{\sigma'^2_y}\right)} \tag{7.55}$$

where

$$a'(z) = \sqrt{\frac{1}{\sigma'^2_x}z^2 + \frac{1}{\sigma'^2_y}} \tag{7.56}$$

$$b'(z) = \frac{\mu'_x}{\sigma'^2_x}z + \frac{\mu'_y}{\sigma'^2_y} \tag{7.57}$$

$$c'(z) = e^{\frac{1}{2}\frac{b'^2(z)}{a'^2(z)} - \frac{1}{2}\left(\frac{\mu'^2_x}{\sigma'^2_x} + \frac{\mu'^2_y}{\sigma'^2_y}\right)} \tag{7.58}$$

$$\Phi(z) = \int_{-\infty}^{z} \frac{1}{\sqrt{2\pi}} e^{-\frac{1}{2}u^2} \, du \tag{7.59}$$

Note that the denominator of Z_t may not be Gaussian due the path-loss exponent. However, using Ito's rule in the small noise regime, one can approximate with time-varying Gaussian processes.

7.6.8.3 Coverage Probability

The coverage probability of a user i who transmits to receiver at $y_{i,t}$ at time t, is the probability that the $mSINR_{i,j,t}$ of the tagged user is larger than β_{SU}. Given the location of the primary receiver, the coverage probability of the primary transmission at time t is the probability that the $mSINR_{PU,PU,t}$ of the tagged user is larger than β_{PU}.

Lemma 7.6.8.4. *The coverage probability of PU at time t with a receiver located y is*

$$C_{PU}(y,t)=1 - F_{Z_t}\left[\frac{\beta_{PU}}{\bar{p}}\left(N_0(y) + \bar{I}_{SU}(y,t)\right)\right]$$

Hence, the density of coverage of PU at time t in a protection zone $\mathcal{A}'$ is

$$\int_{y\in\mathcal{A}'} C_{PU}(y,t)m_{x,t}(dy).$$

Proof. Let $\bar{p} > 0$ be a transmit power of PU. Using the expression of sSINR for PU, the measure of the event $\{mSINR_{PU,PU,t} \geq \beta_{PU}\}$ is equivalent to the measure event

$$\frac{\bar{g}(x_{PU}, y_{PU}, t)}{l(\|x_{PU} - y_{PU,t}\|)} \geq \frac{\beta_{PU}}{\bar{p}}\left(N_0(y) + \bar{I}_{SU}(y,t)\right).$$

The term $\bar{I}_{SU}(y,t)$ is the expectation of the mathematical expectation of a random variable. Thus, the right-hand side term $\frac{\beta_{PU}}{\bar{p}}\left(N_0(y) + \bar{I}_{SU}(y,t)\right)$ is nonrandom. Hence,

$$
\begin{aligned}
C_{PU}(y,t) &= \mathbb{P}\left(mSINR_{PU,PU,t} \geq \beta_{PU}\right) & (7.60)\\
&= 1 - F_{Z_t}\left[\frac{\beta_{PU}}{\bar{p}}\left(N_0(y) + \bar{I}_{SU}(y,t)\right)\right]. & (7.61)
\end{aligned}
$$

$\square$

When the realization of the state of PU and the receiver are given, a similar formula to the above can be used and the coverage probability is given by:
$$COP(s_{PU}, y, t) =$$

$$1 - F_{Z_t}\left[\frac{\beta_{SU}}{\bar{p}}\left(N_0(y) + \bar{I}_{SU}(y,t) + \frac{\bar{p}_{PU}g_{PU,y}}{l(\|x_{PU} - y\|)}\right)\right]$$

For SU, we cannot use the above reasoning because now the denominator is random due to the interference of the primary user (if active) even if $\bar{p}$ is deterministic. However, we can use the distribution of difference of random variables which that stated in the following lemma.

Lemma 7.6.8.5. *In the presence of PU, the coverage probability of a generic SU at a receiver located at y_{SU} is*

$$1 - F_{\bar{Z}_t}\big(\beta_{SU}\left(N_0(y_{SU}) + \bar{I}_{SU}(y_{SU}, t)\right)\big)$$

where $\bar{Z}_t$ is the random variable

$$\frac{\bar{p}_{SU,t}\bar{g}(x_{SU}, y_{SU}, t)}{l(\|x_{SU} - y_{SU,t}\|)} - \beta_{SU}\frac{\bar{p}_{PU,t}\bar{g}(x_{PU}, y_{SU}, t)}{l(\|x_{PU} - y_{SU,t}\|)}.$$

Proof. The coverage inequality event for fixed transmit power gives

$$\frac{\bar{p}_{SU,t}\bar{g}(x_{SU}, y_{SU}, t)}{l(\|x_{SU} - y_{SU,t}\|)} - \beta_{SU}\frac{\bar{p}_{PU,t}\bar{g}(x_{PU}, y_{SU}, t)}{l(\|x_{PU} - y_{SU,t}\|)}$$

$$\geq \beta_{SU}\left(N_0(y_{SU}) + \bar{I}_{SU}(y_{SU}, t)\right).$$

We then take the probability of that event by observing that the right hand side is deterministic. □

The probability density of $\bar{Z}_t$ is obtained by convolution of the two density functions.

Density of interference per area $\mathcal{A} \subset \mathcal{D}$ at time t as

$$d_{SU,t}(\mathcal{A}) = \int_{y \in \mathcal{A}} \bar{I}_{SU}(y, t)m_{x,t}(dy) \tag{7.62}$$

where $m_{x,t}$ is the marginal pdf w.r.t. x. This density allows us to quantify how much interference is around the protection zone of the PU.

Proposition 7.6.8.6. *Consider the SDE (7.32). The distribution of the remaining energy is a weak solution of the following Fokker-Planck-Kolmogorov forward equation:*

$$\partial_t m_{E,t}(\bar{E}) = \partial_{\bar{E}}\left[\bar{p}(.)m_{E,t}(\bar{E})\right] + \frac{\sigma_e^2}{2}\partial^2_{\bar{E}\bar{E}}m_{E,t}(\bar{E}), \tag{7.63}$$

$$m_{E,0}(\bar{E}) = \nu_0(\bar{E}). \tag{7.64}$$

For time-independent $\bar{p}$, the stationary solution is given by

$$m_{st}(\bar{E}) = -C_0 D \exp\left(-D\int_{[0,\bar{E}]}\bar{p}(\xi)d\xi\right) = C'_0 e^{-\Phi(\bar{E})} \tag{7.65}$$

where C_0 is a normalization constant and $D := 2/\sigma_e^2$ and Φ is the potential.

Proof. The statements follows from solving a linear second order ODE. □

Proposition 7.6.8.7. *Consider the SDE (7.33). The distribution of remaining energy is a weak solution of the following Fokker-Planck-Kolmogorov forward equation:*

$$\partial_t m_{E,t}(\bar{E}) = \partial_{\bar{E}}\left[\bar{p}(.)m_{E,t}(\bar{E})\right] + \frac{\sigma_e^2}{2}\partial^2_{\bar{E}\bar{E}}[\bar{E}^2 m_{E,t}(\bar{E})], \qquad (7.66)$$

$$m_{E,0}(\bar{E}) = \nu_0(\bar{E}). \qquad (7.67)$$

For time-independent $\bar{p}$, the stationary solution is given by

$$m_{st}(\bar{E}) = -C_1\frac{D}{\bar{E}^2}\exp\left(-D\int_{[0,\bar{E]}}\frac{\bar{p}(\xi)}{\xi^2}d\xi\right) = C_1'e^{-\Psi(\bar{E})} \qquad (7.68)$$

where C_1 is a normalization factor and $D := 2/\sigma_e^2$ and Ψ is the potential.

Proof. The statement follows from solving a linear second order ODE. $\qquad \square$

The distribution of the remaining energy at time in area $\mathcal{A} \subseteq \mathcal{D}$ is

$$\mu_{E,t}(\mathcal{A}) := \int_{x\in\mathcal{A}}\int_{g\in\mathbb{R}_+} m_t(E, dx, dg) \qquad (7.69)$$

and the distribution of users who have a remaining energy at least $\bar{E}$ is

$$\int_{[0,\bar{E]}} m_{E,t}(de) = F_{E,t}(\bar{E}) \qquad (7.70)$$

Proposition 7.6.8.8. *Consider the SDE (7.32). Then, the function μ : $(t,\bar{E}) \in \mathcal{T} \times \mathbb{R}_+ \longmapsto F_{E,t}(\bar{E})$ is a solution of the following partial differential equation:*

$$\partial_t\mu_t(\bar{E}) - \bar{p}(.,t)\partial_{\bar{E}}\mu_t(\bar{E}) = \frac{\sigma_e^2}{2}\partial^2_{\bar{E},\bar{E}}\mu_t(\bar{E}), \qquad (7.71)$$

$$\mu_0(\bar{E}) = F_0(\bar{E}). \qquad (7.72)$$

where $\bar{p}(.,t)$ is the control of secondary user, and F_0 is a given cumulative function. Similarly, under SDE (7.33), the cumulative distribution of energy is solution of the PDE

$$\partial_t\mu_t(\bar{E}) - \bar{p}(.,t)\partial_{\bar{E}}\mu_t(\bar{E}) = \frac{\sigma_e^2}{2}\partial_{\bar{E}}\left[\bar{E}^2\partial_{\bar{E}}\mu_t\right], \qquad (7.73)$$

$$\mu_0(\bar{E}) = F_0(\bar{E}). \qquad (7.74)$$

Moreover, for time-independent $\bar{p}$, the stationary solutions are given by

$$\mu_{st}(\bar{E}) = \int_{\xi\in[0,\bar{E]}} m_{st}(d\xi) \qquad (7.75)$$

respectively.

Proof. These statements follow by integration of the Fokker-Planck-Kolmogorov forward equation.

$\square$

The total throughput is the mean number of successful transmissions in the network per time slot

$$\frac{1}{T} \int_0^T \int_{(s,y) \in \mathcal{S} \times \mathcal{D}} \mathbb{P}(mSINR(s,y,t) \geq \beta) m_t(ds) m_{x,t}(dy) \, dt$$

Remark 7.6.9. *We have provided the generic analytical expressions of the above analysis for deterministic power of SUs and PU. However, during the interaction between the users correlation and randomness will be involved in these quantities. Thus, we need to examine these performance metrics under optimal behavior of the users. The optimization per user will be interdependent random processes between the powers, and hence independence cannot be assumed.*

7.7 Hybrid Strategic Learning

In this section, we focus on strategic learning for general sum stochastic dynamic games with incomplete information with the following novelties:

- The players do not need to follow the same learning patterns. We propose different learning schemes that the players can adopt. This leads to *heterogeneous learning* [175].

- Each player need not update its strategy at each iteration. The updating time are random and unknown to the players.

- Each player can be in active mode or in sleep mode. When a player is active, he can select among a set of learning patterns to update his strategies and/or estimations. The player can change their learning pattern during the interaction. This leads to *hybrid learning* [199].

- In contrast to the standard learning frameworks developed in the literature which are limited to finite and fixed number of players, we extend our methodology to large systems with multiple classes of populations. This allows us to address the curse of dimensionality problem when the size of the interacting system is very large.

> The players may want to switch between several learning schemes in order to find out the most suitable way for them to adapt their reactions in response to the dynamic, noisy, and interactive environment.

We recall the stochastic approximation theorem: Consider

$$\mathbf{x}_{t+1} = \mathbf{x}_t + \lambda_t(f(\mathbf{x}_t) + M_{t+1}),$$

in $\mathbb{R}^{\sum_i |\mathcal{A}_i|}$ and assume

- **A1** f Lipschitz.
- **A2** $\lambda_t \geq 0$, $\sum_{t \geq 0} \lambda_t = +\infty$, $\sum_{t \geq 0} \lambda_t^2 < \infty$.
- **A31** M_{t+1} is a martingale difference sequence with respect to the increasing family of sigma-fields $\mathcal{F}_t = \sigma(\mathbf{x}_{t'}, \hat{\mathbf{u}}_{t'}, M_{t'}, \ t' \leq t)$ i.e

$$\mathbb{E}\left(M_{t+1} \mid \mathcal{F}_t\right) = 0.$$

- **A32** M_t is square integrable and there is a constant $c > 0$, $\mathbb{E}\left(\| M_{t+1} \|^2 \mid \mathcal{F}_t\right) \leq c(1 + \| \mathbf{x}_t \|^2)$ almost surely, for all $t \geq 0$.
- **A4** $\sup_t \| \mathbf{x}_t \| < \infty$ almost surely.

Then, the asymptotic pseudo-trajectory is given by the ordinary differential equation (ODE)

$$\dot{\mathbf{x}}_t = f(\mathbf{x}_t), \ \mathbf{x}_0 \ \text{fixed}.$$

In Subsection 7.7.1, we focus on the non-zero-sum stochastic game with incomplete information, and present different learning patterns.

7.7.1 Learning in a Simple Dynamic Game

In this section, we formulate a $n-$player non-zero-sum stochastic game model

$$\mathcal{G} = \langle \mathcal{W}, \mathcal{A}_j, \{\tilde{R}_j(w, .)\}_{w \in \mathcal{W}} \rangle,$$

where $\mathcal{A}_j$ is the finite set of actions available to players Pj, respectively, and $\mathcal{W}$ is the set of possible states. We assume that the state space $\mathcal{S}$ and the probability transition on the states are both unknown to the players. A state $w \in \mathcal{W}$ is an independent and identically distributed random variable defined on the set $\mathcal{W}$. We assume that the action spaces are the same in each state, i.e., $\mathcal{A}_j(w) = \mathcal{A}_j$.

7.7.1.1 Learning Patterns

We examine a dynamic game in which each player learns according to a specific learning scheme. Let $\mathcal{L} = \{\mathcal{L}_1, \mathcal{L}_2, \mathcal{L}_3, \mathcal{L}_4, \mathcal{L}_5, \ldots, \mathcal{L}_K\}$ the set of pure learning patterns. At each time, each player can select a learning pattern in $\mathcal{L}$.

We now describe the different learning patterns. Let $r_{j,t}$ be the numerical value of payoff at time t and $\mathcal{B}^n(t)$ be the set of active players at time t.

- $\mathcal{L}_1$: The learning pattern $\mathcal{L}_1$ is given by

$$x_{j,t+1}(s_j) - x_{j,t}(s_j) = \lambda_t \mathbb{1}_{\{j \in \mathcal{B}^n(t)\}} \times$$

$$\frac{r_{j,t} - \Gamma_j}{\sup_{\mathbf{a},w} |R_j(w,a) - \Gamma_j|} \left(\mathbb{1}_{\{a_{j,t} = s_j\}} - x_{j,t}(s_j)\right), \qquad (7.76)$$

$$\hat{r}_{j,t+1}(s_j) - \hat{r}_{j,t}(s_j) = \nu_t \mathbb{1}_{\{a_{j,t} = s_j, j \in \mathcal{B}^n(t)\}} \left(r_{j,t} - \hat{r}_{j,t}(s_j)\right) \qquad (7.77)$$

where Γ_j is a reference level of j.

- $\mathcal{L}_2$

$$x_{j,t+1}(s_j) - x_{j,t}(s_j) = \lambda_t \mathbb{1}_{\{j \in \mathcal{B}^n(t)\}} \times$$

$$\left(\frac{e^{\frac{1}{\epsilon}\hat{r}_{j,t}(s_j)}}{\sum_{s'_j \in \mathcal{A}_j} e^{\frac{1}{\epsilon}\hat{r}_{j,t}(s'_j)}} - x_{j,t}(s_j) \right), \qquad (7.78)$$

$$\hat{r}_{j,t+1}(s_j) - \hat{r}_{j,t}(s_j) =$$
$$\nu_t \mathbb{1}_{\{a_{j,t}=s_j, j \in \mathcal{B}^n(t)\}} \left(r_{j,t} - \hat{r}_{j,t}(s_j) \right) \qquad (7.79)$$

- $\mathcal{L}_3$

$$x_{j,t+1}(s_j) - x_{j,t}(s_j) = \lambda_t \mathbb{1}_{\{j \in \mathcal{B}^n(t)\}} x_{j,t}(s_j) \times$$

$$\left(\frac{e^{\frac{1}{\epsilon}\hat{r}_{j,t}(s_j)}}{\sum_{s'_j \in \mathcal{A}_j} x_{j,t}(s'_j) e^{\frac{1}{\epsilon}\hat{r}_{j,t}(s'_j)}} - 1 \right), \qquad (7.80)$$

$$\hat{r}_{j,t+1}(s_j) - \hat{r}_{j,t}(s_j) =$$
$$\nu_t \mathbb{1}_{\{a_{j,t}=s_j, j \in \mathcal{B}^n(t)\}} \left(r_{j,t} - \hat{r}_{j,t}(s_j) \right) \qquad (7.81)$$

- $\mathcal{L}_4$

$$x_{j,t+1}(s_j) - x_{j,t}(s_j) = \mathbb{1}_{\{j \in \mathcal{B}^n(t)\}} x_{j,t}(s_j) \times$$

$$\left(\frac{(1+\lambda_t)^{\hat{r}_{j,t}(s_j)}}{\sum_{s'_j \in \mathcal{A}_j} x_{j,t}(s'_j)(1+\lambda_t)^{\hat{r}_{j,t}(s'_j)}} - 1 \right), \qquad (7.82)$$

$$\hat{r}_{j,t+1}(s_j) - \hat{r}_{j,t}(s_j) =$$
$$\nu_t \mathbb{1}_{\{a_{j,t}=s_j, j \in \mathcal{B}^n(t)\}} \left(r_{j,t} - \hat{r}_{j,t}(s_j) \right) \qquad (7.83)$$

- $\mathcal{L}_5$

$$x_{j,t+1}(s_j) - x_{j,t}(s_j) \in \mathbb{1}_{\{i \in \mathcal{B}^n(t)\}} \times$$

$$\left((1-\epsilon)\delta_{\arg\max_{s'_j} \hat{r}_{j,t}(s'_j)} + \epsilon \frac{\mathbb{1}}{|\mathcal{A}_j|} - x_{j,t}(s_j) \right), \qquad (7.84)$$

$$\hat{r}_{j,t+1}(s_j) - \hat{r}_{j,t}(s_j) = \nu_t \mathbb{1}_{\{a_{j,t}=s_j, j \in \mathcal{B}^n(t)\}} \left(r_{j,t} - \hat{r}_{j,t}(s_j) \right) \qquad (7.85)$$

7.7.1.2 Description of CODIPAS Patterns

The first equation of $\mathcal{L}_1$ is widely studied in machine learning and has been initially proposed by Bush and Mosteller in 1949-55 [42]. The second equation of $\mathcal{L}_1$ is a payoff estimation for the experimented action by the players. Combining the two equations, one gets a specific combined fully distributed payoff and strategy reinforcement learning based on Bush-Mosteller reinforcement learning.

The strategy learning (7.78) of $\mathcal{L}_2$ is a Boltzmann-Gibbs-based reinforcement learning. Note that the Boltzmann-Gibbs distribution can be obtained from the maximization of the perturbed payoff $R_j + \epsilon H$, where H is the entropy function. It is a smooth best response function.

The strategy learning (7.80) of $\mathcal{L}_3$ is an imitative Boltzmann-Gibbs-based reinforcement learning. The imitation here consists of playing an action with a probability proportional to the previous uses.

The strategy learning (7.82) of $\mathcal{L}_4$ is a learning-rate-weighted imitative reinforcement learning. The main difference with the $\mathcal{L}_2$ and $\mathcal{L}_3$ is that there is no parameter ϵ. The interior outcomes are necessarily exact equilibria of the expected (not approximated equilibria as in $\mathcal{L}_2$).

The last learning pattern $\mathcal{L}_5$ is a combined learning based on the weakened fictitious play. Here a player does not observe the action played by the other at the previous step and the payoff function is not known. Each player estimates its payoff function via the equations (7.85). Equation (7.84) represents a strategy which consists of playing one of the actions with the best estimation $\hat{r}_{j,t}$ with probability $(1 - \epsilon)$ and play an arbitrary action with probability ϵ.

7.7.1.3 Asymptotics of Pure Learning Schemes

Following the multiple time-scale stochastic approximation framework given in appendix, one can write the pure learning schemes in the form

$$\begin{cases} \mathbf{x}_{j,t+1} - \mathbf{x}_{j,t} \in q_{j,t}\left(f_j^{(l)}(\mathbf{x}_{j,t}, \hat{\mathbf{r}}_{j,t}) + M_{j,t+1}^{(l)} \right) \\ \hat{\mathbf{r}}_{j,t+1} - \hat{\mathbf{r}}_{j,t} \in \bar{q}_{j,t}\left(\mathbb{E}_{s,\mathbf{x}_{-j,t},\mathcal{B}^n} R_j - \hat{\mathbf{r}}_{j,t} + \bar{M}_{j,t+1} \right) \end{cases}$$

where $l \in \mathcal{L}$, $q_{j,t}$ is a time-scaling factor function of the learning rates λ_t and the probability activity of player j at time t, $\mathbb{P}(j \in \mathcal{B}^n(t))$, and $\bar{q}_{j,t}$ is the analogous for $\hat{r}_{j,t}$. To establish ODE approximation, we check the assumptions given in appendix.

The term $M_{j,t+1}^{(l)}$ is a bounded martingale difference because the trajectory (generated by the algorithm) belong almost surely in the product of simplice which is non-empty, convex and compact, and the conditional expectation of $M_{j,t+1}$, given the sigma-algebra generated by the random variables $s_{t'}, \mathbf{x}_{t'}, r_{t'}, \hat{\mathbf{r}}_{t'}, \; t' \leq t$, is zero. Similar properties holds for $\bar{M}_{t+1}$. The function f is a regular function (thus Lipschitz over a compact set, which implies linear growth). The parameter λ and ν_t are in $l^2 \backslash l^1$. Note that the case of constant learning rates can be analyzed under the same setting but the convergence result is weaker (convergence in law instead of almost sure convergence). Thus, the asymptotic pseudo-trajectories reduce to

$$\begin{cases} \frac{d}{dt}\mathbf{x}_{j,t} \in g_{j,t}\left(f_j^{(l)}(\mathbf{x}_{j,t}, \hat{\mathbf{r}}_{j,t}) \right) \\ \frac{d}{dt}\hat{\mathbf{r}}_{j,t} = \bar{g}_{j,t}\left(\mathbb{E}_{s,\mathbf{x}_{-j,t},\mathcal{B}^n} R_j - \hat{\mathbf{r}}_{j,t} \right) \end{cases}$$

for the non-vanishing time-scale ratio and,

$$\begin{cases} \frac{d}{dt}\mathbf{x}_{j,t} \in g_{j,t}\left(f_j^{(l)}(\mathbf{x}_{j,t}, \mathbb{E}_{w,\mathbf{x}_{-j,t}}R_j)\right) \\ \hat{\mathbf{r}}_{j,t} \longrightarrow \mathbb{E}_{s,\mathbf{x}_{-j},\mathcal{B}^n}R_j \end{cases}$$

for the vanishing ratio $\frac{\lambda_t}{\mu_t}$.

7.7.1.4 Asymptotics of Hybrid Learning Schemes

Theorem 7.7.1.5. *Consider the hybrid and switching learning*

$$\begin{cases} \mathbf{x}_{j,t+1} - \mathbf{x}_{j,t} \in q_{j,t}\left(\sum_{l\in\mathcal{L}} \mathbb{1}_{\{l_{j,t}=l\}} f_j^{(l)}(\mathbf{x}_{j,t}, \hat{\mathbf{r}}_{j,t}) + M_{j,t+1}^{(l)}\right) \\ \hat{\mathbf{r}}_{j,t+1} - \hat{\mathbf{r}}_{j,t} \in \bar{q}_{j,t}\left(\mathbb{E}_{w,\mathbf{x}_{-j,t}}R_j - \hat{\mathbf{r}}_{j,t} + \bar{M}_{j,t+1}\right) \end{cases}$$

where $l_{j,t}$ is the learning pattern chosen by player j at time t.

Assume that each player j adopts one of the CODIPAS-RLs in $\mathcal{L}$ with probability $\omega \in \Delta(\mathcal{L})$ and the learning rates are in $l^2\backslash l^1$. Then, the asymptotic pseudo-trajectory of the hybrid and switching learning can be written in the form

$$\begin{cases} \frac{d}{dt}\mathbf{x}_{j,t} \in g_{j,t}\left(\sum_{l\in\mathcal{L}} \omega_l f_j^{(l)}(\mathbf{x}_{j,t}, \hat{\mathbf{r}}_{j,t})\right) \\ \frac{d}{dt}\hat{\mathbf{r}}_{j,t} = \bar{g}_{j,t}\left(\mathbb{E}_{s,\mathbf{x}_{-j,t}}R_j - \hat{\mathbf{r}}_{j,t}\right) \end{cases}$$

for the nonvanishing time-scale ratio and

$$\begin{cases} \frac{d}{dt}\mathbf{x}_{j,t} \in g_{j,t}\left(\sum_{l\in\mathcal{L}} \omega_l f_j^{(l)}(\mathbf{x}_{j,t}, \mathbb{E}_{w,\mathbf{x}_{-j,t},\mathcal{B}^n}R_j)\right) \\ \hat{\mathbf{r}}_{j,t} \longrightarrow \mathbb{E}_{w,\mathbf{x}_{-j},\mathcal{B}^n}R_j \end{cases}$$

for the vanishing ratio $\frac{\lambda_t}{\mu_t}$.

7.8 Quiz

7.8.1 What is Wrong in Learning in Games?

The goal of the exercise is to correct the statements. Propose a correct sentence, and justify your statement.

- S1. In learning in games, players use a predefined scheme.

- S2. The outcome of learning in games are Nash equilibria.

- S3. The players follow the same learning scheme.

- S4. The players update at the fixed time slots.

- S5. Each player knows its action space.

- S6. The behavioral strategies of the players converge,

- S7. Having more information is always useful.

- S8. There is a trade-off between information (on the players side) and efficiency of the outcome.

Correction: What is wrong in "learning in games" ?

- Answer One (A1): Different learning schemes with choice of updating rules have been proposed.

- A2: Outcome of learning in games can be non-Nash equilibria (correlated non-Nash equilibrium, bargaining solution, hierarchical solution, etc.).

- cf. A1.

- A4: Random update

- A5: Noisy actions, exploration of action spaces method.

- A6: No.

- A7: No. (cf. this book).

- A8: No. See the robust pseudo-potential counterexample.

- For player j, he needs to learn his action space.

- His payoff function is to be learned

- The optimal value and the equilibrium payoffs are to be learned (if they exist)

- The optimal strategies is to be learned (if there exists)

- The environment state: to be learned.

- The existence of another player in the environment is to be learned.

- The distribution of actions of the others is to be learned.

- The distribution about the environment state is to be learned over the interactions.

To summarize, a player who is in the system wants to know what he can do. What is his payoff function? How can he/she react to the unknown environment? (by only knowing own observations and own noisy measurements in a memoryless way?)

7.8.2   Learning the Action Space

We briefly recall how the action space can be learned.

- *No need to know the action space in advance* Each player should learn his action space (using, for example, exploration techniques). To do so, we need to add an exploration phase or a progressive exploration during the game. The result of Q-learning is that if the all the actions have been explored and sufficiently exploited and if the learning rates are well-chosen then the outcome is "good".

- Example: At each time t, if player j chooses a_{j,k_j} then he automatically learn the existence of the actions $a_{j,k_j-1}, a_{j,k_j}, a_{j,k_j+1}$ i.e., up-down exploration method. (this exploration method is generalized to a neighborhood of an action in a graph).

- Consequence for finite games: An exploration method such that each player learns its action space. (a bad way to do that: taking lot of time to know the action spaces without *exploitation*). Do we need to add an initialization for the action?

7.9   Chapter Review

One of the objectives in distributed interacting multiplayer systems is to enable a collection of different players to achieve a desirable objective. There are two overriding challenges to achieving this objective:

The first one is related to the complexity of finding an optimal solution. A centralized algorithm may be prohibitively complex when there are a large number of interacting players. This motivates the use of adaptive methods that enable players to self-organize into suitable, if not optimal, alternative solutions.

The second challenge is limited information. Players may have limited knowledge of the status of other players, except perhaps for a small subset of neighboring players. The limitations in term of information induce robust stochastic optimization, bounded rationality, and inconsistent beliefs.

We have investigated asymptotic pseudo-trajectories of dynamic robust games under various COmbined fully DIstributed PAyoff and Strategy Reinforcement Learning (CODIPAS-RL) with random updates and hybrid schemes. The expected robust game problem can be formulated as

$$\forall j \in \mathcal{N}, \ ROP_j : \ \sup_{\mathbf{x}_j \in \mathcal{X}_j} \ \mathbb{E}_{\mathbf{w} \sim \mu} R_j(\mathbf{w}, \mathbf{x}_j, \mathbf{x}_{-j}) \qquad (7.86)$$

where $\mathcal{N}$ is the set of players, $\mathcal{X}_j$ is a subset of a finite dimensional space, $\mathbf{x}_{-j} = (\mathbf{x}_{j'})_{j' \neq j}$, $\mathbf{w}$ is a random variable with law μ.

The main issue here is that the mathematical structure of the payoff function $R_j : \mathcal{W} \times \prod_{j' \in \mathcal{N}} \mathcal{X}_{j'} \longrightarrow \mathbb{R}$ is not known to player j. We develop a combined distributed strategic learning in order to learn both expected payoff function as well as the associated optimal strategies. The dynamic game is played as follows.

Time is slotted. We start at some initial state $\mathbf{w}_0$. At each time slot t, the state $\mathbf{w}_t$ is drawn according to its distribution. Each player chooses an action according to some strategy when he is active. Each player is able to observe a measurement of his payoff. Based on the recent measurement, each player updates his strategy and his payoff estimations. The game goes to the next time slot.

Here, the term *dynamic* refers to the time/information dependence and the term *robust game* refers to a *game under uncertainty*. A strategy update of the player is mainly dictated by his learning scheme. We assume each player can choose among a finite set $\mathcal{L}$ of different CODIPAS-RLs. The hybrid learning has the following form:

$$
\begin{cases}
x_{j,t+1}(s_j) = \sum_{l \in \mathcal{L}} \mathbb{1}_{\{l_{j,t}=l\}} f_{j,s_j}^{(l)}(\lambda_{j,t}, \mathbf{x}_{j,t}, a_{j,t}, r_{j,t}, \hat{\mathbf{r}}_{j,t}) \\
\hat{r}_{j,t+1}(s_j) = \sum_{l \in \mathcal{L}} \mathbb{1}_{\{l_{j,t}=l\}} g_{j,s_j}^{(l)}(\mu_{j,t}, a_{j,t}, r_{j,t}, \hat{\mathbf{u}}_{j,t}) \\
s_j \in \mathcal{A}_j, j \in \mathcal{N}, \ t \geq 0 \\
\mathbf{x}_{j,0} \in \mathcal{X}_j, \ \hat{r}_{j,0} \in \mathbb{R}^{|\mathcal{A}_j|}
\end{cases}
$$

where $r_{j,t}$ is a noisy payoff observed by player j at time t, $\mathcal{A}_j$ is a set with the same size as the dimension of $\mathcal{X}_j$, $\mathcal{L}$ is a finite set of learning patterns, $l_{j,t}$ is the learning pattern chosen of player j at time t, $a_{j,t}$ is the action chosen by player j at time t, s_j denotes a generic element of $\mathcal{A}_j$, the real numbers $\lambda_{j,t}$ and $\mu_{j,t}$ are respectively the strategy-learning rate and payoff-learning rate. The vector $\hat{r}_{j,t}$ is the estimated payoffs of player j corresponding to each component. For each $l \in \mathcal{L}$, the function $f^{(l)}$ is well-chosen in order to guarantee the almost sure forward invariance of $\prod_j \mathcal{X}_j$ by the stochastic process $\{\mathbf{x}_t\}_{t \geq 0}$.

Assuming that

- the functions $R_j(., \mathbf{x})$ are integrable with respect to μ.

- the set $\mathcal{X}_j$ is convex, non-empty, and compact, and

- each component of the hybrid CODIPAS-RL learning scheme can be written in the form of Robbins-Monro or Kiefer-Wolfowitz,

we examine the long-run behavior the interacting system. The convergence, non-convergence, and stability/instability properties of the resulting dynamics as well as their connections to different solution concepts are discussed. Those include Lyapunov expected robust games, expected robust pseudo-potential games, S-modular expected robust games, games with monotone expected payoffs, aggregative robust games, and risk-sensitive robust pseudo-potential games.

A particular case of interest for our work is when the size of the system goes to infinity, i.e., $|\mathcal{N}| \longrightarrow +\infty$. Under technical assumptions on indistinguishability per class of learning property, we derive a Fokker-Planck-Kolmogorov equation associated with the hybrid CODIPAS-RL with a large population of players called *mean field learning*. To prove this statement, the steps are as follows:

• Derivation of the asymptotic pseudo-trajectories of the stochastic processes $\mathbf{x}_{j,t}$.

• Identification of ordinary differential equations or Itô's stochastic differential equation approximation via multiple time-scale stochastic approximations

• Time-scaling and classification of payoff functions in different classes.

• Derivation of the law of the empirical measure over the strategies and its associated interdependent systems of partial differential equations.

This chapter extends these results to random updates. To do so, we have eliminated that short-term players and we have assumed finite moments for random payoff with the random number of interacting players and with the respect of the states. The chapter ends with a quiz that includes some results of the previous chapter.

8

Fully Distributed Learning for Global Optima

8.1 Introduction

In the previous chapters, we have studied equilibrium-seeking procedures. In this chapter, we focus on distributed strategic learning for global optima in specific classes of games.

In many problems, distributed strategic learning may have a tendency to converge toward local optima or even arbitrary points rather than the global optimum of the problem. This means that it does not "know how" to sacrifice short-term payoff to gain a longer-term payoff. The likelihood of this occurring depends on the shape of the payoff functions. Certain problems may provide an easy ascent towards a global optimum, whereas others may make it easier for the function to find the local optima. This phenomenon is in part due to the presence of unstable global optima. Then, the objective is to redesign learning schemes that "stabilize" the global optima or that approaches closely global optima of the problem.

The chapter is structured as follows. We provide basic examples of wireless networking problems and illustrate how game dynamics can be used to solve these problems. The first example focuses on frequency selection in wireless environment. The second example examines user-centric network selection. In both case, convergence, non-convergence, stability and instability are carefully investigated. Later, we overview interactive trial-and-error learning based on Markov chain adjustment and evaluate the frequency of visits of Markov chain to Pareto optimal solutions.

8.2 Resource Selection Games

In a resource selection system such as frequency selection in Orthogonal Frequency-Division Multiple Access (OFDMA system), base station (small, macro, femto, pico, etc.), access point selection, route selection etc., there is a set of m resources, $m \geq 1$. Each resource is associated with a certain payoff function $r_a(w, k_a)$, where $r_a(w, k_a)$ is the payoff for every user of resource a if there are k_a users that are active on that resource and the environment state is w. Together with a set of n players, the resource selection system can be seen as a resource selection game. The action set of every player in the game is the set of resources, and the payoff of each player depends on the resource she chooses and on the number of other players who choose this resource. Thus, resource selection games are special types of congestion games. It is also in the class of aggregative game where the aggregative term is the number of users on that action.

8.3   Frequency Selection Games

In this section, we examine a particular resource selection game known as frequency selection game. Consider a multiuser frequency selection (or technology selection) game [175, 164]. Each user can select one of the technologies at a time. When they use the same technology at the same time and same area, there is congestion or a collision (e.g in same frequency), and the data are lost. Table 8.1 represents a simplified example with two-nodes-two-technologies. The frequencies are denote by f_1 and f_2. Node (user) 1 chooses a row and node 2 chooses a column. A successful configuration gives a payoff 1 to the corresponding node. In addition, we look at it in an independent and identically distributed (i.i.d) noisy environment. Let $0 < \alpha \leq 1$. In Table 8.1, n_{ab}^i denotes a real-valued i.i.d noise.

1\ 2	f_1	f_2
f_1	$(0,0)$: collision	(α, α): success
f_2	(α, α): success	$(0,0)$: collision

$n_{1,1}^1, n_{1,1}^2$	$n_{1,2}^1, n_{1,2}^2$
$n_{2,1}^1, n_{2,1}^2$	$n_{2,2}^1, n_{2,2}^2$

+

TABLE 8.1
Strategic form representation of 2 nodes and 2 technologies.

As can be observed, the game has two pure equilibria (f_1, f_2) (f_2, f_1) and one fully mixed equilibrium $(\frac{1}{2}, \frac{1}{2})$ which is also an evolutionarily stable states or strategy in the sense that it is resilient to deviations by small change of investment. The fully mixed equilibrium is less efficient in terms of social welfare (sum of all the performances of all the users). The pure equilibria are *Strong equilibria* (robust to any coalition of any size). The two pure equilibria are maxmin solutions in the sense that the minimum payoffs of the users are maximized. It is easy to see that the two pure equilibria are also global optima.

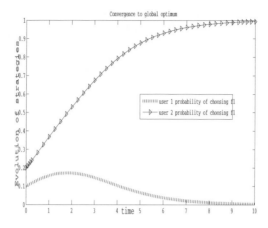

FIGURE 8.1
Convergence to global optimum under imitation dynamics.

In Figures 8.1 8.2,8.3, we represent the evolution of strategies and the vector field: a description of all the possible trajectories starting from the unit square.

8.3.1 Convergence to One of the Global Optima

Since we are observing that the trajectories become close to one of the global optima, it is natural to address the following question:

> *Is there a fully distributed stochastic learning scheme that converges to one of the global optima?*

The answer to this question is positive for most of the initial conditions in the basic setting with two users and two choices. The set of initial conditions under which the convergence to one of the global optima is observed, is of measure 1.

Let x_t be the probability of user 1 choosing the frequency f_1 at time t, and y_t be the probability of user 2 choosing the action f_1 at time t. We use the Borger-Sarin-like startegy learning scheme. In this two player case, the

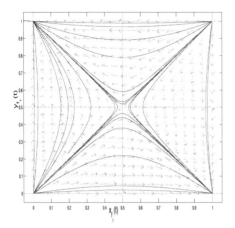
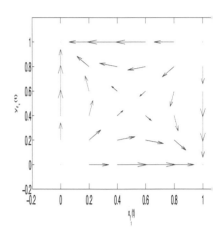

FIGURE 8.2
Vector field of imitation dynamics.

FIGURE 8.3
Vector field of replicator dynamics.

scheme writes

$$x_{t+1} = x_t + \lambda_t r_{1,t+1} \left(\mathbb{1}_{\{a_{1,t+1}=f_1\}} - x_t \right) \tag{8.1}$$

$$y_{t+1} = y_t + \lambda_t r_{2,t+1} \left(\mathbb{1}_{\{a_{2,t+1}=f_1\}} - y_t \right) \tag{8.2}$$

$$(x_0, y_0) \in [0,1]^2. \tag{8.3}$$

One can find the asymptotic pseudo-trajectory for $\lambda_t = \lambda$ constant (convergence in law) or for time-varying λ_t satisfying

$$\lambda_t > 0, \ \sum_{t'} \lambda_{t'} = +\infty, \sum_{t'} \lambda_{t'}^2 < +\infty.$$

The term $\mathbb{1}_{\{a_{1,t}=f_1\}}$ represents the indicator function. It is equal to 1 if user 1 has chosen action f_1 at time t, i.e., $a_{1,t} = f_1$ and 0 otherwise.

Proposition 8.3.1.1. *The algorithm given by the system (8.1) and (8.2) can be tracked asymptotically by a solution of differential equation:*

$$\dot{x} = x(1-x)(1-2y), \tag{8.4}$$

$$\dot{y} = y(1-y)(1-2x), \tag{8.5}$$

$$(x_0, y_0) \in [0,1]^2. \tag{8.6}$$

Proof. By standard stochastic approximations, one can show that the rescaled process from (x_t, y_t) is asymptotically close to a solution of an ordinary differential equation (ODE). Here we identify the exact ODE. To obtain this, we compute the expected change in a one-time slot, also called drift:

$$\mathbb{E}\left(\frac{x_{t+1} - x_t}{\lambda_t} \mid x_t = x, y_t = y\right) \tag{8.7}$$

$$= x(r_{1,f_1}(y) - xr_{1,f_1}(y) - (1-x)r_{1,f_2}(y)) \tag{8.8}$$

$$= x(1-x)(r_{1,f_1}(y) - r_{1,f_2}(y)), \tag{8.9}$$

where $r_{1,f_1}(y)$ is the expected payoff obtained by user 1 when she uses f_1 and user 2 plays a randomized action $(y, 1-y)$. It is easy to see that $r_{1,f_1}(y) = 1-y$ is the probability that user 2 chooses f_2. Similarly, $r_{1,f_2}(y) = y$. Thus,

$$\mathbb{E}\left(\frac{x_{t+1} - x_t}{\lambda_t} \mid x_t = x, y_t = y\right) = x(1-x)(1-2y).$$

We do the same for the process y_t. Since we work in the unit square, the gap between the expected term and the random variable is a martingale difference with bounded second moment. Moreover, the norm of this martingale is bounded by the norm of (x, y). Using stochastic difference framework given in appendix (see Appendix A.4), we deduce the following result:

> *The asymptotic pseudo-trajectories give the replicator dynamics (with two-types).*

If x denotes the probability of user 1 choosing f_1 and y the probability of user 2 choosing f_1, then the ordinary differential satisfied by x and y are:

$$\dot{x} = x(1-x)(1-2y), \tag{8.10}$$

$$\dot{y} = y(1-y)(1-2x), \tag{8.11}$$

$$(x_0, y_0) \in [0,1]^2. \tag{8.12}$$

$$\square$$

Define the rest points (or stationary points) of the system (8.10) as the zeros: $\dot{x} = 0, \dot{y} = 0$.

Proposition 8.3.1.2. *The set of rest points of the dynamics contains both the set of equilibria and the set of global optima.*

Proof. The rest points of the system are obtained by finding the zeros of the right-hand side of the system. The zeros are $(0,0), (1,0), (0,1), (1,1), (\frac{1}{2}, \frac{1}{2})$. Thus, the set of equilibria of the game $\{(1,0), (0,1), (\frac{1}{2}, \frac{1}{2})\}$ is in the set of rest points. The set of global optima $\{(1,0), (0,1)\}$ is also in the set of rest points. $\square$

Proposition 8.3.1.3. *Starting from any point in the unit square $[0,1]^2$ outside the segment $y = x$, the system converges to the set of global optima.*

This result gives a global convergence to an efficient point (global optimum) for almost all initial conditions. We say almost all initial points because the diagonal segment is of Lebesgue measure zero (in two dimensions) compared to the measure of the square $[0,1]^2$.

Proof. By computing the Jacobian at each of the 5 rest points, we check that $(1,0)$ and $(0,1)$ are stable, and the other three rest points are unstable (the Jacobian have a positive eigenvalue). Then, we construct the vector field of the dynamical system. Starting from any point in the unit square $[0,1]^2$ outside the segment $y = x$, the system converges to the corner $(1,0)$ or to $(0,1)$ depending on the initial point. If the starting point is more at the left corner, it goes to the left, otherwise it goes to the right corner. We conclude that the system converges to one of the global optima $\{(1,0),(0,1)\}$. □

As a corollary, we deduce that by carefully choosing the learning parameters, say $\lambda_t = \frac{1}{5+t}$, $\lambda_t = \lambda$, the fully distributed learning algorithm converges almost surely to one of the global optima, which is a very interesting property.

8.3.2 Symmetric Configuration and Evolutionarily Stable State

What happens if the starting points are in the diagonal segment?

The case where the starting points are in the diagonal segment corresponds to a symmetric configuration and the system is reduced to one dynamical equation

$$\dot{x} = x(1-x)(1-2x). \tag{8.13}$$
$$x(0) = x_0 \in [0,1] \tag{8.14}$$

For similar analysis in evolutionary game dynamics, see [166, 153].

Definition 8.3.2.1. *We say that* $x^* = (x_{f_1}^*, x_{f_2}^*)$ *is an evolutionarily stable state (ESS) if for any* $x \neq x^*$ *there exists an* $\epsilon_x > 0$ *such that*

$$\sum_{f \in \{f_1, f_2\}} (x_f^* - x_f) r_f(\epsilon x + (1-\epsilon)x^*) > 0, \ \forall \epsilon \in (0, \epsilon_x). \tag{8.15}$$

The following proposition conducts the analysis for the symmetric case.

Proposition 8.3.2.2. *Consider symmetric configuration.*

- *The symmetric game has a unique evolutionarily stable state which is given by* $(\frac{1}{2}, \frac{1}{2})$.

- *The system goes to the unique evolutionarily stable state starting from any interior point* $x(0) \in (0,1)$.

Proof. In symmetric configurations, the evolutionarily stable states should be symmetric equilibria. Thus, we have to check the stability property among the set of symmetric equilibria which is reduced to the singleton $\{(\frac{1}{2}, \frac{1}{2})\}$. We verify that $(\frac{1}{2}, \frac{1}{2})$ satisfies

$$\left(\frac{1}{2} - x, \frac{1}{2} - (1 - x)\right)\left(\frac{1-x}{x}\right) = \left(\frac{1}{2} - x, -\frac{1}{2} + x\right)\left(\frac{1-x}{x}\right) \tag{8.16}$$

$$= \left(\frac{1}{2} - x\right)(1 - x - x) \tag{8.17}$$

$$= 2\left(\frac{1}{2} - x\right)^2, \tag{8.18}$$

which is strictly greater than 0 for any $x \neq \frac{1}{2}$. We conclude that the vector $(\frac{1}{2}, \frac{1}{2})$ is an evolutionarily stable state or strategy (ESS). Since $\frac{1}{2}$ is a global attractor at the interior, the dynamic system converges exponentially to $\frac{1}{2}$ starting from any point $x_0 \in (0, 1)$. This completes the proof. $\square$

Note that by changing the sign for symmetric configuration, the anti-diagonal segment goes is in the direction of $(1/2, 1/2)$.

The convergence of the above system is exponential, i.e., it takes $O(\log(\frac{1}{\epsilon}))$ time in order to be closer to the rest point with ϵ error gap. This means that when $\epsilon \longrightarrow 0$, the time bound explodes depending on the starting point.

8.3.3   Accelerating the Convergence Time

Since the scenario is a convergent one, we want to know if it is possible to accelerate the convergence time of the system.

Using a time-scale change with factor K in the system, we get

$$\dot{\bar{x}} = Ke^{Kt}\bar{x}(1 - \bar{x})(1 - 2\bar{y}), \tag{8.19}$$

$$\dot{\bar{y}} = Ke^{Kt}\bar{y}(1 - \bar{y})(1 - 2\bar{x}), \tag{8.20}$$

$$(\bar{x}_0, \bar{y}_0) \in [0, 1]^2. \tag{8.21}$$

The solution of this system can be expressed as a function of the nonscaled ordinary differential equation and is given by

$$(\bar{x}(t), \bar{y}(t)) = \left(x[K \int_0^t e^{Ks} ds], y[K \int_0^t e^{Ks} ds]\right).$$

This means that if the first system convergence time to ϵ−close is T_ϵ, the second system will take at most $\frac{1}{K}\log(T_\epsilon + 1)$ which is relatively fast.

8.3.4   Weighted Multiplicative imitative CODIPAS-RL

We now focus on weighted multiplicative imitative CODIPAS-RL. The imitative stochastic iterative procedure is

$$x_{1,t+1}(f_1) - x_{1,t}(f_1) = x_{1,t}(f_1) \times$$

$$\left(\frac{(1 + \lambda_{1,t})^{\hat{r}_{1,t}(f_1)}}{x_{1,t}(f_1)(1 + \lambda_{1,t})^{\hat{r}_{1,t}(f_1)} + x_{1,t}(f_2)(1 + \lambda_{1,t})^{\hat{r}_{1,t}(f_2)}} - 1 \right), \qquad (8.22)$$

$$x_{1,t+1}(f_2) - x_{1,t}(f_2) = x_{1,t}(f_2) \times$$

$$\left(\frac{(1 + \lambda_{1,t})^{\hat{r}_{1,t}(f_2)}}{x_{1,t}(f_1)(1 + \lambda_{1,t})^{\hat{r}_{1,t}(f_1)} + x_{1,t}(f_2)(1 + \lambda_{1,t})^{\hat{r}_{1,t}(f_2)}} - 1 \right), \qquad (8.23)$$

$$\hat{r}_{1,t+1}(f_1) - \hat{r}_{1,t}(f_1) = \nu_{1,t} \mathbb{1}_{\{a_{1,t}=f_1\}} (r_{1,t} - \hat{r}_{1,t}(f_1)) \qquad (8.24)$$

$$\hat{r}_{1,t+1}(f_2) - \hat{r}_{1,t}(f_2) = \nu_{1,t} \mathbb{1}_{\{a_{1,t}=f_2\}} (r_{1,t} - \hat{r}_{1,t}(f_2)) \qquad (8.25)$$

$$x_{2,t+1}(f_1) - x_{2,t}(f_1) = x_{2,t}(f_1) \times$$

$$\left(\frac{(1 + \lambda_{2,t})^{\hat{r}_{2,t}(f_1)}}{x_{2,t}(f_1)(1 + \lambda_{1,t})^{\hat{r}_{1,t}(f_1)} + x_{2,t}(f_2)(1 + \lambda_{1,t})^{\hat{r}_{1,t}(f_2)}} - 1 \right), \qquad (8.26)$$

$$x_{2,t+1}(f_2) - x_{2,t}(f_2) = x_{2,t}(f_2) \times$$

$$\left(\frac{(1 + \lambda_{2,t})^{\hat{r}_{2,t}(f_2)}}{x_{2,t}(f_1)(1 + \lambda_{1,t})^{\hat{r}_{1,t}(f_1)} + x_{2,t}(f_2)(1 + \lambda_{1,t})^{\hat{r}_{1,t}(f_2)}} - 1 \right), \qquad (8.27)$$

$$\hat{r}_{2,t+1}(f_1) - \hat{r}_{2,t}(f_1) = \nu_{2,t} \mathbb{1}_{\{a_{2,t}=f_1\}} (r_{2,t} - \hat{r}_{2,t}(f_1)) \qquad (8.28)$$

$$\hat{r}_{2,t+1}(f_2) - \hat{r}_{2,t}(f_2) = \nu_{2,t} \mathbb{1}_{\{a_{2,t}=f_2\}} (r_{2,t} - \hat{r}_{2,t}(f_2)) \qquad (8.29)$$

The strategy learning (8.22) is a learning-rate-weighted imitative reinforcement learning. We will show this learning is specially well-adapted in the frequency selection game.

We examine two groups of learning rates: constant learning rate and diminishing learning rate.

- $\lambda_{1,t} = \lambda_{2,t} = \epsilon\lambda$ for small $\epsilon > 0$ and a scaling factor $\lambda > 0$.

- $\lambda_{1,t} = \lambda_{2,t} = 1/(t+1)$, $\nu_{1,t} = \nu_{2,t} = (t+1)^{-3/4}$, $t \geq 0$.

The first case is a constant learning rate scaling. In Figures 8.4 and 8.5, we represent one path of the imitative CODIPAS-RL for strategies (frequency 1), estimations for the first frequency and the average payoff. We observe a convergence to one of global optima $(1,0), (0,1)$, i.e., user 1 with f_1 and user 2 will be playing f_2 in the long run. The initial parameters are

$$x_{1,0} = [0.6, 0.4]', x_{2,0} = [0.2, 0.8]';$$

$$\hat{r}_{1,0} = [0.4, 0.3]', \hat{r}_{2,0} = [0.2, 0.4]', \lambda = 0.9.$$

The MATLAB code is provided below:

Initialization

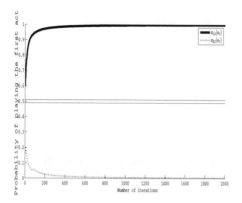

FIGURE 8.4
Strategies.

```
close all;
clear all;

% This program investigates imitative CODIPAS-RL schemes.
P1=[ 0 1
     1 0]

P2=[ 0 1
     1 0]

%P1 is the  column player maximizing
%P2 is the  row player maximizing

N=4000
uhat1=zeros(2,N);
uhat2=zeros(2,N);

a1=zeros(N,1);
a2=zeros(N,1);

f=zeros(2,N);
g=zeros(2,N);

%%%% Initialization
f(:,1)=[0.6,0.4]'
g(:,1)=[0.2,0.8]'

uhat1(:,1)=[0.4 0.3]'
```

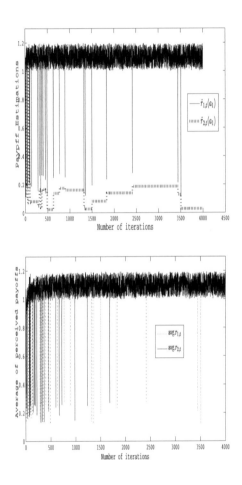

FIGURE 8.5
Estimations and average payoffs.

```
uhat2(:,1)=[0.2 0.4]'

a1(1)=round(rand)+1;
a2(1)=round(rand)+1;

avgP1=[]
avgP2=[]
epsil=1;
lambdat=0.9;
```

Definition of Noisy Payoffs

```
for k=1:N
    S= [1 1; 1 1]-rand(2,2);
    U1 = (1/5)*S + P1;
    U2 = (1/5)*S + P2;

    a1(k)=1+(rand<=f(2,k));
    a2(k)=1+(rand<=g(2,k));
```

Payoff learning

```
uhat1(1,k+1)= uhat1(1,k) + lambdat*(a1(k)==1)*
               (U1(a1(k), a2(k))-uhat1(1,k));
uhat1(2,k+1)= uhat1(2,k) + lambdat*(a1(k)==2)*
               (U1(a1(k), a2(k))-uhat1(2,k));

uhat2(1,k+1)= uhat2(1,k) + lambdat*(a2(k)==1)*
               (U2(a1(k), a2(k))-uhat2(1,k));
uhat2(2,k+1)= uhat2(2,k) + lambdat*(a2(k)==2)*
               (U2(a1(k), a2(k))-uhat2(2,k));
```

Strategy learning

```
f(1,k+1)=f(1,k)+ f(1,k)*(   (1+epsil/(k+1))^(uhat1(1,k)) /
             ((1+epsil/(k+1))^(uhat1(1,k))*f(1,k)+
```

```
                 (1+epsil/(k+1))^(uhat1(2,k))*f(2,k))-1);
   f(2,k+1)=f(2,k)+ f(2,k)*(    (1+epsil/(k+1))^(uhat1(2,k)) /
                  ((1+epsil/(k+1))^(uhat1(1,k))*f(1,k)+
                  (1+epsil/(k+1))^(uhat1(2,k))*f(2,k))-1);


   g(1,k+1)=g(1,k)+ g(1,k)*(    (1+epsil/(k+1))^(uhat2(1,k))/
                  ((1+epsil/(k+1))^(uhat2(1,k))*g(1,k)+
                  (1+epsil/(k+1))^(uhat2(2,k))*g(2,k))-1);
   g(2,k+1)=g(2,k)+ g(2,k)*(    (1+epsil/(k+1))^(uhat2(2,k))/
                  ((1+epsil/(k+1))^(uhat2(1,k))*g(1,k)+
                  (1+epsil/(k+1))^(uhat2(2,k))*g(2,k))-1);


   avgP1(:,k)=f(1,k)*uhat1(1,k)+f(2,k)*uhat1(2,k);
   avgP2(:,k)=g(1,k)*uhat2(1,k)+g(2,k)*uhat2(2,k);

end
figure(1)
hold on
plot(1:N+1, f(1,:), 1:N+1, g(1,:))
ylabel('Probability of playing the first action')
xlabel('Number of iterations')
g2=legend('$x_{1,t}(a_1)$', '$x_{2,t}(a_1)$')
 set(g2,'Interpreter','latex')
hold on
plot(1:N+1,0.51,'r-');
% line to check the speed of convergence to 1/2 if any
hold on
plot(1:N+1,0.49,'r-.');
 hold off
%
figure(2)
plot(1:N+1, uhat1(1,:), 1:N+1, uhat2(1,:))
ylabel('Payoff Estimations')
xlabel('Number of iterations')
 g=legend('$\hat{r}_{1,t}(a_1)$',  '$\hat{r}_{2,t}(a_1)$')
 set(g,'Interpreter','latex')

%Plot the average payoff
figure(3)
plot(1:N, avgP1(1,:), 1:N, avgP2(1,:))
ylabel('Average of perceived payoffs')
xlabel('Number of iterations')
 h=legend('avg.$r_{1,t}$', 'avg.$r_{2,t}$')
  set(h,'Interpreter','latex')
```

> The main reason why the convergence to $(1/2, 1/2)$ is rarely observed is that the uniform strategy is asymptotically unstable in probability.

8.3.5 Three Players and Two Frequencies

Next, we examine the resource selection problem in the case of 3 users and 2 frequencies. There are less resources than users so that there will always be a collision at one of the positions. The case of 3 users and 2 frequencies can be expressed in the matrix form as shown in Table 8.2. The game is the class of minority game in the sense that user who ends up on the minority side is a winner. The minority game, also known as, *El Farol Bar problem* [39] has been extensively studied in the literature of statistical mechanics.

1 \ 2	f_1	f_2
f_1	(0,0,0): collision	(0,1,0)
f_2	(1,0,0)	(0,0,1)

1 \ 2	f_1	f_2
f_1	(0,0,1)	(1,0,0):
f_2	(0,1,0)	(0,0,0): collision

TABLE 8.2
Strategic form representation for 3 users - 2 frequencies

8.3.5.1 Global Optima

The global optima are (f_1, f_1, f_2), (f_2, f_2, f_1) and their permutations, i.e., (f_1, f_2, f_1), (f_2, f_1, f_1), (f_2, f_1, f_2), and (f_1, f_2, f_2). Thus, we have 6 six global optima.

8.3.5.2 Noisy Observation

We assume that a player is able to observe only a noisy version of the entries of these tables. We consider additive noise, i.e., $r_{j,t}$ is a realization of $R_j(.) +$

Noise. We use the following stochastic algorithm

$$
\left\{
\begin{aligned}
& n = 3, \ S = \{f_1, f_2\} \\
& x_{1,t+1}(f_1) = x_{1,t}(f_1) + \lambda_{1,t} r_{1,t} \left(\mathbb{1}_{\{a_{1,t}=f_1\}} - x_{1,t}(f_1) \right) \\
& x_{1,t+1}(f_2) = x_{1,t}(f_2) + \lambda_{1,t} r_{1,t} \left(\mathbb{1}_{\{a_{1,t}=f_2\}} - x_{1,t}(f_2) \right) \\
& \hat{r}_{1,t+1}(f_1) = \hat{r}_{1,t}(f_1) + \nu_{1,t} \mathbb{1}_{\{a_{1,t}=f_1\}} \left(r_{1,t} - \hat{r}_{1,t}(f_1) \right) \\
& \hat{r}_{1,t+1}(f_2) = \hat{r}_{1,t}(f_2) + \nu_{1,t} \mathbb{1}_{\{a_{1,t}=f_2\}} \left(r_{1,t} - \hat{r}_{1,t}(f_2) \right) \\[4pt]
\hline \\[-6pt]
& x_{2,t+1}(f_1) = x_{2,t}(f_1) + \lambda_{2,t} r_{2,t} \left(\mathbb{1}_{\{a_{2,t}=f_1\}} - x_{2,t}(f_1) \right) \\
& x_{2,t+1}(f_2) = x_{2,t}(f_2) + \lambda_{2,t} r_{2,t} \left(\mathbb{1}_{\{a_{2,t}=f_2\}} - x_{2,t}(f_2) \right) \\
& \hat{r}_{2,t+1}(f_1) = \hat{r}_{2,t}(f_1) + \nu_{2,t} \mathbb{1}_{\{a_{2,t}=f_1\}} \left(r_{2,t} - \hat{r}_{2,t}(f_1) \right) \\
& \hat{r}_{2,t+1}(f_2) = \hat{r}_{2,t}(f_2) + \nu_{2,t} \mathbb{1}_{\{a_{2,t}=f_2\}} \left(r_{2,t} - \hat{r}_{2,t}(f_2) \right) \\[4pt]
\hline \\[-6pt]
& x_{3,t+1}(f_1) = x_{3,t}(f_1) + \lambda_{3,t} r_{3,t} \left(\mathbb{1}_{\{a_{3,t}=f_1\}} - x_{3,t}(f_1) \right) \\
& x_{3,t+1}(f_2) = x_{3,t}(f_2) + \lambda_{3,t} r_{3,t} \left(\mathbb{1}_{\{a_{3,t}=f_2\}} - x_{3,t}(f_2) \right) \\
& \hat{r}_{3,t+1}(f_1) = \hat{r}_{3,t}(f_1) + \nu_{3,t} \mathbb{1}_{\{a_{3,t}=f_1\}} \left(r_{3,t} - \hat{r}_{3,t}(f_1) \right) \\
& \hat{r}_{3,t+1}(f_2) = \hat{r}_{3,t}(f_2) + \nu_{3,t} \mathbb{1}_{\{a_{3,t}=f_2\}} \left(r_{3,t} - \hat{r}_{3,t}(f_2) \right)
\end{aligned}
\right.
$$

Below we represent one run of the simulations with constant learning rates. As we can observe the convergence time to global optimum is relatively fast. In Figures 8.7 and 8.6, we illustrate the algorithm with three users. The strategies, estimations, and average payoffs are represented.

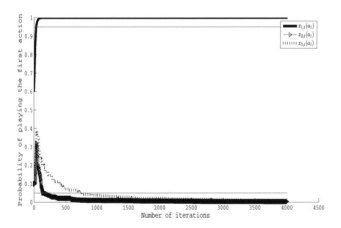

FIGURE 8.6
Three users and two choices.

In Figure 8.8, we represent the impact of the starting point on the outcome. One trajectory of the stochastic algorithm is illustrated. The realization gives different outcomes, but both are global optima.

> How to improve the convergence time of the stochastic algorithm?

We have seen how to accelerate to the convergence time in a certain range of stationary points for continuous time dynamics by using a scaling technique. Now, we ask if it is possible to use a similar technique for the stochastic algorithm in a converging scenario. For CODIPAS-RL algorithms, we examine the effect of ϵ, λ, and μ in order to find an optimal range of parameters with as fast speed as possible while preserving convergence.

8.3.6 Similar Learning Rate

Here we examine the similar learning rate, e.g., $\lambda_{j,t} = \nu_{j,t} = \frac{1}{t+1}$. We propose a differential approach. The asymptotic pseudo-trajectory is given by

$$
\begin{cases}
\quad n = 3, \ S = \{f_1, f_2\} \\
\frac{d}{dt} x_{1,t}(f_1) = x_{1,t}(f_1)(1 - x_{1,t}(f_1)) \times \\
[(1 - x_{2,t}(f_1))(1 - x_{3,t}(f_1)) - x_{2,t}(f_1)x_{3,t}(f_1)] \\
\frac{d}{dt} \hat{r}_{1,t}(f_1) = x_{1,t}(f_1) [(1 - x_{2,t}(f_1))(1 - x_{3,t}(f_1)) - \hat{r}_{1,t}(f_1)] \\
\frac{d}{dt} \hat{r}_{1,t}(f_2) = (1 - x_{1,t}(f_1)) [x_{2,t}(f_1)x_{3,t}(f_1) - \hat{r}_{1,t}(f_2)] \\
\\
\frac{d}{dt} x_{2,t}(f_1) = x_{2,t}(f_1)(1 - x_{2,t}(f_1)) [1 - x_{1,t}(f_1) - x_{3,t}(f_1)] \\
\frac{d}{dt} \hat{r}_{2,t}(f_1) = x_{2,t}(f_1) [(1 - x_{1,t}(f_1))(1 - x_{3,t}(f_1)) - \hat{r}_{2,t}(f_1)] \\
\frac{d}{dt} \hat{r}_{2,t}(f_2) = (1 - x_{2,t}(f_1)) [x_{1,t}(f_1)x_{3,t}(f_1) - \hat{r}_{2,t}(f_2)] \\
\\
\frac{d}{dt} x_{3,t}(f_1) = x_{3,t}(f_1)(1 - x_{3,t}(f_1)) [\hat{r}_{3,t}(f_1) - \hat{r}_{3,t}(f_2)] \\
\frac{d}{dt} \hat{r}_{3,t}(f_1) = x_{3,t}(f_1) [(1 - x_{1,t}(f_1))(1 - x_{2,t}(f_1)) - \hat{r}_{3,t}(f_1)] \\
\frac{d}{dt} \hat{r}_{3,t}(f_2) = (1 - x_{3,t}(f_1)) [x_{1,t}(f_1)x_{2,t}(f_1) - \hat{r}_{3,t}(f_2)]
\end{cases}
$$

Now we modify the algorithm so that $\hat{r}_{j,t}$ replaces $r_{j,t}$.

$$
\begin{cases}
\quad n = 3, \ S = \{f_1, f_2\} \\
\frac{d}{dt} x_{1,t}(f_1) = x_{1,t}(f_1)(1 - x_{1,t}(f_1)) [\hat{r}_{1,t}(f_1) - \hat{r}_{1,t}(f_2)] \\
\frac{d}{dt} \hat{r}_{1,t}(f_1) = x_{1,t}(f_1) [(1 - x_{2,t}(f_1))(1 - x_{3,t}(f_1)) - \hat{r}_{1,t}(f_1)] \\
\frac{d}{dt} \hat{r}_{1,t}(f_2) = (1 - x_{1,t}(f_1)) [x_{2,t}(f_1)x_{3,t}(f_1) - \hat{r}_{1,t}(f_2)] \\
\\
\frac{d}{dt} x_{2,t}(f_1) = x_{2,t}(f_1)(1 - x_{2,t}(f_1)) [\hat{r}_{2,t}(f_1) - \hat{r}_{2,t}(f_2)] \\
\frac{d}{dt} \hat{r}_{2,t}(f_1) = x_{2,t}(f_1) [(1 - x_{1,t}(f_1))(1 - x_{3,t}(f_1)) - \hat{r}_{2,t}(f_1)] \\
\frac{d}{dt} \hat{r}_{2,t}(f_2) = (1 - x_{2,t}(f_1)) [x_{1,t}(f_1)x_{3,t}(f_1) - \hat{r}_{2,t}(f_2)] \\
\\
\frac{d}{dt} x_{3,t}(f_1) = x_{3,t}(f_1)(1 - x_{3,t}(f_1)) [\hat{r}_{3,t}(f_1) - \hat{r}_{3,t}(f_2)] \\
\frac{d}{dt} \hat{r}_{3,t}(f_1) = x_{3,t}(f_1) [(1 - x_{1,t}(f_1))(1 - x_{2,t}(f_1)) - \hat{r}_{3,t}(f_1)] \\
\frac{d}{dt} \hat{r}_{3,t}(f_2) = (1 - x_{3,t}(f_1)) [x_{1,t}(f_1)x_{2,t}(f_1) - \hat{r}_{3,t}(f_2)]
\end{cases}
$$

Another modification to solve a numerical issue is to replace the term

$\nu_{j,t}\mathbb{1}_{\{a_{j,t}=f_1\}}$ by $\frac{\nu_{j,t}\mathbb{1}_{\{a_{j,t}=f_1\}}}{x_{j,t}(f_1)}$. This means the first multiplicative term in the payoff dynamics is omitted.

8.3.7 Two Time-Scales

We look at the case with $n = 3$ users. Each user can choose a frequency in the set $S = \{f_1, f_2\}$. The stochastic algorithm behaves asymptotically as the solutions of the following deterministic system:

$$\begin{cases} \frac{d}{dt}x_{1,t}(f_1) = x_{1,t}(f_1)(1 - x_{1,t}(f_1))\left[1 - x_{2,t}(f_1) - x_{3,t}(f_1)\right] \\ \frac{d}{dt}x_{2,t}(f_1) = x_{2,t}(f_1)(1 - x_{2,t}(f_1))\left[1 - x_{1,t}(f_1) - x_{3,t}(f_1)\right] \\ \frac{d}{dt}x_{3,t}(f_1) = x_{3,t}(f_1)(1 - x_{3,t}(f_1))\left[1 - x_{1,t}(f_1) - x_{2,t}(f_1)\right] \end{cases}$$

- Find the set of rest points of the above system of ODEs.

- Compare with the equilibria, Pareto optimal solutions, and global optima.

- Does the game have an evolutionarily stable strategy?

- What can we say about the convergence and nonconvergence issue?

- What is the speed of convergence?

> *The pure global optima are rest points of the replicator dynamics.*

8.3.8 Three Players and Three Frequencies

Table 8.3 summarizes the case of three players and three frequencies.

The expected payoff can be easily computed. For example, the expected payoff for action f_1 is $r_1(e_{f_1}, x_{-1}) = (1 - x_2(f_1))(1 - x_3(f_1))$, and r_2 is computed in a similar way.

8.3.9 Arbitrary Number of Users

As the throughput of the MAC schemes may significantly affect the overall performance of a wireless network, careful design of MAC schemes is necessary to ensure proper operation of a network. Recall the basic rule of slotted Aloha scheme: *if more than two users transmit, then there is collision.* Following this idea, one can introduce frequency selection case: *if more than two users transmit at the same time with the same frequency, then there is a collision.*

We consider n users and m frequencies. Let $\mathcal{N} := \{1, 2, \ldots, n\}$ be the set of users with n being the total number of users in the system. Denote by

TABLE 8.3

Frequency selection game: 3 players, 3 frequencies

	f_1	f_2	f_3
f_1	$(0,0,0)$	$(0,1,0)$	$(0,1,0)$
f_2	$(1,0,0)$	$(0,0,1)$	$(1,1,1)$
f_3	$(1,0,0)$	$(1,1,1)$	$(0,0,1)$

	f_1	f_2	f_3
f_1	$(0,0,1)$	$(1,0,0)$	$(1,1,1)$
f_2	$(0,1,0)$	$(0,0,0)$	$(0,1,0))$
f_3	$(1,1,1)$	$(1,0,0)$	$(0,0,1)$

	f_1	f_2	f_3
f_1	$(0,0,1)$	$(1,1,1)$	$(1,0,0)$
f_2	$(1,1,1)$	$(0,0,1)$	$(1,0,0)$
f_3	$(0,1,0)$	$(0,1,0)$	$(0,0,0)$

$\mathcal{F} = \{1, 2, \ldots, m\}$ the set of frequencies for the n users. Each user can choose only one among the m frequencies. Denote by $x_{j,t}(f)$ the probability that user j chooses the frequency f at time t. The success probability of user j at time t is given by

$$r_j(x_t) = \sum_{f=1}^{m} x_{j,t}(f) \prod_{j' \neq j} (1 - x_{j',t}(f)).$$

This says that a user j with frequency f has a successful transmission only if no other user is using the same frequency. We distinguish two cases: (i) $m < n$ (ii) $m \geq n$. The state w corresponds to ON/OFF. The state ON means the interface is working and the state OFF means the interface is not working. When the interface is OFF the user cannot access, therefore we look at the probability that the interface to be ON and multiply the performance index by this probability. In the analysis we omit this probability.

8.3.9.1 Global Optimization

The global optimization problem consists of maximizing the probability of successful transmissions of the entire system. The problem can be formulate as follows:

$$\begin{cases} \max_x & \sum_{j \in \mathcal{N}} r_j(x) \\ & \forall j \in \mathcal{N}, \ \sum_{f \in \mathcal{F}} x_j(f) = 1 \\ & \forall j \in \mathcal{N}, \ \forall f \in \mathcal{F}, \ x_j(f) \geq 0 \end{cases}$$

We denote by

$$\Delta(\mathcal{F}) = \{z \mid \sum_{f \in \mathcal{F}} z_j(f) = 1, \ \forall f, \ z_j(f) \geq 0\},$$

the simplex. Then, $\forall j$, $x_j \in \Delta(\mathcal{F})$.

- If $n \leq m$, a direct affectation of frequencies (without repetition) solves the problem. This implies that we have an exponential number of solutions.

- If $n > m$, we affect the first m frequencies to m users. The remaining $n - m$ users remains unaffected. We have again an exponential number of solutions.

8.3.9.2   Equilibrium Analysis

Define a one-shot game by the collection $\mathcal{G} = (\mathcal{N}, (r_j(.))_{j \in \mathcal{N}}, \mathcal{F})$. We say that x is an equilibrium of $\mathcal{G}$ if the following inequalities hold:

$$\forall j, \ r_j(x) \geq r_j(x_1, \ldots, x_{j-1}, y_j, x_{j+1}, \ldots, x_n), \ \forall y_j \in \Delta(\mathcal{F}).$$

We first remark that the above solutions of the optimization problem are pure equilibria of the one-shot game

$$\mathcal{G} = (\mathcal{N}, (r_j(.))_{j \in \mathcal{N}}, \mathcal{F}).$$

In particular, the global optimum value can be obtained as an equilibrium payoff, i.e., the so-called *Price of Stability* is one.

There are many other equilibria of the game $\mathcal{G}$. To see this, consider the case where $n > m$. Any configuration where all the frequencies are used (with probability one) combined with arbitrary strategies of the remaining users leads to an equilibrium of $\mathcal{G}$.

8.3.9.3   Fairness

We briefly introduce the notion of $\alpha-$fairness. Consider the parameterized optimization problem

$$\max \sum_{j \in \mathcal{N}} \frac{1}{1 - \alpha} [g(r_j)^{1-\alpha} - 1].$$

The following classical fairness solution concepts can be obtained:

- $\alpha = 0$ maximization of total payoff problem

- $\alpha \to 1$ proportional fairness (a uniform case of Nash bargaining)

- $\alpha = 2$ cost minimization

- $\alpha \to +\infty$ maxmin fairness (maximize the minimum throughput that a user can have).

When $n > m$, the global optimum payoff (maximum payoff of the system) and the pure equilibrium payoffs are not fair in the sense that some of the users get 1 and some others get 0. More fair solutions can be obtained using mixed strategies. For example if $\forall \ j, \forall \ f$, $x_j^*(f) = \frac{1}{m}$, i.e., a uniform distribution, then the expected payoff of each user is $(1 - \frac{1}{m})^{n-1} > 0$ times the probability of choosing that frequency, and the total system payoff is $\frac{n}{m}(1 - \frac{1}{m})^{n-1}$.

Interestingly, if the proportion of users per frequencies is nonvanishing, the system has a nonzero asymptotic capacity.

Next, we introduce one of the most used solution concepts in cooperative games.

Pareto optimality is a measure of efficiency. An outcome of the game $\mathcal{G}$ is Pareto optimal if there is no other outcome that makes every user at least as well off and at least one user strictly better off. That is, a Pareto Optimal outcome cannot be improved upon without hurting at least one user.

Lemma 8.3.9.4. *The above strategy profile* x^* *is Pareto optimal.*

The proof of this lemma follows from the fact that the strategy maximizes the weighted sum of payoff of the users.

8.4 User-Centric Network Selection

It is envisioned that in the future mobile communication paradigm, the decision of network selection will be delegated to the users to exploit the best available characteristics of different network technologies and network providers, with the objective of increased satisfaction. The consequence of such a user-centric network selection paradigm is *users' short term contractual agreements* with the operators. These contracts will basically be driven by the satisfaction level of users with operator services. In order to more accurately express the user satisfaction, the term Quality of Service (QoS) has been extended to include more subjective and also application specific measures beyond traditional technical parameters, giving rise to the Quality of Experience (QoE) concept. Intuitively this provides the representation and modeling of the *user satisfaction* function, which captures user satisfaction for both technical (delay, jitter, packetloss, throughput, etc.) and economical (service cost, etc.) aspects.

With the evolution toward new multimedia systems and services, user requirements are not limited to requirements on connectivity: users now expect services to be delivered according to their demands on quality. At the same time, audiovisual systems are becoming more and more complex and new possibilities of presenting content are becoming available, including augmented reality and immersive environments. However, for wireless systems the possible limitations due to the characteristics of the transmission channel and of the devices can result in perceivable impairments, originating in the different steps of the value chain from content production to display techniques, that influence the user's perception of quality. In recent years, the concept of quality of service (QoS) has been extended to the new concept of quality of experience (QoE), as the first only focuses on network performance (e.g., packet loss, delay and jitter) without a direct link to the perceived quality, whereas the QoE reflects the overall experience of the consumer accessing and using the provided service. Experience is user- and context-dependent, involving considerations about subjective multimedia quality and users' expectation based on the cost they paid for the service, on their location, on the type of service, on the convenience of using the service, etc. Objective QoE evaluation enables user centric design of novel multimedia systems, including wireless systems based on recent standards, such as WiMAX and 3GPP LTE, through an optimal use of the available resources based on an objective utility index.

The concept of QoE in engineering is also known as Perceived Quality of Service (PQoS), in the sense of the QoS as it is finally perceived by the end-user. The evaluation of the QoE for audiovisual content will provide a user with a range of potential choices, covering the possibilities of low, medium or high quality levels. Moreover the QoE evaluation gives the service provider and network operator the capability to minimize the storage and network resources by allocating only the resources that are sufficient to maintain a specific level of user satisfaction.

It should be noted that, in broader sense, user preferences over different *technical* and *economical* aspects can be translated into user QoE. This motivates the authors to categorize the users into three categories namely *Excellent*, *Good*, and *Fair* users. We define these user types on the basis of the user preferences over different involved parameters for network selection decision making. For instance, an *excellent user* is motivated to pay higher service prices for an excellent service quality and does not care much for service prices. One can think of putting *business users* in this category. On the other hand a *fair user* prefers cheaper services and remains ready to compromise on service quality, an example of such user may be a *student user*. On similar line, a good user stands midway between the two mentioned user types. When it comes to the differentiation of users on the basis of service quality, we mean the users perceived application QoS or QoE. Thus, to differentiate users on these lines for both *real-time* and *non-real-time* traffic types, we need different bounded regions of QoE values e.g., Mean Opinion Score (MOS) values(ranges between $[0-5]$, with *zero* representing the most irritated user and 5 representing the most satisfied user) are the numeric values capturing the user QoE for Voice over IP (VoIP) applications.

We generalize this QoE metric to all the traffic types and set the MOS value bounds for different user types on the similar lines as *Modified E-model* sets its R-factor values to distribute users in *very satisfied, satisfied*, and *some users dissatisfied* etc. categories. In this work the MOS values 4.3 and above, 3.59 ~ 4.3 and 3.1 ~ 3.59 represent the *excellent, good*, and *fair* users respectively. One may object the suitability of MOS values as QoE metric for non-real-time applications e.g., TCP based FTP traffic, and can argue *throughput* or *delivery response time* to be the suitable QoE measurement metric for such traffic. In this case a *transformation or scaling function* may be used to scale the user satisfaction to the MOS value range i.e., $[0 - 5]$. It should be noted that MOS value is the function of QoS measurement metric *delay, jitter*, and *packet loss*. However, we focus that user QoE is the function of both technical and economical parameters. In this connection, the authors have suggested analytical satisfaction function, which takes into account both the mentioned aspects. We validate the QoE prediction of user satisfaction function against the objective measurements (typically for technical parameters [1]). Whereas the user satisfaction for the service cost is captured through the following function.

$$r_k(\pi_k^c) = \tilde{\mu}_k^c - \frac{\tilde{\mu}_k^c}{1 - e^{\tilde{\pi}_k^c}} e^{-\tilde{\pi}_k^c \epsilon},$$

where $\tilde{\mu}_k^c$ represents the maximum satisfaction level of user type k for service type c , and $\tilde{\pi}_k^c$ is the private valuation of service by user, and ϵ represents the price sensitivity of user. Although authors have developed and extensively validated the utility based user satisfaction model against subjective (from experiments) and objective (using network simulator) for different dynamics of the wireless environment. The validation results showed that the proposed user satisfaction function predicts user QoE with the correlation 0.923. However we summarize in Table 8.4 the ranges of technical parameter values attained from user satisfaction function and validated against the subjective and objective measurement results. The range for service costs for different types of user follow the pattern $\pi_{excellent} > \pi_{good} > \pi_{fair}$ and the corresponding user satisfaction from the offered price is computed by the pricing function mentioned earlier.

8.4.1 Architecture for 4G User-Centric Paradigm

We briefly highlight the possible architectural issues associated with the implementation of the proposed user-centric approach and we also propose the

---

[1] As using modified E-model, PESQ etc. for VoIP application one can capture user QoE for delay, packet loss, and jitter parameters, therefore validation could be carried out for these parameters. Similarly for video applications we use PSNR and different codecs for validation.

TABLE 8.4
QoS parameters and ranges from the user payoff function

G 7.11 Codec		(96kbps)	
Parameters	Range	MOS	Category
Delay	$0ms \sim 50ms$	4.3 and above	Excellent
	50ms $\sim$ 200ms	3.59 $\sim$ 4.3	Good
	200ms $\sim$ 300ms	3.1 $\sim$ 3.59	Fair
Packet Loss	0% $\sim$ 3%	4.3 and above	Excellent
	3% $\sim$ 10%	3.59 $\sim$ 4.3	Good
	10% $\sim$ 18%	3.1 $\sim$ 3.59	Fair
	Non-real-time	FTP	
Delay	$0ms \sim 40ms$	4.3 and above	Excellent
	$40ms \sim 50ms$	3.59 $\sim$ 4.3	Good
	$50ms \sim 60ms$	3.1 $\sim$ 3.59	Fair
Packet Loss	0% $\sim$ 3%	4.3 and above	Excellent
	3% $\sim$ 3.5%	3.59 $\sim$ 4.3	Good
	3.4% $\sim$ 4%	3.1 $\sim$ 3.59	Fair
	Video	(x264)	
Delay	$0ms \sim 20ms$	4.3 and above	Excellent
	$20ms \sim 60ms$	3.59 $\sim$ 4.3	Good
	$60ms \sim 90ms$	3.1 $\sim$ 3.59	Fair
Packet Loss	0% $\sim$ 0.4%	4.3 and above	Excellent
	0.4% $\sim$ 1.5%	3.59 $\sim$ 4.3	Good
	1.5% $\sim$ 3.5%	3.1 $\sim$ 3.59	Fair

solution architecture by explaining its functional components and their integration.

Given the basic assumption of users having no long-term contractual agreements with the operators, the natural questions one can think of are:

- When the user mobile is turned on, what will be her default connection operator?

- assuming there exists a default connection operator, how and where in the technical infrastructure is the network selection decision executed?

- Who is responsible for user authentication in the system?

- How does an operator integrated the 3GPP and non-3GPP (trusted and untrusted) technologies providing host based network selection facility?

To address the highlighted issues, the proposed IP Multi-media Subsystem (IMS) based architecture should meet the following requirements;

- It should support the involvement of the third party and extension of services from different IMS core operators.

- It should delegate the service subscription control to the end-users. Hence the operators many implement any Session Initiation Protocol (SIP) services by researching the end user demands. The users should have freedom to subscribe to any service given that it could be delivered using IMS control plane.

- It should enable dynamic partnership of operators with the third party.

- Owing to business contract of User Equipment (UE) with the third party, the complete user profile should be maintained in the Home Subscriber Server (HSS) of third party and each operator should only receive the service specific user data from HSS.

We suggest a model similar to the semi-wall garden business model, where a network provider acts as a bitpipe plus a service broker. This model is open to all parties and its service panoply is as rich as the internet and is a converged business model. We propose IMS(IP Multimedia Subsystems) based on SIP and other IETF protocols to realize the proposed user-centric approach. We assume that there exists a neutral and trusted third party, *Telecommunication Service Provider (TSP)*, the TSP has no *Radio Access Network (RAN)* infrastructure, however it resides the few functional components of IMS architecture namely Proxy Call Session Control Function (P-CSCF), Serving Call Session Control Function (S-CSCF), Interrogating Call Session Control Function (I-CSCF), Application server (AS), and HSS. As discussed in the *OPNET simulation settings* section that operator communication footprint in any geographical area comprises of heterogeneous wireless access technologies (3GPP and non-3GPP).

Figure 8.9 details the integration of operator RANs to the operator core network and operator's integration to the trusted third party operator.

Entity	Stands for	Description
HSS	Home Sub-scriber Server	Central repository to store user information
S-CSCF	Serving call/session control function	SIP server additionally performing session control and also known as SIP registrar
OSA-SCS	Open Service Access Service Capability Server	Provides interference to OSA framework, provides capability to access the IMS security from external networks
IMSSF	IP Multimedia Service Switching Function	It allows to re-use in IMS the customized applications for mobile network enhanced logic services that were developed for GSM. Application servers like IMSSF, OSA SCS provide also the interfaces toward HSS based on Diameter protocol.
AS	Application Server	It is a SIP entity that hosts and executes services.

Sequence of Actions: Users send the service requests to the third party, which then transmits the requests to the available operators. Operators then submit the service offers including QoS indices values and service costs. The third party on behalf of users suggest the best available networks to the users for requested service. The third party takes care service billing.

Architecture functional entities and their interaction:

We now briefly discuss functional entities and their interaction with each other.

Trusted third party functional entity: This entity is basically a SIP application server, which processes SIP messages formulated using SIP MESSAGE method from UE and operators. In the proposed configuration, the SIP application on third party is enabled to understand XML (Extensible Markup Language) messages, which are enclosed in the message body of the proposed SIP MESSAGE method. After the third party receives this message from UE through UE → default operator IMS core network → third party (I,S)-CSCF → third party AS, the registration process is initiated and completed. The third party AS then extracts user service request for the body of SIP message from UE and triggers the network selection decision mechanism. After the computation at the decision maker, the third party generates one of the two possible responses.

- The network selection algorithm has successfully resulted in resource allocation and service price decision.

- These decisions are executed by generating two simultaneous responses; of which one goes to UE indicating successful operation and the other is sent to the operator.

Operator functional entity: Integration of operator technologies

Owing to the maturity of the current communication paradigm, it is needless to highlight the importance of heterogeneous wireless technologies and their co-existence to extend service to end-consumers. When it comes to heterogeneous wireless technologies, one can discern various prevailing standards in the current communication market, such as 3GPP, non-3GPP, 3GPP2, etc. The current communication market is framed to accept the integrated 3GPP and non-3GPP technologies. We follow the 3GPP standard for such integration as shown in Figure-8.9. We now consider the following use cases of the proposed architectural solution; basically non-3GPP technologies can be integrated with 3GPP technologies through one of the three interfaces ($S2a, S2b, S2c$) provided by EPC/SAE(Evolved Packet Core / System Architecture Evolution). The description of each of the interface is as follows:

- S2a - it provides the integration path between trusted non-3GPP IP networks and 3GPP networks. In this case the mobility is handled by the network-based mobility solution e.g., Proxy MIPv6.

- S2b - It provides the integration path between untrusted non-3GPP IP networks and 3GPP networks. In this case mobility is handled by network based mobility solution.

- S2c - It provides the integration between both trusted and un-trusted non-3GPP IP networks and 3GPP networks. In this case the mobility is handled by the host based mobility solution e.g., Dual stack MIPv6.

Interaction of operator with third party

In case the operator wants to participate in the game, she must configure the *Operator functional entity* to register its parameters for indicated time length. This entity can formulate one SIP message for one service or multiple cost and offered quality information for multiple services using one SIP message. However in this case care should be taken that the SIP message size does not exceed the upper bound defined by IETF standards for SIP messages. The *operator functional entity* should maintain a record of all its sent messages because such information cannot be retrieved from the third party application server. This entity, when declared as the chosen operator receives a notification from the third party as a SIP MESSAGE containing the service type, user preferences, user identity etc. It should be noted that in response to a SIP MESSAGE from this entity, a SIP specified acknowledgement response must be sent by the third party that indicates the status of registration process. Basically, the ACK methods in this case may be an OK or any other error message. It keeps track of the registration and their acknowledgements using the Command Sequence (CSeq) header file.

UE functional entity:

Assume that the user has successfully performed SIP registration with IMS platform of the default network. Now, if the UE wants to conduct a session as per proposed mechanism then she must include the type of service, her preferences and identity in regulation of allowed XML syntax and send this information in the body of SIP MESSAGE to the third party. Here we assume that SIP URI (Uniform Resource Identifier) of the third party is known to the user as part of the contract. A user sends only one session request in one SIP message. The UE is also enabled to parse the information received as the part of the response that is sent by the third party against her request. The response can basically lead in accepted or blocked service.

8.4.2   OPNET Simulation Setup

In this section, we describe the simulation setup for the proposed network selection approach. In order to simulate the reference scenario presented in Figure 8.10, the following entities are implemented:

- Impairment entity - we developed an impairment entity that introduces specified packet delays, packet losses and is also able to limit the bandwidth shaping using token bucket algorithm.

- LTE radio access network (eNodeB)).

- User Equipment (UE).

- Serving Gateway (S-GW).

- Packet Data Network Gateway (PDN-GW, whereas the following entities used in the simulation are OPNET standard node models:

  - Wireless LAN access point,

  - Application server,

  - Ethernet link,

– Routers, and

– Mobility model.

Note that the packet delay values in simulation only include codec delays (for real-time applications) and transport network delay excluding fixed delay components e.g., equipment-related delays, compression and decompression delays, etc.

For real-time VoIP applications, we use GSM EFR, G.711, and G.729 codecs in a simulation setup. The purpose of using different codecs is to enable operators to extend offers of different QoE to the users and analyze the users reaction to different offers. For real-time video applications, we use PSNR as video quality metric and make use of EvalVid framework for video quality evaluation.

In this setup, packet losses are injected using *Bernoulli distribution* and we use playout buffer of 250ms during the reconstruction of video file. We consider a reference video sequence called *Highway*. The motivation to use this video sequence its repeated reference in a large number studies in video encoding and quality evaluation e.g., Video Quality Experts Group. This video sequence has been encoded in *H*.264 format using the JM codec with CIF resolution (352×288) using a target bit rate of 256kbps. *H*.264 codec has been selected because its widespread use can be seen in future communication devices. The reference video sequence has total 2000 frames and frame rate of 30fps. Key frame is inserted after every $10th$ frame which provides good error recovery capabilities. An excellent video quality is indicated by 38.9db as an average PSNR value of encoded video sequence. The video file is transmitted over the IP network considering MTU size of 1024 bytes.

For a non-real-time FTP applications simulation setup, file size is considered to be 20MB, which can be downloaded through LTE and WLAN access network. The choice of file size is dictated by the facts; a) slow start effect of TCP can be ignored, b) correlation of TCP throughput and distribution of packetlosses within a TCP can be reduced. Here a bandwidth shaping of 8Mbps is performed. We use TCP flavor *New Reno* with receiver buffer size of 64KB.

As can be viewed in the Figure 8.10 that the users under consideration are covered by the two access networks namely LTE and WLAN of two different operators. The integration of these access technologies follow 3GPP recommendations for integration of 3GPP and non-3GPP access technologies. To have greater control of environment in terms of analysis, impairment entities are placed in the transport networks of each access technology.

Since the mobility is host-based, therefore MIPv6 based mobility management is implemented at PDN-GW, however for network-based mobility PMIPv6 can be implemented, where LMA resides at ePDG in untrusted integration case. User terminals are multi-interface devices, and are capable of simultaneously connecting to multiple access technologies. We also extensively implement the flow management entity, which acts a relay or applies filter rules over the traffic depending on uplink or downlink traffic.

In order to demonstrate user-centric based network selection, and demonstrate the effect of learning in such a telecommunication landscape, we run an extensive round of simulation runs. Service requests of different quality classes (user types) are generated by users. The arrival of requests is modeled by Poisson process, and the service class is chosen randomly among voice, data, and video uniformly. The sizes of requests are assumed to be static and are 60kbps, 150kbps, and 500kbps for voice, data, and video respectively. The capacities of LTE and WLAN network technologies are 32Mbps (downlink)/ 8Mbps (uplink), 8Mbps, respectively. As the network technologies are owned by two different operators, the technical configuration of the technologies owned by both the operators are very similar. However the service pricing scheme is operator specific, which influences the user-centric network selection decision.

8.4.3   Result Analysis

Within the simulation settings, we configure that all the users in the system have the same *initial probability list* i.e., 0.4, 0.3, 0.2, 0.1 for LTE (Operator 1), LTE (Operator 2), WLAN (Operator 1), and WLAN (Operator 2) respectively. We also configure that operator-1 offers lesser service costs when compared with the operator-2, whereas both the operators charge more on LTE than WLAN network technology. The configuration of the technical indices are the same for both the technologies and both the operators, thus the operators offer of technical parameters are influenced by the congestion, available bandwidth, wireless medium characteristics etc. The simulation was run for number of iterations and the convergence of user probabilities of network selection was observed for variable learning schemes. First we analyze the behavior of a *fair user* in the given settings, as can be observed in Figure 8.12 that a *fair* user adjusts its probabilities in the given configuration. As expected the user strategies converge (within relatively small number of iterations) so that she prefers the relatively less costly WLAN (Operator 1) more than any other technology, the probability values of other strategies are the consequences of both technical and non-technical offers of the operators. It should be noted that the Figure 8.12 is result in underloaded system configurations. i.e., both the technologies of both the operators are under utilized.

We now analyze the *fair user* behavior in the congested system configuration (congested system may defined as the system, where most of the resource are already utilized and the option window of user is squeezed), the results for such configuration are presented in Figure 8.13. The impact of congestion over the network selection can be seen by strategy convergence of the user. LTE (Operator 2) turns out to be the only under loaded network technology, this shrinks the options of the user and hence the different convergence result than that of under-loaded configuration even though the simulation settings remain the similar in both the configurations.

These results confirm the superiority of the proposed learning approach in user-centric 4G heterogeneous wireless network selection paradigm. A number of simulations were run and various other results in the similar fashion were taken, where service costs were varied, medium impairments (customized impairments were introduced in the wireless medium with the help of *impairment entity*) were introduced in the wireless access networks of different operators.

The objective of these scenarios was to analyze the behavior of user decision under various dynamics of the system. All the results follow the similar behavior as the ones shown in Figures 8.12,8.13 in different configurations. Thus on the basis of the presented results we can confidently claim that the proposed learning scheme fits well to the future user-centric wireless networks paradigm.

As an illustration, we have implemented the Bush-Mosteller based CODIPAS-RL. In Figure 8.14 we represent the evolution of strategies in a scenario with two users and same action set $m = 2$, $\mathcal{A}_j = \{1, 2\}$ for the two users. As we can observe, the trajectory goes to an equilibrium $(1/2, 1/2)$ which is not efficient. In Figure 8.15, we represent a convergence to an efficient outcome: global optimum using Bush-Mosteller based CODIPAS-RL for different action sets. Note that, in this scenario the convergence time to be arbitrary close is around is around 250 iterations which is relatively fast.

Exercise 8.1. *Examine the convergence of the sequence* $\eta_n^2 = \sum_{t=0}^{n} \frac{1}{(t+1)\log(t+1)}$ *and the sequence given by* $\eta_n^2 = \sum_{t=0}^{n} \frac{1}{(t+1)^2 \log(t+1)^2}$.

On similar lines as discussed in the *user-centric network selection* paradigm, Figures 8.16&8.17, we represent the behavior of the users and their estimated payoff when using variable learning schemes. When the users are active, they can select one of the CODIPAS learning schemes among $\mathcal{L}_1 - \mathcal{L}_5$ with probability distribution $[1/5, 2/5, 1/5, 1/10, 1/10]$. The users are active with probability 0.9. We fix the learning rate $\lambda_t = \frac{1}{(t+1)\log(t+1)}$. Figure 8.16 represents the evolution of strategies and Figure 8.17 represents the estimated payoff evolutions of user 1 and 2. As we can observe, the convergence occurs even for random updating time and hybrid CODIPAS-RLs. Not surprisingly, the convergence time seems very large.

8.5 Markov Chain Adjustment

In this section, a Markov chain adjustment of the interactive trial and error learning is proposed. We consider an interactive system with a finite number of players. The main components of the algorithm are described by the transitions of a Markov chain. Each player has its own Markov chain which is interdependent with the Markov chain of the other players via their decisions. This interdependency leads to interactive learning.

The framework that we present below is based on the work of Young et al. (2009-10). There are four main configurations denoted by $c, c+, c-$, and d. The configuration c represents the mood "content" and d represents a "discontent" state. The other states of the player $c+, c-$ are intermediary states before transitions to discontent d. We denote by $\bar{r}_{j,t}$ the current reference of player j (as a referent-payoff) and $r_{j,t}$ the received payoff, which implicitly depends on the choice of the players.

Each action will be drawn by a player according to her state and strategy. This generates an interactive Markov chain over the product state space. Next, we describe the transitions of the Markov chains.

8.5.1 Transitions of the Markov Chains

- Transitions from content: In "c", we distinguish four cases. The cases $c1 - c3$ are dedicated for experimentation and the state $c4$ is non-experimentation of new action.

  - $c1$: A player can decide whether to experiment a new action or not. The player experiments with probability $\epsilon > 0$ and receives a payoff $r_{j,t+1}$. If $r_{j,t+1} < \bar{r}_{j,t}$, then the reference is the same: $\bar{r}_{j,t+1} = \bar{r}_{j,t}$ and the user mood stays at the configuration $c1$. If $r_{j,t+1} > \bar{r}_{j,t}$, then Markov chain of player j goes to the case $c2$.

  - $c2$: If $r_{j,t+1} > \bar{r}_{j,t}$, player j accepts the new references and actions with probability
    $$q_j = e^{-\beta_j G_j(r_{j,t+1} - \bar{r}_{j,t})},$$
    where $\beta_j > 0$ and G_j is decreasing function which belongs to $[0, \frac{1}{2}]$. The player's chain goes to $c2$.

  - $c3$: If $r_{j,t+1} < \bar{r}_{j,t}$, player j rejects the new references and actions with probability $1 - q_j$ and stays at $c3$.

  - $c4$: No experimentation occurs with probability $(1 - \epsilon)$. If $r_{j,t+1} = \bar{r}_{j,t}$ then the chain stays at c. If $r_{j,t+1} > \bar{r}_{j,t}$ the chain goes to $c+$ and if $r_{j,t+1} < \bar{r}_{j,t}$, then the chain of j goes to $c -$.

- At the state $c+$ the player compares the new payoffs and the references. If $r_{j,t+2} \geq \bar{r}_{j,t}$, the chain goes back to c and the reference $\bar{r}_{j,t+1} = r_{j,t+2}$. If $r_{j,t+2} < \bar{r}_{j,t}$ the chain goes to $c -$.

- At state $c-$ the player compares again the new payoffs and the references. If $r_{j,t+2} > \bar{r}_{j,t}$, the chain goes back to $c+$, if $r_{j,t+2} = \bar{r}_{j,t}$ the chain goes to c and if $r_{j,t+2} < \bar{r}_{j,t}$ then the chain of player j goes to the discontent state d.

- Transitions from discontent state "d": At state d, the player randomly picks an action and receives $r_{j,t}$. The player accepts these new references with probability $p = e^{-\beta_j F_j(r_{j,t})}$ where F_j is a decreasing function such that $nF_j \in [0, 1/2]$, the chain goes to content state c. The user rejects the new references with probability $1 - p_j$ and the chain stay at d.

8.5.2 Selection of Efficient Outcomes

Next, we present the result of the efficient equilibrium in interactive trial and error learning. Recall that a pure Nash equilibrium of a finite game is a configuration such that no user can improve her payoff by deviating unilaterally. A pure equilibrium in this context corresponds to a configuration where

- the states are content,

- the payoffs are the references and are Nash equilibrium payoff.

> We say that Z is a stochastically stable state set if it is the minimal subset of states such that, given any small δ, there is a number $\epsilon_\delta > 0$ such that whenever $\epsilon \in [0, \epsilon_\delta]$, the Markov chain will visit the set Z at least $1 - \delta$ proportion of all time t.

Stochastically stable sets are very important in Markov trees. Our learning process will be most of the time in such a set. The next result shows that the stochastically stable sets are also good for the global optimization problem.

Proposition 8.5.2.1. *If every user follows the above learning process, a pure equilibrium will be visited an arbitrary high proportion of the time. Moreover, every stochastically stable state maximizes the sum function $\sum_j r_j$ i.e the system performance.*

To prove this result, we need to check the interdependency condition, which is the following: given any current choice of actions by the users, any proper subset of users can cause a payoff change for some player not in the subset by a suitable change in their joint actions. This condition is automatically satisfied if each user can make a change in the payoff of other players. Then, the proof of the theorem follows from Theorem 2 in [132].

Remark 8.5.3. *This result is very promising because it selects efficient outcomes that are "stable". However, it does not give the convergence nor the speed of convergence for being arbitrarily close to an equilibrium. Note that the speed of convergence of fully distributed learning algorithms to an approximated equilibrium is a very challenging open problem [69]. The authors have shown that the communication complexity is exponential for zero-error Nash equilibria and polynomial time for correlated equilibria (Nash equilibrium of the game with signal).*

In the next section, we focus on Pareto optimal solutions that are not necessarily equilibria.

8.6   Pareto Optimal Solutions

We now assume that the environment state is reduced to a singleton or empty set. Then, the framework of Young [132] can be extended to Pareto optimal solutions.

At each point in time, a player j's state can be represented as a triple $\bar{s}_{j,t} := (\bar{a}_{j,t}, \bar{r}_{j,t}, m_{j,t})$, where the referent action $\bar{a}_{j,t}$, also called benchmark action is $\bar{a}_{j,t} \in \mathcal{A}_j$, the referent payoff is $\bar{r}_{j,t}$ which is in the range of the $r_j(\prod_j \mathcal{A}_j)$, the current mood is $m_{j,t}$, which can take on two values: content (c) and discontent (d). We normalized the payoff such that $r_j \in [0,1)$.

     Strategy adjustment

We specify how the players' strategies will be changed during the long run interaction.

Fix an experimentation parameter $\epsilon > 0$. We distinguish two different forms of randomized actions depending on the value of the mood which we refer to feeling set. Recall that in this context a simple class of strategies for a player could be the mappings from the set of individual states to the action space of that player. This is what we will do in the description below.

- If $m_{j,t} = content$, the player chooses an action $a_{j,t}$ according to the following probability distribution

$$x_{j,t}(c, s_j) \sim \frac{1}{|\mathcal{A}_j| - 1} \epsilon^{\bar{M}} \delta_{\{a_{j,t} \in \mathcal{A}_j \setminus \{\bar{a}_{j,t}\}\}} + (1 - \epsilon^{\bar{M}}) \delta_{\bar{a}_{j,t}},$$

where the number $\bar{M}$ is chosen that $\bar{M} > \max \sum_j r_j$ and $|\mathcal{A}_j| \geq 2$. It suffices to take $\bar{M} > n$.

- If $m_{j,t} = discontent$, the player chooses an action $a_{j,t}$ according to the following probability distribution

$$x_{j,t}(d, s_j) \sim \frac{1}{|\mathcal{A}_j|} \delta_{a_{j,t} \in \mathcal{A}_j}$$

i.e., the uniform distribution over $\mathcal{A}_j$.

State dynamics

Once the players select actions, each player receives a payoff and the states of all the players change according the following rule:

- In state content (c), we distinguish the following cases: If $(a_{j,t+1}, r_{j,t+1}) = (\bar{a}_{j,t}, \bar{r}_{j,t})$ then the new state is determined by

$$(\bar{a}_{j,t+1}, \bar{r}_{j,t+1}, m_{j,t+1}) = (a_{j,t+1}, r_{j,t+1}, c).$$

  If $(a_{j,t+1}, r_{j,t+1}) \neq (\bar{a}_{j,t}, \bar{r}_{j,t})$ then the new state is determined by

$$(\bar{a}_{j,t+1}, \bar{r}_{j,t+1}, m_{j,t+1}) \sim$$

$$\epsilon^{1-r_{j,t+1}} \delta_{\{(a_{j,t+1}, r_{j,t+1}, c)\}} + (1 - \epsilon^{1-r_{j,t+1}}) \delta_{\{(a_{j,t+1}, r_{j,t+1}, d)\}}.$$

- In state discontent (d), if the selected action and received payoff are $(a_{j,t+1}, r_{j,t+1})$, then the new state is determined by the following rule:

$$(\bar{a}_{j,t+1}, \bar{r}_{j,t+1}, m_{j,t+1}) \sim$$

$$\epsilon^{1-r_{j,t+1}} \delta_{\{(a_{j,t+1}, r_{j,t+1}, c)\}} + (1 - \epsilon^{1-r_{j,t+1}}) \delta_{\{(a_{j,t+1}, r_{j,t+1}, d)\}}.$$

Interdependence condition

Assume in addition that the game satisfies the interdependence condition, i.e., for every action profile $a \in \prod_j \mathcal{A}_j$ and every proper subset of players $J \subset \mathcal{N}$, there exists a player $j' \notin J$ and a choice of actions $a'_J \in \prod_{j \in J} \mathcal{A}_j$ such that

$$r_j(a) \neq r_j(a'_J, a_{-J}).$$

The following result quantifies the frequency of visit of a Pareto optimal point of the game.

Theorem 8.6.0.1. *Let $\mathcal{G}$ be a finite game with single environment state, and assume that $\mathcal{G}$ is an interdependent game. If all players use the dynamics highlighted above, then a state s is stochastically stable if and only if the following conditions are satisfied:*

- *The action profile a optimizes $\sum_{j \in \mathcal{N}} r_j(a)$.*

- *The referent actions and payoffs are aligned, i.e., $\bar{r}_j = r_j(a)$.*

- *The mood of each player is content, i.e., $m_j = c$.*

CODIPAS trial and error for Pareto optimality

Now we add an i.i.d noise to the payoff and associate a payoff learning based on CODIPAS. The scheme is essentially similar. The referent payoff is replaced by the estimated payoff provided by the CODIPAS. The scheme is generic and is well-adapted for cooperative solution concepts such as *Pareto optimality, bargaining solution and global optimum* in finite dynamic robust games. Compared to classical trial and error approach, additional difficulties arise due to the variability of the state of the environment. CODIPAS learning induces a Markov process which can be seen as perturbation of a Markov chain over the finite state space

$$\mathcal{Z} = \prod_j (\mathcal{A}_j \times \mathcal{R}_j \times \mathcal{M}_j)$$

where $\mathcal{R}_j$ denotes the union (over w) of the finite range of $R_j(w, a)$ over all $a \in \mathcal{A} = \prod_j \mathcal{A}_j$, and $\mathcal{M}_j$ is the set of moods $\mathcal{M}_j = \mathcal{M} = \{content, discontent\}$.

We use the theory of resistance trees for regular perturbed Markov decision processes to prove that an action profile is stochastically stable if and only if it maximizes the sum of the expected payoff i.e., a Pareto efficient solution of the expected robust game. Below provide a brief background on the theory of resistance trees.

8.6.1 Regular Perturbed Markov Process

The following lemma holds:

Lemma 8.6.1.1. *The Markov process P^ϵ defined by CODIPAS learning is a regular perturbed Markov process.*

8.6.2 Stochastic Potential

The next theorem is established in [192] and is of great importance in our setting.

Theorem 8.6.2.1. *Let P^ϵ be a regular perturbed Markov process, and for each $\epsilon > 0$ let $(\mu_z^\epsilon)_{z \in \mathcal{Z}}$ be the unique stationary distribution of P^ϵ. The stochastically stable states (i.e., the support of the stationary distribution of P^0) are precisely those states contained in the recurrence classes with minimum stochastic potential.*

Let C^0 be the subset of states in which all the players are content and referent action and payoff are aligned, and D^0 represents the set of states in which everyone is discontent.

The next result provides a characterization of the recurrent classes of the unperturbed process.

The recurrence classes of the unperturbed process P^0 are D^0 and all singletons $z \in C^0$.

The stochastic potential associated with any state $z \in C^0$ is

$$c(|\, C^0 \,| - 1) + \sum_{j \in \mathcal{N}} (\bar{c} - \hat{r}_j(a))$$

where $c > 0$, $\bar{c} = \max_{w \in \mathcal{W}} \max_{a \mathcal{A}} R_j(w, a)$.

The minimizers of the stochastic potential are the maximizers of $\sum_j \hat{r}_{j,t}(a)$ which converges to $\sum_j \mathbb{E}_w R_j(w, a)$ when $t \longrightarrow +\infty$.

Combined learning in large-scale networks

An interacting network can be viewed as a network of decision making nodes, decisions taken and information generated in one part, or node/data center/mobile terminal/agent, rapidly propagate to other nodes, and have impact on the well being (as captured by payoffs) of node at those other nodes. Hence, it is not only the information flow that connects different nodes, but also the cross-impact of individual actions. Individual nodes or a coalition of nodes therefore know that their performance will be affected by decisions taken by at least a subset of other nodes or other coalitions, just as their decisions will affect others. To expect a collaborative effort toward picking the "best" decisions is generally difficult and for various reasons, among which are non-alignment of individual objectives, limits on communication, incompatibility of beliefs, and lack of a mechanism to enforce a stable cooperative solution. Sometimes a node will not even know the objective or payoff functions of other nodes, their motivations, and the possible cross-impacts of decisions (e.g. in cloud, social networks, sensor networks, opportunistic wireless networks, etc.).

Under such constrained situations How can one define a solution concept that will accommodate different elements of such an uncertain strategic decision making environment? How can such a solution be reached when nodes operate under incomplete information? Can nodes learn through an iterative process and with strategic plays the solution-relevant part of the interaction? Would such an iterative process converge, and to the desired solution, when nodes learn at different rates, employ heterogeneous learning schemes, receive information at different rates, and adopt different attitudes toward risk (some being risk-neutral, other being risk-sensitive)? Using control and game theoretic tools, combined learning aims to study inefficiencies in networks, driven by divergence of individual behavior (selfishness) (Nash equilibrium, Stackelberg solutions) from socially optimal allocation. It addresses one of the major challenges in the design of wireless networking and communications, the need for fully distributed strategic learning schemes that consume a minimal amount of information, a minimal amount resources, and very fast convergence time. Each decision maker will try to learn simultaneous the objective function and the associated optimal strategies called combined learning. Combined learning provides deeper understanding of network dynamics and leads to better design of efficient, scalable, stable and robust networks. Combined learning allows us to study quantitative metrics of the cost of selfishness (the Price of Anarchy - PoA), of having limited knowledge of the competitors (the Price of Ignorance - PoI), and of limited time for learning the network environment (the Price of Impatience - PoIm) in dynamically changing radio channels.

Open issues

- Is it possible to extend the above result to robust games and stochastic games with interdependent states and interdependent payoffs?

- What is the performance loss if one of the players does not obey the rules?

- How to establish a convergence result for strategies instead of the frequency of visits of the perturbed Markov chain?

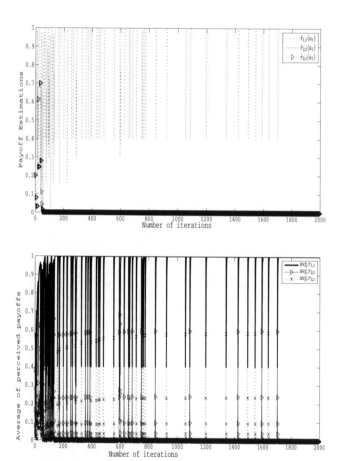

FIGURE 8.7
Three users and two actions.

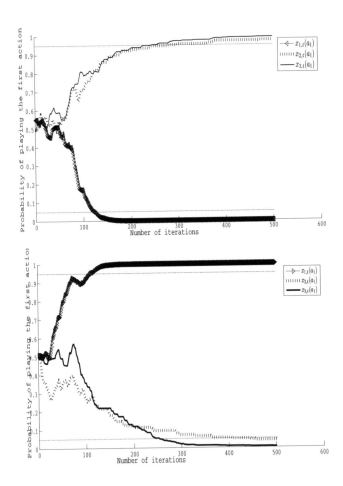

FIGURE 8.8
Impact of the initial condition.

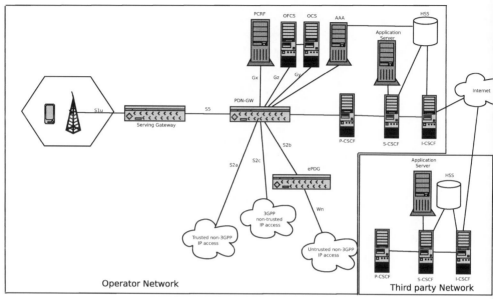

FIGURE 8.9
IMS based integration of operators with trusted third party

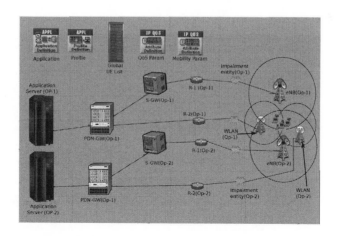

FIGURE 8.10
OPNET simulation scenario

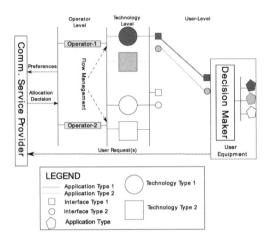

FIGURE 8.11
The scenario

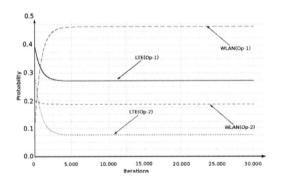

FIGURE 8.12
Evolution of randomized actions for underloaded configuration

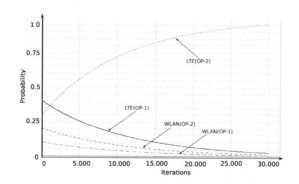

FIGURE 8.13

Evolution of randomized actions for congested configuration.

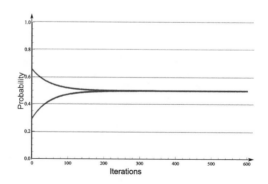

FIGURE 8.14

Convergence to equilibrium.

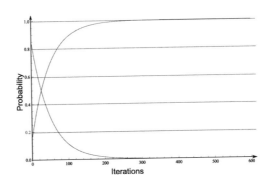

FIGURE 8.15
Convergence to global optimum.

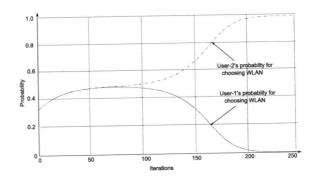

FIGURE 8.16
Evolution of randomized actions.

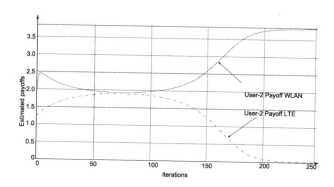

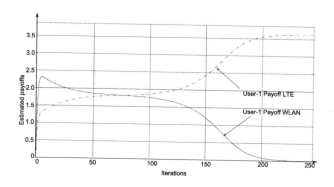

FIGURE 8.17
Evolution of estimated payoffs.

9

Learning in Risk-Sensitive Games

9.1 Introduction

Uncertainty, noise, and time delays are frequently met in wireless networks and communications in multiuser systems. Under such configuration, it is crucial to have adaptive and learning schemes based on the perceived measurements.

This chapter proposes a theoretical framework for learning in risk-sensitive dynamic robust games. The risk-sensitive model began with two key observations: first, that a player chooses between lotteries that the other players generate, and second, that payoffs are not the same as utilities, as models of learning in games often implicitly assume. It then becomes important to study the consequences of risk-aversion, or, more generally, utility functions that are not risk-neutral. A simple way to model a lottery is to approximate its utility with a mean value, a variance, etc.

The central conceptual insight of this chapter is that when a player chooses an action in a game, he/she is choosing a lottery. The feasible payoffs of the lottery are determined by the feasible payoffs associated with the action, and the probabilities are determined by the players' action probabilities.

In many utility models with Boltzmann-Gibbs strategy or quantal response strategy, players always have some non-zero probability of choosing a given action. However, the present chapter deviates from others by making an explicit distinction between rewards and utilities. If the payoffs in games are utilities, then in a sense risk is irrelevant: the expected value of an action is just its mean utility. If, however, the payoffs are rewards, then the fact that utility is not linear in rewards makes a difference and risk matters. In many realistic situations, of course, people earn money or some equivalent, and choosing an action in a game involves risk, so that learning how risky an action is, becomes important for assessing its value.

If a player takes a surprising action, then one's own action must be at least somewhat risky. However, exactly how risky it is depends on how much the other players' actions affects one's payoff. For example, if the surprising action does not affect the payoff of taking an action, then the value of the action should not be significantly affected. If the surprising action does affect the payoff, then a player should respond to the surprise.

Learning about risk would not be important if utility were a linear function of payoffs. However, it is well-known that utility is not a linear function of payoffs. Perhaps the best known version of this is risk aversion in utility calculations. When we incorporate the possibility of a nonlinear utility function, both the expected value of an action and the payoff variance and other moments become important. Therefore, thinking in terms of the basic reinforcement learning problem defined in the previous chapters, it is important to estimate the expected values, the variances, and higher moments.

We develop a risk-sensitive <u>CO</u>mbined fully <u>DI</u>stributed <u>PA</u>yoff and <u>S</u>trategy <u>R</u>einforcement <u>L</u>earning (CODIPAS-RL) to capture realistic behaviors of the players allowing for the possibility of not only risk-sensitivity, limited capabilities, limited observations and computations but also for heterogeneity, adaptive learning scheme, time-delayed, and noisy measurements. We analyze the asymptotic behavior of the dynamic robust game under risk-sensitive ergodic payoffs. In this setting:

- We study convergence and non-convergence properties of the hybrid CODIPAS-RL in Lyapunov dynamic robust games.

- We provide closed-form solutions for specific hybrid learning schemes under a risk-sensitive criterion.

- The convergence time of the hybrid CODIPAS-RL is explicitly given for the payoff-learning,

- In contrast to the standard learning analysis developed in the literature which is limited to a finite number of players, we extend the hybrid-CODIPAS-RL to a class of mean field learning and an explicit mean field dynamics is provided when the size of the system goes to infinity.

Key Additions and Links with the Risk-Neutral Criterion

In this chapter, we focus on hybrid and combined risk-sensitive strategic learning for general-sum stochastic dynamic games with incomplete information and action-independent state-transition with the following points:

- The risk-sensitive criterion is considered. It takes into consideration not only the expectation but also the variance and all the moments. When the risk-sensitive indexes are close to zero, the risk-neutral case is covered. New classes of games: risk-sensitive robust potential games and risk-sensitive monotone games are analyzed and convergence results are provided.

- The players do not need to follow the same learning patterns. We propose different learning schemes that the players can adopt. This leads to *heterogeneous learning*. This is in contrast to standard learning approaches widely studied in the literature. Our motivation for heterogeneous learning follows from the observation that, in heterogeneous systems, the players may not see the environment in the same way and may have different capabilities and different adaptation degrees. Thus, it is important to take into account these differences when analyzing the behavior of the system. *As we will see, the heterogeneity in the learning is crucial in terms of convergence of certain systems.*

- Each player does not need to update her strategy at each iteration. The updating time is random and unknown to the players.

  Usually, in the iterative learning schemes, the time slots at the player updates are fixed. Here, we do not restrict ourselves to fixed updating time. This is because some players come in or exit temporary, and it may be costly [25] to update, or for some other reason the players may prefer to update their strategies at other time slot. One may think that if some players do not update often, the risk-sensitive strategic learning process will be slower in terms of convergence time; this statement is less clear because the off-line players may help indirectly the online players to converge and, when they wake up, they respond to an already converged system, and so on. This suggests a multiple time-scaling scheme.

- Each player can be in active mode or in sleep mode. When a user is active, she can select among a set of learning patterns to update her strategies and estimations. The player can change their learning pattern during the interaction. This leads to an *hybrid learning.*

- In contrast to the standard learning frameworks developed in the literature which are limited to finite and fixed number of players, we extend our methodology to large systems with multiple classes of populations. This allows us to address the curse of dimensionality problem when the size of the interacting system is very large. Finally, different mean field learning schemes are proposed using Fokker-Planck-Kolmogorov equations.

- The case of noisy and time delayed payoffs is also discussed.

The chapter is structured as follows. In Section 9.2, we describe the model of a non-zero-sum dynamic robust game, and present different learning patterns. In Section 9.3, we develop hybrid and delayed learning schemes in a noisy and dynamic environment. Section 9.5 presents mean field learning. Section 9.7 concludes the paper.

9.1.1 Risk-Sensitivity

Mixed (or randomized) actions have been widely examined in game theory and its applications (economics, finance, biology, ecology, engineering, etc.). When at least one the players randomizes her action, the resulting payoff of the players become a random variable. The classical approach resulting from von Neumann and Morgerstern utility theory suggests using the mathematical expectation of the payoff function, also called *expected payoff*. The expected payoff has a natural interpretation in statistics and in learning theory based on the basic observation from the law of large numbers and ergodic theory.

If the game is played infinitely many times and the players always implement a fixed randomized action profile, then the average payoff really obtained by the players converge with probability one to the expected payoff.

Now, consider the following two scenarios:

A player is given the choice between two scenarios, one with a guaranteed payoff and one without.

Scenario 1: In the guaranteed scenario, the player gets 0 surely.

Scenario 2: In the uncertain scenario, a coin is flipped to decide whether the player receives 10^6 or -10^6 Each player gets a -10^6 with probability 1 if she loses and she gets $+10^6$ with probability 1 if she wins. Each player wins or loses with probability $1/2$.

Using the classical approach with the expected payoff criterion, scenarios 1 and 2 lead to the *same outcome*. The expected payoff for both scenarios is 0, meaning that a player who was insensitive to risk would not care whether she took the guaranteed payoff or the gamble. However, players may have different risk attitudes. It is clear that in scenario 2, the player faces a big risk which can be reflected by the variance (or higher moments) of the outcome. A risk-neutral player ignores the risk. However, there is always a risk whenever the action profiles are selected at random. A risk-averse player would accept a payoff of less than 0 (for example, -10), with no uncertainty, rather than taking the gamble and possibly receiving nothing. A risk-seeking player would accept the guaranteed payoff must be more than 0 (for example, 10) to induce her to take the guaranteed option, rather than taking the gamble and possibly winning 10^6.

The first step to take into consideration this phenomenon is to modify the expected payoff criterion to incorporate also variance, third moments, etc.

> We define the term risk as the chances of a huge loss occurring even with the small probability, and propose a risk-sensitive approach that considers distributions not to increase the expected payoff but to mitigate the risks of huge loss. Concretely, instead of the expected payoff, we employ a risk-sensitive metric.

Let

$$var(\tilde{U}_j) = \mathbb{E}[(\tilde{U}_j(w,a))^2] - \left(\mathbb{E}[\tilde{U}_j(w,a)]\right)^2,$$

be the variance of the payoff where $\tilde{U}_j$ is the payoff function of player j. Note that the function $\mathbb{E}[\tilde{U}_j(w,a)] - var(\tilde{U}_j)$, is not necessarily adapted to this scenario. A criterion that takes into consideration all the moments of the random payoff could be

$$\tilde{r}_{j,\mu_j} = \frac{1}{\mu_j} \log\left[\mathbb{E}e^{\mu_j \tilde{U}_j}\right],$$

where $\mu_j \neq 0$ is the risk sensitivity index.

Using Jensen's inequality, for $\mu_j < 0$, the certain equivalent payoff $\tilde{r}_{j,\mu_j} \leq \mathbb{E}\tilde{U}_j$ and when $\mu_j > 0$, the opposition inequality holds. In the first case ($\mu_j <$

0) a player having negative risk factor $\mu_j < 0$ and grading a random payoff according to the certain equivalent payoff $\tilde{r}_{j,\mu_j}$ is referred to as risk-averse and the second case ($\mu_j > 0$) is referred to as risk-seeking .

It is obvious that the certainly equivalent payoff $\tilde{r}_{j,\mu_j}$ takes into consideration all the moments of the payoff $\tilde{U}_j$. By performing a Taylor expansion for μ_j close to zero, one gets,

$$\tilde{r}_{j,\mu_j} \underset{\mu_j \sim 0}{\approx} \mathbb{E}\tilde{U}_j + \frac{\mu_j}{2} var(\tilde{U}_j) + o(\mu_j).$$

A strategic form game with risk-sensitive players (one index μ_j per player) is called a*risk-sensitive game*. We examine such games in a dynamic and uncertain environment.

9.1.2 Risk-Sensitive Strategic Learning

Different from *standard distributed learning optimization*, we use the term *risk-sensitive strategic learning*.

By risk-sensitive strategic learning, we mean how players who are risk-sensitive (averse, seeking, or neutral), are able to learn about the dynamic environment under their complex and interdependent strategies - the convergence of learning of each player depends on the others, and so on.

We provide basic convergence results in risk-sensitive games. These are obtained for a particular structure of the payoffs and action spaces [89, 18].

- Games against nature (with i.i.d random processes).

- Lyapunov risk-sensitive robust games (any finite robust game in which the risk-sensitive payoff leads to hybrid dynamics which have a Lyapunov function). Particular classes of these games are risk-sensitive potential games, common interest games, dummy games, congestion games, etc, under specific dynamics.

- Two-player-two-action games for well-chosen learning patterns and generic risk-sensitive payoffs,

- Particular classes of games with monotone risk-sensitive payoffs,

- Dominant solvable games (games with a dominant strategy) under risk-sensitive criterion,

All the above convergence results for specific classes of games are extended to the class of games under uncertainty (i.i.d).

9.1.3 Single State Risk-Sensitive Game

Risk-neutral problem formulation

The notations are detailed in the next section. For each $j \in \mathcal{N}$, and $\mathbf{x}_{-j}$, one would like to solve the *risk-neutral* optimization problem $RN-OP_j(\mathbf{x}_{-j})$:

$$RN - OP_j(\mathbf{x}_{-j}) : \sup_{\mathbf{x}_j \in \mathcal{X}_j} \quad \mathbb{E}_{\mathbf{x}} \tilde{U}_j(a) \qquad (9.1)$$

Risk-sensitive problem formulation

For each $j \in \mathcal{N}$, and $\mathbf{x}_{-j}$, one would like to solve the *risk-sensitive* robust optimization problem $RS - ROP_j(\mathbf{x}_{-j})$:

$$RS - OP_j(\mu_j, \mathbf{x}_{-j}) : \sup_{\mathbf{x}_j \in \mathcal{X}_j} \quad \frac{1}{\mu_j} \log \left[\mathbb{E}_{\mathbf{x}} e^{\mu_j \tilde{U}_j(a)} \right] \qquad (9.2)$$

The question is how to solve the optimization problems $RS/RN - OP_j(., \mathbf{x}_{-j})$ without knowledge of the mathematical structure of the function $\tilde{U}_j$?

Beyond the specific classes of games mentioned above with risk-neutral payoff, we want to know if it is possible to generalize the convergence results in the class of risk-sensitive games.

9.1.4 Risk-Sensitive Robust Games

A typical problem that we try to solve in this chapter is the following:

Risk-neutral problem formulation of robust games

For each $j \in \mathcal{N}$, and $\mathbf{x}_{-j}$, one would like to solve the *risk-neutral* robust optimization problem $RN - ROP_j(\mathbf{x}_{-j})$:

$$RN - ROP_j(\mathbf{x}_{-j}) : \sup_{\mathbf{x}_j \in \mathcal{X}_j} \quad \mathbb{E}_{\mathbf{w}} U_j(\mathbf{w}, \mathbf{x}_j, \mathbf{x}_{-j}) \qquad (9.3)$$

Risk-sensitive problem formulation of robust games

For each $j \in \mathcal{N}$, and $\mathbf{x}_{-j}$, one would like to solve the *risk-sensitive* robust optimization problem $RS - ROP_j(\mathbf{x}_{-j})$:

$$RS - ROP_j(\mu_j, \mathbf{x}_{-j}) : \ \sup_{\mathbf{x}_j \in \mathcal{X}_j} \quad \frac{1}{\mu_j} \log \left[\mathbb{E}_{\mathbf{w}, \mathbf{x}} e^{\mu_j \tilde{U}_j(\mathbf{w}, \mathbf{a}_j, \mathbf{a}_{-j})} \right] \quad (9.4)$$

The question is how to solve the robust optimization problems $RS/RN - ROP_j(., \mathbf{x}_{-j})$ without knowledge of the mathematical structure of the function $\tilde{U}_j$ or the distribution of $\mathbf{w}$?

> *Beyond these specific classes of games with risk-neutral payoff, we want to know if it is possible to generalize the convergence results into the class of risk-sensitive robust games.*

9.1.5 Risk-Sensitive Criterion in Wireless Networks

To illustrate the relevance of the risk-sensitive criterion in wireless networks, we provide two basic examples.

Security in 802.11 Wireless Networking

802.11 wireless networking is a fast developing technology and new risks are being frequently discovered. Many people feel their home network is at a low risk for attack, but if you have an open Wireless Access Point on your network you are inviting unnecessary risk. As wireless technology has seen wider deployment, many people have realized that they can transfer the risk of spamming, illicit downloads and illegal activity to someone else by doing it through their open wireless Access Point. If the Internet Service Provider detects and tracks the illegal activity to the source address, then that address can be that of the unsuspecting homeowner who will have a lot of questions to answer.

> The question is: Why are the percentages of unsecured wireless networks so high in residential areas even though the risk and damages from having an unprotected network can be catastrophic?

To answer this question, the behavior of the users need to be considered through a risk-sensitivity analysis. Depending on the risk-aversion and risk-seeking of the users, the percentage of unprotected networks changes. Thus, the risk-sensitive criterion is an appropriate criterion in wireless network security systems to analyze the behaviors of the users.

Randomly Varying Channels

Consider a wireless network with an arbitrary varying channel. The payoff function of transmitter j is its mutual information $I(\tilde{s}_j; \tilde{y}_j)(\mathbf{H}, Q)$ where $\tilde{s}_j$ is the input signal, $\tilde{y}_j$ is received signal from j, $\mathbf{H}$ is the channel state and Q is the power allocation scheme. Consider the maximum information rate as the performance of the system. Since $\mathbf{H}$ is a random variable, the classical approach consists of taking the mathematical expectation $\mathbb{E}_{\mathbf{H}} I(\tilde{s}_j; \tilde{y}_j)(\mathbf{H}, Q)$.

Now, consider for simplicity, two channel states: good or bad state. Assume that the channel distribution is $(\frac{1}{2}, \frac{1}{2})$. If the channel is good, the transmitter gets an acceptable payoff $(3/2)\beta > 0$,. If the channel is bad, the transmitter gets nothing. Assume that the minimum guarantee for successful admission is β. With the risk-neutral approach, the expectation is $(3/4)\beta < \beta$, which does not guarantee the minimum requirement. Thus, a transmitter is facing the risk on the variability of the channel, which need, to be considered in the analysis of the performance of the transmitter.

9.2 Risk-Sensitive in Dynamic Environment

9.2.1 Description of the Risk-Sensitive Dynamic Environment

We consider a system with a finite number of *potential players*. The set of players is denoted by $\mathcal{N} = \{1, 2, \ldots, n\}$, $|\mathcal{N}| = n$. The number n can be 10, 10^4, or 10^6. Each player has a finite number of actions denoted by $\mathcal{A}_j$ (which can be arbitrary large). Time is discrete, and the time space is the set of natural numbers $\mathbb{Z}_+ = \{0, 1, 2, \ldots\}$. A player does not necessarily interact at all the time steps. Each player can be in one of the two modes: *active mode* or *sleep mode*. The set of players in interaction at the current time is the set active players $\mathcal{B}^n(t) \subseteq \mathcal{N}$. This time-varying set is unknown to the players. When a player is in an active mode, she does an experiment, and gets a measurement or a reaction of her decision, denoted $u_{j,t} \in \mathbb{R}$ (this may be delayed as we will see). Let $\mathcal{X}_j := \Delta(\mathcal{A}_j)$ be the set of probability distributions over $\mathcal{A}_j$, i.e., the simplex of $\mathbb{R}^{|\mathcal{A}_j|}$. The number $u_{j,t} \in \mathbb{R}$ is the realization of a random variable $\tilde{U}_{j,t}$, which depends on the state of nature $\mathbf{w}_t \in \mathcal{W}$ and the action of the players, where the set $\mathcal{W}$ is a subset of a finite dimensional Euclidean space. Each *active player* j updates her current strategy $\mathbf{x}_{j,t+1} \in \Delta(\mathcal{A}_j)$ based on her experiments and her prediction for her future interaction via the payoff estimation $\hat{\mathbf{r}}_{j,t+1}^{rs} \in \mathbb{R}^{|\mathcal{A}_j|}$. Similarly, we denote the risk-sensitive estimation by $\hat{\mathbf{r}}_{j,t+1}^{rs} \in \mathbb{R}^{|\mathcal{A}_j|}$ and $r_{j,t} = \frac{e^{\mu_j u_{j,t}} - 1}{\mu_j}$.

This leads in the class of dynamic games with unknown payoff function and with imperfect monitoring (the last decisions of the other players are not observed). A risk-neutral payoff in the long-run interaction is the average payoff, which we assume to have a limit. In that case, under the stationary strategies, the limiting of the risk-neutral average payoff $\frac{1}{T+1} \sum_{t=0}^{T} u_{j,t}$ can be expressed as an expected game, i.e., the game with payoff

$$v_j : \prod_{j \in \mathcal{N}} \mathcal{X}_j \longrightarrow \mathbb{R},$$

$$v_j(\mathbf{x}_1, \mathbf{x}_2, \ldots, \mathbf{x}_n) = \mathbb{E}_{\mathbf{x}_1, \mathbf{x}_2, \ldots, \mathbf{x}_n} \left(\mathbb{E} \tilde{U}_j \right).$$

Similarly, we define the *risk-sensitive payoff* in the long run as the limiting of

$$R_{j,\mu_j,T} = \frac{1}{\mu_j(T+1)} \sum_{t=0}^{T} (-1 + e^{\mu_j u_{j,t}})$$

by omitting the logarithmic term, and

$$r_{j,\mu_j}(\mathbf{x}) = \mathbb{E}_{\mathbf{w},\mathbf{x}} \frac{1}{\mu_j} (e^{\mu_j \tilde{U}_j(\mathbf{w},\mathbf{a})} - 1)$$

Assumptions on player's information

The only information assumed is that each player is able to observe or to measure a noisy value of her payoff when she is active and update her strategy based on this measurement.

Note that the players do not need to know their own action space in advance. Each player can learn her action space (using, for example, exploration techniques). In that case, we need to add an exploration phase or a progressive exploration during the dynamic game. It is well-known in machine learning that if the all the actions have been *explored* and sufficiently *exploited* and if the learning rate are well-chosen then the prediction can be "good" enough.

Next, we describe how the risk-sensitive dynamic robust game evolves.

9.2.2 Description of the Risk-Sensitive Dynamic Game

The dynamic robust game is described as follows:

Initialization.

- At time slot $t = 0$, $\mathcal{B}^n(0)$ is the set of active players. The set $\mathcal{B}^n(0)$ is not known to the players. We assume that each player has her internal state in $\{0, 1\}$. The number 1 corresponds to the case where $j \in \mathcal{B}^n(0)$, and 0 otherwise. Each player $j \in \mathcal{B}^n(0)$ chooses an action $a_{j,0} \in \mathcal{A}_j$. The set $\mathcal{A}_j$ is not assumed to be known in advance by player j, we assume that she can explore progressively during the interactions. He measures a numerical noisy value of her payoff which corresponds to a realization of the random variables depending on the actions of the other players and the state of the nature. She initializes her estimation to $\hat{r}_{j,0}^{rs} \in \mathbb{R}$. Each of the non-active players gets zero.

- At time slot t, each player $j \in \mathcal{B}^n(t)$ has an estimation of her payoffs, chooses an action based on her own-experiences and experiments a new strategy. Each player j measures or observes an output $u_{j,t} \in \mathbb{R}$, (eventually after some time delay). Based on her measured value $u_{j,t}$, the player j updates her estimation vector $\hat{\mathbf{r}}_{j,t}^{rs} \in \mathbb{R}^{|\mathcal{A}_j|}$ and builds a strategy $\mathbf{x}_{j,t+1} \in \mathcal{X}_j$ for her next interaction. The strategy at time slot $t+1$, $\mathbf{x}_{j,t+1}$ is a function only of $\mathbf{x}_{j,t}, \hat{\mathbf{r}}_{j,t}^{rs}$ and the most recent target value. Since the players do not interact always, each player has her own clock which counts the activity of that player. At time step t, the clock of player j is $\theta_j(t) = \sum_{t' \leq t} \mathbb{1}_{\{j \in \mathcal{B}^n(t')\}}$.

We assume $\liminf_{t \to \infty} \theta_j(t)/t > 0$. This assumption eliminates short term players in the analysis.

Note that the exact value of the state of the nature at time t, the previous mixed actions $\mathbf{x}_{-j,t-1} := (\mathbf{x}_{k,t-1})_{k \neq j}$ of the other players and their past perceived payoffs $\mathbf{u}_{-j,t-1} := (u_{k,t-1})_{k \neq j}$ are unknown to player j at time t.

- The game moves to $t+1$.

In addition to the above description, we extend the framework to the delayed payoff measurement case. This means that the perceived payoffs at time t are not the instantaneous payoff but the noisy value of the payoff at $t - \tau_j$ i.e., $u_{j,t-\tau_j}$ for a certain time delay $\tau \geq 0$.

In order to define the dynamic robust game, we need some preliminaries. Next, we introduce the notions of histories, strategies and payoffs (performance metrics) as in Chapters 4, 5, and 8. The payoff is associated with a (behavioral) strategy profile which is a collection of mapping from the set of histories to the available actions at the current time.

⚠ Remember that you don't have to know you are playing game to be in a game!

A player's information consists of her past own activities, own actions, and measured own payoffs. A private history up to t for player j is a collection

$$h_{j,t} = (b_{j,0}, a_{j,0}, u_{j,0}, b_{j,1}, a_{j,1}, u_{j,1}, \ldots, b_{j,t-1}, a_{j,t-1}, u_{j,t-1})$$

in the set $H_{j,t} := (\{0,1\} \times \mathcal{A}_j \times \mathbb{R})^t$. where $b_{j,t} = \mathbb{1}_{\{j \in \mathcal{B}^n(t)\}}$, which is 1 if j is active at time t and 0 otherwise.

Behavioral strategies

A behavioral strategy for player j is a mapping

$$\tilde{\tau}_{j,t} : H_{j,t} \longrightarrow \mathcal{X}_j.$$

Let $\tilde{\tau}_j := (\tilde{\tau}_{j,t})_t$. We denote by Σ_j the set of behavioral strategies of player j. The set of complete histories of the dynamic robust game after t stages is $H_t = (2^{\mathcal{N}} \times \mathcal{W} \times \prod_{j \in \mathcal{N}} \mathcal{A}_j \times \mathbb{R}^n)^t$, it describes the set of active players, the states, the chosen actions and the received payoffs for all the players at all past stages before t. The set $2^{\mathcal{N}}$ denotes the set of all the subsets of $\mathcal{N}$ (except the empty set). A behavioral strategy profile $\tilde{\tau} = (\tilde{\tau}_j)_{j \in \mathcal{N}} \in \prod_j \Sigma_j$ and an initial state $\mathbf{w}$ induce a probability distribution $P_{\mathbf{w},\tilde{\tau}}$ on the set of plays $H_\infty = (\mathcal{W} \times \prod_j \mathcal{A}_j \times \mathbb{R}^n)^{\mathbb{Z}_+}$.

Log-run payoffs

Assume that $\mathbf{w}, \mathcal{B}^n$ are independent and independent of the strategy profiles. For a given $\mathbf{w}, \mathcal{B}^n$, we denote

$$U_j^{\mathcal{B}^n}(\mathbf{w}, \mathbf{x}) := \mathbb{E}_{(\mathbf{x}_k)_{k \in \mathcal{B}^n}} \tilde{U}_j^{\mathcal{B}^n}(\mathbf{w}, (a_k)_{k \in \mathcal{B}^n}).$$

Let $\mathbb{E}_{\mathbf{w},\mathcal{B}^n}$ be the mathematical expectation relatively to the measure generated by the random variables $\mathbf{w}, \mathcal{B}^n$. Then, the expected payoff can be written as $\mathbb{E}_{\mathbf{w},\mathcal{B}^n} \tilde{U}_j^{\mathcal{B}^n}(.,.)$. We examine both risk-neutral and risk-sensitive criteria. The *average risk-neutral payoff* is

$$F_{j,T} = \frac{1}{T} \sum_{t=1}^{T} u_{j,t}.$$

The long-term risk-neutral payoff reduces to

$$\frac{1}{\sum_{t=1}^{T} \mathbb{1}_{\{j \in \mathcal{B}^n(t)\}}} \sum_{t=1}^{T} u_{j,t} \mathbb{1}_{\{j \in \mathcal{B}^n(t)\}},$$

when considering only the activity of the player j. We assume that we do not have short-term players or equivalently the probability for a player j to be active is strictly positive. Given an initial state $\mathbf{w}$ and a strategy profile $\tilde{\tau}$, the risk-neutral payoff of player j is the superior limiting of the risk-neutral

Cesaro-mean payoff $\mathbb{E}_{\mathbf{w},\tilde{\tau},\mathcal{B}^n} F_{j,T}$. We assume that $\mathbb{E}_{\mathbf{w},\tilde{\tau},\mathcal{B}^n} F_{j,T}$ has a limit. Then, the risk-neutral expected payoff of an active player j is denoted by

$$v_j(e_{s_j}, \mathbf{x}_{-j}) = \mathbb{E}_{\mathbf{w},\mathcal{B}^n} U_j^{\mathcal{B}^n}(\mathbf{w}, e_{s_j}, \mathbf{x}_{-j}),$$

where e_{s_j} is the vector unit with 1 at the position of s_j and zero otherwise.

Definition 9.2.2.1 (Risk-neutral robust game). *We define the risk-neutral expected robust game as*

$$\left(\mathcal{N}, (\mathcal{X}_j)_{j\in\mathcal{N}}, \mathbb{E}_{\mathbf{w},\mathcal{B}^n} U_j^{\mathcal{B}^n}(\mathbf{w}, .)\right),$$

Definition 9.2.2.2. *A strategy profile* $(\mathbf{x}_j)_{j\in\mathcal{N}} \in \prod_{j=1}^n \mathcal{X}_j$ *is a (mixed) state-independent equilibrium for the risk-neutral expected robust game if* $\forall j \in \mathcal{N}, \forall \mathbf{y}_j \in \mathcal{X}_j$,

$$\mathbb{E}_{\mathbf{w},\mathcal{B}^n} U_j^{\mathcal{B}^n}(\mathbf{w}, \mathbf{y}_j, \mathbf{x}_{-j}) \leq \mathbb{E}_{\mathbf{w},\mathcal{B}^n} U_j^{\mathcal{B}^n}(\mathbf{w}, \mathbf{x}_j, \mathbf{x}_{-j}), \qquad (9.5)$$

The existence of solution of Equation (9.5) is equivalent to the existence of solution of the following *variational inequality problem*: find $\mathbf{x}$ such that

$$\langle \mathbf{x} - \mathbf{y}, V(\mathbf{x}) \rangle \geq 0, \ \forall \mathbf{y} \in \prod_j \mathcal{X}_j$$

where $\langle .,. \rangle$ is the inner product, $V(\mathbf{x}) = [V_1(\mathbf{x}), \ldots, V_n(\mathbf{x})]$,

$$V_j(\mathbf{x}) = [\mathbb{E}_{\mathbf{w},\mathcal{B}} U_j^{\mathcal{B}}(\mathbf{w}, e_{s_j}, \mathbf{x}_{-j})]_{s_j \in \mathcal{A}_j}.$$

Remark 9.2.2.3. *Note that an equilibrium of the expected robust game may not be an equilibrium (of the robust game) at each time slot. This is because* $\mathbf{x}$ *is an equilibrium for the expected robust game does not imply that* $\mathbf{x}$ *is an equilibrium of the game* $\mathcal{G}(\mathbf{w})$ *for some state* $\mathbf{w}$ *and the set of active players may vary.*

Lemma 9.2.2.4. *Assume that* $\mathcal{W}$ *is compact. Then, the expected robust game with unknown state and variable number of interacting players has at least one (state-independent) equilibrium.*

The existence of such equilibrium points is guaranteed since the mappings $v_j : (\mathbf{x}_j, \mathbf{x}_{-j}) \longmapsto \mathbb{E}_{\mathbf{w},\mathcal{B}} U_j^{\mathcal{B}}(\mathbf{w}, \mathbf{x}_j, \mathbf{x}_{-j})$ are jointly continuous, quasi-concave in $\mathbf{x}_j$, the spaces $\mathcal{X}_j$, are non-empty, convex and compact. Then, the result follows by using the Kakutani fixed-point theorem or by applying Nash theorem to the risk-neutral expected robust game.

Definition 9.2.2.5 (Risk-sensitive expected robust game). *We define the risk-sensitive expected robust game as*

$$G^{rs} = \left(\mathcal{N}, (\mathcal{X}_j)_{j\in\mathcal{N}}, \mathbb{E}_{\mathbf{w},\mathcal{B}^n,\mathbf{x}} \frac{1}{\mu_j}(-1 + e^{\mu_j \tilde{U}_j^{\mathcal{B}^n}(\mathbf{w},a)})\right).$$

Consider the robust game $\left(\mathcal{N}, \mathcal{W}, \mathcal{B}^n, \mathcal{X}_j, r_{j,\mu_j}\right)$ where $r_{j,\mu_j} : \prod_{j \in \mathcal{N}} \mathcal{X}_j \longrightarrow \mathbb{R}$ defined by

$$r_{j,\mu_j}(\mathbf{x}) = \mathbb{E}_{w,\mathbf{x}} \frac{1}{\mu_j} \left(e^{\mu_j \tilde{U}_j(w,a)} - 1 \right)$$

Note that $\tilde{r}_j = \frac{1}{\mu_j} \log r_{j,\mu_j}$. The risk-sensitive criterion generalizes the risk-neutral case by the following relations:

$$\lim_{\mu_j \to 0} \frac{(r_{j,\mu_j} - 1)}{\mu_j} = \lim_{\mu_j \to 0} \frac{1}{\mu_j} \left[\mathbb{E}_{w,\mathbf{x}} e^{\mu_j \tilde{U}_j(w,a)} - 1 \right] = \mathbb{E}_{w,\mathbf{x}} \left(\tilde{U}_j(w,a) \right) = v_j(\mathbf{x})$$

This means that the risk-neutral case can be obtained by taking the limit when $\mu_j \longrightarrow 0$. This can be seen by using the fact that the exponential is

$$\mathbb{E}\left(e^Z\right) = \sum_{k \geq 0} \frac{1}{k!} \mathbb{E}\left(Z^k\right),$$

for Z a real or complex-valued random variable. Thus,

$$\frac{(r_{j,\mu_j} - 1)}{\mu_j} = \sum_{k \geq 1} \frac{\mu_j^{k-1}}{k!} \mathbb{E}_{\mathbf{w},\mathcal{B}^n,\mathbf{x}} \left[\left(\tilde{U}_j^{\mathcal{B}^n} \right)^k \right]$$

Definition 9.2.2.6 (Risk-sensitive equilibrium). *A strategy profile* $(\mathbf{x}_j)_{j \in \mathcal{N}} \in \prod_{j=1}^n \mathcal{X}_j$ *is a (mixed) state-independent equilibrium for the risk-sensitive expected robust game if* $\forall j \in \mathcal{N}$, $\forall \mathbf{y}_j \in \mathcal{X}_j$,

$$\mathbb{E}_{\mathbf{w},\mathcal{B}^n,\mathbf{y}_j\mathbf{x}_{-j}} \left[e^{\mu_j \tilde{U}_j^{\mathcal{B}^n}(\mathbf{w},a_j,\mathbf{a}_{-j})} \right] \leq \mathbb{E}_{\mathbf{w},\mathcal{B}^n,\mathbf{x}_j\mathbf{x}_{-j}} \left[e^{\mu_j \tilde{U}_j^{\mathcal{B}^n}(\mathbf{w},a_j,\mathbf{a}_{-j})} \right] \qquad (9.6)$$

This is a Nash equilibrium of the "novel" game where the payoff function is replaced by the risk-sensitive payoff function r_{j,μ_j}, and as a consequence, the following result holds.

Theorem 9.2.2.7 (Existence of risk-sensitive equilibrium). *Assume that $\mathcal{W}$ is compact. Then, the expected robust game with risk-sensitive payoffs, unknown state, and variable number of interacting players has at least one (state-independent) risk-sensitive equilibrium.*

9.2.2.8 Two-by-Two Risk-Sensitive Games

In this subsection, we show that any generic two-by-two games, with orthogonal constraints and risk-sensitive payoff, has at least one evolutionarily stable state or strategy (ESS) under assumptions that the constraint is non-empty (Slater condition is not needed) and the payoff are not constants. In particular, this shows that if a *generic* two-by-two game with symmetric payoff has an interior equilibrium, this interior is an ESS.

Any generic expected symmetric two-by-two game (with non-trivial payoffs) has at least one ESS.

We consider a symmetric two-by-two game. This means that there are precisely two players. Each player has two actions or pure strategies (1 or 2). The payoff to any strategy is independent of which player it is applied to. The payoffs of player 1 is given by $A = \begin{pmatrix} a & b \\ c & d \end{pmatrix}$. and the payoff of player 2 is the transpose of the payoff of first player. There is a constraint for each player $Cx \leq V$ where $C = (c_1, c_2)$, $V \in \mathbb{R}$, $(x_1, x_2) \in \Delta_1$ $\Delta_1 = \{(y, 1-y) \mid 0 \leq y \leq 1\}$. Denote by

$$C := \{x \mid C \begin{pmatrix} x \\ 1-x \end{pmatrix} \leq V\}.$$

The set of constrained best replies to any opponent strategy $mut \in C$ is

$$CBR(mut) = \arg \max_{mut \in C} (x, 1-x)A \begin{pmatrix} mut \\ 1-mut \end{pmatrix}$$

If $c_1 = c_2$ the constraint is independent of the strategies. Therefore, we assume that $c_1 \neq c_2$. We assume that C is non-empty. Denote $\alpha = \frac{V-c_2}{c_1-c_2}$. We transform the matrix A to the following matrix:

$$\bar{A} = \begin{pmatrix} a-c & 0 \\ 0 & d-b \end{pmatrix}$$

The two matrix game $\bar{A}$ and A have the same equilibrium properties.

Denote $\beta_1 = a - c, \beta_2 = d - b$. Suppose that $\beta_1\beta_2 \neq 0$ (if both are zero, the payoffs are constant and any strategy in C is an equilibrium. there is no constrained ESS).

We distinguish two cases:

- Let $c_1 > c_2$. If $V > c_1$, then $C = [0, 1]$ (unconstraint case, there is an ESS). If $V \leq c_1$, then $C = [0, \alpha]$

- $c_1 < c_2$. If $V > c_1$, $C = \emptyset$ (excluded by hypothesis). If $V \leq c_1$, $C = [\alpha, 1]$.

We then have to examine two types of constraints: Type I:

$$c_1 > c_2, \ V < c_1, C = [0, \alpha]$$

and Type II:

$$c_1 < c_2, \ V < c_1, C = [\alpha, 1]$$

9.2.2.9 Type I

- $$\bar{A} = \begin{pmatrix} \beta_1 & 0 \\ 0 & \beta_2 \end{pmatrix}, \beta_1 > 0, \beta_2 \leq 0, \mathcal{C} = [0, \alpha]$$

The first strategy dominates the second one in the unconstrained game. Hence, the mixed strategy $x = \alpha$ is the unique ESS in the constrained game. note that $x = \alpha$ is not an ESS in the unconstrained game.

- $$\bar{A} = \begin{pmatrix} \beta_1 & 0 \\ 0 & \beta_2 \end{pmatrix}, \beta_1 \leq 0, \beta_2 > 0, \mathcal{C} = [0, \alpha]$$

The second strategy dominates the first one in the unconstrained game. Hence, the strategy mixed strategy $x = 0$ is the unique ESS in the constrained game.

- $$\bar{A} = \begin{pmatrix} \beta_1 & 0 \\ 0 & \beta_2 \end{pmatrix}, \beta_1 > 0, \beta_2 > 0, \mathcal{C} = [0, \alpha]$$

the class of Coordination Games . The second strategy is an ESS.

- $$\bar{A} = \begin{pmatrix} \beta_1 & 0 \\ 0 & \beta_2 \end{pmatrix}, \beta_1 < 0, \beta_2 < 0, \mathcal{C} = [0, \alpha]$$

Class of Hawk-Dove Games

If $\frac{\beta_2}{\beta_1 + \beta_2} \geq \alpha$ then $(\frac{\beta_2}{\beta_1 + \beta_2}, \frac{\beta_1}{\beta_1 + \beta_2})$ is an ESS; else if $\frac{\beta_2}{\beta_1 + \beta_2} > \alpha$ then the constrained best response set is

$$CBR(mut) = \arg \max_{mut \in \mathcal{C}} (x, 1-x) A \begin{pmatrix} mut \\ 1 - mut \end{pmatrix} = \begin{cases} \alpha & \text{if } mut < \frac{\beta_2}{\beta_1 + \beta_2} \\ 0 & \text{if } mut > \frac{\beta_2}{\beta_1 + \beta_2} \\ \mathcal{C} & \text{if } mut = \frac{\beta_2}{\beta_1 + \beta_2} \end{cases}$$

Thus, $CBR(\alpha) = \{\alpha\}$ and α is an ESS.

9.2.2.10 Type II

If $\alpha > 1, \mathcal{C}$ is empty. We suppose that $V < c_1$.

- $$\bar{A} = \begin{pmatrix} \beta_1 & 0 \\ 0 & \beta_2 \end{pmatrix}, \beta_1 > 0, \beta_2 \leq 0, \mathcal{C} = [\alpha, 1]$$

The first strategy is the unique ESS in the constrained game.

- $$\bar{A} = \begin{pmatrix} \beta_1 & 0 \\ 0 & \beta_2 \end{pmatrix}, \beta_1 \leq 0, \beta_2 > 0, \mathcal{C} = [\alpha, 1]$$

the strategy mixed strategy $x = \alpha$ is the unique ESS.

-
$$\bar{A} = \begin{pmatrix} \beta_1 & 0 \\ 0 & \beta_2 \end{pmatrix}, \beta_1 > 0, \beta_2 > 0, \mathcal{C} = [\alpha, 1]$$

. The second strategy is an ESS.

-
$$\bar{A} = \begin{pmatrix} \beta_1 & 0 \\ 0 & \beta_2 \end{pmatrix}, \beta_1 < 0, \beta_2 < 0, \mathcal{C} = [0, \alpha]$$

The mixed strategy

$$\min\left(\frac{\beta_2}{\beta_1 + \beta_2}, \alpha\right)$$

is an ESS.

> Note that the risk-sensitive equilibria can be different
> from Nash equilibria (risk-neutral equilibria).

Since we have the existence of a state-independent equilibrium under suitable conditions, we seek heterogeneous and combined algorithms to locate both risk-neutral and risk-sensitive equilibria.

9.3   Risk-sensitive CODIPAS

9.3.1   Learning the Risk-Sensitive Payoff

In this section, we explain how the learning patterns developed for the risk-neutral case in the previous Chapters can be adapted to the risk-sensitive case. We first observe that the classical risk-sensitive criterion

$$\liminf_{T} \frac{1}{T+1} \log\left[\mathbb{E}_{w,\mathbf{x}} e^{\mu_j \sum_{t=0}^{T} u_{j,t}} \mid w_0, \mathbf{x}_0\right]$$

is not necessarily well-adapted and leads to the multiplicative dynamic programming (multiplicative Poisson equation).

To learn the mathematical expectation part, we write the recursive equation satisfied by

$$\tilde{R}_{j,\mu_j,T} = \frac{1}{T+1} \sum_{t=0}^{T} e^{\mu_j u_{j,t}},$$

i.e., $(T+1)\tilde{R}_{j,\mu_j,T} = T\tilde{R}_{j,\mu,T-1} + e^{\mu_j u_{j,T}}$. This means that

$$\tilde{R}_{j,\mu_j,T} = \left(1 - \frac{1}{T+1}\right)\tilde{R}_{j,\mu_j,T-1} + \frac{1}{T+1}(e^{\mu_j u_{j,T}}),$$

which can be rewritten as

$$\tilde{R}_{j,\mu_j,T} = \tilde{R}_{j,\mu_j,T-1} + \frac{1}{T+1}\left(e^{\mu_j u_{j,T}} - \tilde{R}_{j,\mu_j,T-1}\right).$$

Based on this recursive equation, we construct a learning pattern for risk-sensitive payoff as

$$\hat{r}^{rs}_{j,t+1}(s_j) = \hat{r}^{rs}_{j,t}(s_j) + \lambda_{\theta_j(t)}\mathbb{1}_{\{a_{j,t}=s_j, j\in\mathcal{B}^n(t)\}}\left(\frac{(e^{\mu_j u_{j,t+1}} - 1)}{\mu_j} - \hat{r}^{rs}_{j,t}(s_j)\right),\ s_j\in\mathcal{A}_j$$

where $\lambda \in l^2\backslash l^1$.

We propose hybrid, delayed, risk-sensitive COmbined fully DIstributed PAyoff and Strategy Reinforcement Learning in the following form: (CODIPAS-RL)

$$\begin{cases}
\mathbf{x}_{j,t+1}(s_j) - \mathbf{x}_{j,t}(s_j) = \\
\mathbb{1}_{\{j\in\mathcal{B}^n(t)\}}\sum_{l\in\mathcal{L}}\mathbb{1}_{\{l_{j,t}=l\}}K^{1,(l)}_{j,s_j}(\lambda_{j,\theta_j(t)}, a_{j,t}, u_{j,t-\tau_j}, \hat{\mathbf{r}}^{rs}_{j,t}, \mathbf{x}_{j,t}), \\
\hat{\mathbf{r}}^{rs}_{j,t+1}(s_j) - \hat{\mathbf{r}}^{rs}_{j,t}(s_j) = \\
\mathbb{1}_{\{j\in\mathcal{B}^n(t)\}}K^2_{j,s_j}(\nu_{j,\theta_j(t)}, a_{j,t}, u_{j,t-\tau_j}, \hat{\mathbf{r}}^{rs}_{j,t}, \mathbf{x}_{j,t}), \\
j\in\mathcal{N}, t\geq 0, a_{j,t}\in\mathcal{A}_j, s_j\in\mathcal{A}_j, \\
\theta_j(t+1) = \theta_j(t) + \mathbb{1}_{\{j\in\mathcal{B}^n(t)\}}, \\
t\geq 0,\ \mathcal{B}^n(t)\subseteq\mathcal{N}, \\
\mathbf{x}_{j,0}\in\mathcal{X}_j, \hat{\mathbf{r}}^{rs}_{j,0}\in\mathbb{R}^{|\mathcal{A}_j|}.
\end{cases}$$

where

$$\hat{\mathbf{r}}^{rs}_{j,t} = (\hat{r}^{rs}_{j,t}(s_j))_{s_j\in\mathcal{A}_j} \in \mathbb{R}^{|\mathcal{A}_j|},$$

is a vector payoff estimation of player j at time t. Note that when player j uses $a_{j,t} = s_j$, he observes only her measurement corresponding to that action but not those of the other actions $s'_j \neq s_j$. Hence he needs to estimate/predict them via the vector $\hat{\mathbf{r}}^{rs}_{j,t+1}$. The functions K^1 and λ are based on estimated payoffs and perceived measured payoff (delayed and noisy) such that the invariance of simplex is preserved almost surely. The function K^1_j defines the strategy learning pattern of player j and $\lambda_{j,\theta_j(t)}$ is her strategy learning rate. If at least two of the functions K_j are different then we refer to it as *heterogeneous learning* in the sense that the learning schemes of the players are different. If all the K^1_j are identical but the learning rates λ_j are different, we refer to *learning with different speed*: slow, medium, or fast learners. Note that the term $\lambda_{j,\theta_j(t)}$ is used instead of $\lambda_{j,t}$ because the global clock $[t]$ is not known by player j (she knows only how many times she has been active, the activity of others is not known by j). $\theta_j(t)$ is a random variable that determines the local clock of j. Thus, the updates are asynchronous. The functions K^2_j, and ν_j are well-chosen in order to have a good estimation of the payoffs. τ_j is a time delay associated with player j in her payoff measurement. The risk-neutral payoff $u_{j,t-\tau_j}$ at $t - \tau_j$ is perceived at time t. We examine the case where the players can choose different CODIPAS-RL patterns during the dynamic game. They can select among a set of CODIPAS-RLs denoted by $\mathcal{L}_1,\ldots,\mathcal{L}_m, m\geq 1$. The resulting learning scheme is called *hybrid CODIPAS-RL*. The term $l_{j,t}$ is the CODIPAS-RL pattern chosen by player j at time t.

9.3.2 Risk-Sensitive CODIPAS Patterns

We examine the above risk-sensitive dynamic game in which each player learns according to a specific CODIPAS-RL scheme.

9.3.2.1 Bush-Mosteller based RS-CODIPAS

The learning pattern $\mathcal{L}_1$ is given by

$$x_{j,t+1}(s_j) - x_{j,t}(s_j) = \lambda_{\theta_j(t)} \mathbb{1}_{\{j \in \mathcal{B}^n(t)\}} \times$$

$$\frac{\frac{1}{\mu_j}(e^{\mu_j u_{j,t}} - 1) - \Gamma_j}{\sup_{\mathbf{a},w'} |\frac{1}{\mu_j}(-1 + e^{\mu_j U_j(w',a)}) - \Gamma_j|} \left(\mathbb{1}_{\{a_{j,t}=s_j\}} - \mathbf{x}_{j,t}(s_j) \right), \qquad (9.7)$$

$$\hat{r}^{rs}_{j,t+1}(s_j) - \hat{r}^{rs}_{j,t}(s_j) =$$

$$\nu_{\theta_j(t)} \mathbb{1}_{\{a_{j,t}=s_j, j \in \mathcal{B}^n(t)\}} \left(\frac{1}{\mu_j}(e^{\mu_j u_{j,t}} - 1) - \hat{r}^{rs}_{j,t}(s_j) \right) \qquad (9.8)$$

$$\theta_j(t+1) = \theta_j(t) + \mathbb{1}_{\{j \in \mathcal{B}^n(t)\}} \qquad (9.9)$$

where Γ_j is a reference level of j. The first equation of $\mathcal{L}_1$ is widely studied in machine learning and was initially proposed by Bush and Mosteller in 1949-55 [42]. The second equation of $\mathcal{L}_1$ is a payoff estimation for the experimented action by the players. Combined together, one gets a specific combined fully distributed payoff and strategy reinforcement learning based on Bush-Mosteller reinforcement learning.

9.3.2.2 Boltzmann-Gibbs-Based RS-CODIPAS

$$x_{j,t+1}(s_j) - x_{j,t}(s_j) = \lambda_{\theta_j(t)} \mathbb{1}_{\{j \in \mathcal{B}^n(t)\}} \times$$

$$\left(\frac{e^{\frac{1}{\epsilon_j} \hat{\mathbf{r}}^{rs}_{j,t}(s_j)}}{\sum_{s'_j \in \mathcal{A}_j} e^{\frac{1}{\epsilon_j} \hat{r}^{rs}_{j,t}(s'_j)}} - x_{j,t}(s_j) \right), \qquad (9.10)$$

$$\hat{r}^{rs}_{j,t+1}(s_j) - \hat{r}^{rs}_{j,t}(s_j) =$$

$$\nu_{\theta_j(t)} \mathbb{1}_{\{a_{j,t}=s_j, j \in \mathcal{B}^n(t)\}} \left(\frac{1}{\mu_j}(-1 + e^{\mu_j u_{j,t}}) - \hat{r}^{rs}_{j,t}(s_j) \right) \qquad (9.11)$$

$$\theta_j(t+1) = \theta_j(t) + \mathbb{1}_{\{j \in \mathcal{B}^n(t)\}} \qquad (9.12)$$

The strategy learning (9.10) of $\mathcal{L}_2$ is a Boltzmann-Gibbs based reinforcement learning for the risk-sensitive payoff. Note that the Boltzmann-Gibbs distribution can be obtained from the maximization of the perturbed payoff $e^{\mu_j \bar{U}_j} + \epsilon_j H_j$, where H_j is the entropy function i.e

$$H_j(\mathbf{x}_j) = - \sum_{s_j \in \mathcal{A}_j} x_j(s_j) \ln x_j(s_j).$$

It is a smooth best response function to the risk-sensitive criterion. Here the Boltzmann-Gibbs mapping is based on the risk-sensitive payoff estimation (the exact payoff vector is not known, only one component of a noisy value is observed). We denote the risk-sensitive Boltzmann-Gibbs strategy by

$$
\tilde{\beta}_{j,\epsilon_j}(\hat{r}_{j,t}^{rs})(s_j) = \frac{e^{\frac{1}{\epsilon_j}\hat{\mathbf{r}}_{j,t}^{rs}(s_j)}}{\sum_{s_j' \in A_j} e^{\frac{1}{\epsilon_j}\hat{\mathbf{r}}_{j,t}^{rs}(s_j')}}
$$

and the risk-sensitive smooth best response to $\mathbf{x}_{-j,t}$ (also called Logit rule, Gibbs sampling, or Glauber dynamics) is given by

$$
\beta_{j,\epsilon_j}(\mathbf{x}_{-j,t})(s_j) = \frac{e^{\frac{1}{\epsilon_j}r_{j,\mu_j}(\mathbf{e}_{s_j},\mathbf{x}_{-j,t})}}{\sum_{s_j' \in A_j} e^{\frac{1}{\epsilon_j}r_{j,\mu}(\mathbf{e}_{s_j'},\mathbf{x}_{-j,t})}}.
$$

This dynamics is examined in Chapters 4 and 5 for the risk-neutral case.

9.3.2.3 Imitative BG CODIPAS

$$
x_{j,t+1}(s_j) - x_{j,t}(s_j) = \lambda_{\theta_j}(t)\mathbb{1}_{\{j \in \mathcal{B}^n(t)\}} x_{j,t}(s_j) \times
$$
$$
\left(\frac{e^{\frac{1}{\epsilon_j}\hat{r}_{j,t}^{rs}(s_j)}}{\sum_{a_j' \in A_j} x_{j,t}(a_j')e^{\frac{1}{\epsilon_j}\hat{r}_{j,t}^{rs}(a_j')}} - 1 \right), \tag{9.13}
$$
$$
\hat{r}_{j,t+1}^{rs}(s_j) - \hat{r}_{j,t}^{rs}(s_j) =
$$
$$
\nu_{\theta_j}(t)\mathbb{1}_{\{a_{j,t}=s_j, j \in \mathcal{B}^n(t)\}} \left(\frac{1}{\mu_j}(-1 + e^{\mu_j u_{j,t}}) - \hat{r}_{j,t}^{rs}(s_j) \right) \tag{9.14}
$$
$$
\theta_j(t+1) = \theta_j(t) + \mathbb{1}_{\{j \in \mathcal{B}^n(t)\}} \tag{9.15}
$$

The strategy learning (9.13) of $\mathcal{L}_3$ is an imitative Boltzmann-Gibbs based reinforcement learning. The imitation here consists of playing an action with a probability proportional to the previous uses.

9.3.2.4 Multiplicative Weighted Imitative CODIPAS

$$
x_{j,t+1}(s_j) - x_{j,t}(s_j) = \mathbb{1}_{\{j \in \mathcal{B}^n(t)\}} x_{j,t}(s_j) \times
$$
$$
\left(\frac{(1 + \lambda_{\theta_j}(t))^{\hat{r}_{j,t}^{rs}(s_j)}}{\sum_{s_j' \in A_j} x_{j,t}(s_j')(1 + \lambda_{\theta_j}(t))^{\hat{r}_{j,t}^{rs}(s_j')}} - 1 \right), \tag{9.16}
$$
$$
\hat{r}_{j,t+1}^{rs}(s_j) - \hat{r}_{j,t}^{rs}(s_j) =
$$
$$
\nu_{\theta_j}(t)\mathbb{1}_{\{a_{j,t}=s_j, j \in \mathcal{B}^n(t)\}} \left(\frac{1}{\mu_j}(-1 + e^{\mu_j u_{j,t}}) - \hat{r}_{j,t}^{rs}(s_j) \right) \tag{9.17}
$$
$$
\theta_j(t+1) = \theta_j(t) + \mathbb{1}_{\{j \in \mathcal{B}^n(t)\}} \tag{9.18}
$$

The strategy learning (9.16) of $\mathcal{L}_4$ is a learning-rate-weighted imitative reinforcement learning. The main difference with $\mathcal{L}_2$ and $\mathcal{L}_3$ is that there is no parameter ϵ_j. The interior outcomes are necessarily exact equilibria of the expected (not approximated equilibria as in $\mathcal{L}_2$). We will show in appendix that this leads to replicator dynamics (thus, its interior stationary points are Nash equilibria).

9.3.2.5 Weakened Fictitious Play-Based CODIAPS

$$x_{j,t+1}(s_j) - x_{j,t}(s_j) \in \mathbb{1}_{\{j \in \mathcal{B}^n(t)\}} \times$$
$$\left((1 - \epsilon_t)\delta_{\arg\max_{s_j'} \hat{r}_{j,t}^{rs}(s_j')} + \epsilon_t \frac{\mathbb{1}}{|\mathcal{A}_j|} - x_{j,t}(s_j) \right), \qquad (9.19)$$

$$\hat{r}_{j,t+1}^{rs}(s_j) - \hat{r}_{j,t}^{rs}(s_j) =$$
$$\nu_{\theta_j(t)} \mathbb{1}_{\{a_{j,t}=s_j, j\in\mathcal{B}^n(t)\}} \left(\frac{1}{\mu_j}(e^{\mu_j u_{j,t}} - 1) - \hat{r}_{j,t}^{rs}(s_j) \right) \qquad (9.20)$$

$$\theta_j(t+1) = \theta_j(t) + \mathbb{1}_{\{j\in\mathcal{B}^n(t)\}} \qquad (9.21)$$

The last learning pattern $\mathcal{L}_5$ is a combined learning based on the weakened fictitious play. Here a player does not observe the action played by the other at the previous step and the payoff function is not known. Each player estimates her payoff function via the equations (9.20). The equation (9.19) consists of playing one of the actions with the best estimation $\hat{r}_{j,t}^{rs}$ with probability $(1-\epsilon_t)$ and plays an arbitrary action with probability ϵ_t.

Weakened fictitious play-like learning algorithms have been examined in [107, 102, 108]. The work in [102] conducted the convergence analysis of this algorithm for the risk-neutral case without random activities i.e., all the players are active all the time and update their learning patterns, respectively.

9.3.2.6 Risk-Sensitive Payoff Learning

We mention some payoff learning based the idea of risk-sensitive CODIPAS-RL: • $\mathcal{PL}_1$ *No-regret based CODIPAS-RL*

$$x_{j,t+1}(s_j) - x_{j,t}(s_j) = \mathbb{1}_{\{j\in\mathcal{B}^n(t)\}}[\underline{R}_t(s_j) - x_{j,t}(s_j)], \qquad (9.22)$$
$$\hat{r}_{j,t+1}^{rs}(s_j) - \hat{r}_{j,t}^{rs}(s_j) =$$
$$\nu_{\theta_j(t)} \mathbb{1}_{\{a_{j,t}=s_j, j\in\mathcal{B}^n(t)\}} \left(\frac{1}{\mu_j}(e^{\mu_j u_{j,t}} - 1) - \hat{r}_{j,t}^{rs}(s_j) \right) \qquad (9.23)$$
$$\theta_j(t+1) = \theta_j(t) + \mathbb{1}_{\{j\in\mathcal{B}^n(t)\}} \qquad (9.24)$$
$$\underline{R}_t(s_j) = \frac{\phi([\hat{r}_{j,t}^{rs}(s_j) - \frac{1}{\mu_j}(e^{\mu_j u_{j,t}} - 1)]_+)}{\sum_{s_j'} \phi([\hat{r}_{j,t}^{rs}(s_j') - \frac{1}{\mu_j}(e^{\mu_j u_{j,t}} - 1)]_+)} \qquad (9.25)$$

The strategy learning based on no-regret rule is known to converge to the set of correlated equilibria for the risk-neutral case. Here, the non-regret term is based on the risk-sensitive estimations.

- $\mathcal{PL}_2$: *Imitative No-regret-based CODIPAS-RL*

$$x_{j,t+1}(s_j) - x_{j,t}(s_j) = \lambda_{\theta_j(t)} \mathbb{1}_{\{j \in \mathcal{B}^n(t)\}} \left(IR_t(s_j) - x_{j,t}(s_j) \right), \tag{9.26}$$

$$\hat{r}_{j,t+1}^{rs}(s_j) - \hat{r}_{j,t}^{rs}(s_j) =$$

$$\nu_{\theta_j(t)} \mathbb{1}_{\{a_{j,t}=s_j, j \in \mathcal{B}^n(t)\}} \left(\frac{1}{\mu_j}(e^{\mu_j u_{j,t}} - 1) - \hat{r}_{j,t}^{rs}(s_j) \right) \tag{9.27}$$

$$\theta_j(t+1) = \theta_j(t) + \mathbb{1}_{\{j \in \mathcal{B}^n(t)\}} \tag{9.28}$$

$$IR_t(s_j) = \frac{x_{j,t}(s_j)\phi([\hat{r}_{j,t}^{rs}(s_j) - \frac{1}{\mu_j}(e^{\mu_j u_{j,t}} - 1)]_+)}{\sum_{s_j'} x_{j,t}(s_j')\phi([\hat{r}_{j,t}^{rs}(s_j') - \frac{1}{\mu_j}(e^{\mu_j u_{j,t}} - 1)]_+)} \tag{9.29}$$

9.3.3 Risk-Sensitive Pure Learning Schemes

We first recall the standard stochastic approximation result [96, 33]. Consider

$$\mathbf{x}_{t+1} = \mathbf{x}_t + \lambda_t(f(\mathbf{x}_t) + M_{t+1}) \tag{9.30}$$

in $\mathbb{R}^{\sum_{j \in \mathcal{N}} |\mathcal{A}_j|}$ and assume

- **A1** f is a Lipschitz function.
- **A2** The learning rates satisfy $\lambda_t \geq 0$, $\sum_{t \geq 0} \lambda_t = +\infty$, $\sum_{t \geq 0} \lambda_t^2 < \infty$.
- **A31** M_{t+1} is a martingale difference sequence with respect to the increasing family of sigma-fields $\mathcal{F}_t = \sigma(\mathbf{x}_{t'}, M_{t'}, \ t' \leq t)$ i.e

$$\mathbb{E}\left(M_{t+1} \mid \mathcal{F}_t \right) = 0$$

- **A32** M_t is square integrable and there is a constant $c > 0$, $\mathbb{E}\left(\parallel M_{t+1} \parallel^2 \mid \mathcal{F}_t \right) \leq c(1 + \parallel \mathbf{x}_t \parallel^2)$ almost surely, for all $t \geq 0$.
- **A4** $\sup_t \parallel \mathbf{x}_t \parallel < \infty$ almost surely.

Then, the asymptotic pseudo-trajectory is given by the ordinary differential equation (ODE)

$$\dot{\mathbf{x}}_t = f(\mathbf{x}_t), \ \mathbf{x}_0 \ \text{fixed.}$$

Notice that the assumptions **A31** and **A32** imply that the cumulative $\sum_{i=0}^{l-1} \lambda_t M_{t+1}$ goes to zero. We check the assumptions A1-A4 for our strategy-learning patterns in $\mathcal{L}_1, \ldots, \mathcal{L}_4$ and establish their ODE approximations. In Table 9.1, we give the asymptotic pseudo-trajectory of the pure learning when the rate of payoff learning is faster than the strategy learning rate.

In Table 9.1, the replicator dynamics are given by

$$\dot{x}_j(s_j) = g_j x_j(s_j) \left[r_{j,\mu_j}(e_{s_j}, \mathbf{x}_{-j}) - \sum_{s_j' \in \mathcal{A}_j} r_{j,\mu_j}(e_{s_j'}, \mathbf{x}_{-j}) x_j(s_j') \right].$$

TABLE 9.1
Asymptotic pseudo-trajectories of pure learning

Learning patterns	Class of ODE
$\mathcal{L}_1$	Adjusted replicator dynamics
$\mathcal{L}_2$	Smooth best response dynamics
$\mathcal{L}_3$	Imitative BG dynamics
$\mathcal{L}_4$	Time-scaled replicator dynamics
$\mathcal{L}_5$	Perturbed best response dynamics

The smooth best response dynamics are given by

$$\dot{x}_j(s_j) = g_j \left(\frac{e^{\frac{1}{\epsilon_j} r_{j,\mu_j}(e_{s_j}, \mathbf{x}_{-j})}}{\sum_{s'_j} e^{\frac{1}{\epsilon} r_{j,\mu_j}(e_{s'_j}, \mathbf{x}_{-j})}} - x_j(s_j) \right).$$

The imitative Boltzman-Gibbs dynamics are given by

$$\dot{x}_j(s_j) = g_j \left(\frac{x_j(s_j) e^{\frac{1}{\epsilon_j} r_{j,\mu_j}(e_{s_j}, \mathbf{x}_{-j})}}{\sum_{s'_j} x_j(s'_j) e^{\frac{1}{\epsilon_j} r_{j,\mu_j}(e_{s'_j}, \mathbf{x}_{-j})}} - x_j(s_j) \right).$$

The best response dynamics are given by

$$\dot{\mathbf{x}}_j \in g_j(\mathrm{RSBR}_{j,\mu_j}(\mathbf{x}_{-j}) - \mathbf{x}_j), \tag{9.31}$$

where $\mathrm{RSBR}_{j,\mu_j}(\mathbf{x}_{-j}) = \arg\max_{\mathbf{x}'_j \in \mathcal{X}_j} r_{j,\mu_j}(\mathbf{x}'_j, \mathbf{x}_{-j})$, is the best response correspondence of the risk-sensitive payoff. and the *risk-sensitive* payoff dynamics are

$$\frac{d}{dt}\hat{r}_j(s_j) = \bar{g}_j x_j(s_j)(r_{j,\mu_j}(e_{s_j}, \mathbf{x}_{-j}) - \hat{r}_j(s_j)).$$

Following the multiple time-scale stochastic approximation framework developed in [33, 27, 103, 96], one can write the pure learning schemes into the form

$$\begin{cases} \mathbf{x}_{j,t+1} - \mathbf{x}_{j,t} \in q_{j,t} \left(f_j^{(l)}(\mathbf{x}_{j,t}, \hat{\mathbf{r}}_{j,t}^{rs}) + M_{j,t+1}^{(l)} \right) \\ \hat{\mathbf{r}}_{j,t+1} - \hat{r}_{j,t}^{rs} = \bar{q}_{j,t} \left(\mathbb{E}_{\mathbf{w}, \mathbf{x}_{-j,t}, \mathcal{B}^n} \frac{1}{\mu_j} \left(e^{\mu_j \tilde{U}_j} - 1 \right) - \hat{r}_{j,t}^{rs} + \bar{M}_{j,t+1} \right) \end{cases}$$

where $l \in \mathcal{L}$, $q_{j,t}$ is a time-scaling factor function of the learning rates λ_t and the probability activity of the player j at time t, $\mathbb{P}(j \in \mathcal{B}^n(t))$, $\bar{q}_{j,t}$ is the analogous for $\hat{r}_{j,t}^{rs}$. To establish an ODE approximation, we check the assumptions A1-A4. The term $M_{j,t+1}^{(l)}$ is a bounded martingale difference because the strategies are in the product of simplice, which is convex and compact, and the conditional expectation of $M_{j,t+1}$ given the sigma-algebra generated by the random variables $\mathbf{w}_{t'}, \mathbf{x}_{t'}, u_{t'}, \hat{\mathbf{r}}_{t'}$, $t' \leq t$, is zero. Similar properties hold for $\bar{M}_{t+1}$. The function f can be bounded by linear growth in $\mathbf{x}$. The parameters λ_t and ν_t are in $l^2 \backslash l^1$. Thus, the asymptotic pseudo-trajectories reduce to

$$\begin{cases} \frac{d}{dt}\mathbf{x}_{j,t} \in g_{j,t}\left(f_j^{(l)}(\mathbf{x}_{j,t}, \hat{\mathbf{r}}_{j,t}^{rs})\right) \\ \frac{d}{dt}\hat{\mathbf{r}}_{j,t}^{rs} = \bar{g}_{j,t}\left(\mathbb{E}_{s,\mathbf{x}_{-j,t},\mathcal{B}^n} \frac{1}{\mu_j}(e^{\mu_j \tilde{U}_j} - 1) - \hat{\mathbf{r}}_{j,t}^{rs}\right) \end{cases}$$

for the non-vanishing time-scale ratio and,

$$\begin{cases} \frac{d}{dt}\mathbf{x}_{j,t} \in g_{j,t}\left(f_j^{(l)}(\mathbf{x}_{j,t}, \mathbb{E}_{\mathbf{w},\mathbf{x}_{-j,t}} \frac{1}{\mu_j}(e^{\mu_j \tilde{U}_j} - 1))\right) \\ \hat{\mathbf{r}}_{j,t}^{rs} \longrightarrow \mathbb{E}_{\mathbf{w},\mathbf{x}_{-j},\mathcal{B}^n} \frac{1}{\mu_j}(e^{\mu_j \tilde{U}_j} - 1) \end{cases}$$

for the vanishing ratio $\frac{\lambda_{\theta_j(t)}}{\nu_{\theta_j(t)}}$.

Remark 9.3.3.1. *Note that for some pure learning schemes presented here the function f^l does not satisfy the globally Lischpitz condition. However, their asymptotic pseudo trajectory can be described by a differential inclusion.*

Note that the case of constant learning rates can be analyzed under the same setting but the convergence result is weaker (convergence in law instead of almost sure convergence). A positive aspect of the constant learning rate can be the speed of convergence (when it converges) in the weak sense.

9.3.4 Risk-sensitive Hybrid Learning Scheme

Consider the following hybrid and switching learning scheme:

$$\begin{cases} \mathbf{x}_{j,t+1} - \mathbf{x}_{j,t} \in q_{j,t}\left(\sum_{l \in \mathcal{L}} \mathbb{1}_{\{l_{j,t}=l\}} f_j^{(l)}(\mathbf{x}_{j,t}, \hat{\mathbf{r}}_{j,t}^{rs}) + M_{j,t+1}^{(l)}\right) \\ \hat{\mathbf{r}}_{j,t+1}^{rs} - \hat{\mathbf{r}}_{j,t}^{rs} = \bar{q}_{j,t}\left(\mathbb{E}_{\mathbf{w},\mathbf{x}_{-j,t}} \frac{1}{\mu_j}(e^{\mu_j \tilde{U}_j} - 1) - \hat{\mathbf{r}}_{j,t}^{rs} + \bar{M}_{j,t+1}\right) \end{cases}$$

where $l_{j,t}$ is the learning pattern chosen by j at time t.

Assume that each player j adopts one of the CODIPAS-RLs in $\mathcal{L}$ with probability $\omega_j \in \Delta(\mathcal{L})$ and the learning rates are in $l^2 \backslash l^1$. Then, the asymptotic pseudo-trajectory of the hybrid and switching learning can be written in the form

$$\begin{cases} \frac{d}{dt}\mathbf{x}_{j,t} \in g_{j,t}\left(\sum_{l \in \mathcal{L}} \omega_{j,l} f_j^{(l)}(\mathbf{x}_{j,t}, \hat{\mathbf{r}}_{j,t}^{rs})\right) \\ \frac{d}{dt}\hat{\mathbf{r}}_{j,t}^{rs} = \bar{g}_{j,t}\left(\mathbb{E}_{\mathbf{w},\mathbf{x}_{-j,t}} \frac{1}{\mu_j}(e^{\mu_j \tilde{U}_j} - 1) - \hat{\mathbf{r}}_{j,t}^{rs}\right) \end{cases}$$

for the non-vanishing time-scale ratio and

$$\begin{cases} \frac{d}{dt}\mathbf{x}_{j,t} \in g_{j,t}\left(\sum_{l \in \mathcal{L}} \omega_{j,l} f_j^{(l)}(\mathbf{x}_{j,t}, \mathbb{E}_{\mathbf{w},\mathbf{x}_{-j,t},\mathcal{B}^n} \frac{1}{\mu_j}(e^{\mu_j \tilde{U}_j} - 1))\right) \\ \hat{\mathbf{r}}_{j,t}^{rs} \longrightarrow \mathbb{E}_{\mathbf{w},\mathbf{x}_{-j},\mathcal{B}^n} \frac{1}{\mu_j}(e^{\mu_j \tilde{U}_j} - 1) \end{cases}$$

for the vanishing ratio $\frac{\lambda_t}{\nu_t}$.

9.3.5 Convergence Results

Proposition 9.3.5.1. *If the learning rates are well-chosen, the imitative weighted delayed CODIPAS-RLs $\mathcal{L}_1$ (without normalization), $\mathcal{L}_4$, converge almost surely to a nonautonomous adjusted multiple-type replicator dynamics given by*

$$\dot{x}_{j,t}(s_j) = g_j(t) x_{j,t}(s_j) \left[\mathbb{E}_{\mathbf{w}, \mathcal{B}^n, \mathbf{e}_{s_j}, \mathbf{x}_{-j,t}} \frac{1}{\mu_j} (e^{\mu_j \tilde{U}_j^{\mathcal{B}^n}} - 1) \right.$$

$$\left. - \sum_{s_j' \in \mathcal{A}_j} x_{j,t}(s_j') \mathbb{E}_{\mathbf{w}, \mathcal{B}^n, \mathbf{e}_{s_j'}, \mathbf{x}_{-j,t}} \frac{1}{\mu_j} (e^{\mu_j \tilde{U}_j^{\mathcal{B}^n}} - 1) \right] \qquad (9.32)$$

where $g_j(t)$ is the ratio between the strategy learning rate of j and the maximum strategy rate among all the active players at time t. Moreover,

$$\mathbf{x}_j(s_j) > 0 \Longrightarrow \hat{r}_{j,t}^{r,s}(s_j) \longrightarrow \mathbb{E}_{\mathbf{w}, \mathcal{B}^n, \mathbf{e}_{s_j}, \mathbf{x}_{-j}} \frac{1}{\mu_j} (e^{\mu_j \tilde{U}_j^{\mathcal{B}^n}} - 1).$$

In particular, the risk-neutral scheme converges under the same assumptions by taking $\frac{1}{\mu_j} \left(e^{\mu_j \tilde{U}_j} - 1 \right)$ in place of $r_{j,t}$, with vanishing μ_j.

This result says that if the learning rates are chosen as in $\mathcal{L}_1$ and $\mathcal{L}_4$ such that the payoff learning rate is faster than the strategy learning rate, then the payoff estimation converges to the risk-sensitive expected payoff provided that the actions and the state of the nature have been sufficiently explored and the strategy can be approximated by the risk-sensitive version of the well-known replicator dynamics.

Asymptotic pseudo-trajectory of the imitative weighted CODIPAS-RL

Proof of Proposition 9.3.5.1. Denote

$$\lambda_t^* = \max_{j \in \mathcal{B}^n(t)} \lambda_{[\theta_j(t)]}, \ t \geq 0.$$

It can be easily checked that the random learning rate $\{\lambda_t^*\}_{t \in \mathbb{N}}$ satisfies the standard sufficiency conditions: Almost surely, one has:

$$\lambda_t^* \geq 0, \ \sum_{t \in \mathbb{N}} \lambda_t^* = +\infty, \ \text{and} \ \sum_{t \in \mathbb{N}} (\lambda_t^*)^2 < +\infty$$

For any fixed $j \in \mathcal{N}$,

$$+\infty = \sum_{t \in \mathbb{N}} \lambda_t = \sum_{t \in \mathbb{N}} \lambda_{[\theta_j(t)]} \mathbb{1}_{\{j \in \mathcal{B}(t)\}} \leq \sum_{t \in \mathbb{N}} \max_{j \in \mathcal{B}(t)} \lambda_{[\theta_j(t)]} \leq \sum_{t \in \mathbb{N}} \lambda_t^*.$$

Thus, $\sum_{t \in \mathbb{N}} \lambda_t^* = +\infty$, and

$$\sum_{t \in \mathbb{N}} (\lambda_t^*)^2 \leq \sum_{t \in \mathbb{N}} \sum_{j \in \mathcal{N}} \lambda_{[\theta_j(t)]}^2 \mathbb{1}_{\{j \in \mathcal{B}(t)\}} \leq |\mathcal{N}| \sum_{t \in \mathbb{N}} \lambda_t^2 < +\infty$$

We consider the player who has the maximum learning rate among all the active players as a reference clock; this is the random variable λ_t^*. Then, the discrete-time process is written as a function of λ_t^* : denote by $q_{j,t} = \frac{\lambda_{[\theta_j(t)]}}{\lambda_t^*} \mathbb{1}_{\{j \in \mathcal{B}^n(t)\}} \in [0,1]$, $\forall t$. We compute the drift D (the expected changes in one-time slot):

$$D = \frac{x_{j,t+1}(s_j) - x_{j,t}(s_j)}{\lambda_t^*}$$

$$= \mathbb{1}_{\{j \in \mathcal{B}^n(t)\}} \frac{1}{\lambda_t^*} \left[\frac{x_{j,t}(s_j)(1+\lambda_{[\theta_j(t)]})^{\hat{r}_{j,t}^{rs}(s_j)}}{\sum_{s_j' \in \mathcal{A}_j} x_{j,t}(s_j')(1+\lambda_{[\theta_j(t)]})^{\hat{r}_{j,t}^{rs}(s_j')}} - x_{j,t}(s_j) \right]$$

$$= \frac{x_{j,t}(s_j)\mathbb{1}_{\{j \in \mathcal{B}^n(t)\}}}{\sum_{s_j' \in \mathcal{A}_j} x_{j,t}(s_j')(1+q_{j,t}\lambda_t^*)^{\hat{r}_{j,t}^{rs}(s_j')}} \times$$

$$\left[\frac{(1+q_{j,t}\lambda_t^*)^{\hat{r}_{j,t}^{rs}(s_j)}-1}{\lambda_t^*} - \sum_{s_j' \in \mathcal{A}_j} x_{j,t}(s_j') \frac{(1+q_{j,t}\lambda_t^*)^{\hat{r}_{j,t}^{rs}(s_j')}-1}{\lambda_t^*} \right]$$

Using the fact that $\lambda_t^* \longrightarrow 0$, the term $\frac{(1+q_{j,t}\lambda_t^*)^{\hat{r}_{j,t}^{rs}(s_j)}-1}{\lambda_t^*} \longrightarrow q_{j,t}\hat{r}_{j,t}^{rs}(s_j)$ and taking the conditional expectation, we conclude that the drift goes to the multiple-type replicator equation:

$$\dot{x}_{j,t}(s_j) = g_j(t)x_{j,t}(s_j) \left[\mathbb{E}\hat{r}_{j,t}^{rs}(s_j) - \sum_{s_j' \in \mathcal{A}_j} \mathbb{E}\hat{r}_{j,t}^{rs}(s_j')x_{j,t}(s_j') \right]$$

Hence, the ODE is given by

$$\begin{cases} \dot{x}_{j,t}(s_j) = g_j(t)x_{j,t}(s_j) \left[\hat{r}_{j,t}^{rs}(s_j) - \sum_{s_j' \in \mathcal{A}_j} \hat{r}_{j,t}^{rs}(s_j')x_{j,t}(s_j') \right] \\ \frac{d}{dt}\hat{r}_{j,t}^{rs}(s_j) = \tilde{g}_j(t)x_{j,t}(s_j) \left[\mathbb{E}_{\mathbf{w},\mathcal{B}^n,\mathbf{e}_{s_j},\mathbf{x}_{-j,t}} \frac{1}{\mu_j}(e^{\mu_j \tilde{U}_j^{\mathcal{B}^n}(\mathbf{w},s_j,a_{-j})} - 1) - \hat{r}_{j,t}^{rs}(s_j) \right] \end{cases}$$

Since the payoff learning rate ν is faster than the strategy-learning rate λ, one gets the stated first result. The second statement is obtained from the fact that if $\mathbf{x}_j(s_j) > 0$, then

$$\frac{d}{dt}\hat{r}_{j,t}^{rs}(s_j) = \tilde{g}_j(t)\mathbf{x}_j(s_j) \left[\mathbb{E}_{\mathbf{w},\mathcal{B}^n,\mathbf{e}_{s_j},\mathbf{x}_{-j}} \frac{1}{\mu_j}(e^{\mu_j U_j^{\mathcal{B}^n}(\mathbf{w},s_j,a_{-j})} - 1) - \hat{r}_{j,t}^{rs}(s_j) \right]$$

converges globally to $\mathbb{E}_{\mathbf{w},\mathcal{B}^n,\mathbf{e}_{s_j},\mathbf{x}_{-j}} \frac{1}{\mu_j}(e^{\mu_j \tilde{U}_j^{\mathcal{B}^n}(\mathbf{w},s_j,a_{-j})} - 1)$ when $t \longrightarrow \infty$. Now, the limit set is

$$\{(\mathbf{x},\hat{\mathbf{r}}) \mid \forall j, \ \hat{\mathbf{r}}_j(s_j) = \mathbb{E}_{\mathbf{w},\mathcal{B}^n,\mathbf{e}_{s_j},\mathbf{x}_{-j}} \frac{1}{\mu_j}(e^{\mu_j \tilde{U}_j^{\mathcal{B}^n}(\mathbf{w},s_j,a_{-j})} - 1)\}.$$

Thus, $\hat{\mathbf{r}}$ can be incorporated in the dynamics of $\mathbf{x}_t$. We use the multiple time-scale techniques to conclude that the dynamics can be reduced to a composed strategy-dynamics for each of the players. The same technique is used for the other CODIPAS-RLs with particular emphasis on the inclusion case in which an asymptotically stable set argument is used instead of a stable point. Then, the stated result follows. $\qquad \square$

9.3.5.2 Convergence to Equilibria

In the above analysis we have illustrated an example in which the asymptotic behavior of iterative combined learning can be replaced almost surely by differential equations. However, this notion of convergence is not sufficient to show that the learning algorithm converges to equilibria (because, the ODE may not converge). For convergence analysis of the learning algorithm, a deep study of the behavior of the ODE and the stochastic process, is needed. Below, we give a specific class of robust games namely, robust potential games or robust congestion games, with nice convergence properties.

Proposition 9.3.5.3 (Risk-neutral robust potential). *If the dynamic robust game is a risk-neutral robust potential game i.e., if there exists a continuously differentiable function $\tilde{V}$ defined on $\mathbb{R}^{\sum_j |\mathcal{A}_j|}$ such that $\nabla \tilde{V}(\mathbf{x}) = [\mathbb{E}_{\mathbf{w},\mathcal{B}^n} U_j^{\mathcal{B}^n}(\mathbf{w}, ., \mathbf{x}_{-j})]_j$. Then, global convergence to equilibria of the risk-neutral expected robust game holds for almost all initial conditions.*

For $\mathcal{L}_1, \mathcal{L}_4$, we provide a proof below.

Proof of Proposition 9.3.5.3.

$$
\begin{aligned}
\frac{d}{dt}\tilde{V}(\mathbf{x}_t) &= \sum_{j \in \mathcal{N}} \sum_{s_j} [\frac{d}{dt} x_{j,t}][\frac{\partial}{\partial x_{j,t}(s_j)} \tilde{V}(\mathbf{x}_t)] \\
&= \sum_{j \in \mathcal{N}} g_j(t) \sum_{s_j} x_{j,t}(s_j)[\frac{\partial}{\partial x_{j,t}(s_j)} \tilde{V}(\mathbf{x}_t)]^2 \\
&\quad - \sum_{j \in \mathcal{N}} g_j(t)[\sum_{s_j} x_{j,t}(s_j)\frac{\partial}{\partial x_{j,t}(s_j)} \tilde{V}(\mathbf{x}_t)]^2 \geq 0
\end{aligned}
$$

with strict inequality if $\nabla \tilde{V} \neq 0$. In the last inequality (Jensen), we have used the convexity of y^2 and positivity of g. This means that $\tilde{V}$ serves as a Lyapunov function to the non-autonomous dynamics. The convergence to asymptotically stable set follows.

For the Boltzmann-Gibbs based CODIPAS-RLs or smooth best response dynamics, the Lyapunov function of the expected robust game can be constructed from $\tilde{V}$ and it is given by

$$
\tilde{V}_0(\mathbf{x}) = \tilde{V}(\mathbf{x}) + \sum_{j \in \mathcal{N}} H_j(\mathbf{x}_j),
$$

where H_j is the entropy function

$$
H_j(\mathbf{x}_j) = - \sum_{s_j \,|\, x_j(s_j) > 0} x_j(s_j) \log(x_j(s_j)).
$$

$\square$

For CODIPAS-RL such as Boltzmann-Gibbs, smooth best response, weakened fictitious play etc potential games are known to have convergence properties. The same result applies to *risk-neutral robust potential games.*

Definition 9.3.5.4 (Risk-sensitive robust potential). *The dynamic robust game is a risk-sensitive robust potential game i.e., if there exists a continuously differentiable function* $\tilde{V}_{rs}$ *defined on* $\mathbb{R}^{\sum_j |A_j|}$ *such that*

$$\nabla \tilde{V}_{rs}(\mathbf{x}) = \left[\mathbb{E}_{\mathbf{w}, \mathcal{B}^n, ., \mathbf{x}_{-j}} e^{\mu_j U_j^{\mathcal{B}^n}(\mathbf{w},.,a_{-j})} \right].$$

Proposition 9.3.5.5. *Consider* $\mathcal{L}_1$ *and* $\mathcal{L}_4$. *If the dynamic robust game is a risk-sensitive robust potential game, then one has a global convergence to risk-sensitive equilibria for almost all initial conditions.*

Note that the class of risk-sensitive robust potential games differs from the risk-neutral potential games. If a payoff function over pure action is such that $\tilde{U}_j = \frac{1}{\mu_j} \log \tilde{U}_j^*$ where $\{\tilde{U}_j^*\}_j$ has a potential function in the sense of Monderer and Shapley, then $e^{\mu_j \tilde{U}_j}$ is a weighted potential game. This defines an exponential-potential game.

It should be noticed that the replicator dynamics (for the risk-sensitive criterion) may have some rest points that are risk-sensitive non-Nash equilibria (in the faces of the simplex). However, the interior rest points of the replicator dynamics (9.32) are risk-sensitive equilibria of the robust game with variable number of interacting players.

9.3.5.6 Convergence Time

Even if the convergence to equilibria occurs in a specific class of games, the convergence time can be arbitrarily large. Therefore we are interested in how many iterations are needed to be η−close to an equilibrium if the starting point is not too far.

By η−close we mean the distance between the current point and the equilibrium point is less than η, i.e., in a neighborhood of radius ϵ.

Proposition 9.3.5.7. *The convergence time to be close to the risk-sensitive payoff with error tolerance* η *is at most*

$$\max_j (G^j)^{-1} \left(\log[\frac{error(0)}{\eta}] \right)$$

where $(G^j)^{-1}$ *is the inverse of the mapping* $t \longmapsto G^j(t) := \int_0^t g_j(t') \, dt'$ *and* $error_j(0) := \| \hat{r}_{j,0} - \mathbb{E}\frac{1}{\mu_j}(e^{\mu_j \tilde{U}_j} - 1) \|$, $error = \max_j error_j$. *In particular for* $g_j = 1$ *(almost active case), the convergence time is of order of* $\log \left[\frac{error(0)}{\eta} \right]$.

Proof. We verify that the solution of the ODE is $\parallel \hat{r}^{rs}_{j,t} - \mathbb{E}\frac{1}{\mu_j}(e^{\mu_j \tilde{U}_j} - 1) \parallel =$ $\text{error}_j e^{-\int_0^t g_j(s)\, ds}$. From the assumptions, the primitive function G^j is a bijection and $\parallel \hat{r}^{rs}_{j,t} - \mathbb{E}\frac{1}{\mu_j}(e^{\mu_j \tilde{U}_j} - 1) \parallel \leq \eta$ if $t \geq \max_j (G^j)^{-1}\left(\log[\frac{\text{error}(0)}{\eta}]\right)$. The last assertion is obtained for $g_j = 1$. This completes the proof. $\qquad\square$

9.3.5.8   Explicit Solutions

As promised in our introduction of this Chapter, we provide now the explicit solutions for specific cases. For single player case it can be easily checked that

$$x_{j,t}(s_j) = \frac{x_{j,0}(s_j)e^{t\mathbb{E}_{\mathbf{w}}\frac{1}{\mu_j}(e^{\mu_j \tilde{U}_j(\mathbf{w},e_{s_j})}-1)}}{\sum_{s'_j \in \mathcal{A}_j} x_{j,0}(s'_j)e^{t\mathbb{E}_{\mathbf{w}}\frac{1}{\mu_j}(e^{\mu_j \tilde{U}_j(\mathbf{w},e_{s'_j})}-1)}} \tag{9.33}$$

$$= \beta_{j,\frac{1}{t}}(\mathbb{E}_{\mathbf{w}}\frac{1}{\mu_j}(e^{\mu_j \tilde{U}_j(\mathbf{w},\cdot)} - 1))(s_j), \tag{9.34}$$

is the unique solution of the replicator equation starting from $x_{j,0}$. This solution is expressed in terms of the risk-sensitive smooth best response.

   Smooth best response and Boltzmann-Gibbs for n potential players:

   Given trajectory $\{\mathbf{x}_{-j,t'}\}_{t'}$ and an initial condition $\mathbf{x}_{j,0}$, the smooth best response equation

$$\dot{\mathbf{x}}_{j,t} = \beta_{j,\epsilon_j}(\mathbf{x}_{-j,t}) - \mathbf{x}_{j,t}, \; x_{j,0} \in \mathcal{X}_j \text{ fixed} \tag{9.35}$$

has a unique solution given by the vectorial function

$$\xi_j(\mathbf{x}_{-j,t})(s_j) = x_{j,0}(s_j)e^{-t} + e^{-t}\int_0^t z_{j,t'}(s_j)\, e^{t'}\, dt', \; s_j \in \mathcal{A}_j, \tag{9.36}$$

where $z_{j,t'} = \beta_{j,\epsilon_j}(\mathbf{x}_{-j,t'})$. In particular, if the other players are slow learners or if $\mathbf{x}_{-j,t} = \mathbf{x}_{-j}$, constant in time, then the smooth best response equation of player j converges to

$$\xi_j(\mathbf{x}_{-j})(s_j) = (1 - e^{-t})\beta_{j,\epsilon_j}(\mathbf{x}_{-j})(s_j) + e^{-t}x_{j,0}(s_j), \; s_j \in \mathcal{A}_j, \tag{9.37}$$

which goes to $\beta_{j,\epsilon_j}(\mathbf{x}_{-j})$ when $t \longrightarrow +\infty$ and *the rate of convergence is exponential which is very fast.*

   For the replicator equation:

   Given the trajectories of $\{\mathbf{x}_{-j,t'}\}_{t' \geq 0}$ and an interior initial condition $\mathbf{x}_{j,0}$, the replicator equation of player j has a unique solution given by the vectorial function

$$\bar{\xi}_{j,t}(\mathbf{x}_{-j,t})(s_j) = \frac{\mathbf{x}_{j,0}(s_j)e^{\int_0^t r_{j,\mu_j}(e_{s_j},\mathbf{x}_{-j,t'})\, dt'}}{\sum_{s'_j \in \mathcal{A}_j} \mathbf{x}_{j,0}(s'_j)e^{\int_0^t r_{j,\mu_j}(e_{s'_j},\mathbf{x}_{-j,t'})\, dt'}}, \; s_j \in \mathcal{A}_j.$$

In particular, if all the players have already converged or if they are slow learners, i.e., $\mathbf{x}_{-j,t} = \mathbf{x}_{-j}$, constant in time, then the replicator equation of player j converges to

$$\bar{\xi}_{j,t}(\mathbf{x}_{-j})(s_j) = \frac{e^{tr_{j,\mu_j}(e_{s_j},\mathbf{x}_{-j})}}{\sum_{s'_j \in \mathcal{A}_j} e^{tr_{j,\mu_j}(e_{s'_j},\mathbf{x}_{-j})}}, \quad s_j \in \mathcal{A}_j.$$

Note that if the starting point is at the relative interior of the simplex, these solutions remains in the interior of the simplex for finite time t, but the trajectory can be arbitrarily close to the boundary when t goes to infinity. In particular,

$$\bar{\xi}_{j,t}(\mathbf{x}_{-j})(s_j) \to \frac{x_{j,0}(s_j)}{\sum_{s'_j \in RSPBR_j(\mathbf{x}_{-j})} x_{j,0}(s'_j)} \mathbb{1}_{\{s_j \in RSPBR_j(\mathbf{x}_{-j})\}},$$

when $\min_j \epsilon_j \to 0$. The set $RSPBR_j(\mathbf{x}_{-j})$ denotes the set of pure maximizers of $\mathbf{x}_j \longmapsto \mathbb{E}_{\mathbf{w},\mathbf{x}_j,\mathbf{x}_{-j}} \frac{1}{\mu_j}(e^{\mu_j \tilde{U}_j(\mathbf{w},a)} - 1)$.

Consequences:

From these explicit solutions, we get the following: For two potential players with random updates and under uncertainty, the time-average of the asymptotic pseudo-trajectory of the learning processes leading to the replicator equation and the best response dynamics are asymptotically close. This is because for $\mathcal{N} = \{1,2\}$,

$$\bar{\xi}_{1,t}(s_1) = \frac{\mathbf{x}_{1,0}(s_1)e^{\int_0^t r_{1,\mu_1}(e_{s_1},\mathbf{x}_{2,t'}) \, dt'}}{\sum_{s'_1 \in \mathcal{A}_1} \mathbf{x}_{1,0}(s'_1)e^{\int_0^t r_{1,\mu_1}(e_{s'_1},\mathbf{x}_{2,t'}) \, dt'}}$$

can be written as $\bar{\xi}_{1,t} = \tilde{\beta}_{1,\frac{1}{t}}(\mathbb{U}_{1,\mu_1,t})$ where

$$\mathbb{U}_{1,\mu_1,t}(s_1) := \frac{1}{t} \int_0^t r_{1,\mu_1}(e_{s_1},\mathbf{x}_{2,t'}) \, dt' = r_{1,\mu_1}(e_{s_1},\bar{\mathbf{x}}_{2,t}),$$

with $\bar{\mathbf{x}}_{2,t}(s_2) = \frac{1}{t} \int_0^t \bar{\mathbf{x}}_{2,t'}(s_2)dt'$ is the time average trajectory of player 2.

For more than two potential players, the limiting behavior of the cross terms can be different than the product of the limiting time averages:

$$\lim_t \frac{1}{t} \int_0^t x_{j,t'}(s_j)x_{k,t'}(s_k) \, dt' \neq$$

$$\left(\lim_t \frac{1}{t} \int_0^t x_{j,t'}(s_j) \, dt' \right) \times \left(\lim_t \frac{1}{t} \int_0^t x_{k,t'}(s_k) \, dt' \right)$$

This property is known as *violation of the Decoupling Assumption* in the context of the mean field limit [160, 163, 158].

9.3.5.9   Composed Dynamics

The following result follows from the above analysis.

Proposition 9.3.5.10 (Slow and fast learning in CODIPAS-RL). *We consider two potential players.* • *Assume that player 1 is a slow learner of $\mathcal{L}_2$ and player 2 is a fast learner of $\mathcal{L}_2$, then almost surely, $\|\mathbf{x}_{2,t} - \beta_{2,\epsilon}(\mathbf{x}_1)\| \longrightarrow 0$, as t goes to infinity and,*

$$\dot{x}_{1,t} = \beta_{1,\epsilon}(\beta_{2,\epsilon}(\mathbf{x}_{1,t})) - \mathbf{x}_{1,t}$$

is the asymptotic pseudo-trajectory of $\{\mathbf{x}_{1,t}\}_{t\geq 0}$.

   • *Assume that player 1 is a slow learner of $\mathcal{L}_4$ and player 2 is a fast learner of $\mathcal{L}_4$, then almost surely,*

$$\|\mathbf{x}_{2,t} - \xi_2(\mathbf{x}_1)\| \longrightarrow 0,$$

as t goes to infinity and,

$$\dot{x}_{1,t}(s_1) = x_{1,t}(s_1)\left[\mathbb{E}_{\mathbf{w},\mathbf{e}_{s_1},\xi_2(\mathbf{x}_{1,t})}\frac{1}{\mu_1}(e^{\mu_1\tilde{U}_1(w,a)} - 1)\right.$$

$$\left. - \sum_{s_1'\in\mathcal{A}_1} x_{1,t}(s_1')\mathbb{E}_{\mathbf{w},\mathbf{e}_{s_1'},\xi_2(\mathbf{x}_{1,t})}\frac{1}{\mu_1}(e^{\mu_1\tilde{U}_1(w,a)} - 1)\right],$$

is the asymptotic pseudo-trajectory of $\{\mathbf{x}_{1,t}\}_{t\geq 0}$ where

$$\xi_j(\mathbf{x}_{-j})(s_j) = \frac{e^{\int_0^t \mathbb{E}_{\mathbf{w},\mathbf{e}_{s_j},\mathbf{x}_{-j,t'}}\frac{1}{\mu_j}(e^{\mu_j\tilde{U}_j(\mathbf{w},s_j,a_{-j})}-1)\ dt'}}{\sum_{s_j'\in\mathcal{A}_j} e^{\int_0^t \mathbb{E}_{\mathbf{w},\mathbf{e}_{s_j'},\mathbf{x}_{-j,t'}}\frac{1}{\mu_j}(e^{\mu_j\tilde{U}_j(\mathbf{w},s_j',a_{-j})}-1)\ dt'}}.$$

   • *Assume that player 1 is a slow learner of imitative Boltzmann-Gibbs $\mathcal{L}_3$ and player 2 is a fast learner of $\mathcal{L}_2$. Then, almost surely, $\|\mathbf{x}_{2,t} - \beta_{2,\epsilon}(\mathbf{x}_1)\| \longrightarrow 0$, as t goes to infinity and, $\dot{x}_{1,t} = \sigma_{1,\epsilon}(\beta_{2,\epsilon}(\mathbf{x}_{1,t})) - \mathbf{x}_{1,t}$ is the asymptotic pseudo-trajectory of $\{\mathbf{x}_{1,t}\}_{t\geq 0}$ where*

$$\sigma_{j,\epsilon}(\hat{\mathbf{r}}_{\mathbf{j,t}}^{\mathbf{rs}})(s_j) = \frac{x_{j,t}(s_j)e^{\frac{\hat{r}_{j,t}^{rs}(s_j)}{\epsilon_j}}}{\sum_{s_j'\in\mathcal{A}_j} x_{j,t}(s_j')e^{\frac{\hat{r}_{j,t}^{rs}(s_j')}{\epsilon_j}}}.$$

   • *Assume that player 1 is a slow learner of $\mathcal{L}_4$ and player 2 is a fast learner of $\mathcal{L}_2$. Then, almost surely, $\|\mathbf{x}_{2,t} - \beta_{2,\epsilon}(\mathbf{x}_1)\| \longrightarrow 0$, as t goes to infinity and,*

$$\dot{x}_{1,t}(s_1) = x_{1,t}(s_1)\left[\mathbb{E}_{\mathbf{w},\mathbf{e}_{s_1},\beta_{2,\epsilon}(\mathbf{x}_{1,t})}\frac{1}{\mu_1}(e^{\mu_1\tilde{U}_1} - 1)\right.$$

$$\left. - \sum_{s_1'\in\mathcal{A}_1} x_{1,t}(s_1')\mathbb{E}_{\mathbf{w},\mathbf{e}_{s_1'},\beta_{2,\epsilon}(\mathbf{x}_{1,t})}\frac{1}{\mu_1}(e^{\mu_1\tilde{U}_1} - 1)\right]$$

is the asymptotic pseudo-trajectory of $\{\mathbf{x}_{1,t}\}_{t\geq 0}$.

• *Assume that player 1 is a slow learner of $\mathcal{L}_4$ and player 2 is a fast learner of $\mathcal{L}_3$. Then almost surely, $\|\mathbf{x}_{2t} - \sigma_{2,\epsilon}(\mathbf{x}_1)\| \longrightarrow 0$, as t goes to infinity and,*

$$\dot{x}_{1,t}(s_1) = x_{1,t}(s_1)\left[\mathbb{E}_{\mathbf{w},\mathbf{e}_{s_1},\sigma_{2,\epsilon}(\mathbf{x}_{1,t})}\frac{1}{\mu_1}(e^{\mu_1\tilde{U}_1} - 1)\right.$$

$$\left. - \sum_{s_1'\in\mathcal{A}_1} x_{1,t}(s_1')\mathbb{E}_{\mathbf{w},\mathbf{e}_{s_1'},\sigma_{2,\epsilon}(\mathbf{x}_{1,t})}\frac{1}{\mu_1}(e^{\mu_1\tilde{U}_1} - 1)\right]$$

is the asymptotic pseudo-trajectory of $\{\mathbf{x}_{1,t}\}_{t\geq 0}$.

9.3.5.11   Non-Convergence to Unstable Rest Points

Proposition 9.3.5.12. *Let $\mathbf{x}^*$ be a rest point for the asymptotic of Bush-Mosteller CODIPAS-RL, or equivalently the adjusted replicator dynamics. If either*

• $\mathbf{x}^*$ *is not an equilibrium of the expected robust game,*

• $\mathbf{x}^*$ *is a Nash equilibrium linearly unstable under the adjusted replicator dynamics,*

then, for the Bush-Mosteller based CODIPAS-RL process from any relative interior initial condition, that is $x_{j,0}(s_j) > 0$ for each player j and all action s_j, $\mathbb{P}(\lim_t \mathbf{x}_t = \mathbf{x}^*) = 0$.

Proof of Proposition 9.3.5.12. The proof of non-convergence to linearly unstable point of a Bush-Mosteller based CODIPAS-RL follows the same line as in [76, 36, 29]. □

Following the non-convergence to linearly unstable equilibria developed in [27], the Proposition 9.3.5.12 extends to the CODIPAS-RLs $\mathcal{L}_1, \ldots, \mathcal{L}_4$.

9.3.5.13   Dulac Criterion for Convergence

For robust games with two potential players and two actions per player, i.e., $\mathcal{A}_j = \{s_j^1, s_j^2\}, j \in \{1,2\}$, one can transform the system of ODEs of the strategy-learning into a planar system in the form

$$\dot{\alpha}_1 = Q_1(\alpha_1, \alpha_2), \ \dot{\alpha}_2 = Q_2(\alpha_1, \alpha_2), \tag{9.38}$$

where we let $\alpha_j = x_j(s_j^1)$. The dynamics for player j can be expressed in terms of α_1, α_2 only as $x_1(s_1^2) = 1 - x_1(s_1^2)$, and $x_2(s_2^2) = 1 - x_2(s_2^2)$.

We use the Poincaré-Bendixson lemma and the Dulac criterion [67] to establish a convergence result for (9.38).

Proposition 9.3.5.14 ([67]). *For an autonomous planar vector field as in (9.38), the Dulac's criterion states as follows. Let $\gamma(.)$ be a scalar function defined on the unit square $[0,1]^2$. If*

$$\frac{\partial[\gamma(\alpha))\dot{\alpha}_1]}{\partial \alpha_1} + \frac{\partial[\gamma(\alpha)\dot{\alpha}_2]}{\partial \alpha_2},$$

is not identically zero and does not change sign in $[0,1]^2$, then there are no cycles lying entirely in $[0,1]^2$.

Corollary 9.3.5.15. *Consider a two-action robust game with two potential players. Assume that each of the adopt the Boltzmann-Gibbs CODIPAS-RL with*

$$\frac{\lambda_{j,t}}{\nu_{j,t}} = \frac{\lambda_t}{\nu_t} \longrightarrow 0.$$

Then, the asymptotic pseudo-trajectory reduces to a planar system in the form $\dot{\alpha}_1 = \tilde{\beta}_{1,\epsilon_1}(r_{1,\mu_1}(e_{s_1}, \alpha_2)) - \alpha_1; \dot{\alpha}_2 = \tilde{\beta}_{2,\epsilon_2}(r_{2,\mu_2}(\alpha_1, e_{s_2})) - \alpha_2$. Moreover, the system satisfies the Dulac criterion.

Proof. We apply Proposition 9.3.5.14 with $\gamma(\cdot) \equiv 1$. One obtains that the divergence is -2, which is strictly negative. Hence, the result follows. $\square$

Note that for the replicator dynamics, the Dulac criterion reduces to $(1 - 2\alpha_1)(r_{1,\mu_1}(e_{s_1^1}, \alpha_2) - r_{1,\mu_1}(e_{s_1^2}, \alpha_2)) + (1 - 2\alpha_2)(r_{2,\mu_2}(\alpha_1, e_{s_2^1}) - r_{2,\mu_2}(\alpha_1, e_{s_2^2}))$ which vanishes for $(\alpha_1, \alpha_2) = (1/2, 1/2)$. It is possible to have limit cycles in the replicator dynamics, and hence the Dulac criterion does not apply. However, the stability of the replicator dynamics can be directly studied in the two-action case by simply identifying the game in one of the following types: coordination, anti-coordination, prisoner's dilemma, and hawk-and-dove [190]. For the best response dynamics the same methodology is used for generic two-player-two-action games.

The next corollary follows from Proposition 9.3.5.14.

Corollary 9.3.5.16.   **Heterogeneous learning:** *If one of the potential player (say player 1) is with Boltzmann-Gibbs CODIPAS-RL and player's 2 learning process leads to replicator dynamics, then the convergence condition reduces to*

$$(1 - 2\alpha_2)(r_{2,\mu_2}(\alpha_1, e_{s_2^1}) - r_{2,\mu_2}(\alpha_1, e_{s_2^2})) < 1,$$

for all (α_1, α_2).

Hybrid learning: *If all the potential players use hybrid learning [153] obtained by combining Boltzmann-Gibbs CODIPAS-RL with weight $\omega_{j,1}$ and the replicator dynamics with weight $1 - \omega_{j,1}$ then the Dulac criterion reduces to $\omega_{1,2}[(1 - 2\alpha_1)(r_{1,\mu_1}(e_{s_1^1}, \alpha_2) - r_{1,\mu_1}(e_{s_1^2}, \alpha_2))] + \omega_{2,2}[(1 - 2\alpha_2)(r_{2,\mu_2}(\alpha_1, e_{s_2^1}) - r_{2,\mu_2}(\alpha_1, e_{s_2^2}))] < w_{1,1} + w_{2,2}$ for all (α_1, α_2).*

Remark 9.3.5.17 (Symmetric games with three actions). *If the expected game is a symmetric game with three actions per player, then the symmetric game dynamics reduce to a two-dimensional dynamical system. This allows us to apply the Dulac criterion.*

9.4 Risk-Sensitivity in Networking and Communications

Risk-Sensitivity in Reliability and Network Services

In today's provision of network services, a rapid development of many new technologies (like virtualization, cloud computing, or advanced long-term-evolution) and applications (e.g., social networking) can be seen. They have a considerable engineering, commercial and societal impact and pose numerous challenges to engineers, designers, and researchers. One class of issues is the continuous offering of services in spite of various types of failures that are everyday events in networks. The increasing complexity, the participation of many different autonomous entities in the service provision, the lack of information, coordinated actions as well as governance makes this challenge increasingly difficult. In addition to a design of reliable network segments, server clusters etc., new technological issues like stability and failure management across heterogeneous autonomous systems arise. They also involve novel fields at the border of technology, business and society, like quality of experience (QoE), risk-awareness or influence of public policy.

Next we examine a generic frequency selection interaction.

Example 9.4.1 (Frequency selection games: two-players-two-frequencies). *Consider two potential players in a wireless network. Each player can choose one of the frequencies $\{f_1, f_2\}$. If the players choose at the same frequency during the same frame, there is a collision and the packets are lost and each of the players gets 0. If they choose at different frequencies, they have successful transmissions and the payoffs are 1. The following table summarizes this scenario. In Table 9.2 player 1 uses a row, and player 2 chooses a column.*

TABLE 9.2
Frequency selection games

	f_1	f_2
f_1	0,0	1,1
f_2	1,1	0,0

The game has two pure equilibria and one fully mixed equilibrium which is also an evolutionarily stable strategy in the sense of Maynard-Smith and Price. The fully mixed equilibrium is less efficient in terms of social welfare. The pure equilibria are Strong equilibria (robust to any coalition of any size). The two pure equilibria are maxmin solutions in the sense that the minimum of the payoffs of the players is maximized. It is easy to see that the two pure equilibria are also global optima. Now, a natural question is,

> *Is there a fully distributed learning scheme to converges to global optima?*

The answer to this question is positive for some initial conditions. The set of initial conditions under which convergence to global optima is observed is of measure 1.

Using the risk-neutral CODIPAS-RL $\mathcal{L}_1, \mathcal{L}_4$ one has that the asymptotic pseudo-trajectories give the replicator dynamics. If x denotes the probability of player 1 choosing action f_1 and y the probability of player 2 choosing action f_1, then the ordinary differential equation satisfied by x and y are:

$$\dot{x} = x(1-x)(1-2y), \tag{9.39}$$

$$\dot{y} = y(1-y)(1-2x), \tag{9.40}$$

The rest points of the system are $(0,0),(1,0),(0,1),(1,1),(\frac{1}{2},\frac{1}{2})$. One can easily check that $(1,0)$ and $(0,1)$ are stable. Starting from any point in the unit square $[0,1]^2$ outside the segment $y = x$, the system converges to the set of global optima, which is very interesting. Now, if we consider symmetric configuration, then the system goes to the unique evolutionarily stable strategy starting from any interior point. Next, we add the activity of the two players. There are four configurations (see Table 9.3): $\mathcal{B}^2 \in \{\{1,2\},\{1\},\{2\},\emptyset\}$. It easy to see that the above analysis to the i.i.d state case where the set $\{1,2\}$ is in the support of the random variable $\mathcal{B}_t^2$.

Risk-sensitive case Let $\mu_j \neq 0$. The two player case is summarized in Table 9.4.

Notice that the global optimum value $\frac{e^{\mu_1}-1}{\mu_1} + \frac{e^{\mu_2}-1}{\mu_2}$ can be greater the risk-neutral global optimum value, which is 2. Moreover, the risk-sensitive CODIPAS-RL based on replicator dynamics selects one of the risk-sensitive global optima for almost all initial conditions.

Using the risk-sensitive CODIPAS-RL $\mathcal{L}_1$ and $\mathcal{L}_4$, one has that the asymptotic pseudo-trajectories give the rescaled replicator dynamics. If x denotes the probability of player 1 choosing action f_1 and y the probability of player 2 choosing action f_1, then the ordinary differential equation satisfied by x and y are:

$$\dot{x} = \frac{e^{\mu_1}-1}{\mu_1}x(1-x)(1-2y), \tag{9.41}$$

$$\dot{y} = \frac{e^{\mu_2}-1}{\mu_2}y(1-y)(1-2x), \tag{9.42}$$

$$(x_0, y_0) \in [0,1]^2 \tag{9.43}$$

The rest points of the system are $(0,0),(1,0),(0,1),(1,1)$, and $(\frac{1}{2}\frac{1}{2})$. One can easily check that $(1,0)$ and $(0,1)$ are stable points. Starting from any point in the unit square $[0,1]^2$ outside the segment $y = x$, the system converges

TABLE 9.3
Frequency selection games: random activity

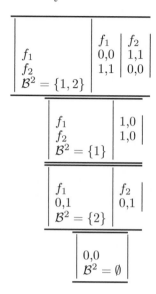

TABLE 9.4
Risk-sensitive frequency selection games

	f_1	f_2
f_1	0,0	$\frac{e^{\mu_1}-1}{\mu_1}, \frac{e^{\mu_2}-1}{\mu_2}$
f_2	$\frac{e^{\mu_1}-1}{\mu_1}, \frac{e^{\mu_2}-1}{\mu_2}$	0,0

to the set of risk-sensitive global optima, which is very interesting. Now, if
we consider a symmetric configuration, then the system goes to the unique
evolutionarily stable state (ESS) starting from any interior point.

Notice that for $\mu_1, \mu_2 > 0$ the risk-sensitive replicator dynamics converges
faster than the risk-neutral.

Next, we add the activity of the two players. There are four configurations
(see table 9.5): $\mathcal{B}^2 \in \{\{1,2\}, \{1\}, \{2\}, \emptyset\}$. It easy to see that the above analysis
to the i.i.d state case where the set $\{1,2\}$ is in the support of the random
variable $\mathcal{B}_t^2$.

Three users:

We now examine the three-user case: $\mathcal{B}^3 = \{1,2,3\}$. We prove that the
pure global optima are linearly stable under the time-scaled replicator dynamics. The proof is as follows. We first check that the global optima are

TABLE 9.5
Frequency selection games: random activity

	f_1	f_2
f_1	0,0	$\frac{e^{\mu_1}-1}{\mu_1}, \frac{e^{\mu_2}-1}{\mu_2}$
f_2 $\mathcal{B}^2=\{1,2\}$	$\frac{e^{\mu_1}-1}{\mu_1}, \frac{e^{\mu_2}-1}{\mu_2}$	0,0

	$\frac{e^{\mu_1}-1}{\mu_1},0$
f_1	
f_2 $\mathcal{B}^2=\{1\}$	$\frac{e^{\mu_1}-1}{\mu_1},0$

	f_1	f_2
$\mathcal{B}^2=\{2\}$	$0,\frac{e^{\mu_2}-1}{\mu_2}$	$0,\frac{e^{\mu_2}-1}{\mu_2}$

0,0
$\mathcal{B}^2=\emptyset$

$(1,0,0),(0,1,0),(0,0,1),(0,1,1),(1,0,1),$ and $(1,1,0).$ *We compute the risk-sensitive version of the replicator equation associated with the game:*

$$\dot{x} = \frac{e^{\mu_1}-1}{\mu_1}x(1-x)(1-y-z), \tag{9.44}$$

$$\dot{y} = \frac{e^{\mu_2}-1}{\mu_2}y(1-y)(1-x-z), \tag{9.45}$$

$$\dot{z} = \frac{e^{\mu_3}-1}{\mu_3}z(1-z)(1-x-y), \tag{9.46}$$

$$(x_0,y_0,z_0)\in[0,1]^2. \tag{9.47}$$

We verify that the global optima are also rest points of the replicator equation. Now, to establish a local stability result, we compute the Jacobian of the system which is

$$\begin{pmatrix} (1-2x)(1-y-z) & -x(1-x) & -x(1-x) \\ -y(1-y) & (1-2y)(1-x-z) & -y(1-y) \\ -z(1-z) & -z(1-z) & (1-2z)(1-x-y) \end{pmatrix}$$

The Jacobian evaluated at the global optima reduces to the matrices

$$J_{global} = \begin{pmatrix} -1 & 0 & 0 \\ 0 & 0 & 0 \\ 0 & 0 & 0 \end{pmatrix},$$

and the permutation over the diagonal elements. Recall that Lyapunov stability of the linearized system holds if and only if all eigenvalues of the matrix J_{global} have negative or zero real parts and for each eigenvalue with zero real part there is an independent eigenvector. Here, the independent eigenvectors are obtained by taking two of the elements of the canonical basis of $\mathbb{R}^3$. However, the fact that this matrix is degenerated, it is difficult to reach completely pure strategies.

For non-hyperbolic stationary points, the local stability of the non-linear system can be different than the local stability of linear systems. In other words, Hartman-Grobman's theorem (Theorem A.1.0.3 in Appendix) cannot be used here directly. We need to a find a Lyapunov function (if any) of the system.

As we can see $(1, y, 0)$ is a stationary point for any $y \in [0, 1]$. Thus, with this scheme we observe a convergence to the set

$$\{(1, y, 0) \mid y \in [0, 1]\},$$

and all the possible permutations. However, convergence to pure strategies can be observed with the (stochastic) CODIPAS-RL with constant learning rate. This can be seen from the stationary points of the strategy-learning scheme under constant learning rate. Below we detail these statements using numerical examples. In Figures 9.1, 9.2, 9.3 and 9.4, we illustrate the imitative CODIPAS-RL and observe the convergence to one global optimum starting from a non-symmetric point.

Details of Figures 9.1 and 9.2: These figures represent the evolution of the strategy-learning, its zoom, and the payoff estimations. We choose a window of size 3000iterations, $\lambda = 0.1$, $\nu = 0.6$, and initialize the strategies at $x_{1,0} = [0.6, 0.4]$, $x_{2,0} = [0.1, 0.9]$, $x_{3,0} = [0.2, 0.8]$. The payoff estimations are initialized at $\hat{r}_{1,0} = [0.40.3], \hat{r}_{2,0} = [0.20.4], \hat{r}_{3,0} = [0.20.3]$. All the users have negative risk-sensitivity index $\mu_1 = \mu_2 = \mu_3 = -1$. We observe that the payoffs are around $1 - \frac{1}{e}$ which is less than 1. This is due to the negativity of the risk-sensitivity indexes. We plot two lines 0.1 and 0.9 as reference thresholds to check the convergence time. As we can observe that the curves of the evolution of strategies in time across the referent lines after around 1000 iterations (maximum time of all the actions to be $0.005-$close to the global optimum.)

In Figure 9.3, we run the imitative CODIPAS-RL for positive risk-sensitivity index $\mu_j = 1$. The other parameters are the same as in Figure 9.2. We observe that the payoffs are around $e - 1$ which is greater than 1. We plot two lines 0.1 and 0.9 as reference thresholds to check the convergence time. As we can observe the curves across these lines after around 500 iterations (maximum time of all the actions to be 0.005 close to the global optimum.)

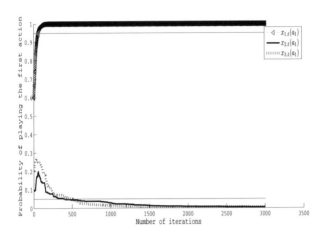

FIGURE 9.1

Global optima: $\mu_j < 0$.

Interestingly, the convergence time of the strategy learning is improved com-
pared the one with negative risk-sensitive index. This is because asymptotically,
when $\mu_j > 0$, the multiplicative factor of the equation is $(e^{\mu_j} - 1)/\mu_j > 1$ which
accelerate the convergence time to be $0.05 - close$.

    In Figure 9.4, we represent the evolution of strategies of a mixed sys-
tem: Two risk-averse users and one risk-seeking user. $\mu_1 = \mu_2 = -1 < 0$, $\mu_3 =$
$6 > 0$. The CODIPAS converges to one of the global optima. We observe that
the convergence of the strategy-learning of user 3 is faster but the the variance
is very large compared to the one with $\mu_3 \in \{1, -1\}$.

    Figures 9.5 and 9.6 illustrate the impact of initial conditions in the outcome
of the stochastic algorithm. In figure 9.5, the initial conditions for strategies
are $x_{1,0} = [0.51, 0.49]$, $x_{2,0} = [0.5, 0.5]$, $x_{3,0} = [0.49, 0.51]$. We observe that the
imitative CODIPAS-RL converges to $(1, 0, 0)$.

    In Figure 9.6, we fix $x_{1,0} = [0.52, 0.48]$, $x_{2,0} = [0.5, 0.5]$, $x_{3,0} = [0.49, 0.51]$
and $\hat{r}_{j,0} = [0.1, 0.1], j \in \{1, 2, 3\}$. We choose the risk-sensitivity parameters
$\mu_j = 0.1$, $j \in \{1, 2, 3\}$. For one run of the stochastic algorithm, we observe
that the imitative CODIPAS-RL converges to $(0, 1, 1)$, which is another global
optimum corresponding to the case where user 1 chooses frequency 2 and users
2 and 3 choose the same frequency 1. This scenario tell us that the stochastic
nature of the imitative CODIPAS-RL plays an important role in the outcome
and is affected by both starting point and risk-sensitivity parameters. This ob-
servation suggests also the instability of the point $(1/2, 1/2, 1/2)$ in the asym-
metric behavior. This can be proved by computing the Jacobian at the point
$(1/2, 1/2, 1/2)$, which has 1 as eigenvalue.

    Figure 9.7 is three dimensional plot that illustrates a convergence to a pure
equilibrium under imitative CODIPAS.

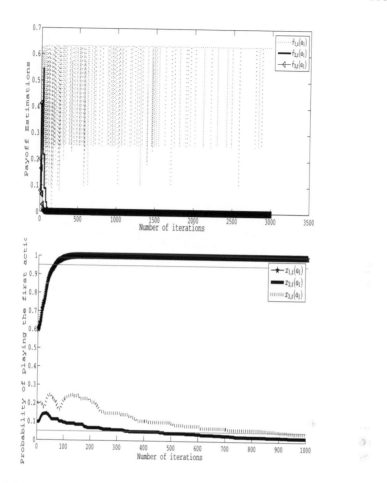

FIGURE 9.2
Convergence to global optima, $\mu_j < 0$.

Example 9.4.2. *Dynamic channel selection is an important component of wireless systems. It allows a transmitter to adaptively choose for transmission the channel offering the best radio conditions and to avoid interference created by other transmitters. In the absence of interference, the channel selection problem can be simply interpreted as a Multi-Armed Bandit (MAB) problem for which low-regret learning algorithms can be found. Now, the problem is much more complicated because of interference due to the interaction of transmitters, and the fact that a given transmitter experiencing a transmission failure cannot identify whether the failure is due to a channel error or to interference. A transmitter has then to learn not only which channel is best but also to share channels with other transmitters. For the dynamic channel*

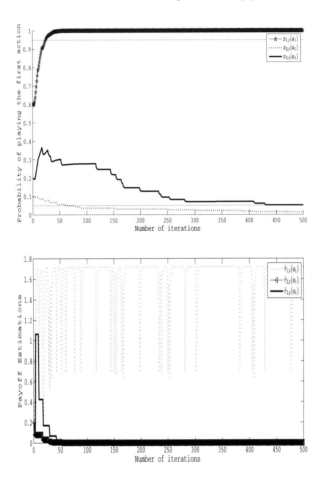

FIGURE 9.3
Convergence to global optimum $(1, 0, 0)$. $\mu_i > 0$.

selection problem with multiple users we propose the following algorithm: Let the imitative strategy to be

$$\tilde{\sigma}_{j,\lambda_t,s_j}(\hat{d}_{j,t}) = \frac{x_{j,t}(s_j)(1 + \lambda_t)^{-\hat{d}_{j,t}(s_j)}}{\sum_{s'_j \in \mathcal{A}} x_{j,t}(s'_j)(1 + \lambda_{j,t})^{-\hat{d}_{j,t}(s'_j)}}$$

The algorithm is described as follows.

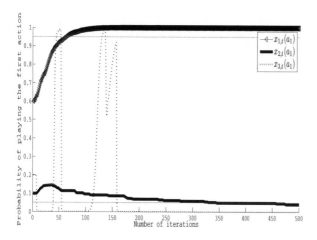

FIGURE 9.4
Two risk-averse users and one risk-seeking user. $\mu_1 < 0, \mu_2 < 0, \mu_3 > 0$.

Algorithm 8: Imitative CODIPAS based Multi-Armed Bandit

Fix $\mathbf{x}_0$;

Initialize $\hat{d}_0 = 0 \in \mathbb{R}^K$;

for $t=1$ **to** T **do** *do*;

foreach *User* j **do**

Draw an action $a_{j,t} = s_{j_t}$ *with random* j_t *distributed according to* $x_{j,t}$.;
Observe its current payoff $d_{j,t}$;
Update its estimation via
$\hat{d}_{j,t+1}(s_{j_t}) = \hat{d}_{j,t}(s_{j_t}) + \nu_t \left(\frac{1}{x_{j,t}(s_{j_t})}[d_{j,t} + \delta_t]_+ \mathbb{1}_{\{a_{j,t}=s_{j_t}\}} \right)$;
Update via Imitative CODIPAS: $\mathbf{x}_{j,t+1}(s_{j_t}) = \tilde{\sigma}_{j,\lambda_t,s_{j_t}}(\hat{d}_{j,t})$;

Example 9.4.3 (Learning the ergodic capacity). *We consider a multi-player wireless MIMO system with independent and identically distributed (i.i.d) channel states. We consider* $n-link$ *communication network that can be modeled by a MIMO Gaussian interference channel. Each link is associated with a transmitter-receiver pair. Each transmitter and receiver are equipped with* $\bar{n}_t$ *and* $\bar{n}_r$ *antennas, respectively.*

The sets of transmitters is denoted by $\mathcal{N}$. *The cardinality of* $\mathcal{N}$ *is* n, *and transmitter* j *transmits a complex signal vector* $\tilde{s}_{j,t} \in \mathbb{C}^{\bar{n}_t}$ *of dimension* $\bar{n}_t$. *Consequently, a complex baseband signal vector of dimension* $\bar{n}_r$ *denoted by* $\tilde{y}_{j,t}$ *is received as output. The vector of received signals from* j *is defined by*

$$\tilde{y}_{j,t} = H_{j,j,t}\tilde{s}_{j,t} + \sum_{j' \neq j} H_{j,j',t}\tilde{s}_{j',t} + z_{j,t},$$

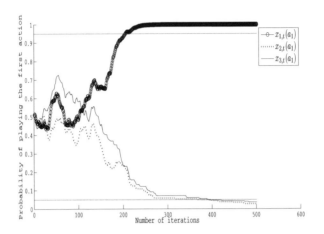

FIGURE 9.5
Imitative CODIPAS-RL. Impact initial condition. $\mu_i = -0.01$.

where t is the time index, $\forall j \in \mathcal{N}$, $H_{j,j',t}$ is the complex channel matrix of dimension $\bar{n}_r \times \bar{n}_t$ from the transmitter j to the receiver j', and the vector $z_{j,t}$ represents the noise observed at the receivers; it is a zero mean circularly symmetric complex Gaussian noise vector with an arbitrary nonsingular covariance matrix R_j.

For all $j \in \mathcal{N}$, the matrix $H_{j,j,t}$ is assumed to be non-zero. We denote $\mathbf{H}_t = (H_{j,j',t})_{j,j'}$. The vector of transmitted symbols $\tilde{s}_{j,t}$, $\forall \, j \in \mathcal{N}$ is characterized in terms of power by the covariance matrix $Q_{j,t} = \mathbb{E}\left(\tilde{s}_{j,t}\tilde{s}_{j,t}^{\dagger}\right)$, which is Hermitian (self-adjoint) positive semi-definite matrix. Now, since transmitters are power-limited, we have that

$$\forall \, j \in \mathcal{N}, \; \forall t \geq 0, \quad tr(Q_{j,t}) \leq p_{j,\max}. \tag{9.48}$$

Note that the constraint at each time slot can be relaxed to a long-term time-average power budget. We define a transmit power covariance matrix for transmitter $j \in \mathcal{N}$ as a matrix $Q_j \in \mathcal{M}_+$ satisfying (9.48), where $\mathcal{M}_+$ denotes the Hermitian positive matrix. The payoff function of j is its mutual information $I(\tilde{s}_j; \tilde{y}_j)(\mathbf{H}, Q_1, \ldots, Q_n)$. Under the above assumption, the maximum information rate [50] is

$$\underline{R}_j(\mathbf{H}, Q) = \tilde{U}_j(\mathbf{H}, Q_1, \ldots, Q_n) = \log \det \left(I + H_{jj}^{\dagger}\Gamma_j^{-1}(Q_{-j})H_{jj}Q_j\right),$$

where $\Gamma_j(Q_{-j}) = \underline{R}_j + \sum_{j' \neq j} H_{jj'}Q_{j'}H_{jj'}^{\dagger}$ is the multi-player interference plus noise observed at j and $Q_{-j} = (Q_k)_{k \neq j}$ is the collection of players' covariance matrices, except the j-th one. The robust individual optimization problem of

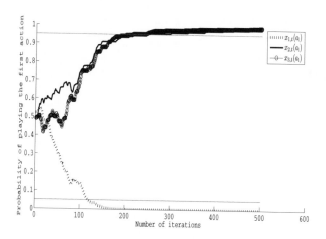

FIGURE 9.6
Imitative CODIPAS-RL, $\mu_j > 0$. Impact of initial condition.

player j is then

$$j \in \mathcal{N}, \quad \sup_{Q_j \in \mathcal{Q}_j} \; \inf_{\mathbf{H}} \; I(\tilde{s}_j; \tilde{y}_j)(\mathbf{H}, Q_1, \ldots, Q_n)$$

where

$$\mathcal{Q}_j := \{Q_j \in \mathbb{C}^{\bar{n}_t \times \bar{n}_t} \mid Q_j \in \mathcal{M}_+, \; tr(Q_j) \le p_{j,\max}\}.$$

Assume that each player does not know the structure of the payoff function and wants to learn it. Note that no channel state information is used in this model. The transmitter does not have knowledge of the distribution over $\mathbf{H}$ in $I(\tilde{s}_j; \tilde{y}_j)(\mathbf{H}, Q_1, \ldots, Q_n)$. We assume that player employs a state-independent stationary strategy. We use the payoff learning of CODIDAS-RL under a fixed law of $\mathbf{w}_t = \mathbf{H}_t$. Using Proposition 9.3.5.7, we deduce that the payoff dynamics (ODE) of CODIPAS-RL converges exponentially to the expected payoff, i.e., the ergodic capacity is almost learned.

Example 9.4.4 (Learning the outage probability). *The action $s_j = Q_j$ is now a finite matrix. The matrix is fixed in a way such that $tr(Q_j) \le c_j$, $\forall j$. The constant c_j is a fixed power constraint for player j. Let $u_{j,t} = \mathbb{1}_{\{\underline{R}_j(H_t, Q_t) \ge R_j^*\}}$. When t goes to infinity, the above algorithm converges to $\mathbb{P}(\underline{R}_j(H, Q) \ge R_j^*)$. This result is important since it allows to estimate the ergodic capacity without knowing the law of the channel H_t, provided that all the matrices Q have been sufficiently explored. Thus, we deduce the outage probability by taking $1 - \mathbb{P}(\underline{R}_j(H, Q) \ge R_j^*)$.*

Example 9.4.5 (Power allocation games with i.i.d channels). *We reconsider*

Convergence to a pure strategy by imitation

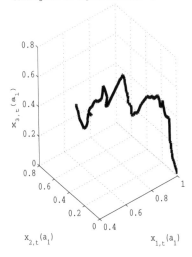

FIGURE 9.7
Imitative CODIPAS-RL: 3D plot.

the above power allocation game where each transmitter player employs a hybrid CODIPAS-RL scheme. We assume that the players have a minimal amount of information in the sense that each player observes a numerical, noisy and delayed value of her own payoff when she is active.

In particular, transmitter j does not know the mathematical structure of her payoff function, or her own channel state. The actions and payoffs of the other players are unknown too.

> *It is important to notice that the players do not need to know their own channel state in order to update the CODIPAS-RL.*

Under such conditions the question becomes: Is there a fully distributed learning scheme to achieve an expected payoff and an equilibrium in MIMO interference dynamic environment?

We say that the family of games indexed by $\mathcal{H}$ is an expected robust pseudo-potential game if there exists a function ϕ defined on $\mathcal{H} \times \prod_{j \in \mathcal{N}} \mathcal{Q}_j$ such that

$$\forall\ j,\ \forall \mathbf{Q}_{-j},\ \arg\max_{Q_j} \mathbb{E}_{\mathbf{H}}\phi(\mathbf{H}, \mathbf{Q}) \subseteq \arg\max_{Q_j} \mathbb{E}_{\mathbf{H}}U_j(\mathbf{H}, \mathbf{Q}) \qquad (9.49)$$

The work in [162] analyzed in detail the heterogeneous CODIPAS-RL for the ergodic Shannon rate case. By rewriting the risk-neutral payoff as

$$\tilde{U}_j(\mathbf{H}, \mathbf{Q}) = \phi(\mathbf{H}, \mathbf{Q}) + \tilde{B}_j(\mathbf{H}, \mathbf{Q}_{-j}),\ \ where, \qquad (9.50)$$

$$\phi(\mathbf{H}, \mathbf{Q}) = \log \det \left(\tilde{R} + \sum_{j \in \mathcal{N}} H_{j,t} Q_{j,t} H_{j,t}^\dagger \right),$$

$$\tilde{B}_j(\mathbf{H}, \mathbf{Q}_{-j}) = -\log \det \left(\underline{R} + \sum_{l \neq j} H_{l,t} Q_{l,t} H_{l,t}^\dagger \right),$$

we deduce that the risk-neutral power allocation game is an expected robust pseudo-potential game with potential function $\mathbb{E}_{\mathbf{H}} \phi$.

To see this, we verify that

$$\xi: \ \mathbf{Q} \longmapsto \mathbb{E}_{\mathbf{H}} \phi(\mathbf{H}, \mathbf{Q}) = \int_{\mathbf{H}} \phi(\mathbf{H}, \mathbf{Q}) \ \nu(d\mathbf{H}),$$

where ν is the law of $\mathbf{H}$. By taking the expectation of Equation (9.50), one has that

$$\forall j \in \mathcal{N}, \ \forall \mathbf{Q}_{-j}, \ \arg\max_{Q_j} \mathbb{E}_{\mathbf{H}} I(\tilde{s}_j; \tilde{y}_j)(\mathbf{H}, \mathbf{Q}) = \arg\max_{Q_j} \mathbb{E}_{\mathbf{H}} \phi(\mathbf{H}, \mathbf{Q}) = \arg\max_{Q_j} \xi(\mathbf{Q})$$

Using Proposition 9.3.5.3, we deduce global convergence result to the set of risk-neutral equilibria for the mixed extension of the expected robust potential function.

9.5 Risk-Sensitive Mean Field Learning

When the number of players is large, the system of ODEs is difficult to analyze. This leads to a *curse of dimensionality* problem. It is important to reduce the number of parameters without losing so many equilibrium/stability properties. This reduction is crucial in particular when the size of the population grows (one gets an exponential number of random processes to analyze). The mean field approach allows us to reduce drastically the number of parameters in the analysis of large systems by using a dynamical system description of the aggregate of the individual states and/or actions. One of the questions that the mean field approach addresses is the derivation of the *mean field game dynamics*. In this section, we derive the mean field game dynamics corresponding to the risk-sensitive CODIPAS-RL.

We consider a finite number of subsystem called subpopulations. We denote by $\mathcal{A}^p$ the set of all the possible actions in the subpopulation p. We assume that these sets are finite. We establish a relation between the CODIPAS-RL with a random set of interacting players and the mean field game dynamics $K-$large subpopulations. Usually, in population games, the players play only in pure actions[1] The consequence of this assumption is that the mass

---
[1] Revision protocols in pure actions.

behavior is in finite dimension if the sets of actions are finite. The standard evolutionary game dynamics do not cover the scenario that we have described above in which a player may use a mixed action $\mathbf{x}_t$. We aim to study the mean field limit when the players are allowed to use mixed actions (as in the above learning schemes). We derive the corresponding evolutionary game dynamics for risk-sensitive CODIPAS-RL called *mean field game dynamics*.

We assume that players from different sub-populations can adopt different learning patterns. We denote the set of subpopulations by $\mathcal{P} = \{1, \ldots, K\}$ where K is the total number of learning schemes. The set of learning schemes is denoted by $\mathcal{L} = \{\mathcal{L}_1, \ldots, \mathcal{L}_K\}$. A player from subpopulation p adopts a learning scheme $\mathcal{L}_p$ and her set of pure actions is a finite set $\mathcal{A}^p$. The drift of the learning scheme $\mathcal{L}_p$ is denoted by $\tilde{F}^p$, i.e.,

$$\mathbf{x}_{t+1}^p = \mathbf{x}_t^p + \lambda_t \tilde{F}_t^p(\mathbf{x}_t^p, \hat{\mathbf{u}}_t^p), \ p \in \mathcal{P}.$$

Denote by F_t^p the drift limit of the learning governed by $\tilde{F}_t^p$. Therefore, the mixed actions evolution is given by the following dynamical system

$$\dot{\mathbf{x}}_t^p = F_t^p(\mathbf{x}_t), \ p \in \mathcal{P}, \ \mathbf{x}_0^p \in \Delta(\mathcal{A}^p).$$

We define the population profile of mixed actions as

$$M_t^{p,n_p} = \frac{1}{n_p} \sum_{j \in \mathcal{N}_p} \delta_{\mathbf{x}_{j,t}}, t \geq 0$$

$\mathcal{N}_p$ is the set of players in subpopulation p and $n_p = |\mathcal{N}_p|$. Given a subset B_p of $\Delta(\mathcal{A}^p)$, the measure of B_p relatively to M_t^{p,n_p} is

$$M_t^{p,n_p}(B_p) = \frac{1}{n_p} \sum_{j \in \mathcal{N}_p} \delta_{\{\mathbf{x}_{j,t}^p \in B_p\}}.$$

This says that the mass of players in subpopulation p using a mixed action in B_p.

The process $M_t^{p,n_p} \in \Delta(\Delta(\mathcal{A}^p))$. Hence,

$$M^{p,n_p} \in \Delta(\Delta(\mathcal{A}^p))^{\mathbb{N}}.$$

It is known that the space $\Delta(\Delta(\mathcal{A}^p))^{\mathbb{N}}$ is a Polish space. Under the assumptions of exchangeability per subpopulation (for learning scheme, strategies, and payoffs), we can derive the weak convergence of the mean process (see [164] and the references therein), i.e., the weak convergence of the random process $M^n = (M^{1,n_1}, \ldots, M^{p,n_p}, \ldots, M^{K,n_K})$. We refer to $m = (m^p)_p$ as the limit. Sufficient conditions for deterministic mean field limit can be found in [164, 163]. Let ϕ^p be a test function that vanishes at the boundary of the space $X^p := \Delta(\mathcal{A}^p)$. Here, we do not restrict ourselves to pure actions. Our objective is to derive the mean field game dynamics, i.e., the evolution of the density of the subpopulation playing mixed actions. The mean field dynamics for *mixed actions* are derived from the different learning dynamics. The mean field measure is defined by

$$m^p : \Delta(\mathcal{A}^p) \times \mathbb{R}_+ \longrightarrow [0,1],$$

$$\int_{\Delta(\mathcal{A}^p)} m_t^p(\mathbf{x}^p) \, d\Pi(\mathbf{x}^p) = 1,$$

$$m_t^p(\mathbf{x}^p) \geq 0, \ \forall(t,p),$$

where Π is a measure of the set of mixed actions. In the single subpopulation case (homogeneous population), the population of players is represented by a probability measure $m_t(d\mathbf{x})$. The total mass of players using mixed strategies in the open set $O \subset int\Delta(A)$ is $\int_O m_t(\mathbf{x}) \, d\Pi(\mathbf{x})$. The rate of changes of mass in the set O is

$$\frac{d}{dt} \int_O m_t(\mathbf{x}) \, d\Pi(\mathbf{x}) = \int_O \frac{\partial}{\partial t} m_t(\mathbf{x}) \, d\Pi(\mathbf{x}).$$

Let us examine the incoming flux and the outgoing flux of O across the boundary ∂O. The population learning process is represented by

$$\dot{\mathbf{x}}_t = F_t(\mathbf{x}_t).$$

The flux of mass out of O is given by the vector field $\langle F_t, \vec{n} \rangle m_t$ on ∂O ($\vec{n}$ normal of ∂O. The aggregate flux of mass into O across its boundary is

$$-\int_{\partial O} \langle F_t, \vec{n} \rangle m_t(\mathbf{x}) \, d\text{Area},$$

where *Area* is an elementary measure at ∂O. Recall that the *Divergence theorem* states:

$$\int_{\partial O} \langle F_t, \vec{n} \rangle m_t(\mathbf{x}) \, d\text{Area} = \int_O \text{div}[F_t(\mathbf{x}) m_t(\mathbf{x})] \, d\Pi(\mathbf{x}).$$

Thus,

$$\int_O \frac{\partial}{\partial t} m_t(\mathbf{x}) \, d\Pi(\mathbf{x}) = -\int_O \text{div}(F_t m_t) \, d\Pi(\mathbf{x})$$

i.e

$$\int_O \left[\frac{\partial}{\partial t} m_t(\mathbf{x}) + \text{div}(F_t m_t) \right] d\Pi(\mathbf{x}) = 0.$$

The mean field dynamics for a single population is

$$\frac{\partial}{\partial t} m_t(\mathbf{x}) + \text{div}(F_t(\mathbf{x}) m_t(\mathbf{x})) = 0, \tag{9.51}$$

$$\dot{\mathbf{x}}_t = F_t(\mathbf{x}_t), \tag{9.52}$$

$$\mathbf{x}_0 \in \Delta(\mathcal{A}), m_0(.) \in \Delta(\Delta(\mathcal{A})). \tag{9.53}$$

We now focus on the multiple subpopulations case. Let

$$m^p : \Delta(\mathcal{A}^p) \times \mathbb{R}_+ \longrightarrow [0,1],$$

be the time-dependent probability density function over the space of mixed strategies for subpopulation p. If B_p is a measurable set, then

$$m_t^p(B_p) = \int_{\mathbf{x}^p \in B_p} m_t^p(\mathbf{x}^p) \, d\Pi(\mathbf{x}^p),$$

is the mass of players using their mixed action in B_p.

Proposition 9.5.0.1. *The mean field game dynamics is given by*

$$\begin{cases} \frac{\partial}{\partial t} m_t^p(\mathbf{x}^p) + div[\Gamma_t^p(\mathbf{x}^p, m_t^{-p}) m_t^p(\mathbf{x}^p)] = 0 \\ \dot{\mathbf{x}}_t^p = F_t^p(\mathbf{x}_t^1, \ldots, \mathbf{x}_t^K) \\ \mathbf{x}_0^p \in \Delta(\mathcal{A}^p), \ m_0^p \in \Delta(\Delta(\mathcal{A}^p)), \end{cases}$$

where

$$\Gamma_t^p(\mathbf{x}^p, m_t^{-p}) = \int_{\mathbf{x}^{-p}} \sum_{(s^1, \ldots, s^K)} \left(\prod_{p=1}^K \mathbf{x}^p(s^p) \right) F_{t,s}^p(\mathbf{x}_t) \times$$
$$\prod_{p' \neq p} m_t^{p'}(\mathbf{x}^{p'}) d\Pi(\mathbf{x}^{p'})$$

and $m_t^{-p} = (m_t^{p'})_{p' \neq p}.$

The proof of Proposition 9.5.0.1 follows similar lines as in the above analysis. The above mean field game dynamics is related to the transport equations (see [146] and the references therein), also called continuity equations. These structures are particular cases of Fokker-Planck-Kolmogorov equations which contain not only a divergence term but also a diffusion term. The computation of $\Gamma_t^p(\mathbf{x}^p, m_t^{-p})$ for a revision protocol of mixed actions leading to the replicator dynamics is easily obtained by incorporating

$$F_{t,s}^p(\mathbf{x}_t) = \mathbf{x}_t^p(s^p) \left[\hat{\mathbf{r}}_t^p(s^p) - \sum_{s'^p \in \mathcal{A}^p} \hat{\mathbf{r}}_t^p(s'^p) \mathbf{x}_t^p(s'^p) \right].$$

How useful is mean field learning?

In large-scale systems, the differential equations representing the learning patterns of each player lead to a high a dimensional dynamical system, for example, the asymptotic pseudo-trajectories could be a system of $\sum_{j=1}^n |\mathcal{A}_j|$

ODEs. When the number of players goes to infinity, we cannot use the standard numerical methods to solve these systems. One of the contributions of mean field learning is to reduce the complexity of these systems to condensate in density form, and under the above assumptions, it leads to the continuity equations given in Proposition 9.5.0.1. Note that in the presence of non-vanishing noises, we will get a second order equation known as Fokker-Planck-Kolmogorov forward equation.

9.6 Extensions

9.6.1 Risk-Sensitive Correlated Equilibria

Definition 9.6.1.1. *Following the same methodology as above, we say that a probability distribution $p^* \in \Delta(\prod_j \mathcal{A}_j)$ is a risk-sensitive correlated equilibrium if for any $a_j \in \mathcal{A}_j$,*

$$\sum_{a_{-j}} p^*(a_j, a_{-j}) \left[\frac{1}{\mu_j} (e^{\mu_j \tilde{U}_j(a)} - 1) - \frac{1}{\mu_j} (e^{\mu_j \tilde{U}_j(s_j, a_{-j})} - 1) \right] \geq 0 \ \forall s_j \in \mathcal{A}_j.$$

Conditioned on the j-th component of a mixed action profile drawn from x being a_j, the risk-sensitive payoff for j of playing a_j is no smaller than that of playing b_j.

The set of risk-sensitive correlated equilibria is non-empty and contains the set of risk-sensitive Nash equilibria. Moreover, using the fact that the set of risk-sensitive correlated equilibria can be described by linear inequalities and has convexity properties, a regret-based learning scheme with low complexity can be developed following the idea developed in [70]. Using the works in [58, 70] for finite games, we deduce that *risk-sensitive regret* minimizing procedures can cause the empirical frequency distribution of play to converge to the set of *risk-sensitive correlated equilibria*. Extension to games with continuous action space can found in [149]. The stability of correlated equilibria have been studied in [169] using the notion *correlated evolutionary stable strategies* (CESS).

Conditioned on the j-th component of a mixed action profile drawn from x being a_j, the risk-sensitive payoff for j of playing a_j is no smaller than that of playing b_j.

Definition 9.6.1.2. *We say that probability distribution $p^* \in \Delta(\prod_j \mathcal{A}_j)$ is a risk-sensitive state-independent correlated equilibrium if for any $a_j \in \mathcal{A}_j$,*

$$\sum_{a_{-j}} \mathbb{E}_w p^*(a_j, a_{-j}) \left[e^{\mu_j \tilde{U}_j(w,a)} - e^{\mu_j \tilde{U}_j(w,s_j,a_{-j})} \right] \geq 0 \ \forall s_j \in \mathcal{A}_j.$$

> The risk-sensitive no-regret algorithm generates a frequency of play that becomes close to the set of *risk-sensitive state-independent correlated equilibria*.

9.6.2 Other Risk-Sensitive Formulations

Below we list two alternative formulations that are related to the risk-sensitive analysis:

- For each $j \in \mathcal{N}$, and $\mathbf{x}_{-j}$, one would like to solve the *risk-sensitive* robust optimization problem 1 $RS - ROP1_j(\mu_j, \mathbf{x}_{-j})$:

$$\sup_{\mathbf{x}_j \in \mathcal{X}_j} \left[\mathbb{E}_{\mathbf{w},\mathbf{x}} \tilde{U}_j(\mathbf{w}, \mathbf{a}_j, \mathbf{a}_{-j}) - \lambda var[\tilde{U}_j(\mathbf{w}, \mathbf{a}_j, \mathbf{a}_{-j})] \right]. \qquad (9.54)$$

for a certain $\lambda \in \mathbb{R}$.

- For each $j \in \mathcal{N}$, and $\mathbf{x}_{-j}$, one would like to solve the constrained robust optimization problem 2 $RS - ROP2_j(\mu_j, \mathbf{x}_{-j})$: maximizing the expected payoff while maintaining the variance below some threshold.

$$\sup_{\mathbf{x}_j \in \mathcal{X}_j} \mathbb{E}_{\mathbf{w},\mathbf{x}} \tilde{U}_j(\mathbf{w}, \mathbf{a}_j, \mathbf{a}_{-j}), \qquad (9.55)$$

$$var\left[\tilde{U}_j(\mathbf{w}, \mathbf{a}_j, \mathbf{a}_{-j}) \right] \leq \overline{variance}. \qquad (9.56)$$

Both formulations take into consideration the expectation and the variance which play important roles in the risk-sensitive systems. A closely related problem to $RS - ROP2$ could be a multiobjective view point: Each user maximizes her expected payoff and minimizes the variance of her payoff.

9.6.3 From Risk-Sensitive to Maximin Robust Games

In this subsection, we explain how the risk-sensitive payoff is related to a pessimistic/optimistic payoff. To establish this connection, we use large deviation theory. Let $(E, \mathcal{P})$ be a measurable space and $u(.)$ a bounded mapping E into real values, and $\tilde{\nu}$ a probability measure over E.

The **relative entropy** is defined by

$$\tilde{D}(\tilde{\nu}|\tilde{\nu}) = \int_E \log\left(\frac{d\tilde{\nu}}{d\tilde{\nu}} \right) d\tilde{\nu}$$

whenever $\log\left(\frac{d\tilde{\nu}}{d\tilde{\nu}}(w) \right)$ is integrable with the respect to $\tilde{\nu}$. In all other cases, it is defined to be $+\infty$.

Lemma 9.6.3.1. *For every $\tilde{\nu}, \tilde{\nu}$ probability measures over a complete separable metric space E, $\tilde{D}(\tilde{\nu}|\tilde{\nu})$ with equality if and only if $\tilde{\nu}$ and $\tilde{\nu}$ are identical.*

Proof. The proof is simple. First, observe that for any $x \geq 0$, $\log(x) \leq x - 1$ with equality if and only if $x = 1$. We deduce that $\log(1/x) \geq 1 - x$, $x > 0$. Similarly, $x \log(x) \geq x - 1$ if and only if $x = 1$. All these inequalities come from the concavity of the function $\log(.)$. Thus,

$$\tilde{D}(\tilde{\nu}|\tilde{\nu}) = \int_E \log\left(\frac{d\tilde{\nu}}{d\tilde{\nu}}\right) d\tilde{\nu} \tag{9.57}$$

$$\geq \int_E \left[1 - \frac{d\tilde{\nu}}{d\tilde{\nu}}\right] d\tilde{\nu} \tag{9.58}$$

$$= 0 \tag{9.59}$$

with equality if and only if $\frac{d\tilde{\nu}}{d\tilde{\nu}} = 1$ for $\tilde{\nu}-$almost everywhere. □

Note that the nonnegative of the relative entropy holds also for the discrete state space.

$$\tilde{D}(\tilde{\nu} \mid \tilde{\nu}) = \sum_{s \in E} \tilde{\tilde{\nu}}_s \log\left(\frac{\tilde{\tilde{\nu}}_s}{\tilde{\nu}_s}\right) \tag{9.60}$$

$$\geq \sum_{s \in E} \tilde{\tilde{\nu}}_s \left(1 - \frac{\tilde{\nu}_s}{\tilde{\tilde{\nu}}_s}\right) \tag{9.61}$$

$$= \sum_{s \in E} \tilde{\tilde{\nu}}_s - \sum_{s \in E} \tilde{\nu}_s \tag{9.62}$$

$$= 0 \tag{9.63}$$

with equality if and only if $\forall s \in E$, $\frac{\tilde{\tilde{\nu}}_s}{\tilde{\nu}_s} = 1$.

Note that the relative entropy measure (which is not a distance) quantifies how different two distributions are.

The following theorem holds.

Theorem 9.6.3.2. • *We have the variational formula:*

$$-\log\left(\int_E e^{-u} d\tilde{\nu}\right) = \inf_{\tilde{\nu}} \left\{ \tilde{D}(\tilde{\nu}|\tilde{\nu}) + \int_E u d\tilde{\nu} \right\} \tag{9.64}$$

where $\tilde{D}(\tilde{\nu}|\tilde{\nu})$ denotes the relative entropy (also called, Kullback-Leibler measure).

• *Let $\tilde{\nu}_0$ denote the probability measure on E which is absolute continuous with the respect to $\tilde{\nu}$ and satisfies*

$$\frac{d\tilde{\nu}_0}{d\tilde{\nu}}(w) = \frac{e^{-u(w)}}{\int_E e^{-u(w)} d\tilde{\nu}(w)} \tag{9.65}$$

Then, the infimum in the variational formula is uniquely attained at $\tilde{\nu}_0$.

- *The risk-sensitive payoff can be written as a robust payoff i.e*

$$\epsilon \log \left(\int_E e^{\frac{1}{\epsilon} u} d\tilde{\nu} \right) = \sup_{\tilde{\nu}} \left\{ \int_E u(w) d\tilde{\nu}(w) - \epsilon \tilde{D}(\tilde{\nu}|\tilde{\nu}) \right\}$$

Proof.

$$\tilde{D}(\tilde{\nu}|\tilde{\nu}) + \int_E u d\tilde{\nu} \qquad (9.66)$$

$$= \int_E \log \left(\frac{d\tilde{\nu}}{d\tilde{\nu}} \right) d\tilde{\nu} + \int_E u d\tilde{\nu} \qquad (9.67)$$

$$= \int_E \log \left(\frac{d\tilde{\nu}}{d\tilde{\nu}_0} \right) d\tilde{\nu} + \int_E \log \left(\frac{d\tilde{\nu}_0}{d\tilde{\nu}} \right) d\tilde{\nu} + \int_E u d\tilde{\nu} \qquad (9.68)$$

$$= \tilde{D}(\tilde{\nu}|\tilde{\nu}_0) + \int_E \left[\log e^{-u} - \log \left(\int_E e^{-u} d\tilde{\nu} \right) \right] d\tilde{\nu} + \int_E u d\tilde{\nu} \qquad (9.69)$$

$$= \tilde{D}(\tilde{\nu}|\tilde{\nu}_0) - \log \left(\int_E e^{-u} d\tilde{\nu} \right) \qquad (9.70)$$

Now we use the fact that $\tilde{D}(\tilde{\nu}|\tilde{\nu}_0) \geq 0$ with equality if and only if $\tilde{\tilde{\nu}} = \tilde{\nu}_0$. The other statements follow. $\qquad \square$

> The consequence of Theorem 9.6.3.2 is that the risk-sensitive payoff can be seen as robust payoff obtained by perturbation of the risk-neutral payoff with the relative entropy among all the possible distributions.

9.6.4 Mean-Variance Approach

In this subsection, we propose an alternative risk-sensitive approach that is widely used in financial markets.

> How to learn the mean-variance payoff from the measurements?

We decompose the mean-variance payoff-learning problem in two parts. We know how to learn the expected payoff with the risk-neutral CODIPAS-RL developed in Chapter 3. We combine this approach with a learning scheme for the variance of the payoffs.

Learning the variance of the payoffs:

The variance learning scheme is constructed as follows. The estimator of the variance is

$$\hat{v}_{j,t} = \frac{1}{t} \sum_{t'=1}^{t} \left(r_{j,t'} - \hat{r}_{j,t'} \right)^2 .$$

Using the same methodology as in Lemma 2.2.1.1 of Chapter 2, we get the recursive stochastic equation:

$$\hat{v}_{j,t+1} = \hat{v}_{j,t} + \tilde{\mu}_{j,t}\left((r_{j,t} - \hat{r}_{j,t})^2 - \hat{v}_{j,t}(s_j)\right).$$

The mean-variance learning $\hat{r}_{var,t}$ is then given by

$$\begin{cases} \hat{v}_{j,t+1}(s_j) = \hat{v}_{j,t}(s_j) + \tilde{\mu}_{j,t}\mathbb{1}_{\{a_{j,t}=s_j\}}\left((r_{j,t} - \hat{r}_{j,t}(s_j))^2 - \hat{v}_{j,t}(s_j)\right). \\ \hat{r}_{j,t+1}(s_j) = \hat{r}_{j,t}(s_j) + \mu_{j,t}\mathbb{1}_{\{a_{j,t}=s_j\}}\left(r_{j,t} - \hat{r}_{j,t}(s_j)\right) \\ \hat{r}_{var,t}(s_j) = \hat{r}_{j,t}(s_j) - \hat{v}_{j,t}(s_j) \end{cases}$$

Combining together, we obtain a new CODIPAS called *Mean-Variance CODIPAS*:

$$\begin{cases} \mathbf{x}_{j,t+1}(s_j) - \mathbf{x}_{j,t}(s_j) = \\ \mathbb{1}_{\{j \in \mathcal{B}^n(t)\}}\sum_{l \in \mathcal{L}}\mathbb{1}_{\{l_{j,t}=l\}}K_{j,s_j}^{1,(l)}(\lambda_{j,\theta_j}(t), a_{j,t}, u_{j,t-\tau_j}, \hat{\mathbf{r}}_{j,t}, \hat{v}_{j,t}, \mathbf{x}_{j,t}), \\ \hat{\mathbf{r}}_{j,t+1}(s_j) - \hat{\mathbf{r}}_{j,t}(s_j) = \\ \mathbb{1}_{\{j \in \mathcal{B}^n(t)\}}K_{j,s_j}^2(\nu_{j,\theta_j}(t), a_{j,t}, u_{j,t-\tau_j}, \hat{\mathbf{r}}_{j,t}, \hat{v}_{j,t}, \mathbf{x}_{j,t}), \\ \hat{v}_{j,t+1}(s_j) = \hat{v}_{j,t}(s_j) + \tilde{\mu}_{j,t}\mathbb{1}_{\{j \in \mathcal{B}^n(t)\}}\left((r_{j,t} - \hat{r}_{j,t}(s_j))^2 - \hat{v}_{j,t}(s_j)\right) \\ j \in \mathcal{N}, t \geq 0, a_{j,t} \in \mathcal{A}_j, s_j \in \mathcal{A}_j, \\ \theta_j(t+1) = \theta_j(t) + \mathbb{1}_{\{j \in \mathcal{B}^n(t)\}}, \\ t \geq 0, \ \mathcal{B}^n(t) \subseteq \mathcal{N}, \\ \hat{\mathbf{v}}_{j,0} \in \mathbb{R}^{|\mathcal{A}_j|}. \\ \mathbf{x}_{j,0} \in \mathcal{X}_j, \hat{\mathbf{r}}_{j,0} \in \mathbb{R}^{|\mathcal{A}_j|}. \end{cases}$$

9.7 Chapter Review

9.7.1 Summary

In this chapter we have studied hybrid learning for both payoffs and strategies under risk-sensitive criterion. We provided convergence and nonconvergence results.

A number of further issues are under consideration. It would be of great interest to develop theoretical bounds for the rate of convergence of hybrid risk-sensitive CORDIPAS-RLs. Also, we aim to extend the analysis of our CODIPAS-RL algorithms to more classes of dynamic games including schemes described by Itô's stochastic differential equation (SDE). Typically, the case where the strategy learning has the following form:

$$\mathbf{x}_{t+1} = \mathbf{x}_t + \lambda_t(f(\mathbf{x}_t, \hat{\mathbf{r}}_t) + M_{t+1}) + \sqrt{\lambda_t}\sigma(\mathbf{x}_t, \hat{\mathbf{r}}_t)\xi_{t+1},$$

can be seen as an Euler scheme of Itô's SDE:

$$d\mathbf{x}_{j,t} = f_j(\mathbf{x}_t, \hat{\mathbf{r}}_t)dt + \sigma_j(\mathbf{x}_t, \hat{\mathbf{r}}_t)d\mathbb{B}_{j,t},$$

where $\mathbb{B}_{j,t}$ is a standard Brownian motion in $\mathbb{R}^{|\mathcal{A}_j|}$. Finally, we point out the extension of the stochastic approximation framework to correlated noises (non-i.i.d states) and random updating times.

Exercise 9.1 (Risk-sensitive solution concepts). *You can think of this question as "true or false and explain why" discussion of the italicized statements, although parts of it are somewhat more specific.*

- *Risk-sensitive games study risk as chances of huge loss occurring even with small probability, and propose a risk-sensitive approach that considers distributions not to increase the expected payoff but to mitigate the risks of a huge loss.*

  - *Risk-sensitive equilibria are not necessarily Nash equilibria.*
  - *Risk-sensitive solutions can be global optima.*
  - *Risk-sensitive equilibria can be characterized as solutions of variational inequalities.*

- *Risk-sensitive correlated equilibrium may improve the performance of the system compared to risk-neutral equilibrium.*

Exercise 9.2 (Risk-sensitive learning). *Consider a risk-sensitive learning framework.*

- *Give an example of a networking game in which a risk-sensitive CODIPAS-RL converges. Propose a non-convergent game under CODIPAS-RL.*

- *Give a basic two-by-two example of a risk-sensitive robust potential game and provide a convergent risk-sensitive CODIPAS.*

- *Provide a convergent CODIPAS-RL for aggregative potential games.*

- *Explain how to learn the risk-sensitive payoff.*

We summarize some comparative properties of the different learning schemes in Table 9.6.

where the abbreviations are summarized below.

- LC: limit cycle

- PD: Partially distributed, FD: fully distributed,

- RL: reinforcement learning,

- NE: Nash equilibrium, CE: Correlated equilibrium, SS: Satisfactory solution,

- boundary: the boundary of the simplex.

- RS: risk-sensitive, FP: Fictitious play

- BM - RS-CODIPAS: Bush-Mosteller-RS-CODIPAS, $\mathcal{L}_1$

TABLE 9.6
Summary

	PD	FD	Outcome
Fictitious play	yes	no	NE
Better reply	yes	no	NE
Best response	yes	no	NE
Smooth best response	yes	no	NE
Cost-to-Learn	yes		NE
RL: Arthur's model	no	yes	NE or boundary, LC
RL: Borger & Sarin	no	yes	NE or boundary, LC
Boltzmann-Gibbs	no	yes	NE, LC
No-regret	no	yes	CE
Trial and error	no	yes	NE
Bargaining	yes	no	Pareto optimal
Satisfactory learning	no	yes	SS
BM -CODIPAS	no	yes	NE, LC, boundary
BM - RS-CODIPAS	no	yes	NE, LC, boundary
BG-CODIPAS	no	yes	$\epsilon-$NE, LC
RS-BG-CODIPAS	no	yes	RS-NE, LC
IBG-CODIPAS	no	yes	ϵ-NE, boundary, LC
RS-IBG-CODIPAS	no	yes	RS-NE, Boundary, LC
Weakened FP-CODIPAS	no	yes	NE, LC
RS-Weakened FP-CODIPAS	no	yes	RS-NE, LC
MW-IBG-CODIPAS	no	yes	NE, boundary,LC
RS-MW-IBG-CODIPAS $\mathcal{L}_4$	no	yes	RS-NE, boundary,LC

- BG-CODIPAS: Boltzmann-Gibbs CODIPAS-RL, $\mathcal{L}_2$,

- IBG: Imitative Boltzmann-Gibbs, IBG-CODIPAS $\mathcal{L}_3$

- MW: Multiplicative weight;

9.7.2 Open Issues

- Stochastic iterative estimations

  - Fast convergence of iterative learning patterns

  - Hitting time to a set

  - Frequency of visits

- Strategy learning and payoff estimations

  - Best learning algorithms for stability/performance tradeoff,

– How to learn global optima in a fully distributed way?

- Wireless communications and networking

    – Why should we care about distributed strategic learning in wireless networks?

    – How to extract useful information from outdated and noisy measurements?

- Economics of networks

    – Simultaneous, coalitional and mutual learning

    – Learn how to flatten the peaks in large-scale wireless networks

- Cross-layer learning

    – How to learn in cross-layer interactive systems (e.g. PHY and MAC)?

    – Evolutionary coalitional learning in multilevel interacting networks

A

Appendix

A.1   Basics of Dynamical Systems

The concept of a dynamical system has its origins in mathematical physics. There, as in other sciences and engineering, the evolution rule of dynamical systems is given implicitly by a relation that gives the state of the system only a short time into the future. The relation is either a differential equation, difference equation, or other time scale. Determining the state for all future times requires iterating the relation many times, each advancing time a small step called *step-size, learning rate, intensity of interaction*, etc. The iteration procedure is referred to as solving the system or integrating the system. Once the system can be solved, given an initial point it is possible to determine all its future positions, a collection of points known as a trajectory or orbit.

Definition A.1.0.1. *A dynamical system is a manifold called the state space endowed with a family of smooth evolution functions ϕ_t that for any element of $t \in \mathcal{T}$, the time, maps a point of the phase space back into the phase space. The notion of smoothness changes with applications and the type of manifold. There are several choices for the set $\mathcal{T}$.*

- *When $\mathcal{T}$ is taken to be the reals, the dynamical system is called a flow; and*

- *If $\mathcal{T}$ is restricted to the nonnegative reals, then the dynamical system is a semi-flow.*

- *When $\mathcal{T}$ is taken to be the integers, it is a cascade or a map;*

The evolution function ϕ_t is often the solution of a "differential equation of motion"

$$\dot{x} = f(x). \tag{A.1}$$

The equation gives the time derivative, represented by the dot, of a trajectory x_t on the state space starting at some point x_0. The *vector field $f(x)$* is a smooth function that at every point of the state space $\mathcal{X}$ provides the velocity vector of the dynamical system at that point. These vectors are not vectors in the state space $\mathcal{X}$, but in the *tangent space $T_x\mathcal{X}$* of the point x.

There is no need for higher-order derivatives in the equation, nor for time

417

dependence in $f(x)$ because these can be eliminated by considering systems of higher dimensions.

Below we give a formal definition.

Definition A.1.0.2. *A dynamical system is a semigroup $\bar{G}$ acting on a space $\mathcal{X}$. That is, there is a map*

$$\tilde{T}: \ \bar{G} \times \mathcal{X} \longrightarrow \mathcal{X}$$

defined by $(g, x) \longmapsto \tilde{T}_g(x)$ such that

$$\tilde{T}_{g_1 o g_2} = \tilde{T}_{g_1} o \tilde{T}_{g_2}.$$

In 1892, the Russian mathematician Alexander Mikhailovitch Lyapunov introduced his famous stability theory for nonlinear and linear systems. A complete English translation of Lyapunov's doctoral dissertation was published in the *International Journal of Control* in March 1992. In this chapter, stability means *stability in the sense of Lyapunov*.

We consider the in/stability question in more detail, and explicitly answer it in the case of linear systems. Throughout, we will restrict our attention to a certain class of systems, defined as follows:

> A first-order system is called *autonomous* if it can be expressed in the form $\dot{x} = f(x)$; that is, if there is no explicit dependence on the time variable t.

We now define more precisely what we mean by the stability of a solution to an autonomous system:

> ### Lyapunov stability

The solution $x = \phi_1$ to the autonomous equation $\dot{x} = f(x)$ is said to be stable if for every $\epsilon > 0$, there exists $\delta_\epsilon > 0$ such that whenever $\tilde{x} = \phi_2$ is another solution with

$$\| \phi_1(0) - \phi_2(0) \|_\infty = \max_k | \phi_{1,k}(0) - \phi_{2,k}(0) | < \delta_\epsilon$$

then we are guaranteed that

$$\| \phi_1(t) - \phi_2(t) \|_\infty < \epsilon.$$

In words we are saying that a solution x is stable if we can force nearby solutions to stay arbitrarily close to the solution y merely by requiring their initial values to be sufficiently close to the initial values of x. On the contrary with an unstable solution x we are never guaranteed that a solution with initial values very close to x will behave in any way at all like x in the long run.

According to Lyapunov, one can check the stability of a system by finding some function called the Lyapunov function, which for time-invariant systems satisfies

- $V(x)$ is positive: $V(x) > 0, \forall x \neq 0$, and $V(0) = 0$

-
$$\dot{V}(x) = \frac{d}{dt}[V(x)] = \langle \nabla V(x), \dot{x} \rangle \leq 0$$

⚠ There is no general procedure for finding the Lyapunov functions for nonlinear systems.

Linearization and local stability

In the study of dynamical systems, the Hartman-Grobman theorem, also called linearization theorem is an important theorem about the local behaviour of dynamical systems in the neighbourhood of a hyperbolic equilibrium point (i.e., all eigenvalues of the linearization have nonzero real part).

Basically, the theorem states that the behavior of a dynamical system near a hyperbolic equilibrium point is qualitatively the same as the behavior of its linearization near this equilibrium point provided that no eigenvalue of the linearization has it's real part equal to 0. Therefore when dealing with such fixed points we can use the simpler linearization of the system to analyze its behavior.

Theorem A.1.0.3 (Hartman-Grobman theorem). *Let $f : \mathbb{R}^d \to \mathbb{R}^d$ be a smooth map with a hyperbolic equilibrium point x, that is, $f(x) = 0$ and no eigenvalue of the linearization M of f at point x has its real part equal to 0. Then there exists a neighborhood $\mathcal{N}(x)$ of x and a homeomorphism (a continuous function between two topological spaces that has a continuous inverse function),*

$$h : \mathcal{N}(x) \to \mathbb{R}^d$$

such that $h(x) = 0$, and such that in a neighborhood $\mathcal{N}(x)$ of x, the flow of f is topologically conjugate by h to the flow of its linearization M. In general, even for infinitely differentiable maps f, the homomorphism h need not to be smooth, nor even locally Lipschitz.

The above result is very important and is used in this book to study the local stability of game dynamics under uncertainties.

How to analyze the stability of linear systems in continuous time?

Consider the dynamical system $\dot{x} = Mx$, $x(0) = x_0$. We aim to study the stability/instability of this system. Assume that all the eigenvalues are nonzero.

If all the eigenvalues of M have strictly negative real parts, then the system is stable.

The theorem below answers the stability question of linear systems in continuous time.

Theorem A.1.0.4. *Consider the linear system $\dot{x} = Mx$.*

- *If all of the eigenvalues of M have negative real part, then every solution is stable.*

- *If any eigenvalue of M has positive real part, then every solution is unstable.*

- *If some but not all of the eigenvalues of M have negative real part and the rest have zero real part, then let $\lambda \in \{i\mu_1, i\mu_2, \ldots, i\mu_K\}$ be the eigenvalues with zero real part (so that the numbers $\mu_1, \ldots, \mu_K$ are reals). Let l_j be the multiplicity of the eigenvalue $i\mu_j$ as a root of the polynomial $\det(M - \lambda I)$. If $i\mu_j$ has l_j linearly independent vectors for all j, then all solutions are stable. Otherwise all solutions are unstable.*

Note that to use this theorem for non-linear systems, we need to check that the rest point is hyperbolic i.e does not have it eigenvalues in the imaginary axis.

Fixed-Point Iterations

Let $\mathcal{X}$ be a non-convex, compact and non-empty subset of an Hilbert space.

Picard iteration

The Picard iteration is given by

$$x_{t+1} = f(x_t), \ t \geq 1 \tag{A.2}$$
$$x_0 \in \mathcal{X}. \tag{A.3}$$

Definition A.1.0.5. *We say that* $f : \mathcal{X} \longrightarrow \mathcal{X}$ *is contractive if there exists* $k < 1$

$$\| f(x) - f(y) \| \leq k \| x - y \|, \ \forall (x, y) \in \mathcal{X}^2.$$

Definition A.1.0.6. *We say that* $f : \mathcal{X} \longrightarrow \mathcal{X}$ *is non-expansive if*

$$\| f(x) - f(y) \| \leq \| x - y \|, \ \forall (x, y) \in \mathcal{X}^2.$$

For nonexpansive map, then Picard iteration does not converge, generally, or even if it converges, its limit is not a fixed point of f.

Definition A.1.0.7. *We say that* $f : \mathcal{X} \longrightarrow \mathcal{X}$ *is pseudo-contractive if*

$$\| f(x) - f(y) \| \leq \| x - y \| + \| x - y + f(y) - f(x) \|, \ \forall (x, y) \in \mathcal{X}^2.$$

Krasnoselskij iteration

$$x_{t+1} = (1 - \lambda)x_t + \lambda f(x_t), \ t \geq 1 \tag{A.4}$$
$$x_0 \in \mathcal{X}. \tag{A.5}$$
$$\lambda \in (0, 1) \tag{A.6}$$

Mann iteration

$$x_{t+1} = (1 - \lambda_t)x_t + \lambda_t f(x_t), \ \forall t \geq 1 \tag{A.7}$$
$$x_0 \in \mathcal{X}. \tag{A.8}$$
$$\lambda_t \in (0, 1), \ \forall t \geq 1 \tag{A.9}$$

Ishikawa iteration

$$x_{t+1} = x_t + \lambda_t[f(y_t) - x_t] \tag{A.10}$$
$$y_t = x_t + \mu_t[f(x_t) - x_t], \tag{A.11}$$
$$x_0 \in \mathcal{X}, \tag{A.12}$$
$$0 \le \lambda_t \le \mu_t < 1 \tag{A.13}$$
$$\lim_{t \longrightarrow +\infty} \mu_t = 0 \tag{A.14}$$
$$\sum_{t \ge 1} \lambda_t \mu_t = +\infty \tag{A.15}$$

Definition A.1.0.8 (Deterministic algorithms, How to distinguish fast and slow?). *Let x_t and y_t be two convergent sequences generated by deterministic iterates. Assume that $\lim_t x_t = \lim_t y_t = x^*$ and the ratio $\frac{\|x_t - x^*\|}{\|y_t - y^*\|}$ has a limit denoted by δ (finite). We say that x is faster than y if $\delta = 0$. If $\delta > 0$ then we say that the two sequences have similar rate.*

Definition A.1.0.9. *We say f be strongly pseudo-contractive i.e there exists $t' > 1$ such that*

$$\| x - y \| \le \| (1 + s)(x - y) - st'(f(x) - f(y)) \|, \forall x, y,$$

and $s > 0$

Theorem A.1.0.10 ([142], Fastest algorithm with constant rate). *Let f be strongly pseudo-contractive over a non-empty, convex compact set $\mathcal{X}$. If f is Lipschitz with constant L and has at least one fixed point then the algorithm*

$$x_{t+1} = x_t + \lambda(f(x_t) - x_t), \ \lambda \in (0, \bar{\lambda}), \tag{A.16}$$

where $\bar{\lambda} = \frac{t'}{(L+1)(L+2-t')}$, converges to a fixed-point. Moreover there is a fastest algorithm among this class and the fastest one is obtained for the learning rate $\lambda = -1 + \sqrt{1 + \bar{\lambda}}$.

Theorem A.1.0.11 (Ishikawa, 1974, pseudo-contractive map, [80]). *Let $\mathcal{X}$, a non-empty, convex, compact, subset of Hilbert space. Let $f : \mathcal{X} \longrightarrow \mathcal{X}$ be a Lipschitz and pseudo-contractive map*

$$x_{t+1} = x_t + \lambda_t[f(y_t) - x_t] \tag{A.17}$$
$$y_t = x_t + \mu_t[f(x_t) - x_t], \tag{A.18}$$
$$x_0 \in \mathcal{X}, \tag{A.19}$$
$$0 \le \lambda_t \le \mu_t < 1 \tag{A.20}$$
$$\lim_{t \longrightarrow +\infty} \mu_t = 0 \tag{A.21}$$
$$\sum_{t \ge 1} \lambda_t \mu_t = +\infty \tag{A.22}$$

Then x_t converges to a point-fixed point of f.

The above result can be used to approximate an equilibrium if the best-response is a function (not a correspondence) and satisfies the above assumptions.

Exercise A.1. *Apply the result A.1.0.11 for $f(x) = 1/x$; $\mathcal{X} = [\frac{1}{2}, 2]$. Find the fixed points. Show that the Picard iteration does not converge. Is the Ishikawa's algorithm convergent to a fixed-point?*

A.2   Basics of Stochastic Approximations

The basic stochastic approximation algorithms introduced by Robbins and Monro and by Kiefer and Wolfowitz in the 1950s have been the subject of huge literature, both theoretical and applied. This is due to the large number of applications and the interesting theoretical issues in the analysis of stochastic processes. The basic paradigm is a stochastic difference equation such as

$$x_{t+1} = x_t + \lambda_t y_t,$$

, where x_t takes its values in some Euclidean space, y_t is a random variable, and the step size $\lambda_t > 0$ is small and might go to zero as t goes to zero. In its simplest form, x is a parameter of a system, and the random vector y_t is a function of noise-corrupted observations taken on the system when the parameter is set to x_t. One recursively adjusts the parameter so that some goal is met asymptotically.

We review some standard concepts for the study of dynamical systems. The following notions are taken from Kushner (1978), Borkar (1996, 2008), and Benaïm (1999). Let $\mathcal{A}$ be a nonempty subset of $\mathbb{R}^n$ with the Euclidean metric d.

Definition A.2.0.12 (semi-flow). *A semi-flow is a continuous map $\Phi : \mathcal{A} \times \mathbb{R}_+ \longrightarrow \mathcal{A}$ such that $\Phi(a, 0) := \Phi_0(a) = a$ and $\Phi_t(\Phi_{t'}(a)) = \Phi_{t+t'}(a)$, for every $(t, t') \in \mathbb{R}_+^2$, $a \in \mathcal{A}$.*

We extend Φ to subset of $\mathcal{A}$ by defining $\Phi_t(A_1) = \{\Phi_t(a), \ a \in A_1\}$ for $A_1 \subseteq \mathcal{A}$.

Definition A.2.0.13 (invariant subset). *We say that a subset A_1 of $\mathcal{A}$ is invariant of $\Phi_t(A_1) = A_1$ for all $t \in \mathbb{R}_+$.*

Definition A.2.0.14 (attracting set). *We say that A_1 is an attracting subset if A_1 is nonempty and compact, and there exists some open neighborhood $\mathcal{O}_1$ of A_1 such that $\lim_{t \longrightarrow \infty} d(\Phi_t(x), A_1) = 0$ uniformly in $\mathcal{O}_1$. An attractor is an invariant attracting set. $t \in \mathbb{R}_+$.*

Every attracting set contains an attractor. A globally attracting set is an attracting set such that

$$\lim_{t \longrightarrow \infty} d(\Phi_t(a), A_1) = 0, \ \forall a \in \mathcal{A}$$

Definition A.2.0.15 (Omega-limit set). *The $\omega-limit$ set of a point a is the set*

$$\omega(a) = \{a' \in \mathcal{A} \mid \Phi_{t_k}(a) \longrightarrow a', \text{ for some } t_k \longrightarrow +\infty.\}$$

Definition A.2.0.16 (Internally-Chain-Transitive). *For a nonempty invariant set A_1, a $(\delta, T)-pseudo-orbit from a to a' in A_1 is a finite series of partial trajectories, $\{\Phi_t(a_k), \ 0 \leq t \leq t_k\}, k \in \{1, 2, \ldots, l-1\}, t_k \geq T$ such that*

- $a_k \in \mathcal{A}, \ \forall k.$

- $d(a_0, a) < \delta, \ d(\Phi_{t_k}(a_k), a_{k+1}) < \delta, \ \forall k \in \{1, 2, \ldots, l-1\}$ and $x_l = a'.$

A nonempty compact invariant set A_1 is internally chain-transitive if, for every $a, a' \in \mathcal{A}$ and every $\delta > 0, T > 0$, there is a $(\delta, T)-pseudo-orbit in $\mathcal{A}$ from a to a'.

Definition A.2.0.17. *A subset of a topological space $\mathcal{A}$ is precompact in $\mathcal{A}$ if its closure in $\mathcal{A}$ is compact.*

Every internally chain-transitive set is connected. Moreover, a nonempty compact invariant set is internally chain-transitive if and only if the set has no proper attracting set. The internally chain-transitive sets provide a characterization of the possible long-run behavior of the system; Benaïm has shown that this particular concept of long-run behavior is useful when working with stochastic approximations. The classical stochastic approximation algorithm is a discrete-time stochastic process whose step size decreases with time, so that asymptotically the system converges to its deterministic continuous-time limit. The early work on stochastic approximation was done by Robbins and Monro (1951,[134]) and Kiefer and Wolfowitz (1952, [93]) and has since been applied and extended by a number of authors. For most of our learning mechanism purposes, the main results are the following

Theorem A.2.0.18. *Consider the discrete time process on a non-empty convex subset $\mathcal{A}$ of $\mathbb{R}^n$ defined by the recursion,*

$$\mathbf{x}_{t+1} = \mathbf{x}_t + \frac{1}{t+1} \left[f(\mathbf{x}_t) + \epsilon_{t+1}^1 + \epsilon_t^2 \right] \tag{A.23}$$

and the corresponding continuous time semi-flow Φ induced by the system of ordinary differential equations (ODEs)

$$\dot{x}(t) = f(x(t)) \tag{A.24}$$

where

- $f : \mathcal{A} \longrightarrow \mathbb{R}^n$ *is continuous and the ODE (A.23) is globally integrable;*

- *The* $\{\epsilon_t^j\}$, $j \in \{1, 2\}$ *are stochastic processes adapted to filtration* $\mathcal{F}_t$, *i.e., for each time slot* $t \in \mathbb{N}$, ϵ_t^1 *and* ϵ_t^2 *are random variables that are measurable with respect to* $\mathcal{F}_t$, *where* $\mathcal{F}_t$ *is the* $\sigma-$*algebra corresponding to the history of the system up through the end of period* t,

- $\mathbb{E}\left(\epsilon_{t+1}^1 | \mathcal{F}_t\right) = 0$ *almost surely (a.s), and* $\mathbb{E}\left(\parallel \epsilon_{t+1}^1 \parallel^2\right) < \infty$, $\lim_{t \longrightarrow \infty} \epsilon_t^2 = 0$ *almost surely,*

- *The set* $\{\mathbf{x}_t, \ t \geq 0\}$ *is precompact in* $\mathcal{A}$ *almost surely.*

Then, with probability 1, every $\omega-$*limit of the process* $\{\mathbf{x}_t\}_t$ *lies in an internally chain-transitive set for* Φ.

If $\mathcal{A}$ is compact and f is Lipschitz continuous, then ODE is globally integrable and $\{\mathbf{x}_t\}_t$ is precompact.

Recall that a rest point x^* of the ODE (A.23) (i.e., $f(x^*) = 0$) is linearly unstable if the Jacobian matrix of f at x^*, $Df(x^*)$, has some eigenvalue with a positive real part. Let $R^m = E_+ \oplus E_-$, where $E+$ and E_- are the generalized eigenspaces of $Df(x^*)$ corresponding to eigenvalues with positive and nonpositive real parts, respectively. If x^* is linearly unstable rest point, then E_+ has at least one dimension. We are now ready for the result of Brandière and Duflo (1996).

Theorem A.2.0.19 (Brandière and Duflo (1996,[36])). *Consider the stochastic process (A.23) on a nonempty open subset* $\mathcal{A}$ *of* $\mathbb{R}^n$. *Let* x^* *be a linearly unstable rest point of* f *and* $\epsilon_{t,+}^1$ *be the projection of* ϵ_t^1 *on* E_+ *in the directions of* E_-. *Assume that*

- f *is continuously differentiable and its derivative is Lipschitz continuous on a neighborhood of* x^*,

- *The* $\{\epsilon_t^j\}$, $j \in \{1, 2\}$ *are stochastic processes adapted to filtration* $\mathcal{F}_t$, *i.e., for each time slot* $t \in \mathbb{N}$, ϵ_t^1 *and* ϵ_t^2 *are random variables that are measurable with respect to* $\mathcal{F}_t$, *where* $\mathcal{F}_t$ *is the* $\sigma-$*algebra corresponding to the history of the system up through the end of period* t,

- $\mathbb{E}\left(\epsilon_{t+1}^1 | \mathcal{F}_t\right) = 0$ *almost surely (a.s), and*

$$\limsup_{t \longrightarrow \infty} \mathbb{E}\left(\parallel \epsilon_{t+1}^1 \parallel^2\right) < \infty,$$

$$\liminf_{t \longrightarrow \infty} \mathbb{E}\left(\parallel \epsilon_{t+1,+}^1 \parallel^2 \mid \mathcal{F}_t\right) > 0 \ (a.s)$$

- $\sum_{t \geq 1} \parallel \epsilon_t^2 \parallel^2 < \infty$ *almost surely,*

Then, $\lim_{t \longrightarrow \infty} \mathbf{x}_t = x^*$ *with probability 0.*

Robbins-Monro-like algorithms

Let $F : \mathbb{R}^{\sum_j n_j} \longrightarrow \mathbb{R}^{\sum_j n_j}$ be a continuous map. Consider a discrete time process $\{\mathbf{x}_t\}_{t \in \mathbb{N}}$ elements of $\mathbb{R}^{\sum_j n_j}$ whose general form can be written as

$$\mathbf{x}_{t+1} = \mathbf{x}_t + \lambda_t [F(\mathbf{x}_t) + M_{t+1}] \qquad (A.25)$$

where

- $\{\lambda_t\}_t$ is a sequence of non-negative numbers such that $\lambda \notin l^1$ and $\lim_t \lambda_t = 0$.

- $M_t \in \mathbb{R}^{\sum_j n_j}$ is a perturbation.

The iterative equation (A.25) can be considered as a perturbation version of a variable step-size Euler-scheme for numerical implementation of the ordinary differential equation

$$\dot{\mathbf{x}}_t = F(\mathbf{x}_t) \qquad (A.26)$$

i.e., the sequence

$$\mathbf{y}_{t+1} = \mathbf{y}_t + \lambda_t F(\mathbf{y}_t) \qquad (A.27)$$

We compare the behavior of a sample path $\{\mathbf{x}_t\}_{t \geq 0}$ with the trajectories of the flow induced by the vector field F. To do this, we denote by,

$$T_0 := 0, \ T_k := \sum_{t=1}^{k} \lambda_t, \ k \geq 1.$$

We define the continuous time affine and piecewise interpolated process $X : \mathbb{R}_+ \longrightarrow \mathbb{R}^{\sum_j n_j}$,

$$X(T_k + s') = \mathbf{x}_k + s' \frac{\mathbf{x}_{k+1} - \mathbf{x}_k}{T_{k+1} - T_k}, \ \forall \ k \in \mathbb{N}, \ 0 \leq s' \leq \lambda_{k+1}$$

Define $\tilde{T} : t \longmapsto \sup\{k, \ T_k \leq t\}$ the left-hand-inverse of the map $k \longmapsto T_k$. Denote by $\tilde{M}, \tilde{\lambda}$ the continuous time processes defined by

$$\begin{cases} \tilde{M}(T_k + s') = M_k, \ \forall \ k \in \mathbb{N}, \ 0 \leq s' \leq \lambda_{k+1} \\ \tilde{\lambda}(T_k + s') = \lambda_k, \ \forall \ k \in \mathbb{N}, \ 0 \leq s' \leq \lambda_{k+1} \end{cases}$$

Equation (A.25) can be written as

$$X(t) - X(0) = \int_0^t \left[F(\tilde{X}(s)) + \tilde{M}(s) \right] ds \qquad (A.28)$$

Recall that if F is bounded and locally Lipschitz, then F has a unique integral curve. Next we give a sufficient condition for convergence of the Robbins-Monro's like algorithms.

Proposition A.2.0.20. *Let F be a locally Lipschitz vector field, bounded on the simplex. Assume that,*

$$(H1) \quad \forall T, \quad \lim_{t \longrightarrow \infty} \sup_{0 \leq \Delta t \leq T} \| \int_t^{t+\Delta t} \tilde{M}(s) \, ds \| = 0 \qquad (A.29)$$

Then, the interpolated process X is an asymptotic pseudo-trajectory of the flow Φ induced by F i.e

$$\forall \, T > 0, \quad \lim_{t \longrightarrow +\infty} \sup_{0 \leq \Delta t \leq T} d(X(t + \Delta t), \Phi_{\Delta t}(X(t))) = 0.$$

Moreover, for t sufficiently large enough, we have that

$$\sup_{0 \leq \Delta t \leq T} \| X(t + \Delta t) - \Phi_{\Delta t}(X(t)) \|$$

$$\leq C_T \left[\sup_{s \in [t, t+T]} \tilde{\lambda}(s) + \sup_{0 \leq \Delta t \leq T+1} \| \int_{t-1}^{t-1+\Delta t} \tilde{M}(s) \, ds \| \right],$$

where C_T depends only on T and F.

In this Proposition, d denotes the metric on the space of continuous map from $\mathbb{R}$ to $\prod_j \mathcal{X}_j$. The process X is extended to $\mathbb{R}_-$ by setting $X(t) = X(0)$ if $t \leq 0$.

$$d(f, g) = \sum_{k \in \mathbb{N}} \frac{1}{2^k} \min\{1, d_k(f, g)\} \qquad (A.30)$$

$$d_k(f, g) = \sup_{t \in [-k, k]} \tilde{d}(f(t), g(t)) \qquad (A.31)$$

Denote by

$$\tilde{\epsilon}_{t,T} = \sup_{0 \leq \Delta t \leq T} \| \int_t^{t+\Delta t} \tilde{M}(s) \, ds \| .$$

The assumption $H1$ says that $\lim_{t \longrightarrow \infty} \tilde{\epsilon}_{t,T} = 0$. The discrete time version of the assumption $H1$ is exactly

$$\forall T, \quad \lim_{k \longrightarrow \infty} \sup \left\{ \| \sum_{t=k}^{L'-1} \lambda_{t+1} M_{t+1} \| : \; L' \in \{k+1, k+2, \ldots, \tilde{T}(T_k + T)\} \| \right\} = 0$$

$$(A.32)$$

Note that this assumptions is automatically satisfied if M_{t+1} is a martingale difference with the respect to the filtration $\mathcal{F}_t = \sigma(\mathbf{x}_s, \ s \leq t)$ and $\lambda \succeq 0$, $\lambda \in l^2$. A formal proof of these statements can be found in Kushner (1978), Benaim (1999), Borkar (2008).

We are now ready for a sketch of proof of Proposition A.2.0.20.

Proof. By continuity of F and the fact that $\{\mathbf{x}_t\}_t \subset \prod_j \mathcal{X}_j$ which a convex and compact set, there exists a constant $c > 0$ such that

$$\| F(\mathbf{x}_t) \| \leq c, \ \forall \ t \geq 0.$$

From Equation (A.28), one has that

$$\sup_{\Delta t \in [0,T]} \| X(t + \Delta t) - X(t) \| \leq \int_{[0,T]} \sup_{s \in [0,T]} \| F(X(s)) \| + \tilde{\epsilon}_{t,T}$$

This implies by H1 that

$$\lim_{t \longrightarrow +\infty} \sup_{\Delta t \in [0,T]} \| X(t + \Delta t) - X(t) \| \leq cT.$$

Hence, the interpolated process is uniformly continuous.

We now rewrite (A.28) as

$$X(t + \Delta t) \quad = \quad X(t) + \int_0^{\Delta t} F(X(t + s')) \, ds' \tag{A.33}$$

$$+ \int_t^{t+\Delta t} [F(\tilde{X}(s')) - F(X(s'))] \, ds' \tag{A.34}$$

$$+ \int_t^{t+\Delta t} \tilde{M}(s') \, ds' \tag{A.35}$$

By Assumption H1, $\lim_{t \to \infty} \int_t^{t+\Delta t} \tilde{M}(s') \, ds' \equiv 0$ where the equality is in the sense of the space of continuous function from $\mathbb{R}$ to $\mathbb{R}^{\sum_j n_j}$. Again, by (A.28), one gets that, for any $T > 0$ and any $s \in [t, t + T]$,

$$\| X(s) - \tilde{X}(s) \| = \| \int_{T_{\tilde{T}(s)}}^s F(\tilde{X}(s')) + \tilde{M}(s') \, ds' \|$$

$$\leq c\tilde{\lambda}(s) + \| \int_{T_{\tilde{T}(s)}}^s \tilde{M}(s') \, ds' \| .$$

Since $\tilde{\lambda}(.) \longrightarrow 0$, one has that $\tilde{\lambda}(s) < 1$ for s large enough.

$$\| \int_{T_{\tilde{T}(s)}}^s \tilde{M}(s') \, ds' \| \leq \| \int_{t-1}^{T_{\tilde{T}(s)}} \tilde{M}(s') \, ds' \| + \| \int_{t-1}^s \tilde{M}(s') \, ds' \| \leq 2\tilde{\epsilon}_{t-1,T+1}.$$

Thus, $\sup_{s \in [t,t+T]} \| X(s) - \tilde{X}(s) \| \leq 2\tilde{\epsilon}_{t-1,T+1} + \sup_{s \in [t,t+T]} c\tilde{\lambda}(s)$. Since F is uniformly continuous on $\prod_j \mathcal{X}_j$, therefore the second term (A.34) satisfies

$$\lim_{t \longrightarrow \infty} \int_t^{t+\Delta t} \left[F(\tilde{X}(s')) - F(X(s')) \right] \, ds' \equiv 0$$

where the equality is in the sense of the space of continuous function from $\mathbb{R}$ to $\mathbb{R}^{\sum_j n_j}$. Let X^* denote an omega-limit of $\{X(t + .)\}$. Then,

$$X^*(s) = X^*(0) + \int_0^s F(X^*(s'))\, ds'$$

By unique integral curve, this implies that $X^* = \Phi(X^*)$ and X is an asymptotic pseudo-trajectory of Φ.

It remains to establish the bound of convergence. First, we estimate the norm of $\int_t^{t+\Delta t} [F(\tilde{X}(s')) - F(X(s'))]\, ds'$. Let $\Delta t \in [0, T]$ and L is Lipschitz constant of F. Then,

$$a_0(\Delta t) := \left\| \int_t^{t+\Delta t} [F(\tilde{X}(s')) - F(X(s'))]\, ds' \right\| \leq LT \sup_{\Delta t \in [0,T]} \| \tilde{X}(s')) - X(s') \|$$

$$\leq LT \left[c \sup_{s \in [t, t+T]} \tilde{\lambda}(s) + 2\tilde{\epsilon}_{t-1, T+1} \right]$$

Thus,

$$\| X(t + \Delta t) - \Phi_{\Delta t}(t) \| \leq L \int_0^{\Delta t} \| X(t + s') - \Phi_{s'}(X(t)) \|\, ds' + a_0(\Delta t)$$

$$+ \left\| \int_t^{t+\Delta t} \tilde{M}(s')\, ds' \right\|$$

$$\leq L \int_0^{\Delta t} \| X(t + s') - \Phi_{s'}(X(t)) \|\, ds'$$

$$+ LT \left[c \sup_{s \in [t, t+T]} \tilde{\lambda}(s) + 2\tilde{\epsilon}_{t-1, T+1} \right] + \tilde{\epsilon}_{t-1, T+1}$$

We now use Gronwall's inequality to get the announced bound. Let f, g, h be a continuous function on the compact interval $[a, \Delta t]$ It states that if g is nonnegative and if h satisfies the inequality

$$f(t) \leq h(t) + \int_0^t g(s')f(s')\, ds', \ \forall t \in [0, \Delta t]$$

Then,

$$f(t) \leq h(t) + \int_a^t h(s')g(s')e^{\int_{s'}^t g(s'')\, ds''}\, ds'$$

In addition, if h is constant then,

$$f(t) \leq h e^{\int_a^t g(s')\, ds'}, \ t \in [0, \Delta t].$$

This completes the proof. $\qquad\qquad\qquad\qquad\qquad\qquad\qquad\qquad\square$

Error bound for constant step-size

- $\mathcal{N} = \{1, 2, \ldots, n\}$, set of players

- $\mathcal{A}_j$ set of actions of j (finite) with cardinality $=k_j < +\infty$.

- s_j is a generic element of $\mathcal{A}_j$

$$x_{j,t+1} = x_{j,t} + br_{j,t}\left(e_{a_{j,t}} - x_{j,t}\right), \quad \forall j, \; \forall t, \; x_{j,t} \in \mathbb{R}^{k_j} \tag{A.36}$$

where $e_{a_{j,t}}$ is the unit vector with one at the position of $a_{j,t}$ and zero otherwise.
- The invariance of the simplex $\Delta(\mathcal{A}_j)$ is preserved if $br_{j,t} \in [0, 1]$.
- This iterative scheme can be written into the form of Robbins-Monro (1951).

- Let $F : \mathbb{R}^{\sum_{j \in \mathcal{N}} k_j} \longrightarrow \mathbb{R}^{\sum_j k_j}$, continuously differentiable

$$x_{j,t+1} = x_{j,t} + \lambda_t(F_j(x_t) + M_{j,t+1}),$$

- we examine the case where $\lambda_t = b, \; \forall t$.

How to prove the weak convergence to ODE

We summarize the methodology of proving convergence in stochastic approximation, fluid approximation or ODE approach.

- First write the difference equation in the form of Robbins-Monro scheme

- Derive the re-scaled drift

- Show that the drift-ODE is well-posed. For example, has the ODE a unique solution?

- Prove that the noise terms are controlled

- Compare with Euler schemes

- Finally, prove that the asymptotic weak-convergence is in of order of $O(b)$.

In order to write the learning scheme into the form of Robbins-Monro, we compute the drift (the expected in one step-size).

$$
\begin{aligned}
F_{j,s_j}(x) \;\; &:= \;\; \lim_{b \longrightarrow 0} \frac{1}{b} \mathbb{E}\left(x_{j,t+1}(s_j) - x_{j,t}(s_j) \mid x_t = x\right) && \text{(A.37)} \\
&= \;\; \lim_{b \longrightarrow 0} \frac{1}{b} \sum_{s'_j \in \mathcal{A}_j} \mathbb{E}\left(x_{j,t+1}(s_j) - x_{j,t}(s_j) \mid x_t = x, a_{j,t} = s'_j\right) x_{j,t}(s'_j) \\
&= \;\; \sum_{s'_j \neq s_j} x_{j,t}(s'_j) \mathbb{E}\left(-r_{j,t} \mid x_t = x, a_{j,t} = s'_j\right) x_{j,t}(s_j) \\
&\qquad + x_{j,t}(s_j) \mathbb{E}\left(r_{j,t} \mid x_t = x, a_{j,t} = s_j\right)(1 - x_{j,t}(s_j)) && \text{(A.38)} \\
&= \;\; x_{j,t}(s_j) \Big[\mathbb{E}\left(r_{j,t} \mid x_t = x, a_{j,t} = s_j\right) \\
&\qquad\quad - \sum_{s'_j \in \mathcal{A}_j} x_{j,t}(s'_j) \mathbb{E}\left(r_{j,t} \mid x_t = x, a_{j,t} = s'_j\right) \Big] \\
&= \;\; x_{j,t}(s_j) \left[r_j(e_{s_j}, x_{-j,t}) - \sum_{s'_j \in \mathcal{A}_j} x_{j,t}(s'_j) r_j(e_{s'_j}, x_{-j,t}) \right] && \text{(A.39)}
\end{aligned}
$$

The autonomous ODE $\dot{x}_{j,t}(s_j) = F_{j,s_j}(x_t)$ is a (multitype) replicator equation.

Now write, Equation (A.36) as

$$
\begin{aligned}
x_{j,t+1} \;\; &= \;\; x_{j,t} + b(F_j(x_t) + r_{j,t}\left(e_{a_{j,t+1}} - x_{j,t}\right) - F_j(x_t)) && \text{(A.40)} \\
&= \;\; x_{j,t} + b(F_j(x_t) + M_{j,t+1}) && \text{(A.41)}
\end{aligned}
$$

where $M_{j,t+1} = r_{j,t}\left(e_{a_{j,t+1}} - x_{j,t}\right) - F_j(x_t)$ is a martingale difference (with null expectation).

- $\forall t > 0,\ \mathbb{E}(M_{t+1} \mid \mathcal{F}_t) = 0$

We want to know if this iterative scheme can be seen as a noisy approximation of the ODE $\dot{x} = F(x)$

Lemma A.2.0.21. *Property 1: The mapping F is continuously differentiable.*

Proof. $\forall j \in \mathcal{N},\ \forall s_j \in \mathcal{A}_j$, the mapping

$$
x \longmapsto x_j(s_j)[r_j(e_{s_j}, x_{-j}) - \sum_{s'_j \in \mathcal{A}_j} x_j(s'_j) r_j(e_{s'_j}, x_{-j})],
$$

is C^∞. $\qquad\qquad\square$

Lemma A.2.0.22. *Property 2: Given an initial condition $x_0 \in \mathbb{R}^{\sum_j k_j}$, the ODE $\dot{x} = F(x)$ has unique solution.*

To prove this result, we will use a fixed point theorem (Banach-Picard, Tykonoff, Schauder, etc) and the Gronwall inequalities. We first establish the Gronwall inequalities that are useful for the estimates.

Gronwall inequality: continuous time

Lemma A.2.0.23. *Gronwall inequality: continuous version Let $f(.), g(.) \geq 0$ be continuous mapping satisfying*

$$(ineq1) \quad f(t) \leq h + \int_0^t g(s)f(s) \, ds, \ t \in [0,T], h \in \mathbb{R}.$$

Then,

$$f(t) \leq h e^{\int_0^T g(s) \, ds}, \ \forall t \in [0,T]$$

Proof. Let

$$B(t) := \int_0^t f(s)g(s) \, ds, \ \forall t \in [0,T].$$

By inequality (ineq1) and the positivity of the function g one gets, $f(t)g(t) \leq hg(t) + g(t)B(t)$ i.e $B'(t) \leq hg(t) + g(t)B(t)$.

$$e^{-\int_0^t g(s) \, ds}[B'(t) - g(t)B(t)] \leq hg(t)e^{-\int_0^t g(s) \, ds} \iff$$

$$[B'(t)e^{-\int_0^t g(s) \, ds} - g(t)e^{-\int_0^t g(s) \, ds}B(t)] \leq hg(t)e^{-\int_0^t g(s) \, ds} \iff$$

$$\frac{d}{dt}\left[B(t)e^{-\int_0^t g(s) \, ds}\right] \leq hg(t)e^{-\int_0^t g(s) \, ds}$$

By integration between 0 and t, one has,

$$B(t)e^{-\int_0^t g(s) \, ds} - B(0) \leq h[-e^{-\int_0^t g(s) \, ds}]_0^t = h(1 - e^{-\int_0^t g(s) \, ds}), \ B(0) = 0.$$

Thus,

$$f(t) \leq h + B(t) \leq h + h(1 - e^{-\int_0^t g(s) \, ds})e^{\int_0^t g(s) \, ds} = h e^{\int_0^t g(s) \, ds}.$$

$\square$

Discrete Gronwall inequality

Lemma A.2.0.24. *Gronwall inequality: discrete case*

$$f_{t+1} \leq h + \sum_{s=0}^t g_s f_s, \ g_s \geq 0, f_s > 0,$$

implies that

$$f_{t+1} \leq he^{\sum_{s=0}^{t} g_s}.$$

Proof. $B_t := \sum_{s=0}^{t} g_s f_s$. Multiplying by g_{t+1}, one gets,

$$B_{t+1} - B_t \leq hg_{t+1} + g_{t+1}B_t$$

i.e $B_{t+1} \leq hg_{t+1} + (1 + g_{t+1})B_t$. By induction,

$$B_t \leq h \sum_{k=0}^{t} \prod_{k'=k+1}^{t} (1+g_{k'})g_k \leq h \int_0^{\sum_{s'=0}^{t} g_{s'}} e^{\sum_{s'=0}^{t} g_{s'} - s} \, ds = h[-1 + e^{\sum_{s'=0}^{t} g_{s'}}].$$

Thus,

$$f_{t+1} \leq h + B_t \leq he^{\sum_{s'=0}^{t} g_{s'}}.$$

$\square$

Lemma A.2.0.25 (Uniqueness). *The autonomous ODE has a unique solution starting from any point $x_0 \in \mathbb{R}^{\sum_j k_j}$.*

Proof. Let

$$\Psi : \ x(.) \longrightarrow y(.), \ y(t) = x_0 + \int_0^t F(x(s)) \, ds, \ t \in [0, T].$$

Then, x is a solution of ODE if and only if $\Psi(x(.)) = x(.)$. (fixed point). Denote by

$$\| \, y(.) \, \|_c = \sup_{s \in [0,T]} \| \, y(s) \, \|, \ y(.) \in C^0([0,T], \mathbb{R}^{\sum_j k_j}).$$

Clearly,

$$\| \, \Psi(x_1) - \Psi(x_2) \, \|_c \ \leq \ \int_0^T \| \, F(x_1(s)) - F(x_2(s)) \, \| \ ds \quad (A.42)$$

$$\leq \ L \int_0^T \| \, x_1(s) - x_2(s) \, \| \ ds \quad (A.43)$$

$$\leq \ LT \| \, x_1 - x_2 \, \|_c \quad (A.44)$$

where L is a Lipschitz constant of F over the simplex. Taking $TL < 1$ and using the Banach space, Ψ has a unique fixed point solution on $[0,T], T < \frac{1}{L}$. By repeating a similar argument the interval in the form $\bigcup_{k \geq 0} [kT, (k+1)T]$, the solution is defined over any compact interval. Moreover, the solution $x_0 \longrightarrow x(.) \in C^0([0,T], \mathbb{R}^{\sum_j k_j})$ is Lipschitz:

$$\| \, x_1(t) - x_2(t) \, \| \leq \| \, x_1(0) - x_2(0) \, \| + L \int_0^t \| \, x_1(s) - x_2(s) \, \| \ ds.$$

By Gronwall inequality,

$$\| \, x_1(t) - x_2(t) \, \| \leq e^{LT} \| \, x_1(0) - x_2(0) \, \|.$$

$\square$

Now we establish a connection between the solution of the ODE and the interpolated process from the discrete time evolution $\{x_t\}_t$. Consider the interpolated process $X(.)$. Denote

$$T_0 = 0, \ T_k = kb, \ X(T_k) = x_k, \ X(T_k + s) = x_k + s\frac{x_{k+1} - x_k}{b}, \ 0 \le s < b.$$

Let $\phi_{\delta t}(t), \ t \le \delta t$ be the trajectory of $\dot{x} = F(x)$ starting from $X(\delta t)$ at time δt. In particular,

$$\phi_{\delta t}(t) = X(\delta t) + \int_0^t F(\phi_{\delta t}(s)) \ ds, \ \delta t \le t.$$

This implies

$$\| \phi_{\delta t}(t) \| \le \| X(\delta t) \| + \int_{\delta t}^t L_1(1 + \| \phi_{\delta t}(s) \|) \ ds$$

where the last term follows from the fact that Lipschitz continuity implies at most linear growth and

$$L_1 = \max(L, L \| X(\delta t) \| + \| F(X(\delta t)) \|).$$

We now use the Gronwall inequality:

$$\| \phi_{\delta t}(t) \| \le [\| X(\delta t) \| + L_1 T]e^{L_1 T} \le L_2(1 + \| X(\delta t) \|)$$

where $L_2 = L_1 T e^{L_1 T}$. Using this result, we determine the variation of ϕ between two times $t, t', \ t > t'$ is bounded by:

$$\| \phi_{\delta t}(t) - \phi_{\delta t}(t') \| (t - t') = \| \int_{t'}^t F(\phi_{\delta t}(s) \ ds \|$$

$$\le (t - t') \sup_{[\delta t, \delta t + s]} \| F(\phi_{\delta t}(.)) \| \le (t - t')L_2(1 + \| X(\delta t) \|) \ (*t1)$$

We are now ready to establish the main result.

Theorem A.2.0.26 (Weak convergence). *For any $T >$*

$$\mathbb{E}\left(\sup_{s \in [0,T]} \| \phi_{\delta t}(\delta t + s) - X(\delta t + s) \|^2 \right) \le C_T b = O(b)$$

where C_T depends only on T and the payoff function $r(.)$.

Proof. Without loss of generality, consider $T = Kb$ for some $K > 0$. For $t \ge 0$ let

$$\tilde{T}(t) = \sup\{kb, \ k \ge 0, \ kb \le t\}, \ N_k = b \sum_{k'=0}^k M_{k'}.$$

Rewriting the above equation, we have

$$X((k + k')b) - X(kb) = \sum_{k''=1}^{k'} X((k'' + k)b) - X((k'' + k - 1)b)$$

$$= b \sum_{k''=1}^{k'} F(X((k + k'' - 1)b)) + b \sum_{k''=1}^{k'} M_{k+k''}$$

$$= \int_{kb}^{(k+k')b} F(X(\tilde{T}(s))) \, ds + N_{k+k'} - N_k$$

$\bullet$

$$(*a) \; X((k + k')b) = X(kb) + \int_{kb}^{(k+k')b} F(X(\tilde{T}(s))) \, ds + N_{k+k'} - N_k$$

$\bullet$

$$(*b) \; \phi_{kb}((k + k')b) = X(kb) + \int_{kb}^{(k+k')b} F(\phi_{kb}(\tilde{T}(s))) \, ds$$

$$+ \int_{kb}^{(k+k')b} \left[F(\phi_{kb}(s)) - F(\phi_{kb}(\tilde{T}(s))) \right] ds \; (**)$$

The norm of second term $(**)$ is bounded by

$$(*3) \; Lb \sum_{k''=0}^{k'-1} \| X((k + k'')b) - \phi_{kb}((k + k'')b) \|$$

One has,

$$\left\| \int_{T_{k+k'}}^{T_{k+k'+1}} \left(F(X(s)) - F(X(\tilde{T}(s))) \right) \, ds \right\| \tag{A.45}$$

$$\leq \int_{T_{k+k'}}^{T_{k+k'+1}} \| F(X(s)) - F(X(\tilde{T}(s))) \| \, ds \tag{A.46}$$

$$\leq L \int_{T_{k+k'}}^{T_{k+k'+1}} \| X(s) - X(\tilde{T}(s)) \| \, ds \tag{A.47}$$

$$\leq L(T_{k+k'+1} - T_{k+k'})L_2(1 + \| x_{kb} \|^2) = LbL_2(1 + \| x_{kb} \|^2) \tag{A.48}$$

where the last inequality is obtained via $(*t1)$.

Combining $(*a)$ $(*b)$:

$$\sup_{0 \leq l \leq k'} \| X((k + l)b) - \phi_{kb}((k + l)b) \| \tag{A.49}$$

$$\leq bL \sum_{k''=0}^{k'-1} \sup_{0 \leq l \leq k''} \| X((k + l)b) - \phi_{kb}((k + l)b) \|$$

$$+ LbL_2(1 + \| x_{kb} \|^2)$$

$$+ \sup_{1 \leq l \leq k'} \| N_{k+l} - N_k \| \tag{A.50}$$

The noise term is controlled in expectation by

$$\mathbb{E}\sup_{1\le l\le k'}\| N_{k+l} - N_k \|^2 \le b^2\mathbb{E}(\sum_{0\le l\le k'}\| M_{k+l} \|) \le b^2TC$$

Again, by discrete Gronwall:

$$\mathbb{E}(\sup_{1\le l\le k'}\| X(T_{k+l}) - \phi_{kb}(T_{k+l}) \|^2)^{1/2} \qquad (A.51)$$

$$\le bL\sum_{l\le k-1}\mathbb{E}(\sup_{1\le l\le k'}\| X(T_{k+l}) - \phi_{kb}(T_{k+l}) \|^2)^{1/2} \qquad (A.52)$$

$$+bLL_2\mathbb{E}(1+\| X(T_{k+l}) \|^2)^{1/2} + \mathbb{E}\left(\sup_{l\le k'}\| N_{k+l} - N_k \|^2\right)^{1/2} \qquad (A.53)$$

Thus,

$$\mathbb{E}(\sup_{l\in[k,k+T]}\| X(T_l) - \phi_{kb}(T_l) \|^2)^{1/2} \le C_T\sqrt{b}$$

We have

$$\mathbb{E}(\sup_{s\in[kb,(k+1)b]}\| X(s) - X(kb) \|^2)^{1/2} = O(\sqrt{b}),$$

and

$$\mathbb{E}(\sup_{s\in[kb,(k+1)b]}\| \phi_{kb}(s) - \phi_{kb}(kb) \|^2)^{1/2} = O(\sqrt{b}).$$

We conclude that

$$\mathbb{E}(\sup_{s\in[0,T]}\| X(T_k + s) - \phi_{T_k}(s) \|^2)^{1/2} \le C_T\sqrt{b}.$$

This completes the proof. □

To summarize, we have shown

- weak convergence (convergence in law) of the algorithm when $b \longrightarrow 0$,

$$\lim_{b\longrightarrow 0}\mathbb{E}(\sup_{s\in[0,T]}\| X(T_k + s) - \phi_{T_k}(s) \|^2)^{1/2} = 0$$

- the error is in order of b.

Note that the proof extends to

- finite robust games, see Chapter 7.

- weak convergence to *stochastic differential equation* (the case where the noise does not vanish) can be found in [150], chapters 10-11.

Multiple time-scales stochastic approximations

In this subsection, we follow the work by Kushner, Borkar, Leslie and Collins on multiple time scale stochastic approximations.

Consider L interdependent stochastic approximation iterates as follows

$$(MTSA) \begin{cases} x_{t+1}^{(l)} = x_t^{(l)} + \lambda_t^{(l)} \left[f^{(l)}(x_t^{(1)}, \dots, x_t^{(L)}) + M_{t+1}^{(l)} \right], \\ x_t^{(l)} \in \mathbb{R}^{d_l}, \\ d_l \geq 1, \ L \geq 1, \\ l \in \{1, \dots, L\} \end{cases}$$

- **A01** We assume the learning rates are ordered and satisfy: for $l \in \{1, 2, \dots, L\}$, $\lambda^{(l)} \in l^2 \backslash l^1$.
- **A02** For $l \in \{1, 2, \dots, L-1\}$, we have that

$$\frac{\lambda_t^{(l)}}{\lambda_t^{(l+1)}} \longrightarrow 0,$$

when t goes to infinity.
- **A03** The functions $f^{(l)}$ are globally Lipschitz continuous.
- **A04** For any $l \in \{1, 2, \dots, L\}$,

$$\sup_t \| x_t^{(l)} \| < +\infty$$

- **A05** For any $l \in \{1, 2, \dots, L\}$, the random process $\sum_{t=0}^{n} \lambda_t^{(l)} M_{t+1}^{(l)}$ converges almost surely. This says that the cumulative noises remains convergent.
- **A06** Denote by

$$l' > l, \ \underline{x}_t^{(l \to l')} = (x_t^{(l)}, \dots, x_t^{(l')})$$

We start by the stochastic approximation equation with the highest learning rate. Consider the system of ordinary differential equations:

$$(M - SA)_L \begin{cases} \underline{x}_0^{(1 \to L-1)} = \phi_0 \\ \underline{\dot{x}}_t^{(1 \to L-1)} = 0 \\ \dot{x}_t^{(L)} = f^L(\underline{x}_t^{(1 \to L-1)}, x_t^{(L)}) \end{cases}$$

Then we assume that there exists a Lipschitz continuous function $y^{(L)}(\phi_0)$ such that the system $(M - SA)_L$ converges to a point $(\phi_0, y^{(L)}(\phi_0))$. Recursively we define the l-th assumption: $l \in \{1, 2, \dots, L-1\}$

$$(M - SA)_l \begin{cases} \underline{x}_0^{(1 \to l-1)} = \phi_0^{(l-1)} \\ \underline{\dot{x}}_t^{(1 \to l-1)} = 0 \\ \dot{x}_t^{(l)} = f^l(\underline{x}_t^{(1 \to l-1)}, x_t^{(l)}, y^{(l+1)}(\phi_0^{(l-1)}, x_t^{(l)})) \end{cases}$$

where $y^{(l+1)}(\phi_0^{(l-1)}, x_t^{(l)})$ is determined by $(l+1)$ iterates of $(M-SA)_{l+1}$, then we assume that there exists a Lipschitz continuous function $y^{(l)}(\phi_0^{(l-1)})$ such that the system of ordinary differential equations $(M-SA)_l$ converges to a point $(\phi_0^{(l-1)}, y^{(l)}(\phi_0^{(l-1)}))$.

Theorem A.2.0.27 ([103, 97, 34]). *Consider the multiple time scale stochastic approximation in (MTSA). Assume that $A01 - A06$ hold. Then the asymptotic behavior of the iterates converges to the invariant set of the dynamical system*

$$\dot{x}_t^{(1)} = f^1(x_t^{(1)}, y^{(2)}(x_t^{(1)}))$$

where $y^{(2)}(.)$ is obtained from assumption A06.

A.3   Differential Inclusion

A set-valued map (or correspondence) $F : \mathbb{R}^d \rightrightarrows \mathbb{R}^d$ defines a differential inclusion with

$$\frac{d}{dt}x_t \in F(x_t).$$

Good upperhemicontinuous differential inclusion

The differential inclusion is said a good upperhemicontinuous differential inclusion if the following conditions are satisfied:

- *Nonemptiness:* The set $F(x)$ is nonempty for all $x \in \mathbb{R}^d$, ;
- *Convex valued:* The set $F(x)$ is convex for all $x \in \mathbb{R}^d$;
- *Boundedness:* There exists a $K > 0$ such that

$$\sup\{\| y \| \mid y \in F(x)\} \leq K$$

for all $x \in \mathbb{R}^d$;

- *Upper hemicontinuous:* The graph of F,

$$graph(F) = \{(x, y) \mid y \in F(x)\},$$

is a closed set.

Tangent Cone

Let $\mathcal{X}$ be a compact convex subset of $\mathbb{R}^d$, and let

$$TX(x) = closure(\{z \in \mathbb{R}^d \mid z = \mu(y - x), \text{ for some} \mu \geq 0, \ y \in \mathcal{X}\}),$$

be the tangent cone of $\mathcal{X}$ at the point x.

Sufficiency condition for existence of a solution

Theorem A.3.0.28. *Let F defining a good upperhemicontinuous differential inclusion. Suppose that $F(x) \in TX(x)$, $\forall x$. Then, there exists an absolutely continuous mapping $x : \mathbb{R}_+ \longrightarrow \mathcal{X}$ with $x(0) = x_0$ satisfying $\dot{x}_t \in F(x_t)$, for almost every time $t \in \mathbb{R}_+$.*

Note that F is not assumed to be Lipschitz.

⚠ Given an initial condition x_0, the uniqueness of the trajectory is not guaranteed!

Stochastic approximation and differential inclusion

We now introduce a class of stochastic approximation processes with constant step size defined on the probability space $(\Omega, \mathcal{F}, \mathbb{P})$. For a sequence of values of λ that are close to zero, let $(M_{\lambda,t})_{t\geq 0}$ be a sequence of $\mathbb{R}^d$-valued random variables and $F_\lambda(x)$ be a family of set-valued maps on $\mathbb{R}^d$. Let $y \in F_\lambda(x)$ be a good upper hemicontinuous differential inclusion. Assume that the following are satisfied:

- $\lambda > 0$,

- $\forall t,\ x_{\lambda,t} \in \mathcal{X}$,

- $x_{\lambda,t}$ satisfies the recursive formula

$$x_{\lambda,t+1} - x_{\lambda,t} - \lambda M_{\lambda,t} \in F_\lambda(x_{\lambda,t}),$$

- For any $\alpha > 0$, there exists an $\lambda_0 > 0$ such that for all $\lambda \leq \lambda_0$ and $x \in \mathcal{X}$,

$$F_\lambda(x) \subset \{z \in \mathbb{R}^d \mid \exists y \in B(x; \alpha) \text{ such that } distance(z, F_\lambda(y)) < \alpha\}$$

- Let $\lfloor t \rfloor$ be the integer part of t. For all horizon $T > 0$ and all $\delta > 0$, we have

$$(C)\ \lim_{\lambda \to 0}\ \mathbb{P}\left(\max_{l \leq \lfloor \frac{T}{\lambda} \rfloor} \|\sum_{i=0}^{l-1} \lambda M_{\lambda,i+1}\| > \delta\,\Big|\, x_{\lambda,0} = x\right) = 0$$

uniformly in x

Sufficient condition

Note that if $\mathbb{E}(M_{\lambda,t+1}\,|\,\mathcal{F}_t) = 0$, $\forall t$ and if

$$\sup_{\lambda > 0} \sup_{t \geq 0} \mathbb{E}(\| M_{\lambda,t+1} \|^p) < \infty,$$

for some $p \geq 2$, then the condition (C) given above is satisfied, where $\mathcal{F}_t$ is the filtration generated by the process up to t.

The following theorem gives a very useful track of the stochastic process.

Theorem A.3.0.29. *Suppose the above hypothesis holds. Then for any $T > 0$ and any $\delta > 0$, we have*

$$\lim_{\lambda \to 0}\ \mathbb{P}\left(\inf_{x \in S_\phi} \sup_{t \in [0,T]} \| \bar{x}_{\lambda,t} - x_t \| > \delta\,\Big|\, x_{\lambda,0} = x\right) = 0,$$

where S_ϕ the set of all solution curves of the differential inclusion and $\bar{x}_{\lambda,t}$ is the (affine) interpolated process from $x_{\lambda,t}$.

Vanishing learning rates

Theorem A.3.0.30. *Consider the stochastic recursive equation:*

$$x_{t+1} = x_t + \lambda_t(y_t + M_{t+1}),$$

where

- $\lambda \in l^2 \backslash l^1$,

- $y_t \in F(x_t), \ t \geq 0$.

- M_{t+1} *is is a martingale difference sequence with the respect to the increasing σ−field $\mathcal{F}_t = \sigma(x_{t'}, y_{t'}, M_{t'}, \ t' \leq t)$.*

Assume in addition that

- *Nonemptiness and convexity: the set $F(x)$ is non-empty and convex for for all $x \in \mathbb{R}^d$, ;*

- *Linear growth: There exists a $K' > 0$ such that*

$$\sup_{y \in F(x)} \| y \| \leq \ K'(1+ \| x \|)\},$$

for all $x \in \mathbb{R}^d$;

- *Upper hemicontinuous: The graph of F,*

$$graph(F) = \{(x, y) \mid y \in F(x)\},$$

is a closed set.

- *Almost sure boundedness of the scheme*

$$a.s : \sup_t \| x_t \| < +\infty$$

Define a piecewise constant $\bar{y}_t$ such that

$$\bar{y}_t = y_t, \ if \ t \in [\sum_{k=0}^{t-1} \lambda_k, \lambda_t + \sum_{k=0}^{t-1} \lambda_k]$$

and $y_t \in F(x_t)$. Let ϕ_s be the solution of the solution of

$$\dot{\phi}_t = \bar{y}_t, \ t \geq t_0, \ \phi_{t_0} = \bar{x}_{t_0},$$

where $\bar{x}_s$ is the interpolated process from x. Then, for any $T > 0$

$$\lim_{t_0 \longrightarrow +\infty} \sup_{t \in [t_0, t_0+T]} \| \bar{x}_t - \phi_t \| = 0$$

A.4 Markovian Noise

We now introduce a class of stochastic approximation processes with Markovian noise defined on the probability space $(\Omega, \mathcal{F}, \mathbb{P})$.

For a sequence of values of λ_t that converges to zero, let $(M_t)_{t \geq 0}$ be a sequence of $\mathbb{R}^d$-valued random variables and $F(x, w)$ be a mapping on $\mathbb{R}^d \times \mathcal{W}$.

Suppose
$$x_{t+1} = x_t + \lambda_t(F(x, w) + M_{t+1} + \epsilon_t),$$

where

- b_t is a deterministic sequence that converges to zero,

- w_{t+1} is a Markov decision process with kernel

$$q_{wxw'} := \mathbb{P}(w_{t+1} = w' \mid w_t = w, x_t = x),$$

- For any fixed $x \in \mathcal{X}$, the Markov chain has a unique stationary distribution $\pi_x \in \Delta(\mathcal{W})$

- $Q_x = (q_{wxw'})_{w,w'}$ is geometrically ergodic with stationary distribution π_x.

In addition to the above, the classical assumption holds for the learning rate.

Then, x_t converges to stable sets of the ODE

$$\dot{x} = \mathbb{E}_{w \sim \pi_x} F(x, w).$$

Geometrically Ergodic Markov Chain

Recall that a Markov chain with transition kernel Q_x and stationary distribution π_x is π-a.e. geometrically ergodic, i.e., there is $\rho < 1$, and constants $0 < c_x < +\infty$ for each $x \in \mathcal{X}$, such that for π-a.e. $x \in \mathcal{X}$,

$$\| Q_x^t - \pi_x \|_{var} \leq c_x \rho^t$$

where

$$\| \mu \|_{var} = \sup_{A \subseteq \mathcal{W}} |\mu(A)|.$$

Bibliography

[1] M. Aghassi and D. Bertsimas. Robust game theory. *Mathematical Programming*, 107(1):231–273, 2006.

[2] C. Alós-Ferrer and N. Netzer. The logit-response dynamics. *Games and Economic Behavior*, 68(2):413–427, March 2010.

[3] T. Alpcan and T. Basar. *Network Security: A Decision and Game Theoretic Approach*. Cambridge University Press, 2010.

[4] E. Altman, T. Başar, I. Menache, and H. Tembine. A dynamic random access game with energy constraints. *In Proceedings 7th International Symposium on Modeling and Optimization in Mobile, Ad Hoc, and Wireless Networks, (WiOpt'09)*, 2009.

[5] E. Altman, T. Boulogne, R. El-Azouzi, T. Jimenez, and L. Wynter. A survey on networking games in telecommunications. *Computers and Operations Research*, 2006.

[6] E. Altman, R. El-Azouzi, Y. Hayel, and H. Tembine. Evolutionary power control games in wireless networks. In *Proceedings of the 7th international IFIP-TC6 networking conference on AdHoc and sensor networks, wireless networks, next generation internet*, NETWORKING'08, pages 930–942, Berlin, Heidelberg, 2008. Springer-Verlag.

[7] E. Altman, R. ElAzouzi, Y. Hayel, and H. Tembine. An evolutionary game approach for the design of congestion control protocols in wireless networks. *6th International Symposium on Modeling and Optimization in Mobile, Ad Hoc, and Wireless Networks and Workshops, WiOPT*, 2008.

[8] Eitan Altman, Rachid El Azouzi, Yezekael Hayel, and H. Tembine. The evolution of transport protocols: An evolutionary game perspective. *Computer Networks*, 53(10):1751–1759, 2009.

[9] K. B. Ariyur and M. Krstic. *Real Time Optimization by Extremum Seeking*. Hoboken, NJ: Wiley, 2003.

[10] Kenneth J. Arrow and Gerard Debreu. Existence of an equilibrium for a competitive economy. *Econometrica*, 22:265–290, July 1954.

[11] W.B. Arthur. On designing economic agents that behave like human agents. *Journal Evolutionary Econ. 3*, pages 1–22, 1993.

[12] P. Auer, N. Cesa-Bianchi, and P. Fischer. Finite-time analysis of the multiarmed bandit problem. *Machine Learning*, 47:235–256, 2002.

[13] R. J. Aumann. Agreeing to disagree. *The Annals of Statistics*, 4:1236–1239, 1976.

[14] R. J. Aumann. Rationality and bounded rationality. *Games and economic behavior*, 21:2–14, 1997.

[15] Robert Aumann and Sergiu Hart. *Handbook of Game Theory with Economic Applications*, volume 1. Handbooks in Economics, North-Holland, 1992.

[16] Robert Aumann and Sergiu Hart. *Handbook of Game Theory with Economic Applications*, volume 2. Handbooks in Economics, North-Holland, 1994.

[17] Robert Aumann and Sergiu Hart. *Handbook of Game Theory with Economic Applications*, volume 3. Handbooks in Economics, North-Holland, 2002.

[18] A. Azad and H. Tembine. Dynamic routing games: An evolutionary game theoretic approach. *CDC-ECC, 50th IEEE Conference on Decision and Control and European Control Conference, Orlando, Florida,* December 12-15 2011.

[19] A.G. Barto, R.S. Sutton, and C. Anderson. Neuron-like adaptive elements that can solve difficult learning control problems. *IEEE Transactions on Systems, Man, and Cybernetics, SMC*, 13:834–846, 1983.

[20] T. Basar and G. J. Olsder. Dynamic noncooperative game theory. *SIAM Series in Classics in Applied Mathematics, Philadelphia,* January 1999.

[21] M. Beckmann, C. B. McGuire, and C. B. Winston. Studies in the economics of transportation. *Yale University Press*, 1956.

[22] R. Bellman. On the theory of dynamic programming. *Proceedings of the National Academy of Sciences of the U.S.A*, 38:716–719, 1952.

[23] R.E. Bellman. A markov decision process. *Journal of Mathematical Mech.*, 6:679–684, 1957.

[24] E.V. Belmega, H. Tembine, and S. Lasaulce. Learning to precode in outage minimization games over mimo interference channels. *The IEEE Asilomar Conference on Signals, Systems, and Computers, California, US*, November 2010.

[25] E. Ben-Porath and M. Kahneman. Communication in repeated games with costly monitoring. *Games and Economic Behavior*, 44:227–250, 2003.

[26] M. Benaïm. Dynamics of stochastic approximations. *Le Seminaire de Probabilites. Lectures Notes in Mathematics*, 1709:1–68, 1999.

[27] M. Benaïm and M. Faure. Stochastic approximations, cooperative dynamics and supermodular games. *Preprint*, 2010.

[28] M. Benaïm, J. Hofbauer, and S. Sorin. Stochastic approximations and differential inclusions. *Part II: Applications, Mathematics of Operations Research*, 2006.

[29] A. Benveniste, P. Priouret, and M. Metivier. *Adaptive algorithms and stochastic approximations*. Springer-Verlag New York, Inc. New York, NY.

[30] Ulrich Berger. Fictitious play in 2×n games. *Journal Econo. Theory*, pages 139–154, 2005.

[31] D.T. Bishop and C. Cannings. The war of attrition with random rewards. *Journal of Theoretical Biology*, 74:377–388, 1978.

[32] T. Borgers and R. Sarin. Learning through reinforcement and replicator dynamics. *Mimeo, University College London.*, 1993.

[33] V. S. Borkar. *Stochastic approximation: a dynamical systems viewpoint*. New Delhi: Hindustan Publishing Agency, and Cambridge, UK.

[34] V. S. Borkar. Stochastic approximation with two timescales. *Systems Control Lett.*, 29:291–294, 1997.

[35] Diaconis P. Xiao L. Boyd, S. Fastest mixing markov chain on a graph. *SIAM Rev.*, 46:667–689, 2004.

[36] O. Brandière and M. Duflo. Les algorithmes stochastiques contournent-ils les pièges? *Annales de l'institut Henri Poincaré (B) Probabilites et Statistiques*, 32:395–427, 1996.

[37] Haim Brezis. *Functional analysis, Sobolev spaces and partial differential equations*. Universitext. Springer, New York, ISBN 978-0-387-70913-0, 2011.

[38] Skyrms Brian and Robert Pemantle. A dynamic model of social network formation. *In Proceedings of the National Academy of Sciences*, 97(16):9340–9346, 2000.

[39] W. Brian Arthur. Inductive reasoning and bounded rationality. *American Economic Review*, 84:406–411, 1994.

[40] G. W. Brown. Some notes on computation of games solutions. *Report P-78, The Rand Corporation,* 1949.

[41] G.W. Brown. Iterative solutions of games by fictitious play,. *In Activity Analysis of Production and Allocation, T.C. Koopmans (Ed.), New York: Wiley.,* 1951.

[42] R. Bush and F. Mosteller. Stochastic models of learning. *Wiley Sons, New York.,* 1955.

[43] A. Cabrales. Stochastic replicator dynamics. *International Economic Review,* 41:451–481, 2000.

[44] U. O. Candogan, I. Menache, A. Ozdaglar, and P. A. Parrilo. Near-optimal power control in wireless networks: A potential game approach. *Annual IEEE International Conference on Computer Communications,* 2010.

[45] R.G. Cassidy, C.A. Field, and M.J.L. Kirby. Solution of a satisficing model for random payoff games. *Management Science,* 19:266–271, 1972.

[46] N. Cesaro-Bianchi and G. Lugosi. *Prediction, Learning and Games.* Cambridge University Press, 2006.

[47] R. Cominetti, E. Meloy, and S. Sorin. A simple model of learning and adaptive dynamics in traffic network games. 2007.

[48] P. Coucheney, C. Touati, , and B. Gaujal. Selection of efficient pure strategies in allocation games. *In Proceedings of the International Conference on Game Theory for Networks,* 2009.

[49] Augustin Cournot. Recherches sur les principes mathematizues de la theorie des richesses. *Paris, France: L. Hachette, 1838; Italian translation in Biblioteca DelliEcon., 1875; English translation by N. T. Bacon in Economic Classics, New York, NY: Macmillan, 1897; reprinted by Augustus M. Kelly, 1960.,* 1838.

[50] T. M. Cover and J. A. Thomas. *Elements of Information Theory.* New York, Wiley, 1991.

[51] G. Debreu. A social equilibrium existence theorem. *Proceedings of the National Academy of Science,* 38:886–893, 1952.

[52] P. Diaconis and D. Stroock. Geometric bounds for eigenvalues of markov chains. *Ann. Appl. Probab.,* 1:36–61, 1991.

[53] M. Dindos and C. Mezzetti. Better-reply dynamics and global convergence to nash equilibrium in aggregate games. *Games and Economic Behavior,* 54:261–292, 2006.

[54] S. Ellner. Ess germination strategies in randomly varying environments. i. logistic-type models. *Theoretical Population Biology*, 28:50–79, 1985.

[55] Behdis Eslamnour, S. Jagannathan, and Maciej Zawodniok. Dynamic channel allocation in wireless networks using adaptive learning automata. *International Journal of Wireless Information Networks*, pages 1–14, May 2011.

[56] S. Fischer, H. Räcke, and B. Vöcking. Fast convergence to wardrop equilibria by adaptive sampling methods. *Proceedings of the thirty-eighth annual ACM symposium on Theory of computing*, pages 653–662, 2006.

[57] J. Flesch, G. Schoenmakers, and K. Vrieze. Stochastic games on a product state space. *Math. Oper. Res.*, 33:403–420, 2008.

[58] D. Foster and R. V. Vohra. Calibrated learning and correlated equilibrium. *Games and Economic Behavior*, 21:40–55, 1997.

[59] D. Fudenberg and D. Kreps. A theory of learning, experimentation, and equilibrium in games. 1988.

[60] D. Fudenberg and D. Levine. *Learning in Games*. MIT Press, Cambridge, MA, 1998.

[61] D. Fudenberg and S. Takahashi. Heterogeneous beliefs and local information in stochastic fictitious play. 2008.

[62] J. Willard Gibbs. On the equilibrium of heterogeneous substances. *Connecticut Acad. Sci.*, 1875-1878.

[63] I. Gilboa and A. Matsui. Social stability and equilibrium. *Econometrica*, 59:859–867, 1991.

[64] D.B. Gillies. Some theorems on n-person games. *Ph.D. Thesis, Princeton University Press, Princeton*, 1953.

[65] AJ Goldman. *The probability of a saddle point*, 64:729–730, 1957.

[66] Sanjeev Goyal. Learning in networks: a survey. *Group Formation in Economics: Networks, Clubs, and Coalitions edited by G. Demange and M. Wooders. Cambridge University Press.*, 2003.

[67] J. Guckenheimer and P. Holmes. *Nonlinear Oscillations, Dynamical Systems, and Bifurcations of Vector Fields*. Springer-Verlag, New York, 1983.

[68] M.A. Haleem and R. Chandramoulli. Adaptive downlink scheduling and rate selection: a cross-layer design. *IEEE Journal On Selected Areas in Communications*, 23(6), June 2005.

[69] S. Hart and Y. Mansour. How long to equilibrium? the communication complexity of uncoupled equilibrium procedures. *Games and Economic Behavior*, 69:107–126, 2010.

[70] S. Hart and A. Mas-Colell. A simple adaptive procedure leading to correlated equilibrium. *Econometrica*, 68:1127–1150, 2000.

[71] Y. Hayel, H. Tembine, E. Altman, and R. ElAzouzi. A markov decision evolutionary game for individual energy management. *Annals vol. XI of ISDG International Society of Dynamic Games, Advances in Dynamic Games: Theory, Applications, and Numerical Methods, Editors: Michele Breton, Krzysztof Szajowski*, pages 313–335, 2010.

[72] G. He, H. Tembine, and M. Debbah. Deployment analysis of cooperative ofdm base stations. *IEEE International Conference on Game Theory for Networks (GameNets), Istanbul, Turkey*, 2009.

[73] J. Hofbauer and L. Imhof. Time averages, recurrence and transience in the stochastic replicator dynamics. *Annals of Applied Probability*, 19:1347–1368, August 2009.

[74] J. Hofbauer and K. Sigmund. *Evolutionary Games and Population Dynamics*. Cambridge University Press, 1998.

[75] J. Hofbauer, S. Sorin, and Y. Viossat. Time average replicator and best-reply dynamics. *Mathematics of Operations Research*, 34(2):263–269, May 2009.

[76] Ed Hopkins and M. Posch. Attainability of boundary points under reinforcement learning. *Games and Economic Behavior*, 53:110–125, 2005.

[77] A.I. Houston and J. McNamara. Fighting for food: a dynamic version of the hawk-dove game. *Evolutionary Ecology*, 2:51–64, 1988.

[78] J. W. Huang and V. Krishnamurthy. Transmission control in cognitive radio as a markovian dynamic game: Structural result on randomized threshold policies. *IEEE Transactions on Communications*, 58(1), Jan. 2010.

[79] Kaplansky I. A contribution to von neumann's theory of games. *Ann. of Math.*, 2(46):474–479, 1945.

[80] S. Ishikawa. Fixed points by a new iteration method. *Proceedings Amer. Math. Soc. 44*, (1):147–150, 1974.

[81] Luis R. Izquierdo, Segismundo S. Izquierdo, Nicholas M. Gotts, and J. Gary Polhill. Transient and asymptotic dynamics of reinforcement learning in games. *Games and Economic Behavior*, 61:259–276, 2007.

[82] Matthew Jackson and Watts Allison. On the formation of interaction networks in social coordination games. *Games and Economic Behavior*, 41(2):265–291, 2002.

[83] D. H. Jacobson and D. Q. Mayne. Differential dynamic programming. *American Elsevier Pub. Co., New York, NY*, 1970.

[84] T. Joshi, D. Ahuja, D. Singh, and D. Pagrawal. Sara: Stochastic automata rate adaptation for ieee 802.11 networks. *IEEE Transactions on Parallel and Distributed Systems*, 19(10):1579–1590, October 2008.

[85] Shizuo Kakutani. A generalization of brouwer's fixed point theorem. 8(3):457–459, 1941.

[86] W. Karush. Minima of functions of several variables with inequalities as side constraints. *M.Sc. Dissertation. Dept. of Mathematics, Univ. of Chicago, Chicago, Illinois.*, 1939.

[87] G. Kasbekar and A. Proutiere. Opportunistic medium access in multi-channel wireless systems: A learning approach. *in Proceedings of Allerton Conference on Communications, Control, and Computing*, September 2010.

[88] M. Khan and H. Tembine. Energy-efficiency networking games. *Green It: Technologies and Applications, Springer Verlag*, 2011.

[89] M. Khan, H. Tembine, and A. Vasilakos. Game dynamics and cost of learning in heterogeneous 4g networks. *IEEE Journal on Selected Areas in Communications*, 30(1):198–213, January 2012.

[90] M. Khan, H. Tembine, and T. Vasilakos. Evolutionary coalitional games: Design and challenges in wireless networks. *IEEE Wireless Communications Magazine*, 2012.

[91] Manzoor Ahmed Khan and H Tembine. Evolutionary coalitional games in network selection. *Wireless Advanced 20th-22nd*, June 2011.

[92] Manzoor Ahmed Khan, H. Tembine, and Stefen Marx. Learning in user-centric iptv services selection in heterogeneous wireless network. *IEEE INFOCOM 2011 International Workshop on Future Media Networks and IP-based TV*, 2011.

[93] J. Kiefer and J. Wolfowitz. Stochastic estimation of the maximum of a regression function. *Annals of Mathematical Statistics*, 23:462–466, 1952.

[94] Alan Kirman. The economy as an evolving network. *Journal of Evolutionary Economics*, 7:339–353, 1997.

[95] H. W. Kuhn and A. W. Tucker. Nonlinear programming. *In the Proceedings of 2nd Berkeley Symposium. Berkeley: University of California Press*, pages 481–492, 1951.

[96] H. J. Kushner and D. S. Clark. Stochastic approximation methods for constrained and unconstrained systems. *Springer, New York*, 1978.

[97] H. J. Kushner and G. Yin. *Stochastic Approximation and Recursive Algorithms and Applications.* 2nd Edition, Springer-Verlag, New York, 2003, [Applications of Mathematics, Volume 35], xxii+474 pp. Stochastic Approximation Algorithms and Applications, 1st Edition, 1997. xxi+417 pp.

[98] S. H. Low L. Chen, T. Cui and J. C. Doyle. A game-theoretic model for medium access control. *IEEE Journal on Selected Areas in Communications*, 26(7), September 2008.

[99] S. Lasaulce and H. Tembine. *Game Theory and Learning for Wireless Networks: Fundamentals and Applications.* Academic Press, Elsevier, ISBN: 978-0-12-384698-3, 336 pages, 2011.

[100] M. Le Treust, H. Tembine, S. Lasaulce, and M. Debbah. Coverage games in small cells networks. *IEEE Proceedings of the Future Network and Mobile Summit (FUNEMS'10)*, 2010.

[101] D. Leslie and E. Collins. Generalised weakened fictitious play. *Games and Economic Behavior*, 56:285–298, 2005.

[102] D. Leslie and E. Collins. Individual q-learning in normal form games. *SIAM Journal Control Optim.*, 44:495–514, 2005.

[103] D. S. Leslie and E. J. Collins. Convergent multiple timescales reinforcement learning algorithms in normal form games. *The Annals of Applied Probability*, 13(4):1231–1251, 2003.

[104] Shu-Jun Liu and Miroslav Krstic. Stochastic nash equilibrium seeking for games with general non-linear payoffs. *SIAM Journal CONTROL OPTIMIZATION*, 49(4):1659–1679, 2011.

[105] L. Ljung. Analysis of recursive stochastic algorithms. *IEEE Transactions on Automatic Control*, 22(4):551–575, 1977.

[106] Chengnian Long, Z. Qian, Li Bo, Y. Huilong, and G. Xinping. Non-cooperative power control for wireless ad hoc networks with repeated games. *IEEE JSAC*, 25(6):1101–1112, August 2007.

[107] J. R. Marden, G. Arslan, and J. S. Shamma. Joint strategy fictitious play with inertia for potential games. *In Proceedings 44th IEEE Conf. Decision Control*, pages 6692–6697, Dec. 2005.

[108] J. R. Marden, G. Arslan, and J. S. Shamma. Joint strategy fictitious play with inertia for potential games. *IEEE Transactions on Automatic Control*, 54(2), February 2009.

[109] J. R. Marden, H. Peyton Young, G. Arslan, and J. S. Shamma. Payoff-based dynamics for multi-player weakly acyclic games. *SIAM Journal on Control and Optimization*, 2009.

[110] L. W. McKenzie. On equilibrium in graham's model of world trade and other competitive systems. *Econometrica*, 22:147–161, 1954.

[111] P. Mehta and S. Meyn. Q-learning and pontryagin's minimum principle. *in IEEE Proceedings CDC*, 2009.

[112] I. Menache and N. Shimkin. Efficient rate-constrained nash equilibrium in collision channels with state information. *IEEE Journal on Selected Areas in Communications, special issue on Game Theory*, 267:1070–1077.

[113] P. Mertikopoulos and A. L. Moustakas. Correlated anarchy in overlapping wireless networks. *IEEE Journal on Selected Areas in Communications*, 26(7):1160 – 1169, 2008.

[114] Mihail Mihaylov, Karl Tuyls, and Ann Nowé. Decentralized learning in wireless sensor networks. *In the Proceedings of the Adaptive and Learning Agents Workshop (ALA 2009), Taylor and Tuyls (eds.), Budapest, Hungary.*

[115] K. Miyasawa. On the convergence of learning processes in a 2x2 non-zero-person game. *Princeton University Research Memo 33*, 1961.

[116] D. Monderer and L. S. Shapley. Potential games. *Games and Economic Behavior*, 14:124–143, 1996.

[117] Anthony P. Morse. The behaviour of a function on its critical set. *Annals of Mathematics*, 40:62–70, January 1939.

[118] J. Nachbar. Evolutionary selection dynamics in games: Convergence and limit properties. *International Journal of Game Theory*, 19:59–89, 1990.

[119] John Nash. Non-cooperative games. *The Annals of Mathematics*, 54:286–295, 1951.

[120] John Von Neumann and Oskar Morgenstern. *Theory of Games and Economic Behavior*. Princeton University Press, 1944.

[121] P. Nicopolitidis, G. I. Papadimitriou, P. Sarigiannidis, Mohammad S. Obaidat, and A. S. Pomportsis. Adaptive wireless networks using learning automata. *IEEE Wireless Communications*, pages 75 – 81, 2011.

[122] P. Nicopolitidis, G.I. Papadimitriou, , and A.S. Pomportsis. Learning-automata-based polling protocols for wireless lans. 51(3):453–463, March 2003.

[123] H. Nikaido and K. Isoda. Note on non-cooperative convex games. *Pacific Journal of Mathematics*, 5:807–815, 1955.

[124] R. Pemantle. Nonconvergence to unstable points in urn models and stochastic approximations. *Annals of Probability*, 18:698–712, 1990.

[125] S. M. Perlaza, V. Q. Florez, H. Tembine, and S. Lasaulce. On the convergence of fictitious play in channel selection games. *IEEE Latin America Transactions*, 9, April 2011.

[126] S. M. Perlaza, H. Tembine, S. Lasaulce, , and M. Debbah. Satisfaction equilibrium for quality-of-service provisioning in decentralized networks. *IEEE Journal on Selected Topics in Signal Processing. Special Issue in Game Theory for Signal Processing*, 2012.

[127] S. M. Perlaza, H. Tembine, and S. Lasaulce. How can ignorant but patient cognitive terminals learn their strategy and utility? *In IEEE Proceedings of the 11th International Workshop on Signal Processing Advances for Wireless Communications (SPAWC)*, 2010.

[128] S. M. Perlaza, H. Tembine, S. Lasaulce, and M. Debbah. Satisfaction equilibrium: A general framework for qos provisioning in self-configuring networks. *IEEE Globecom*, 2010.

[129] Samir Medina Perlaza, H. Tembine, Samson Lasaulce, and V. Quintero-Florez. On the fictitious play and channel selection games. *In Proceedings of the Latin-American Conference on Communications (LATIN-COM), Bogota, Colombia*, September 2010.

[130] J. Ponstein. Existence of equilibrium points in non-product spaces. *SIAM Journal on Applied Mathematics*, 14(1):181–190, Jan. 1966.

[131] A. Poznyak and K. Najim. Learning through reinforcement for n-person repeated constraint games. *IEEE Trans. on Systems, Man, and Cybernetics: Part B Cybernetis*, 32:759–771, 2002.

[132] Bary S. R. Pradelski and H. Peyton Young. Efficiency and equilibrium in trial and error learning. 2010.

[133] H. Robbins. Some aspects of the sequential design of experiments. *Bulletin of the American Mathematical Society*, 55:527–535, 1952.

[134] H. Robbins and S. Monro. A stochastic approximation method. *Annals of Mathematical Statistics*, 22:400–407, 1951.

[135] David P. Roberts. Kernel sizes for random matrix games. *Working paper*, 2006.

[136] David P. Roberts. Nash equilibria in cauchy-random zero-sum and coordination matrix games. *International Journal of Game Theory*, 34:167–184, 2006.

[137] J. Robinson. An iterative method of solving a game. *Annals of Mathematics*, 54:296–301, 1951.

[138] J. B. Rosen. Existence and uniqueness of equilibrium points for concave n-person games. *Econometrica*, 33(3):520–534, 1965.

[139] A. Roth and I. Erev. Learning in extensive form games: Experimental data and simple dynamic models in the intermediate term. *Games and Economic Behavior*, 8(1):164–212, 1995.

[140] W. H. Sandholm. Population games and evolutionary dynamics. *MIT Press*, 2010.

[141] Arthur Sard. The measure of the critical values of differentiable maps. *Bulletin of the American Mathematical Society*, 48:883–890, 1942.

[142] K. P. R. Sastry and G. V. R. Babu. Approximation of fixed points of strictly pseudo-contractive mappings on arbitrary closed convex sets in a banach space. *Proceedings of Amer. Math. Soc.*, *128*, (10):2907–2909, 2000.

[143] L. Shapley. Some topics in two-person games. *In Advances in Game Theory M. Drescher, L.S. Shapley, and A.W. Tucker (Eds.), Princeton: Princeton University Press*, 1964.

[144] L. S. Shapley. Stochastic games. *Proceedings Nat. Acad. Sciences*, 39:1095–1100, 1953.

[145] Alonso Silva, Eitan Altman, Mérouane Debbah, H. Tembine, and Chloé Jimenez. Optimal mobile association on hybrid networks: centralized and decentralized case. *CoRR*, abs/0911.0257, 2009.

[146] Alonso Silva, H. Tembine, Eitan Altman, and Mérouane Debbah. Spatial games and global optimization for the mobile association problem: the downlink case. *49th IEEE Conference on Decision and Control, Atlanta, GA, USA*, December 15-17 2010.

[147] V. Srivastava, J. Neel, A. B. MacKenzie, R. Menon, L. A. DaSilva, J. E. Hicks, J. H. Reed, and R. P. Gilles. Using game theory to analyze wireless ad hoc networks. *IEEE Communications Surveys and Tutorials*, 7(4):46–56, 2005.

[148] W. Stanford. On the number of pure strategy nash equilibria in finite common payoffs games. *Economics Letters*, 62:29–34, 1999.

[149] G. Stoltz and G. Lugosi. Learning correlated equilibria in games with compact sets of strategies. *Games and Economic Behavior*, 59:187–208, April 2007.

[150] D.W. Stroock and SRS Varadhan. *Multidimensional Diffusion Processes*. Springer, 1979.

[151] P. D Taylor and L. Jonker. Evoltionarily stable strategies and game dynamics. *Mathematical Bioscience*, 40:145–156, 1978.

[152] Emre Telatar. Capacity of multi-antenna Gaussian channels. *European Transactions on Telecommunications*, 10(6):585–596, 1999.

[153] H. Tembine. Hybrid mean field game dynamics in large populations. *American Control Conference, ACC , San Francisco, California, US.*

[154] H. Tembine. Stochastic optimization: the non-differentiable case. *Master thesis dissertation, INRIA*, 2004.

[155] H. Tembine. Folk theorem for stochastic games with public signals. *Master dissertation*, 2006.

[156] H. Tembine. Repeated games with imperfect monitoring. *Master Thesis Dissertation, Ecole Polytechnique, France*, 2006.

[157] H. Tembine. Evolutionary network formation games and fuzzy coalition in heterogeneous networks. *2nd IFIP Wireless Days (WD)*, 2009.

[158] H. Tembine. Population games with networking applications. *Ph.D Dissertation, University of Avignon*, September 2009.

[159] H. Tembine. Distributed strategic learning for wireless engineers. *Notes*, 2010.

[160] H. Tembine. Population games in large-scale networks. *LAP, 250 pages, ISBN 978-3-8383-6392-9*, 2010.

[161] H. Tembine. Codipas: Combined fully distributed payoff and strategy learning. *Notes*, January 2011.

[162] H. Tembine. Dynamic robust games for mimo systems. *IEEE Transactions Systems, Man, Cybernetics*, 99(41):990 – 1002, August 2011.

[163] H. Tembine. Mean field stochastic games. *Notes*, 2011.

[164] H. Tembine. Mean field stochastic games: convergence, q/h learning, optimality. *American Control Conference, ACC, San Francisco, California, US*, 2011.

[165] H. Tembine, E. Altman, R. El-Azouzi, and H. Hayel. Evolutionary games with random number of interacting players with application to access control. *IEEE/ACM Proceedings 6th International Symposium on Modeling and Optimization in Mobile, Ad Hoc, and Wireless Networks (WiOpt) March 31 - April 4, Berlin, Germany,* 2008.

[166] H. Tembine, E. Altman, and R. ElAzouzi. Delayed evolutionary game dynamics applied to medium access control. *In Proceedings 4th IEEE International Conference on Mobile Ad-Hoc and Sensor Systems (MASS), Pisa, Italy.,* 2007.

[167] H Tembine, E. Altman, R. ElAzouzi, and Y. Hayel. Stable networking games. *In Proceedings of Forty-Sixth Annual Allerton Conference on Communication, Control, and Computing, Allerton Retreat Center, Monticello, Illinois,* September 2008.

[168] H. Tembine, E. Altman, R. ElAzouzi, and Y. Hayel. Stochastic population games with individual independent states and coupled constraints. *ACM Proceedings of the Second International Workshop on Game Theory in Communication Networks (GameComm), Athens, Greece,* October 2008.

[169] H. Tembine, E. Altman, R. ElAzouzi, and Y. Hayel. Correlated evolutionarily stable strategies in random medium access control. *International IEEE Conference on Game Theory for Networks, Gamenets,* 2009.

[170] H. Tembine, E. Altman, R. ElAzouzi, and Y. Hayel. Evolutionary games in wireless networks. *IEEE Trans. on Systems, Man, and Cybernetics, Part B, Special Issue on Game Theory,* June 2010.

[171] H. Tembine, E. Altman, R. ElAzouzi, and Y. Hayel. Bio-inspired delayed evolutionary game dynamics with networking applications. *Telecommunication Systems Journal, DOI: 10.1007/s11235-010-9307-1.,* 47:137–152(16), 2011.

[172] H. Tembine, E. Altman, R. ElAzouzi, and W. H. Sandholm. Evolutionary game dynamics with migration for hybrid power control in wireless communications. *47th SIAM/IEEE CDC,* December 2008.

[173] H. Tembine and M. Assaad. Hybrid mean field learning in large-scale dynamic robust games. *International Conference on Control and Optimization with Industrial Applications, COIA,* 2011.

[174] H. Tembine and A. Kobbane. Robust power allocation games under channel uncertainty and time delays. *IFIP Wireless Days,* 2010.

[175] H. Tembine, A. Kobbane, and M. El Koutbi. Dynamic robust power allocation games under channel uncertainty and time delays. *Elsevier Computer Communications,* 12, 34:1529–1537, 2011.

[176] H. Tembine, J. Y. Le Boudec, R. El-Azouzi, and E. Altman. Mean field asymptotics of markov decision evolutionary games and teams. *in Proceedings of IEEE Gamenets*, May 2009.

[177] H. Tembine, P. Vilanova, and M. Debbah. Noisy mean field game model for malware propagation in opportunistic networks. *2rd International Conference on Game Theory for Networks, Gamenets*, 2011.

[178] M. A. L. Thathachar and P. S. Sastry. Learning automata in stochastic games with incomplete information. *Systems and Signal Processing, (R. N. Madan, N. Viswanadham and R. L. Kashyap, eds.), (New Delhi)*, pages 417–434, 1991.

[179] M.A.L. Thathachar, P.S. Sastry, and V.V. Phansalkar. Decentralized learning of nash equilibria in multiperson stochastic games with incomplete information. *IEEE transactions on system, man, and cybernetics*, 24(5), 1994.

[180] E.L. Thorndike. Animal intelligence: An experimental study of the associative processes in animals. *Psychological Review, Monograph Supplements, vol. 8. MacMillan, New York*, 1898.

[181] J. Akbari Torkestani and M.R. Meybodi. Mobility-based multicast routing algorithm for wireless mobile ad-hoc networks: A learning automata approach. *Computer Communications, Elsevier*, 33(6):721–735, April 2010.

[182] D. N. C. Tse and S. V. Hanly. Multiaccess fading channels. i. polymatroid structure, optimal resource allocation and throughput capacities. *IEEE Trans. on Info. Theory*, 44(7):2796–2815, 1998.

[183] M. van der Schaar and F. Fu. Spectrum access games and strategic learning in cognitive radio networks for delay-critical applications. *Proceedings IEEE*, 2009.

[184] S. Verdú and V. Poor. On minimax robustness: A general approach and applications. *IEEE Transactions Information Theory*, 30:328–340, Mar. 1984.

[185] Y. Viossat. The replicator dynamics does not lead to correlated equilibria. *Games and Economic Behavior*, 59:397–407, 2007.

[186] P. Viswanath, D.N.C. Tse, and V. Anantharam. Asymptotically optimal water-filling in vector multiple-access channels. *IEEE Transactions on Information Theory*, 47(1):241–267, Jan 2001.

[187] J. Wardrop. Some theoretical aspects of road traffic research. *Proceedings of the Institution of Civil Engineers, Part II*, 1(36):352–362, 1952.

[188] C.I.C.H. Watkins. Learning from delayed rewards. *PhD thesis, University of Cambridge, Cambridge, UK*, 1989.

[189] C.I.C.H. Watkins and P. Dayan. Q-learning. *Machine Learning*, 8:279–292, 1992.

[190] J. Weibull. Evolutionary game theory. *MIT Press*, 1995.

[191] Y. Xing and R. Chandramouli. Stochastic learning solution for distributed discrete power control game in wireless data networks. *IEEE/ACM Transactions on Networking*, 16:932–944, august 2008.

[192] H. P. Young. The evolution of conventions. *Econometrica*, 61:57–84, 1993.

[193] H. P. Young. *Strategic Learning and Its Limits*. Oxford University Press, 2004.

[194] H. P. Young. Learning by trial and error. *Games and Economic Behavior, Elsevier*, 65(2):626–643, March 2009.

[195] Q. Zhu, H. Tembine, and T. Başar. A constrained evolutionary gaussian multiple access channel game. *in the IEEE proceedings of GameNets*, 2009.

[196] Q. Zhu, H. Tembine, and T. Başar. Evolutionary game for hybrid additive white gaussian noise multiple access control. *in IEEE Proceedings of Globecom*, 2009.

[197] Q. Zhu, H. Tembine, and T. Başar. Heterogeneous learning in zero-sum stochastic games with incomplete information. *in 49th IEEE Conference on Decision and Control, Atlanta, GA, USA*, 2010.

[198] Q. Zhu, H. Tembine, and T. Başar. Network security configuration: A nonzero-sum stochastic game approach. *in IEEE Proceedings of American Control Conference*, June 2010.

[199] Q. Zhu, H. Tembine, and T. Başar. Distributed strategic learning with application to network security. *American Control Conference, ACC 2011, San Francisco, California, US*, 2011.

Index

Printed and bound by CPI Group (UK) Ltd, Croydon, CR0 4YY

24/10/2024

01778720-0001